BIOLOGY and PHYSIOLOGY of the OSTEOCLAST

BIOLOGY and PHYSIOLOGY of the OSTEOCLAST

Edited by

Barry R. Rifkin, D.D.S., Ph.D.
Head, Division of Basic Sciences
and Professor of Oral Medicine
and Pathology
New York University College of Dentistry
New York, New York

Carol V. Gay, Ph.D.
Professor of Molecular and Cell Biology
and of Poultry Science
The Pennsylvania State University
University Park, Pennsylvania

CRC Press
Boca Raton Ann Arbor London Tokyo

Library of Congress Cataloging-in-Publication Data

Biology and Physiology of the osteoclast / editors, Barry R. Rifkin,
Carol V. Gay.
p. cm.
Includes bibliographical references and index.
ISBN 0-8493-5437-4
1. Osteoclasts. I. Rifkin, Barry R. II. Gay, Carol V.
QP88.2.B54 1992
599′.01852--dc20 92-7698
CIP

Book cover illustration from Collin-Osdoby, P., Oursler, M., Webber, D., and Osdoby, P., *J. Bone Miner. Res.*, 6, 1353, 1991. With permission.

Direct all inquiries to CRC Press, Inc., 2000 Corporate Blvd., N.W., Boca Raton, Florida 33431.

International Standard Book Number 0-8493-5437-4

Library of Congress Card Number 92-7698

Printed in the United States of America 2 3 4 5 6 7 8 9 0

Printed on acid-free paper

DEDICATION

To the memory of my brother Howard and my father Samuel.

Barry R. Rifkin

To my mother, Charlotte Mariam Stuart Lovejoy, who made sure I had a chance.

Carol V. (Lovejoy) Gay

PREFACE

The last decade has witnessed unprecedented progress in the understanding of the biology and physiology of the osteoclast. The role of the osteoclast in shaping and remodeling the skeleton, in well-regulated mineral homeostasis, and in the development of skeletal-eroding diseases has progressed significantly because many investigators from a wide variety of disciplines have become intrigued by this strange cell which sometimes moves like an amoeba and other times behaves like an immobile oxyntico-peptic cell that secretes both acid and proteolytic enzymes. Prior to the early 1980's, our knowledge of the osteoclast was substantially descriptive and had not changed much through the years. In the late 1970's, studies reporting the isolation of osteoclasts emerged. The Nelson and Bauer (1977) report describing the isolation of rabbit osteoclasts must be considered a landmark paper which launched the current productive period of osteoclast investigation. This was followed in the early 1980's by several laboratories reporting on the isolation of avian osteoclasts in large numbers. Later, the bone slice assay to monitor osteoclast function was introduced. The advances in the isolation of osteoclasts, coupled with the use of modern cell and molecular biological techniques, have led to a substantial increase in our knowledge of the development, regulation, and function of the osteoclast. Thus, it seems prudent at this juncture to collect in one volume much of what has been discovered in recent years. This volume, consisting of 21 chapters, has been written by many of the scientists who have made significant contributions to our understanding of the biology and physiology of this unique cell. The overall plan was to fully cover the biology of osteoclasts. In order to accomplish this goal, chapter topics were purposely selected to have some overlap. This allowed contributing authors to fully present their perspectives, thereby enriching the total effort. Some chapter topics were selected to focus on experimental systems for studying osteoclasts and to evaluate unique osteoclast characteristics such as membrane markers, acid phosphastase content, ultrastructure, histochemical distributions of enzymes, and the nature of the cytoskeleton. Other chapters focus on origin and development. Mechanisms of osteoclast responses including signal transduction, electrophysiology, acidification, oxygen radical production, mineral dissolution, and matrix dissolution are explored. The regulation of osteoclasts is considered from several directions, including the influence of systemic factors, local factors, extracellular matrix, mechanical forces, and osteoclast/osteoblast coupling. Finally, osteoclast related diseases and external ways of altering osteoclast action are discussed.

This volume is intended as resource material for scientists, clinicians, and students. It should be of interest to cell and molecular biologists, skeletal researchers, dental scientists, endocrinologists, orthopedic researchers, rheumatologists, and investigators in the field of aging.

Completion of this volume was no easy task. The editors would like to extend their sincere thanks to the contributing authors. Their willingness to prepare timely and significant chapters, despite grant deadlines, teaching responsibilities, study section duties, seminars, committee meetings, publication deadlines, and research experiments, is truly appreciated.

It is our hope that this work will spark even greater progress in this exciting field.

Barry R. Rifkin
Carol V. Gay

THE EDITORS

Barry R. Rifkin, D.D.S., Ph.D., is Head, Division of Basic Sciences, and Professor of Oral Medicine and Pathology, New York University College of Dentistry, New York, New York.

Dr. Rifkin graduated in 1961 from Ohio State University with a B.S. degree in Anatomy. He obtained an M.S. degree in Physiology from the University of Illinois, Champaign-Urbana, in 1964. In 1968 he received a D.D.S. degree from Temple University, and in 1974 a Ph.D. in Experimental Pathology from the University of Rochester.

In 1973 Dr. Rifkin was appointed Assistant Professor of Pathology and Dental Research at the University of Rochester School of Medicine and Dentistry, and in 1974 Associate Pathologist at Strong Memorial Hospital. Since 1980 he has been at the New York University College of Dentistry. He first served as Associate Professor and Chairman and then in 1984 became Professor and Chairman, Department of Oral Medicine. In 1987 Dr. Rifkin was appointed Chairman of the newly created Department of Oral Medicine and Pathology, and in 1991 he was named Head, Division of Basic Sciences.

Dr. Rifkin is a member of the American Association of Dental Research, the International Association of Dental Research, the American Society of Bone and Mineral Research, the American Society of Cell Biology, the American Association for the Advancement of Science, the American Association of Oral Biologists, the New York Academy of Sciences, and the Honor Society Sigma XI. He has served as Chairman of the Constitution Committee of the American Association for Dental Research and is presently a member of the Electorate Nominating Committee and a member at large for Dentistry of the American Association for Advancement of Science. Dr. Rifkin is currently President-Elect of the American Association of Oral Biologists. He was elected to Omicron Kappa Upsilon and to fellowship in the American College of Dentists. In 1974 he was appointed an Andrew Mellon Faculty Fellow at the University of Rochester.

Dr. Rifkin has been the recipient of research grants from the National Institutes of Health and private industry. He has published more than 90 papers and abstracts, given over 25 invited lectures, and has served as a peer reviewer for NIH. His current research interests include tetracycline effects on bone cells and osteoclast resorption mechanisms.

Carol V. Gay, Ph.D., is Professor of Molecular and Cell Biology and of Poultry Science at The Pennsylvania State University, University Park, Pennsylvania.

Dr. Gay graduated in 1962 from the University of Maine with a B.A. degree in Zoology. She obtained an M.S. in Biophysics in 1967 and a Ph.D. in Physiology in 1972, both at The Pennsylvania State University. From 1972 to 1975 she was an NIH Postdoctoral Fellow and from 1975 to 1988 a Research Associate and a Senior Research Associate in Biophysics at Penn State. In 1988, she became an Associate Professor in Molecular and Cell Biology, and in 1991, she assumed her present position.

Dr. Gay is a member of the American Society for Bone and Mineral Research, the American Association for the Advancement of Science, the American Physiology Society, the American Society for Cell Biology, and the Phi Beta Kappa and Sigma Xi honor societies. She is an affiliate member of the Royal Society of Medicine. She serves on the Editorial Boards of the Journal of Histochemistry and Cytochemistry and the Journal of Bone and Mineral Research and has served as peer reviewer for granting agencies. She has been the recipient of many NIH and USDA grants, including a Research Career Development Award.

Dr. Gay has presented over 40 invited lectures and has authored over 60 research papers and 10 book chapters. Her current major research interests relate to bone cell structure and function.

CONTRIBUTORS

Göran Andersson, Ph.D.
Department of Pathology
Karolinska Institutet
Huddinge University Hospital
Huddinge, Sweden

Stuart A. Arkett, Ph.D.
Department of Physiology
The University of Western Ontario
London, Ontario, Canada

Harry C. Blair, M.D.
Department of Pathology
University of Alabama at Birmingham
Birmingham, Alabama

Elisabeth H. Burger, Ph.D.
Department of Oral Cell Biology
ACTA, Free University
Amsterdam, The Netherlands

Timothy J. Chambers, M.D., Ph.D.
Department of Histopathology
St. George's Hospital Medical School
London, United Kingdom

Patricia Collin-Osdoby, Ph.D.
Department of Biology
Washington University
St. Louis, Missouri

Ricardo de Grooth
Department of Cell Biology and Histology
Faculty of Medicine
University of Leiden
Rijnsburgerweg, The Netherlands

Jean-Marie Delaissé, Dr. Sci.
Department of Biochemistry and Cell Biology and International Institute of Cellular and Molecular Pathology
University of Louvain
Brussels, Belgium

S. Jeffrey Dixon, D.D.S., Ph.D.
Division of Oral Biology and Department of Physiology
Faculty of Dentistry
The University of Western Ontario
London, Ontario, Canada

Henry J. Donahue, Ph.D.
Musculo-Skeletal Research Laboratory
Department of Orthopaedics
State University of New York
Stony Brook, New York

Barbro Ek-Rylander, B.Sc.
Department of Oral Pathology and Department of Pathology
Karolinska Institutet
Huddinge Hospital
Huddinge, Sweden

Carol V. Gay, Ph.D.
Department of Molecular and Cell Biology and Department of Poultry Science
The Pennsylvania State University
University Park, Pennsylvania

Julie Glowacki, Ph.D.
Department of Orthopedic Surgery
Harvard Medical School and Brigham and Women's Hospital
Boston, Massachusetts

Robert A. Greenwald, M.D.
Department of Medicine/
Rheumatology
Long Island Jewish Medical
Center
New Hyde Park, New York

Johan N. M. Heersche, Ph.D.
MRC Group in Periodontal
Physiology
Faculty of Dentistry
University of Toronto
Toronto, Ontario, Canada

Miep H. Helfrich, Ph.D.
Department of Haematology
Imperial Cancer Research Fund
St. Bartholomew's Hospital
London, United Kingdom

Michael A. Horton, F.R.C.Path.
Department of Haematology
Imperial Cancer Research Fund
St. Bartholomew's Hospital
London, United Kingdom

Motomi Ishibe, M.D.
Department of Orthopaedic
Surgery
Hokkaido University School of
Medicine
Sapporo, Japan

Melanie E. M. Kelly, Ph.D.
Department of Pharmacology
Dalhousie University
Halifax, Nova Scotia, Canada

Marilyn Krukowski, Ph.D.
Department of Biology
Washington University
St. Louis, Missouri

Peter Lakatos, M.D.
1st Department of Internal
Medicine
Semmelweis University Medical
School
Budapest, Hungary and
Department of Pharmacology
Northwestern University Medical
School
Chicago, Illinois

Sandy C. Marks, Jr., D.D.S., Ph.D.
Department of Cell Biology
University of Massachusetts
Medical School
Worcester, Massachusetts

Cedric Minkin, Ph.D.
Department of Basic Sciences/
Physiology
University of Southern California
School of Dentistry
Los Angeles, California

Gregory R. Mundy, M.D.
Department of Medicine/
Endocrinology
University of Texas Health
Science Center
San Antonio, Texas

Peter J. Nijweide, Ph.D.
Department of Cell Biology and
Histology
Faculty of Medicine
University of Leiden
Rijnsburgerweg, The Netherlands

Philip Osdoby, Ph.D.
Department of Biology
Washington University
St. Louis, Missouri

Lois K. Osier, M.D.
Department of Orthopedic Surgery
University of Massachusetts Medical School
Worcester, Massachusetts

J. Edward Puzas, Ph.D.
Department of Orthopaedics
University of Rochester School of Medicine
Rochester, New York

Barry R. Rifkin, D.D.S., Ph.D.
Division of Basic Sciences/Oral Medicine and Pathology
New York University College of Dentistry
New York, New York

G. David Roodman, M.D., Ph.D.
Department of Medicine/Hematology
University of Texas Health Science Center and Audie Murphy Memorial Veterans Administration Hospital
San Antonio, Texas

Paul H. Schlesinger, M.D., Ph.D.
Department of Cell Biology and Physiology
Washington University School of Medicine
St. Louis, Missouri

Stephen M. Sims, Ph.D.
Department of Physiology
The University of Western Ontario
London, Ontario, Canada

Paula H. Stern, Ph.D.
Department of Pharmacology
Northwestern University Medical School
Chicago, Illinois

Anna Teti, Ph.D.
Department of Human Anatomy
School of Pharmacy
University of Bari
Bari, Italy

Gilbert Vaes, M.D., Ph.D.
Department of Biochemistry and Cell Biology and International Institute of Cellular and Molecular Pathology
University of Louvain
Brussels, Belgium

Alberta Zambonin-Zallone, Ph.D.
Department of Human Anatomy
School of Medicine
University of Bari
Bari, Italy

TABLE OF CONTENTS

Chapter 1

EXPERIMENTAL SYSTEMS FOR STUDYING OSTEOCLAST BIOLOGY

Philip Osdoby, Marilyn Krukowski, and Patricia Collin-Osdoby

TABLE OF CONTENTS

I. INTRODUCTION

The exploration of bone has captured the interest of investigators in a variety of biological fields, including morphology, physiology, developmental and cell biology, and recently, immunology and hematology, owing to new-found discoveries concerning immune modulation of bone and the hematopoietic origin of the bone-resorbing osteoclast. An increased understanding of the skeleton is of continuing benefit to the clinical disciplines of orthopedics, rheumatology, endocrinology, dentistry, and geriatrics. Contributions from this broad spectrum of investigators have steadily expanded our knowledge regarding the skeleton at all levels: holistic, organismic, cellular, and molecular. The complexity of bone cell and matrix interactions, along with the multitude of signals regulating and coordinating the development and activity of cells involved in bone formation and turnover, elicits intrigue and at times confusion in those endeavoring to understand these processes.

Bone is a remarkable multifunctional organ. The cells and matrix comprising skeletal tissue form a specialized architecture uniquely adapted for weight bearing, motion, and marrow development. Bone also serves as a mineral reservoir for the organism and is, by necessity, highly responsive to complicated endocrine signals responsible for maintaining and modifying mineral homeostasis. In addition, the complex bone matrix may function as a growth factor repository and thereby operate to influence hematopoietic events and contribute to the regenerative properties of bone. Normal bone development and remodeling depend upon intricate temporal interactions between precursor cells, mature bone cells, extracellular matrix molecules, and signals generated by growth factors, immune cytokines, and humoral modulators. Bone modeling (during development) and remodeling (throughout life) are dependent upon factors that regulate the number and activity of both bone-forming osteoblasts and bone-resorbing osteoclasts. During much of adult life, overall bone mass remains unchanged due to the net balance between formative and degradative processes. Deciphering the biochemical mechanisms governing bone cell precursor recruitment, differentiation, and activity is therefore of paramount importance for understanding both normal and pathological processes in bone. A major challenge that has confronted skeletal researchers in general, and osteoclast investigators in particular, relates to the development of appropriate systems and assays for examining the physiology, cellular and molecular biology, and responsiveness to humoral, local, or physical cues which affect these bone-resorbing cells.

In this chapter, we will attempt to review many of the experimental systems currently available for studying osteoclast development, regulation, and activity. Although a summary of any reasonable length cannot be all inclusive, an effort has been made to review a variety of examples from different methods and ones that appear to be particularly valuable or novel. The relative strengths and limitations of various approaches have been evaluated from our viewpoint and presented for consideration. The complexity of bone and the inherent difficulties associated with analysis of such a richly vascularized, mineralized tissue have tested the creativity and resources of many investigators interested in understanding skeletal behavior. Despite the complications, including those posed by the interactive nature of cells residing within bone and by the overlapping or multitiered cascades of signal regulation among these cells, rapid progress has been achieved in elucidating the autocrine, paracrine, endocrine, and physical signals that influence normal and abnormal bone modeling and remodeling. Contributions have derived from every area of bone research, including whole animal studies coupled with morphological and pharmacological manipulation, organ and cell culture systems, transformed and immortalized cell lines, and naturally occurring genetic lesions.

II. HISTORICAL OVERVIEW

Three nineteenth-century anatomists, Robin, Rolet, and Kolliker, are credited with the identification of large multinucleated cells associated with bone.[1] Kolliker[2] christened these cells osteoclasts, since he believed that they were responsible for bone resorption. Osteoclast research during the last quarter of the nineteenth century and the first half of the twentieth century was primarily descriptive. According to Hancox[1] in his 1972 review, osteoclast investigation had already passed through three major phases of inquiry that corresponded to advances in microscopic technique. He stated: "The subject has been reviewed several times in recent years and there seems little point in considering the historical aspects of the subject any further here. It would be chiefly a narrative of various ways in which a similar histological picture has been interpreted . . . ." Hancox's review, then, focused on the third phase of inquiry in which structure/function relationships for osteoclasts were probed using up-to-date methodology such as autoradiography, cytochemistry, electron microscopy, and cinemaphotography.

Although Jordan's 1928[3] and Jaffee's 1930[4] studies identified phagocytosed matrix particles in osteoclasts when routine tissue sections were viewed, there was still doubt as to whether osteoclasts were the primary cells responsible for bone resorption. Accepted experimental evidence for osteoclast degradation of bone did not arise until 1957, at which time an autoradiographic study by Arnold and Jee[5] demonstrated incorporation of plutonium-labeled bone matrix material into osteoclasts. The next two decades saw major strides in elucidating osteoclast ultrastructure and function in parallel with the development of advanced methods for electron microscopy, histochemistry, and immunohistochemistry. A description of the osteoclast with its rich cytoplasmic complement of lysosomes and mitochondria, and the distinctive clear zone and ruffled border adjacent to the bone surface, was forthcoming (for reviews, see References 6 to 8). Numerous enzymes associated with osteoclasts were also defined by histochemical and immunohistochemical procedures, revealing an array of acid hydrolases[9] and carbonic anhydrase[10] activities (for reviews, see References 7 and 11).

Histomorphometric analyses of tissues from whole animals pharmacologically manipulated with calcium-regulating hormones, most notably calcitonin[12] and parathyroid hormone,[13] enabled determination of endocrine cue influences on bone remodeling. Such descriptive and histomorphometric analysis of osteoclasts in whole animals subjected to physiological perturbation continues to serve an important role in understanding osteoclast development, regulation, and activity, and is, in fact, critical for the adoption of therapeutic strategies to treat a variety of bone disorders. Despite revealing an effect upon the osteoclast and/or its precursors, this approach does not permit evaluation of whether the response to a pharmacological, mechanical, or other signal originates at the level of the osteoclast (or precursor cell) itself or, alternatively, acts via one or more intermediary cascading pathways which are ultimately manifested in altered osteoclast development or behavior.

Bone organ culture systems obviate many of the complex systemic variables in the whole animal while maintaining some aspects of the local skeletal microenvironment. Although Fell[14] demonstrated the utility of bone organ culture in the 1920s, its use in investigating osteoclasts was limited prior to 1965, at which time the rat long-bone organ culture studies of Raisz[15] and Raisz and Niemann[16] were reported. Early-1960 studies of cultured bone pieces or calvaria demonstrated that osteoclasts were impressively motile cells capable of movement even while attached to the bone surface.[17-21] With the development and utilization of the Raisz and Niemann[15,16] bone organ culture system, the modern era of inquiry aimed at clarifying osteoclast regulation and function *in vitro* was launched. However, the artificial environment created in organ culture requires that investigators exercise caution when they interpret results and attempt to extrapolate findings to *in vivo* situations. In addition, despite the diminished complexity when compared with whole animal systems, organ cultures still contain heterogeneous cell populations with diverse phenotypes. This hinders the extrapolation of final results to presumed effects upon individual cells or cell types, and the ability to selectively target an agent toward a chosen cell type.

Studies on the origin of the osteoclast were for many years mired in controversy since most initial studies were limited to microscopic analyses. One popular view was that osteoclasts shared a common ancestry with osteoblasts (monophyletic theory), whereas opposing evidence suggested that they derived from some other precursor population. Resolution of this issue was not forthcoming until studies were performed using naturally occurring genetic mutants which have impaired osteoclasts and consequently develop osteopetrosis (for a review, see Reference 22). This approach was originally pioneered by Walker, who demonstrated that the osteopetrotic condition, which resulted from an osteoclast defect in the mutant mice, could be cured either through parabiotic (circulatory) union with normal littermates[23] or by marrow transplantation from normal mice.[24] Strong evidence for the marrow origin of the osteoclast was also garnered from a human marrow transplant performed between an opposite-sex sibling donor and recipient, wherein markers associated with newly formed osteoclasts matched those of the donor.[25] Together with earlier ^{3}H-thymidine labeling studies during newt limb regeneration,[26] and later chick-quail chimera cell marker experiments,[27,28] these studies clearly established osteoclast precursors to be of hematopoietic origin and to derive from cells related to mononuclear phagocytes (for a review, see Reference 29).

Elucidation of the relationship between osteoclasts and cells of the mononuclear phagocyte family precipitated new avenues of osteoclast inquiry. Taking one direction, Horton et al.[30,31] demonstrated that a product of the immune system (osteoclast activating factor, OAF) could modulate osteoclast activity, and possibly also development, thereby laying the foundation for current studies of immune and local cytokine influences on osteoclast function in particular, and bone remodeling in general (for reviews, see References 32 and 33). Another direction encouraged the use of monocyte and macrophage cultures as surrogate cell populations to investigate cellular aspects of osteoclast activity.[34-37] Questions arose with the latter studies, however, as to whether such systems were, in fact, reliable tools for understanding the unique bone-resorbing activity and regulating events associated with the osteoclast.[38,39] Such concern fostered renewed efforts to develop methods for the isolation, purification, and cultivation of authentic osteoclasts[40-45] (for reviews, see Reference 46 and Section III.D.1 of this chapter.)

III. EXPERIMENTAL SYSTEMS IN CURRENT USE

A. INTRODUCTION

Knowledge concerning osteoclast activity and development has accumulated rapidly in recent years. In spite of this progress, numerous gaps remain in our understanding of osteoclast function, differentiation, and fate. Further, conflicting observations have fueled controversy relative to the origin, regulation, and cell physiology of the osteoclast. The wide range of diverse systems currently available to investigate these areas represents both a strength and a weakness for resolving the outstanding issues pertaining to these bone-resorbing cells. Before embarking on a survey of the systems now in use for researching osteoclasts, a few caveats deserve mention. First, it is important when choosing a system to keep in mind the source of the starting material for either *in vivo* or *in vitro* analyses. Whether cells of different species develop, function, and are regulated in a comparable manner can affect the final results and should be considered. For instance, much debate has been generated concerning the presence or absence of calcitonin receptors on avian osteoclasts compared to mammalian osteoclasts. Additionally, the age and sex of the animal used, or the organ or cell culture material from which the osteoclasts were obtained, must be taken into account. Further, for organ and cell culture experiments, the results may be sensitive to and dependent upon numerous extraneous factors, such as the particular culture conditions, feeding regimen, and batch of serum employed. Finally, the relative purity of cell cultures can markedly bias results, as can the nature of the substrate with which the cells are associated. It should be stressed that the cell populations purported to be phenotypically "homogeneous" may consist of cells in various stages of differentiation and/or senescence, as well as, in the case of dividing cells, in various stages of the cell cycle. Clearly, all these parameters, and more, have the potential to elicit subtle to major variations in systemic, organ, and cellular behavior.

B. WHOLE ANIMAL SYSTEMS

1. General

Whole animal systems have proven to be valuable for the study of bone modeling and remodeling, and remain an important component in the repertoire of methods available for the investigation of osteoclast development and behavior in multiple species, including humans, rats, cats, mice, rabbits, birds, and cartilagenous and bony fishes.[47-54] Frequently, in man and other vertebrates, histological assessment of osteoclast numbers is correlated with a profile of physical measurements — dual photon bone densitometry, blood calcium and osteocalcin levels, and urine hydroxyproline and deoxypyridinoline — as indicators of bone turnover.[55-61] Pharmacological studies designed to examine the influence of hormones and effectors on bone remodeling, and the actions of inhibitors such as fluorides and diphosphonates, have yielded general mechanistic insights as well as potentially effective therapeutic protocols.[62-64] A number of systems for evaluating physical and mechanical forces have been developed. For example, the advent of the space program has spurred investigation into bone remodeling under conditions of weightlessness.[65] In a ground-based system, the rat tail suspension model[66] has been used to study the consequences of disuse and/or weightlessness on particular components of the skeleton. Weightlessness produces effects similar to those observed in disuse osteoporosis, and the data suggest that the osteoblast, not the osteoclast, is the cell primarily affected.[67] Orthodontists and orthopedists have substantial interest in learning how mechanical forces lead to changes in bone, and so, experimental systems have been devised to assay alterations in bone remodeling induced by physical forces or electromagnetic stimuli.[68-72] Perhaps one of the more novel systems used to examine osteoclast development and function is the pressurized middle ear chamber model of Chole and colleagues.[73,74] This approach permits controlled pressure regulation, enabling measurements to be recorded for changes in osteoclast number and activity which occur as a consequence of pressure alone or of pressure in combination with locally administered cytokines. For most of the above studies, analysis after such manipulations involves histological and histomorphometric measurements. Findings are thus restricted to reporting alterations in osteoclast number and resorptive activity.

2. Osteoclast Development

Other *in vivo* systems have been primarily utilized for investigating osteoclast development. These systems can be manipulated in such a manner as to enable an initial incident to precipitate a series of reproducible events, culminating in the recruitment and differentiation of giant multinucleated cells that may possess osteoclast-like characteristics. Three *in vivo* systems that share this ability include the synchronized remodeling sequence initiated by tooth removal in the rat,[75,76] the subcutaneous implantation of devitalized bone or other test material in rats and mice,[77-79] and the implantation of such materials onto the extraembryonic chorioallantoic membrane of the developing chick.[80,81] Baron and colleagues[75] have used the synchronous tooth removal system to characterize (enzymatically and histomorphometrically) the cellular changes that take place in advance of the formation of osteoclasts and the subsequent bone remodeling events. Glowacki and colleagues[77-79] have used the subcutaneous implant system to document that bone matrix (and osteocalcin as a critical matrix constituent) is essential for the recruitment and formation of osteoclasts displaying calcitonin receptors.[78] Krukowski and Kahn[80] and Webber et al.[81] took advantage of the chick chorioallantoic membrane implant system to demonstrate that bone mineral, in concert with the organic component of bone matrix, was required for osteoclast differentiation, as evidenced by the appearance of giant cells displaying ruffled borders and an osteoclast-specific membrane antigen (Figure 1). Groessner-Schreiber et al.[82,83] also used the chorioallantoic membrane model to show that with both rat and human bone implants, older bone recruited fewer osteoclasts, suggesting that impaired bone remodeling in older individuals may be partly a consequence of some properties of the aged bone matrix. Some investigators have challenged the utility of the subcutaneous implant system, since they did not observe ruffled borders on the giant cells which had formed.[84,85] In contrast, with the chorioallantoic membrane implant system, giant cells with ruffled borders have been consistently described[80,81] and Webber et al.[86] have recently demonstrated that six days after implantation of cortical bone slices onto the extraembryonic membrane, the bone slices display characteristic osteoclast-mediated resorption pits (Figure 1). Much has already been learned through the use of the synchronous tooth removal model and the subcutaneous and chorioallantoic membrane implant systems. It is expected that these systems will continue to provide valuable insights into understanding cellular events and factors which influence osteoclast recruitment and function.

3. Genetic Variants

As described earlier, mouse osteopetrotic mutants have been exploited for delineating factors involved in osteoclast development (for a review, see

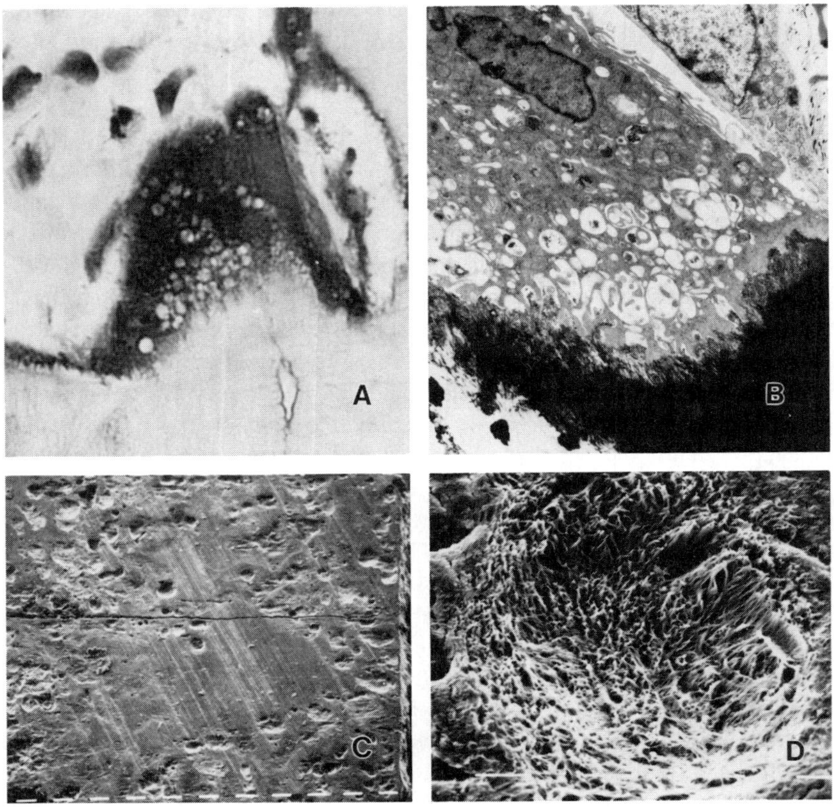

FIGURE 1. (A) Light photomicrograph from mineralized bone implant that was placed onto the chick chorioallantoic membrane (CAM) for 8 d. A large multinucleated cell can be observed in what appears to be a resorption bay. (B) Transmission electron photomicrograph from mineralized bone particle implant placed on the CAM for 8 d. Note ruffled border on multinucleated giant cell. (From Webber, D., Osdoby, P., Hauschka, P., and Krukowski, M., *J. Bone Miner. Res.*, 5, 401, 1990. With permission.) (C, D) Scanning electron photomicrographs of bovine cortical bone slab implant placed on the CAM and recovered after 8 d. Cells have been removed to reveal numerous resorption pits (C) and exposed collagen (D). (From Webber, D., Menton, D., and Osdoby, P., *Bone Miner.*, 11, 127, 1990. With permission.)

Reference 22). More recently, *op/op* osteopetrotic mice have been shown to contain a point mutation in the gene encoding the growth factor cytokine M-CSF.[87,88] Osteoclasts possess high levels of the receptor, c-*fms*,[89] to which this cytokine binds, and the lack of M-CSF in *op/op* mice may account for this osteopetrotic condition. Other pathological conditions, such as osteoporosis, Paget's disease, hypercalcemia of malignancy, and inflammatory states associated with arthritis and periodontal disease,[90-93] have been extensively probed to better understand events linked to osteoclast development and action. The use of genetically engineered or transgenic animals promises advances in recognizing how specific gene products modulate the formation and behavior of osteoclasts. Recent findings by Soriano et al.[94] demonstrated that transgenic mice containing a major disruption of the proto-oncogene c-*src* develop osteopetrosis. Subsequently, it was discovered that high levels of c-*src* are detectable within the bone environment only in the osteoclast.[95] This unexpected observation has generated excitement in the field, and many laboratories have initiated experiments to determine whether osteoclasts from such animals (which are present in normal numbers but lack ruffled borders and resorptive capability) are themselves defective or are responding to an altered local microenvironment.

4. Histochemical and Immunohistochemical Studies

Although Hancox's comments[1] suggested that descriptive microscopic evaluation of the osteoclast was, for the most part complete, he was aware that substantial improvements in tissue preparations and histochemical methods could provide answers to new and unresolved issues. In recent years, the osteoclast has been carefully scrutinized with an array of immunological techniques, using light or electron microscopic methods. In addition, imprints of osteoclasts prepared by pressing pieces of bone onto agar-coated slides,[96,97] following a procedure originally described by Walker,[98] have been informative. Monoclonal antibodies have provided researchers with an expanded panel of potential enzymes, receptors, differentiation markers, and other determinants to explore the relationship of osteoclasts to other cells of the mononuclear phagocyte family.[99-103] The prime activity and defining attribute of the osteoclast is its ability to resorb bone. From immunohistochemical and immunocytochemical studies, it has been shown that osteoclasts are polarized cells on bone that are endowed with a rich complement of acid hydrolase and carbonic anhydrase activities.[104-110] Although no longer considered exclusive to the osteoclast, these cells also possess a distinguishing tartrate-resistant acid phosphatase.[111,112] Immunohistochemical probing has led to the discovery of a group of osteoclast plasma membrane proteins, including a mannose 6-phosphate receptor,[113] and ATPase ion transport systems of the K^+,[114] H^+,[115] vacuolar H^+,[116-118] and Ca^{2+} [119] types. These enzymes are postulated to contribute to the acidification, proteolytic activation, mineral dissolution, and calcium translocation associated with the resorption process. The importance of osteoclast polarity with respect to its functional capabilities has been underscored in immunological experiments by Baron and co-workers.[113] Although it is evident that immunohistochemical methods provide a powerful tool for examining osteoclast characteristics *in vivo,* it is important to remember that tissue processing and fixation protocols may, under some circumstances, interject artifacts that can lead to false positive or negative results. Finally, as with all immunological procedures, investigators need to be cognizant of antibody cross-reactions, for in some cases such reactions may indicate antigenic components shared between the cross-related species, whereas in others the cross-reaction may merely describe a shared epitope (typically five to seven amino acids or carbohydrate moieties) in common between disparate molecules. Distinction can be made by Western blot probing or purification of the antibody-reactive antigen(s). At present, monoclonal antibody technology, in conjunction with light and transmission electron microscopy, remains one of the most powerful approaches for studying osteoclast development and behavior, and for the discovery of new properties of these cells.

C. ORGAN CULTURE SYSTEMS

1. General

As described earlier, the fetal long-bone organ culture system of Raisz and Niemann[15,16] prompted research geared toward deciphering the mode of action and regulation of the osteoclast. That system and its progeny have served as the workhorse paradigm for learning about the effects of specific agents upon bone resorption. Stern and Raisz[120] have reviewed and discussed the approaches, relative merits, and possible pitfalls of organ culture systems for investigating osteoclast activity and development. The Raisz-Niemann model requires that pregnant mice or rats receive ^{45}Ca- and/or ^{3}H-proline to label fetal long bones and calvaria. After a typical 24-h labeling period, the fetal bones are removed and placed in defined media on either Millipore filters or stainless steel rafts. Following equilibration of the organ cultures, test substances are added, and the subsequent release of radiolabeled calcium or proline is monitored as an indication of resorptive activity. Bones are usually paired, with one serving as a control relative to the test material added. The radiolabel analysis is usually compared with the histological profile of the organ culture to look for parallel alterations in osteoclast number or resorption. In this way, individual hormonal effects, coordinated endocrine influences, immune modulation of bone resorption, and cytokine regulation of resorptive activity have been investigated. The strengths of this system lie in its reproducibility, inherent balance toward resorptive activity, potential for retaining intact many of the original cell interactions, and its simplicity of design. However, counterbalancing weaknesses have sometimes led to confusion in the literature with regard to interpretation of organ culture results. For instance, direct effects cannot be distinguished from potential indirect actions on a target cell population. This was apparently the case when it was reported that PTH stimulated bone resorption by direct action on osteoclasts.[121] The more widely accepted general view, based on other studies, is that PTH may exert its stimulatory action through an effect on intermediary cells, possibly osteoblasts, which in turn influence the osteoclast. The organ culture system does not allow one to discriminate between these options, nor to necessarily extrapolate from results obtained with fetal or neonatal animals to predicted responses that might occur in older organisms. Despite these limitations, the bone organ culture systems represent an exceedingly valuable means of studying bone formation and resorption events.

2. Developmental

Bone organ cultures have also been instrumental in refining our understanding of osteoclast development and fate. Both the calvarial organ culture system by Ko and Bernard[122] and the mouse metatarsal model designed by Burger and co-workers[123,124] have enabled probing of the cell lineage and cell interaction events occurring during osteoclast development. More recently, the mouse metatarsal system has been coupled to a hydrostatic mechanism eliciting intermittent changes in pressure, so that alterations in bone modeling and remodeling induced by physical force may be examined.[123] Klein-Nulend et al.[125] demonstrated that the application of intermittent force to mouse calvaria led to an increase in bone formation concomitant with diminished bone resorption. Taylor et al.[126] have shown in the mouse metatarsal system that calcitonin receptors on preosteoclasts and osteoclast-like cells could be detected in autoradiographs of cells preincubated with radiolabeled calcitonin. Braidman et al.[127] cultured day 19 fetal rat calvaria (with a few osteoclasts present) in liquid media or on collagen gels in order to monitor osteoclast differentiation by neutral red staining or by acid phosphatase histochemical localization. The findings revealed that, whereas the morphology and enzymatic activity of the cells of the calvaria cultured in liquid media deteriorated, the calvaria cultured on type I collagen gels supplemented with calcium, β-glycerophosphate, proline, and ascorbate supported osteoclast differentiation. From these examples and additional studies, it is clear that the mouse metatarsal and calvarial organ culture systems have been useful models for defining precursor populations for the osteoclast and for determining that osteoblast products, both diffusible and matrix-associated, are required for osteoclast cytodifferentiation. Due to the small amounts of material under study in these systems, it has not been feasible to identify the source of any bone-derived factors nor to purify the agents responsible for the activation or differentiation of the osteoclast. Instead, analysis is more appropriately directed toward histological assessments and radiolabel release studies for discerning resorptive changes in the explants.

D. ISOLATED CELL SYSTEMS

There are great benefits to be realized by working with purified isolated osteoclasts, but these can be offset by limitations imposed by removal of osteoclasts and their precursors from normal interactive influences with neighboring cells and complex local and systemic modulation. A major advantage of dealing with isolated purified cell populations is the ability to attribute a specific response to the cell under study. Receptor binding, signal transduction, and cell-matrix interaction studies are amenable to close scrutiny in pure populations of isolated cells. Moreover, if sufficient material can be obtained, biochemical and molecular analyses become possible. A considerable effort has been directed toward isolating and purifying osteoclasts and their precursor cells in order to study individual cell function, factors influential in osteoclast development and differentiation, the role of cell-cell and cell-matrix interactions in osteoclast formation and behavior, and the interfacing of these processes with endocrine, immune, mechanical, and other signals. This discussion will be divided into systems utilizing authentic and surrogate osteoclasts, and systems designed to investigate osteoclast formation and cytodifferentiation. Recent articles[7,46] reviewing these systems are recommended as additional reading.

1. Authentic Osteoclasts

The number of osteoclasts is low when compared to other cell types present in bone. Osteoclasts represent a morphologically and functionally heterogeneous cell population. This heterogeneity stems from their wide variation in size (20 to 100 μm in diameter), nuclear content (anywhere from 2 to 100 nuclei per cell), age, stage of differentiation, and physiological status — parameters which affect bone-resorbing activity, hormonal responsiveness, and adherence to bone.[128] Osteoclast tight attachment to bone, inaccessibility within resorption bays, high frequency of proteolytic activities, and size heterogeneity complicate attempts to isolate and purify sufficient numbers of viable cells to perform biochemical, physiological, and molecular analyses. Multiple methods have been developed over the years to isolate small populations of mammalian osteoclasts for single-cell analysis or for procedures not requiring large numbers of cells. Mammalian osteoclasts have been successfully isolated from rabbits, rats, mice, and humans. For research on much larger populations of osteoclasts, chicken and quail have provided a convenient, reliable source.[46]

During the 1970s, reports appeared that described studies on isolated authentic osteoclasts.[41,42] The pioneering approach used by Nelson and Bauer in 1977[41] took advantage of the large size of osteoclasts to isolate these cells from rabbits using unit gravity sedimentation. Although the number of isolated cells was low and their viability poor, this technique served as the foundation for many of the current protocols which have been subsequently developed for osteoclast isolation. In 1979, Chambers[42] reported successful culture of a small number of viable osteoclasts from rabbit bone. In 1982, four papers were published that described various procedures for isolating and culturing rabbit,[40] rat,[45] embryonic chicken,[43] and hen osteoclasts,[44] although only small numbers of osteoclasts could be obtained from the mammalian sources. Chambers and Magnus[40] used curetting procedures to remove rabbit osteoclasts from the bone surfaces, which were then selectively enriched for osteoclasts based on the latter's rapid sedimentation and adhesion properties. Longer sedimentation times introduced osteoblasts and various other cell types. This method was valuable for determining whether osteoclast hormonal and cytokine responses were elicited either directly or via other intermediary cell types, such as osteoblasts. The limitations of this strategy relate to the number of cells obtained and to concerns that the cell heterogeneity of the system is not fully defined. In an alternative procedure, Helfly and Stern[45] treated bones with hyaluronidase and/or calcitonin to effect cell release; however, they noted that viability was poor following the use of calcitonin. In avian osteoclast methods, Osdoby and co-workers[43] used sequential incubations with trypsin and collagenase to release osteoclasts from day 19 embryonic chick bones. Comparisons of the final cell yield and purity obtained after filtration through Nitex mesh screens and Percoll density gradient separations were reported. Osteoclasts purified according to this scheme had therefore been separated on the basis of both their size and buoyant density. Zambonin-Zallone and colleagues[44] introduced the use of the low-calcium modified diet for birds. Bones from these animals exhibited significantly increased osteoclast numbers and provided an important strategy for investigators to obtain large numbers of osteoclasts. In the Zambonin-Zallone et al. method, osteoclasts were then isolated by physical scraping of the bone surface. Released cells were isolated and purified by unit gravity sedimentation in the presence of protease inhibitors at low temperature to minimize potential proteolytic damage resulting from disrupted cells. Following their purification, the cells were cultured in the presence of cytosine arabinoside to inhibit overgrowth by rapidly proliferating cells. In a later report, Gay and co-workers[129] used a brief calcitonin incubation to aid osteoclast release from chick bone, followed by sequential filtrations through nylon mesh screens. These multiple reports precipitated a workshop at the 1983 annual meeting of the American Society for Bone and Mineral Research that was entitled "Isolation and Properties of Osteoclasts". Gay presented a review of the methods available at that time to isolate osteoclasts and provided a summary table outlining the approaches in use. This table is presented here with Gay's permission (Table 1).[130] It can be seen from this summary that the procedures listed vary dramatically in osteoclast population yields and purity, with the avian systems consistently providing the greatest number of cells. The 1983 workshop elicited discussion on a range of topics related to osteoclast identification, yield, purity, and viability. Apart from osteoclast identification and yield, the issues of greatest concern for experimental design and data interpretation continue to be those pertaining to cell purity, viability, and, most recently, species of origin.

Purity and viability — The Zambonin-Zallone et al.[44] method allowed purity to be increased by repeating the sedimentation step. Usually, greater than 90% purity could be achieved after three sedimentations. However, as a consequence of repeated trials, both yield and viability were heavily compromised. In 1985, Oursler and co-workers[103] reported using the low-calcium diet with chick hatchlings to increase osteoclast yield. Sufficient numbers of osteoclasts were produced in this way to enable screening of monoclonal antibodies raised against isolated osteoclasts. Blair and colleagues[131,132] used

TABLE 1
Outline of Osteoclast Isolation Methods

Source of osteoclasts	Method of releasing cells	Enrichment method	Purity	Yield	Ref.
Rabbit femur, 12—14 d	Curette	Unit gravity sedimentation	30% of volume 73% of volume	1.5—1.8 × 10^4/8 bones	41
Rabbit femur, 12—14 d	Vigorous pipetting	Adherence to glass	40% of nuclei	Millions	42
Rabbit femur, 1—4 d	Curette, pipetting	Adherence to glass		Tens of osteoclasts	40
Fetal rat, 19 d long bones	Hyaluronidase	Culturing with 1,25 $(OH)_2D_3$	15% of cells	50 osteoclasts/bone	45
Chick embryo, 19 d tibia	Trypsin-EDTA; collagenase	Sequential filtration, Percoll gradients	50—75% of cells	Millions	43
Laying hen medullary bone	Scraping	Low-calcium diet, unit gravity sedimentation	90—95% of cells	1—1.5 × 10^6/hen	44
Chick metatarsi, 2 weeks	Calcitonin; trypsin; collagenase-EDTA	Sequential filtration	62% of cells, 78% of volume	3.5 × 10^5/10 bones	129

Table courtesy of Dr. Carol Gay from the 1983 Osteoclast Workshop, American Society for Bone and Mineral Research. References from this chapter.

a modification of the Zambonin-Zallone et al.[44] method in conjunction with selective osteoclast attachment to small bone particles to improve osteoclast purity. These isolated cells were capable of releasing radiolabeled proline from prelabeled rat bone particles, suggesting that at least some portion of the isolated cell population was actively resorbing bone. However, neither study documented cell viability of their osteoclast population. In 1988, de Vernejoul et al.[133] reported that only 10% of the avian osteoclasts isolated by a modification of the methods of Zambonin-Zallone et al.[44] and Oursler et al.[103] were viable when stained with fluorescein diacetate and ethidium bromide. More recently, Oursler et al.[134] confirmed this finding and proceeded to develop methods for selectively isolating viable osteoclasts from chick hatchlings. The strategy entailed maintaining the hatchlings on a low-calcium diet for three to four weeks. Cells were then removed from isolated bones by a sequential series of mechanical and enzymatic procedures, yielding three subpopulations. The first and second populations were high in osteoclast yield but low in viability, whereas the third bone-derived cell population contained fewer total osteoclasts but a substantially larger proportion of viable cells (1 to 2 × 10^6 viable osteoclasts per 15 chicks) (Table 2). Owing to the varying sizes and buoyant densities of the viable osteoclasts, this purification regimen preferentially selected for the largest osteoclasts, while smaller viable osteoclasts were not routinely recovered. Collin-Osdoby et al.[135] have overcome this limitation, while increasing both cell yield and purity, by employing osteoclast-specific monoclonal antibodies coupled to small magnetic beads. Using this procedure, they were able to generate viable osteoclast populations exhibiting purity on the order of 85 to 95%, viability as high as 99%, and

TABLE 2
Yield, Purity, and Viability of Percoll-Fractionated Population 1, 2, and 3 Osteoclasts

Population	6% Percoll location[a]	% OC purity	% viable OCs[b]	OC yield[c]	
				Total	Viable
1	Top	40	0.5	4.0×10^7	2.0×10^5
1	Bottom	40	2.5	8.0×10^6	2.0×10^5
2	Top	30	0	6.0×10^6	0
2	Bottom	40	0	4.0×10^6	0
3	Top	5	64	2.5×10^6	1.6×10^6
3	Bottom	9	50	2.8×10^6	1.4×10^6

Note: Osteoclasts were sequentially removed from the bones of 15 chicks fed a low-calcium diet by shaking (population 1), collagenase/shaking (population 2), and trypsin/shaking (population 3). Each subpopulation was fractionated by Percoll density centrifugation.

[a] Top refers to cell suspension overlaid onto a 6% Percoll column and removed after 60 min. Bottom refers to cells that have penetrated the 6% Percoll within 60 min of sedimentation.
[b] Viability was assessed by trypan-blue dye exclusion. Dye exclusion was highly correlated with adherence in culture, since at least 95% of the adherent cells were viable (dye excluded), and less than 5% of the unattached cells (after 5 h in culture) were viable.
[c] Yield reported is total derived from tibiae and humeri of 15 chick hatchlings.

From Oursler, M., Collin-Osdoby, P., Anderson, R., Li, L., Webber, D., and Osdoby, P., *J. Bone Miner. Res.*, 6, 375, 1991. With permission.

yields in the range of 5 to 7 $\times$ 10^6 cells from 15 chicks (Table 3, Figure 2). The osteoclasts collected by this approach have been shown to resorb bone and to respond to modulators of bone resorption. In this context, this immunomagnetic osteoclast isolation procedure has been used to obtain cells to purify RNA for production of an osteoclast cDNA library[136] and for Northern blot analysis. One series of studies has already demonstrated that osteoclasts synthesize mRNA coding for estrogen receptors and that osteoclastic resorption[137] and proteolytic enzyme gene expression[138] are downregulated by exposure to this sex steroid. The high purity and yield of immunomagnetically isolated osteoclasts could potentially facilitate sensitive physiological studies, such as those designed to determine whether a hormone, cytokine, or other modulator of bone remodeling acts directly on the osteoclast. In addition, this method may provide cellular material suitable for biochemical and molecular characterization of the osteoclast phenotype, and for protein profiles of structural, metabolic, and secretory proteins involved in osteoclast activity, and cell-cell or cell-matrix interactions. Since at least one of the chick osteoclast-specific monoclonal antibodies (121F) also recognizes rat and human osteoclasts, the immunomagnetic method possesses the added promise of being valuable for the isolation of relatively pure rat, human, and

TABLE 3
Viability and Purity of Osteoclasts Immunoisolated from Bone Cell Subpopulations

Population	Total OCs	Viable OCs	% viable OCs	% OC purity
2	3.20×10^6	3.20×10^5	10	92
3	4.86×10^6	4.80×10^6	99	86

Note: Population 2, collagenase-released osteoclasts (one sixth of an osteoclast preparation from 15 chicks fed a low-calcium diet), and population 3, trypsin-released osteoclasts (three fifths of a preparation), were incubated ice-cold with 2×10^7 and 1×10^7 121F antibody-conjugated beads, respectively. Bead-bound cells were magnetically sorted, washed, and cultured for 5 h prior to light microscopic determinations of total osteoclast number (yield of immunomagnetically sorted OCs), and osteoclast purity and viability (trypan blue dye exclusion, adherence in culture, and fluorscein diacetate staining).

From Collin-Osdoby, P., Oursler, M., Webber, D., and Osdoby, P., *J. Bone Miner. Res.*, 6, 1353, 1991. With permission.

perhaps other mammalian osteoclast populations. Alternatively, negative selection by immunomagnetic removal of osteoclasts might be performed to generate osteoclast-deficient bone cell populations. One caveat to the use of this procedure concerns the potential for the cell surface-bound antibodies, required for the cell isolation, to alter cell activity. This issue deserves careful consideration with respect to performing physiological and cell interaction studies.

Species considerations — It is clear from the previous discussion that investigators have experienced difficulty in obtaining sufficient numbers of mammalian osteoclasts. Many researchers have used the methods of Chambers and Magnus[40] and others to isolate, culture, and study mammalian osteoclasts. Although at the present time, large numbers of authentic osteoclasts can only be prepared from birds, their use has been circumscribed by a series of concerns. First, there is some uncertainty as to whether avian and mammalian osteoclasts display equivalent activity and hormonal responsiveness. Second, there is concern over the low-calcium diet that many avian isolation methods use, since along with elevating osteoclast number, this diet might generate osteoclasts different from those of animals maintained under normal dietary conditions. Foremost, there have been conflicting reports as to whether chicken osteoclasts possess calcitonin receptors and can respond to this hormone in a manner similar to mammalian osteoclasts.[139-144] Since calcitonin is a prime regulator of mammalian osteoclast activity, the failure of chicken osteoclasts to possess these receptors and to respond to this hormone would signify a serious drawback for those wishing to extrapolate information derived from chickens to osteoclast pathologies associated with human disease. At this time, there are probably as many reports documenting the presence of calcitonin receptors on chicken osteoclasts[139-141] as there are those which refute

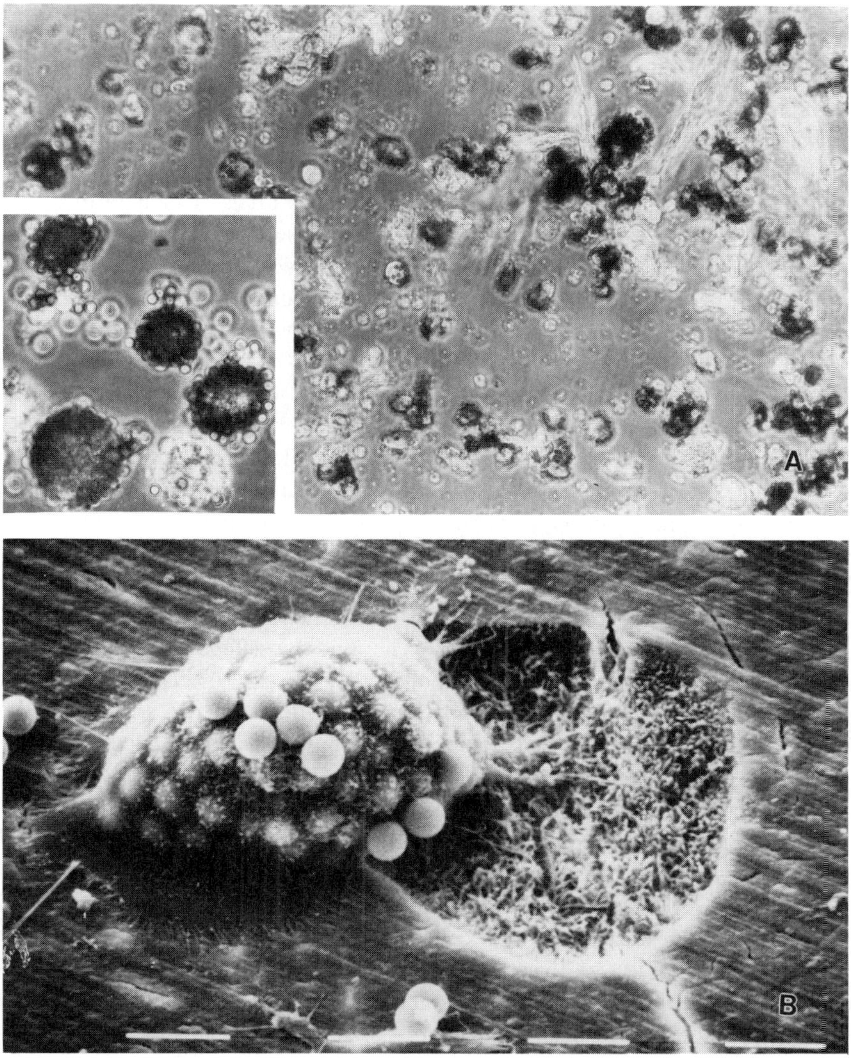

FIGURE 2. (A) Low-power light photomicrograph of immunomagnetically purified avian osteoclasts. The dark material represents magnetic beads coating the numerous osteoclasts. Insert illustrates a higher magnification to illustrate beads bound to osteoclasts. (B) Scanning electron photomicrograph of an immunomagnetically purified avian osteoclast cultured on a bovine cortical bone slice. Note many beads on and within the cell and that the osteoclast has formed a characteristic resorption pit. (From Collin-Osdoby, P., Oursler, M., Webber, D., and Osdoby, P., *J. Bone Miner. Res.*, 6, 1353, 1991. With permission.)

it.[142-144] What appears clear is that calcitonin receptors which are present on chicken osteoclasts must be so at levels that are much reduced from those of the rat. On the whole, avian and mammalian osteoclasts demonstrate more similarities than differences. Aside from the controversy over avian calcitonin receptors, several reports have suggested that there may be subtle differences in osteoclast resorptive activity among different vertebrates. Jones et al.[145,146] demonstrated that mammalian and avian osteoclasts can resorb similar mineralized substrates, but that bone-resorbing capability was highest in chick osteoclasts, followed next by those obtained from rabbits, and then rats.[147] Therefore, there appear to be differences in resorptive activity among mammalian species as well as between birds and mammals, which further emphasizes the need for caution when extrapolating results from a study using one species to that of another. The ideal situation would permit similar studies to be performed using cells from a variety of species, including humans[148] and higher primates.

2. Surrogate Cell Populations

Since mammalian osteoclasts are such a limited commodity, investigators have at times used surrogate cell populations. Two general categories of surrogate cells are currently in use: giant cells and transformed cells. The first type, represented by giant cells derived from naturally occurring giant cell tumors of bone, has been probed for characteristic osteoclast features, such as the presence of calcitonin receptors, tartrate-resistant acid phosphatase activity, the ability to resorb bone, and an immunological profile consistent with the osteoclast phenotype.[149-151] The findings suggest that giant cells derived from giant cell tumors of bone possess many osteoclast attributes, especially their unique ability to resorb bone. Unfortunately, the giant cells from these tumors are not a transformed population and are, therefore, incapable of replenishing themselves. Furthermore, it is difficult to obtain a regular supply of fresh tumor material with which to continue or initiate new studies. An additional complication pertains to the unusually large size and excessive multinuclearity of these tumor-derived giant cells when compared with osteoclasts derived from normal bone. As with any surrogate system, experimental results using such cells could potentially be nonrepresentative of osteoclasts. Despite these caveats, the giant cells obtained from human bone tumors have allowed for substantial advances in our understanding of human osteoclasts. The second category of surrogate cells is the transformed cell lines which have expedited experimental description of the regulation and activity of bone-forming osteoblasts. To date, however, there have not been any reports of similarly transformed differentiated osteoclast cell lines. Since mature osteoclasts are nondividing, fully differentiated, multinucleated cells, it has not been possible to immortalize or transform them. At best, researchers have utilized transformed human or other mammalian cells related to cells of the monocytic family as a source of continually propagating cells from which osteoclast-like cells may be induced to form *in vitro*. Yoneda et al.,[152] using the human promyelocytic HL-60 cell line, cultured cells in methylcellulose, and treated them with conditioned media obtained from MH85 tumor cells followed by 1,25-dihydroxyvitamin D_3, to form a subpopulation of multinucleated cells that were capable of resorbing bone in culture. The lengthy culture time and multiple manipulations involved in this system have, however, discouraged its widespread acceptance for general use. Recently, another human promyelocytic cell line, FLG 29.1, has been shown to give rise to osteoclast-like cells when treated with the signal-generating, differentiation-inducing phorbol ester, TPA.[153] After TPA exposure, cells become multinucleated, resorb bone particles, express distinctive osteoclast phenotypic markers, and respond to calcitonin. However, unlike mature osteoclasts, these cells fail to form resorption pits on cortical bone slices. Additional manipulation of this system may prove necessary to elicit formation of mature osteoclast-like cells capable, under appropriate conditions, of producing well-excavated resorption pits on mineralized substrates.

3. Bone Marrow Cultures and Osteoclast Development

Knowledge that osteoclasts are of hematopoietic origin, and early reports indicating that avian[154,155] or feline[156,157] marrow cells form multinucleated giant cells *in vitro,* suggested that osteoclasts could be produced *in vitro* from precursor populations. If this proved to be the case, then investigators could potentially circumvent the need for the isolation of sufficient numbers of mature osteoclasts for biochemical and physiological analysis. These marrow and monocyte cultures provide the added potential for identifying osteoclast precursors and for monitoring their development under the influence of factors or events suspected of modulating osteoclast differentiation. Chicken, cat, mouse, baboon, and human marrow systems[158-163] have all been used to generate osteoclast-like cells *in vitro.* A number of excellent reviews describe these systems in detail,[46,164] including Chapters 4 and 5 of this volume. Therefore, only a few relevant comments will be made here. Species differences have been noted with respect to the ease with which multinucleated cells can be encouraged to form in culture. Chicken and mouse appear to pose the fewest problems, whereas rat marrow has proven to be problematic. Although the originally nonadherent or less differentiated cell population from mammalian sources[161] may contain most of the osteoclast precursors, this may not be true for avian marrow systems.[158] The mouse system, under optimal conditions, seems to give rise to a greater proportion of multinucleated cells active in pit resorption on cortical bone than do other systems.[160] Aside from species variations, investigators employ diverse protocols in their preparation and culture of marrow cells. Some researchers use whole marrow,[165] whereas others, in order to select specific subpopulations, fractionate the marrow by Ficoll-Hypaque separation,[158] Percoll density centrifugation,[156] or antibody reaction.[166] Therefore, the starting marrow populations of these systems can range in heterogeneity from highly mixed to relatively restricted, which may well impact on the outcome of studies aimed at analyzing humoral, cytokine, and cell-interactive parameters. It has been evident that not all the giant cells formed in these cultures express the full profile of distinguishing osteoclast features. In fact, cells at varying stages of osteoclastic differentiation and activity can exist simultaneously, especially in whole-marrow culture systems, due to the presence of other mononuclear cells, including marrow stromal cells and osteoblasts. Complications resulting from contaminating cells therefore make certain marrow systems less desirable than others for particular experimental purposes. One way to partially overcome the problem of cell heterogeneity for inquiry into the regulatory and behavioral aspects of these cells is to separate the giant cells from the mononuclear cells prior to the experimental study. In addition, it is recommended that osteoclastic phenotypic traits be assessed both before and after any experimental manipulation of the cells in order to evaluate whether the modulator altered the activity of the fully differentiated cell or, alternatively, acted to promote or suppress further differentiation of the giant cells. Particularly noteworthy are a series of careful studies performed by Suda and co-workers,[160,167,168] as well as those of other researchers (for reviews, see References 164 and 169), using marrow cultures in combination with osteoblasts and stromal cell populations, since these studies have provided valuable insights into how cell interactive events interface with humoral regulation. In a similar vein, the elegant studies conducted by Roodman, Mundy,[160-162,164] and others[166,169] have paved the way for understanding endocrine and paracrine control of osteoclast formation and activity. No doubt, as we learn more about osteoclasts, we will be able to clarify how closely these marrow-derived, osteoclast-like cells resemble bona fide osteoclasts. A challenge before us is to document the similarities and differences between these cell populations so that we may better exploit them to promote an improved understanding of the process of bone remodeling. Finally, it should be noted that some differences between organ and cell cultures in response to cytokines and osteotropic hormones have been observed. Careful scrutiny of these disparities could reveal the basis for their occurrence and, ultimately, help untangle the individual and interactive events that participate in osteoclast regulation.

E. METHODS FOR ASSESSING OSTEOCLAST DEVELOPMENT AND ACTIVITY

Osteoclast function and regulation can be examined in either individual cells or populations of isolated osteoclasts. Enzyme activity, cytoskeletal modifications, motility, and response to hormones or pharmacological agents have been monitored using histochemistry, immunocytochemistry, autoradiography, cinematography, time-lapse photography, and bone resorption assays. Boyde[170] and Chambers[171] are credited with the development of the resorption pit assay, whereby isolated osteoclasts and accompanying contaminating cells are cultured on thin slices of mineralized dentin, cortical bone, or other mineralized substrates, such as egg shell, for varying periods of time.[172] This material is later processed, with or without the removal of the cells, for scanning electron microscopic examination. Osteoclast excavation sites can be visualized and the amount of resorbed bone quantitated by several methods. In early studies, Boyde analyzed pit number, size, depth, and area by means of stereophotogrammetry.[172] More recently, confocal microscopy has provided comparable data analysis.[173] The formation of resorption lacunae on mineralized substrates has become the standard assay by which osteoclast function is measured, and the protocol of choice for determining whether marrow precursor cells have differentiated into active mature osteoclasts. There are, unfortunately, a number of problems associated with this technique. For example, the quantitative analysis of resorption lacunae requires expensive and specialized equipment, including computer software, to achieve statistically relevant findings. The method is also labor intensive, time consuming, and requires well-trained, specialized investigators. For many purposes, it is sufficient to demonstrate that a cell preparation can produce resorption pits on a mineralized substrate. An alternative approach pioneered by Arnett and Dempster uses the technique of visualizing resorption pits stained with toluidine blue by light microscopy.[144,174,175] When integrated with computer-based data analysis, this method reduces preparation time and simplifies instrument requirements for resorption quantitation, and as a result, investigators can accumulate reliable estimates of pit number and area quickly. The trade-offs include uncertainty with regard to the validity of depth and volume estimates measured this way, so that precise values for the quantity of bone resorbed may not be attainable.[176] Alternatives to toluidine blue staining of pits include immunodetection of collagen fibrils exposed by osteoclast degradation or antibody localization of other matrix components unveiled in the resorptive process.[177]

A more quantitative, although perhaps less definitive method of assessing bone resorptive activity is provided by the bone particle resorption assay.[36,131,132] For this assay, radiolabeled, uniformly small bone particles are prepared from fetal bones metabolically labeled *in utero* with either ^{45}Ca- or ^{3}H-proline. Osteoclasts, or other test cells, are then plated prior to or along with the bone particles, and measurements of radiolabel release over time in culture are taken as a reflection of bone resorption. In general, particles prelabeled with ^{3}H-proline, in contrast to ^{45}Ca, engender less background release. In the bone particle assay, it is of prime importance to ensure that the "resorption" being monitored represents extracellular matrix degradation, as opposed to intracellular phagocytic degradation, since the latter can be accomplished by many nonosteoclastic cells. For example, monocytes and macrophages have been shown to release radiolabel from such particles.[36,37] One way to alleviate this concern is to employ sufficiently large bone particles to minimize phagocytosis. Even so, it is expected that the debate will continue over whether this assay actually simulates osteoclastic bone resorption.

Improved techniques for monitoring single-cell physiology by microelectrode probes,[178] patch-clamp analysis,[179] fluorescent dye binding,[180-182] or advanced image analysis[182,183] have provided valuable procedural ways to examine individual osteoclast function in response to specific modulators. Changes in pH,[178] ion fluxes,[181-183] and signal transduction[184,185] have all been studied using these methods and have yielded insights into osteoclast activity and regulation. Despite the obvious strengths of such techniques, there are also counterbalancing limitations restricting their utility or the conclusions drawn from studies employing these strategies. Inaccuracies can arise and be magnified when results obtained from individual cases are extrapolated to a population as a whole, making it necessary to examine a statistically significant number of cells under identical experimental conditions. This, then, leads to discussion of a further problem typical of the majority of isolated cell systems currently available, namely, the effects of the heterogeneity of the isolated cell population (as discussed earlier). In addition, many studies have been performed using tissue culture plastic as a substratum, and it should be recognized that the absence of a natural extracellular matrix could markedly alter many of the dynamic events under study.

The large yield of osteoclasts possible from avian sources, along with improved cell fractionation procedures and immunological reagents, have fostered many studies to characterize osteoclast structural and functional proteins. Even with avian sources and advanced methods, the starting material is never homogeneous and is generally quite limited, in contrast to other biological systems where propagating cells or cell lines are available. Furthermore, the high proteolytic content of the osteoclast must be considered, and an array of protease inhibitors should be included in any biochemical study unless, of course, specific proteases are the targets of investigation.

Until now, experiments at the molecular level on osteoclasts or osteoclast surrogate cells have been few. This is mainly due to the relatively low number of primary isolated osteoclasts which can be obtained. Moreover, problems with osteoclast purity, viability, and damage have made it difficult to extract pure undegraded RNA for use in either Northern blot analysis or construction of a cDNA library. Selection for viable osteoclasts and high-purity immunomagnetic isolation methods[135] have enabled us to, in part, overcome these obstacles. However, from these studies, it has been noted that less than 1% (150 ng) of the total RNA extracted from viable osteoclasts (10^6) was polyadenylated mRNA.[136,137] Therefore, apart from *in situ* hybridization studies, it would appear that high yields of viable osteoclasts will generally be required for many molecular biological studies. Polymerase chain reaction methods might alleviate this limitation, if applicable to this cell type. However, concern over contaminating cell-derived RNA complicating data interpretation is warranted, given the low quantity of mRNA that our group has repeatedly found to be extractable from chicken osteoclasts. Investigators may circumvent some of the potential problems through the use of marrow-derived osteoclast-like cells or transformed cell-derived giant cells for sources of protein or RNA. Since, however, the similarities and differences between osteoclasts and these other giant cell populations await clarification, researchers should clearly acknowledge their source of material.

IV. CONCLUSIONS AND FUTURE DIRECTIONS

This is, without doubt, an exciting and productive period in osteoclast research, with great promise for the future. The last decade witnessed the development of advanced methods for isolating and culturing osteoclasts, and for generating osteoclast-like cells from marrow precursors *in vitro*. Development of an *in vitro* biological resorption pit assay and a series of immunological reagents, along with new microscopic and single-cell analytical techniques, have rapidly expanded our understanding of osteoclast development, behavior, and mechanisms of regulation. These efforts have provided a foundation upon which future researchers can build. Increasingly sophisticated osteoclast isolation schemes and marrow-derived, osteoclast-like culture systems will likely be forthcoming. Identification of a specific osteoclast precursor may be achieved, which, through cloning and immortalization, would provide a powerful tool for further examination of osteoclast developmental events. Many of the signals that influence osteoclast differentiation and activity may be defined, and signal origin and interplay described. New insights into how osteoclasts affect osteoblasts and other cells within the bone environment will be illuminated.

Immunological and single-cell analytical techniques were the hallmarks of approaches in the last decade; molecular biology approaches may be more prevalent during this decade. The cloning of the calcitonin receptor gene is a case in point, and the construction of cDNA libraries from osteoclasts and related cells should catalyze this avenue of research. However, since bone represents a complex tissue with a multitude of interactions, it will be necessary to continue to use whole animal, organ, and cell systems in concert with ever-evolving analytical tools, including molecular biology. This combination of approaches, along with carefully designed experimental strategies, should help resolve outstanding issues in osteoclast biology in particular, and bone remodeling in general. Questions which have not yet been entertained will probably arise in the course of future studies, again necessitating clever and innovative improvements in methods and analysis. Future progress in this field should unveil new aspects of osteoclast basic biology and holds promise for clinical therapies to alleviate pathologies resulting from abnormalities in the development, behavior, and regulation of the bone-degrading osteoclast.

ACKNOWLEDGMENTS

The authors thank Dr. Carol Gay for permitting us to use the table prepared for the 1983 Osteoclast Workshop. We also thank Mrs. Cheryl Genovese for preparing this manuscript.

REFERENCES

1. **Hancox, N. M.,** The osteoclast, in *Biology of Bone,* Vol. 1, 2nd ed., Harrison, R. J. and McMinn, R. M., Eds., Academic Press, New York, 1972, 45.
2. **Kolliker, A.,** *Die normale resorption des knochengewebes und ihre Bedentung fur die Entstehung der typischen Knochenformen,* Vogel, Leipzig, 1873.
3. **Jordan, H. E.,** Histological preparation of jaw-bone of newborn rat, showing osteoclasts containing globules of osseous material, *Anat. Rec.,* 18, 268, 1928.
4. **Jaffee, H. L.,** The absorption of bone, *Chicago Arch. Surg.,* 20, 355, 1930.
5. **Arnold, J. and Jee, W.,** Bone growth and osteoclastic activity as indicated by radioautographic distribution of plutonium, *Am. J. Anat.,* 101, 367, 1957.
6. **Holtrop, M.,** Light and electron microscopic structure of osteoclasts, in *The Osteoclast,* Vol. 2, Hall, B., Ed., CRC Press, Boca Raton, FL, 1991, chap. 1.
7. **Vaes, G.,** Cellular biology and biochemical mechanism of bone resorption, *Clin. Orthop. Relat. Res.,* 231, 239, 1988.
8. **Gothlin, G. and Ericsson, J.,** The osteoclast, *Clin. Orthop. Relat. Res.,* 120, 201, 1976.
9. **Delaisse, J., Ledent, P., Eackhout, Y., and Vaes, G.,** Cysteine proteases and bone resorption, in *Cysteine Proteinases and Their Inhibitors,* Turk, V., Ed., de Gruyter, Berlin, 1986, 259.
10. **Gay, C. and Mueller, W.,** Carbonic anhydrase and osteoclasts: localization by labeled inhibitor autoradiography, *Science,* 183, 432, 1974.
11. **Doty, S.,** Histochemistry and enzymology of osteoclasts, in *The Osteoclast,* Vol. 2, Hall, B., Ed., CRC Press, Boca Raton, FL, 1991, chap. 3.
12. **Holtrop, M., King, G., Cox, K., and Reit, B.,** Time-related changes in the ultrastructure of osteoclasts after injection of parathyroid hormone in young rats, *Calcif. Tissue Int.,* 27, 129, 1979.
13. **Lucht, U.,** Effects of calcitonin on osteoclasts *in vivo.* An ultrastructural and histochemical study, *Z. Zellforsch. Mikrosk. Anat.,* 145, 75, 1973.
14. **Fell, H. and Robinson, R.,** The growth, development, and phosphatase activity of embryonic avian femora and limb buds cultivated in vitro, *Biochem. J.,* 23, 767, 1929.
15. **Raisz, L.,** Bone resorption in tissue culture. Factors influencing the response to parathyroid hormone, *J. Clin. Invest.,* 44, 103, 1965.
16. **Raisz, L. and Niemann, I.,** Early effects of parathyroid hormone and thyrocalcitonin in bone in organ culture, *Nature (London),* 214, 486, 1967.
17. **Hancox, N.,** Motion picture observations on osteoclasts *in vitro, J. Physiol.,* 110, 205, 1949.
18. **Hancox, N. and Boothroyd, B.,** Motion picture and electron microscope studies on embryonic avian osteoclasts, *J. Biophys. Biochem. Cytol.,* 11, 651, 1961.
19. **Goldhaber, P.,** Behavior of bone in tissue culture, in *Calcification in Biological Systems,* Sognnges, R., Ed., American Association for the Advancement of Science, Washington, D.C., 1960, 349.
20. **Goldhaber, P.,** The effect of hyperoxia on bone resorption in tissue culture, *AMA Arch. Pathol.,* 66, 635, 1958.
21. **Gaillard, P.,** Parathyroid gland and bone *in vitro,* in *The Parathyroids,* Greep, R. and Talmage, R., Eds., Charles C Thomas, Springfield, IL, 1961, 20.
22. **Marks, S.,** Osteoclast biology: lessons from mammalian mutations, *Am. J. Med. Genet.,* 34, 1989, 43.
23. **Walker, D.,** Congenital osteopetrosis in mice cured by parabiotic union with normal siblings, *Endocrinology,* 91, 916, 1972.
24. **Walker, D.,** Osteopetrosis cured by temporary parabiosis, *Science,* 180, 875, 1973.

25. **Coccia, P., Krivit, W., Cervenka, J., Clawson, C., Kershey, J., Kim, T., Nesbit, M., Ramsay, N., Warkentin, P., Teitelbaum, S., Kahn, A., and Brown, D.,** Successful bone-marrow transplantation for infantile malignant osteopetrosis, *N. Engl. J. Med.*, 302, 701, 1980.
26. **Fischman, D. and Hay, D.,** Origin of osteoclasts from mononuclear leucocytes in regenerating new limbs, *Anat. Rec.*, 143, 329, 1962.
27. **Joterau, F. and LeDouarin, N.,** The developmental relationship between osteocytes and osteoclasts: a study using the quail-chick nuclear marker in endochondral ossification, *Dev. Biol.*, 63, 253, 1978.
28. **Kahn, A. and Simmons, D.,** Investigation of cell lineage in bone using a chimaera of chick and quail embryonic tissue, *Nature (London)*, 258, 325, 1975.
29. **Krukowski, M., Simmons, D., and Kahn, D.,** Cell lineage studies, in *Skeletal Research*, Vol. 2, Simmons, D., Ed., Academic Press, New York, 1983, chap. 4.
30. **Horton, J., Raisz, L., Simmons, H., Oppenheim, J., and Mergenhagen, S.,** Bone resorbing activity in supernatant fluid from cultured human peripheral blood leucocytes, *Science*, 177, 793, 1972.
31. **Horton, J., Oppenheim, J., Mergenhagen, S., and Raisz, L.,** Macrophage-lymphocyte synergy in the production of osteoclast activating factor, *J. Immunol.*, 113, 1278, 1974.
32. **Baron, R., Vignery, A., and Horowitz, M.,** Lymphocytes, macrophages and the regulation of bone remodeling, in *Bone and Mineral Research 2*, Peck, W., Ed., Elsevier, Amsterdam, 1984, 175.
33. **Mundy, G. and Bonewald, L.,** Effects of immune cell products on bone, in *Macrophage-Derived Cell Regulatory Factors*, Vol. 1, Sorg, C., Ed., S. Karger, Basel, 1989, 38.
34. **Rifkin, B., Baker, R., and Coleman, S.,** An ultrastructural study of macrophage-mediated resorption of calcified tissue, *Calcif. Tissue Res.*, 202, 125, 1979.
35. **Mundy, G., Altman, A., Gondek, M., and Bandelin, J.,** Direct resorption of bone by human monocytes, *Science*, 196, 1109, 1977.
36. **Teitelbaum, S., Stewart, C., and Kahn, A.,** Rodent peritoneal macrophages as bone resorbing cells, *Calcif. Tissue Int.*, 27, 255, 1979.
37. **Kahn, A., Stewart, C., and Teitelbaum, S.,** Contact-mediated bone resorption by human monocytes in vitro, *Science*, 199, 988, 1978.
38. **Mundy, G.,** Monocyte-macrophage system and bone resorption, *Lab. Invest.*, 49, 119, 1983.
39. **Hogg, N., Shapiro, I., Jones, S., Slusarenko, M., and Boyde, A.,** Lack of Fc receptors on osteoclasts, *Cell Tissue Res.*, 212, 509, 1980.
40. **Chambers, T. and Magnus, C.,** Calcitonin alters behavior of isolated osteoclasts, *J. Pathol.*, 136, 27, 1982.
41. **Nelson, R. and Bauer, G.,** Isolation of osteoclasts by velocity sedimentation at unit gravity, *Calcif. Tissue Res.*, 22, 303, 1977.
42. **Chambers, T.,** Phagocytosis and trypsin-resistant glass adhesion by osteoclasts in culture, *J. Pathol.*, 127, 55, 1979.
43. **Osdoby, P., Martini, M., and Caplan, A.,** Isolated osteoclasts and their presumed progenitor cells, the monocyte, in culture, *J. Exp. Zool.*, 244, 331, 1982.
44. **Zambonin-Zallone, A., Teti, A., and Primavera, M.,** Isolated osteoclasts in primary culture: first observation on structure and survival in culture media, *Anat. Embryol.*, 165, 405, 1982.
45. **Helfley, T. and Stern, P.,** Isolation of osteoclasts from fetal rat long bones, *Calcif. Tissue Int.*, 34, 480, 1982.
46. **Zambonin-Zallone, A. and Teti, A.,** Isolation and behavior of cultured osteoclasts, in *The Osteoclast*, Vol. 2, Hall, B., Ed., CRC Press, Boca Raton, FL, 1991, chap. 4.
47. **Addison, W.,** The effect of parathyroid hormone on the numbers of nuclei in feline osteoclasts *in vivo*, *J. Anat.*, 130, 479, 1980.
48. **Tinkler, S., Williams, D., and Johnson, N.,** Osteoclast formation in response to intraperitoneal injection of 1α hydroxycholecalciferol in mice, *J. Anat.*, 133, 91, 1981.

49. **Miller, S.,** Osteoclast cell-surface specialization and nuclear kinetics during egg-laying in Japanese quail, *Am. J. Anat.*, 162, 35, 1981.
50. **Rifkin, B., Brand, J., Cushing, J., Coleman, S., and Sanavi, F.,** Fine structure of fetal rat calvarian provisional identification of preosteoclasts, *Calcif. Tissue Int.*, 31, 21, 1980.
51. **Key, L., Carnes, D., Cole, S., Holtrop, M., Bar-Shavit, Z., Shapiro, F., Acceci, R., Steinberg, J., Gundberg, C., Kahn, A., Teitelbaum, S., and Anast, C.,** Treatment of congenital osteopetrosis with high dose calcitriol, *N. Eng. J. Med.*, 310, 410, 1984.
52. **Parfitt, A.,** The actions of parathyroid hormone on bone: relation to bone remodeling and turnover, calcium homeostasis, and metabolic bone disease. I. Mechanisms of calcium transfer between blood and bone and their cellular basis: morphological and kinetic approaches to bone turnover, *Metabolism,* 25, 809, 1976.
53. **Powell, J. and DeMark, B.,** Clinical pharmacokinetics of diphosphonates, in *Bone Resorption, Metastasis and Diphosphonates,* Garattini, S., Ed., Raven Press, New York, 1985, 41.
54. **Glowacki, J., Cox, K., O'Sullivan, J., Wilkie, D., and Deftos, L.,** Osteoclasts can be induced in fish having an acellular bony skeleton, *Proc. Natl. Acad. Sci. U.S.A.*, 83, 4104, 1986.
55. **Epstein, S., Posee, J., McClintock, R., Johnston, C., Bryce, G., and Hui, S.,** Differences in serum bone gla protein with age and sex, *Lancet,* 1, 307, 1984.
56. **Riggs, B. and Nelson, K.,** Effect of long term treatment with calcitriol on calcium absorption and mineral metabolism in postmenopausal women, *J. Clin. Endocrinol. Metab.*, 61, 457, 1985.
57. **Uebelhart, D., Gineyts, E., Chapuy, M., and Delmas, P.,** Urinary excretion of pyridinium crosslinks; a new marker of bone resorption in metabolic bone disease, *Bone Miner.*, 8, 87, 1990.
58. **Carniggia, A., Nuti, R., Galli, M., Lore, F., Turchetti, V., and Righi, G.,** Effect of long-term treatment with 1,25-dihydroxy vitamin D_3 on osteocalcin in postmenopausal osteoporosis, *Calcif. Tissue Int.*, 38, 328, 1986.
59. **Krolner, B. and Nielson, S.,** Bone mineral content of the lumbar spine in normal and osteoporotic women: cross-sectional and longitudinal studies, *Clin. Sci.*, 62, 329, 1982.
60. **Genant, H., Cann, C., Ettingen, B., and Gordon, G.,** Quantitative computed tomography of vertebral spongiosa, a sensitive method for detecting early bone loss after oophorectomy, *Ann. Int. Med.*, 97, 699, 1982.
61. **Kanders, B., Lindsay, R., Dempster, D., Markhard, L., and Valiquetto, L.,** Determinants of bone mass in young healthy women, in *Osteoporosis: Proceedings of the Copenhagen Symposium on Osteoporosis,* Christiansen, C., Arnaud, C., Nordin, B., Parfitt, A., Peck, W., and Riggs, B., Eds., Department of Clinical Chemistry, Glostrup Hospital, Denmark, 1984, 337.
62. **Riggs, B., Seeman, E., Hodgson, S., Taves, D., and O'Fallon, W.,** Effect of the fluoride/calcium regimen on vertebral fracture occurrence in postmenopausal osteoporosis, *N. Engl. J. Med.*, 306, 446, 1982.
63. **Fleisch, J., Russell, R., and Francis, M.,** Diphosphonates inhibit hydroxyapatite dissolution *in vitro* and bone resorption in tissue culture *in vivo, Science,* 165, 1262, 1969.
64. **Johnston, C., Altman, R., Canfield, R., Finnerman, G., Taulbee, J., and Ebert, M.,** Review of fracture experience during treatment of Paget's disease with etidronate disodium (EMDP), *Clin. Orthop.*, 172, 186, 1983.
65. **Tilton, F., Degioanni, J., and Schneider, V.,** Long-term follow-up of skylab bone demineralization, *Aviat. Space Environ. Med.*, 51, 1209, 1980.
66. **Globus, R., Bikle, D., Halloran, B., and Morey-Horton, E.,** Skeletal response to dietary calcium in a rat model simulating weightlessness, *J. Bone Miner. Res.*, 1, 191, 1986.
67. **Morey, E. and Baylink, D.,** Inhibition of bone formation during space flight, *Science,* 201, 1138, 1978.

68. **Mostafa, Y., Weaks-Dybvig, M., and Osdoby, P.,** Orchestration of tooth movement, *Am. J. Orthod.*, 83, 245, 1983.
69. **Storey, E.,** Bone changes associated with tooth movement. A histological study of the effect of force in the rabbit, guinea pig and rat, *Austr. J. Dent.*, 59, 147, 1955.
70. **Roberts, W., Goodwin, W., and Heinen, S.,** Cellular response to orthodontic force, *Dent. Clin. North Am.*, 25, 3, 1981.
71. **Norton, L., Hanley, and Turkewics, J.,** Bioelectric perturbations of bone angle, *Orthodontics*, 54, 73, 1984.
72. **Brighton, C.,** Current concepts review — the treatment of non-union with electricity, *J. Bone J. Surg. Am. Vol.*, 63, 847, 1981.
73. **Chole, R., McGinn, M., and Tinling, S.,** Pressure induced bone resorption in the middle ear, *Ann. Otol. Rhinol. Laryngol.*, 94, 165, 1985.
74. **Wolfman, D. and Chole, R.,** Osteoclast induction by positive middle ear pressure, *Arch. Otolaryngol.*, 112, 1037.
75. **Tran Van, P., Vignery, A., and Baron, R.,** Cellular kinetics of the bone remodeling sequence in the rat, *Anat. Rec.*, 202, 445, 1982.
76. **Baron, R., Neff, L., Tran Van, P., Nefussi, J. R., and Vignery, A.,** Kinetic and cytochemical identification of osteoclast precursors and their differentiation into multinucleated osteoclasts, *Am. J. Pathol.*, 122, 363, 1986.
77. **Glowacki, J., Jasty, M., and Goldring, S.,** Comparison of multinucleated cells elicited in rats by particulate bone, polyethylene or polymethylmethacrylate, *J. Bone Miner. Res.*, 1, 327, 1986.
78. **Glowacki, J. and Lian, J.,** Impaired recruitment and differentiation of osteoclast progenitors by osteocalcin-deplete bone implants, *Cell Differ.*, 21, 247, 1987.
79. **Glowacki, J., Cox, K., and Wilson, S.,** Impaired osteoclast differentiation in subcutaneous implants of bone particles in osteopetrotic mutants, *Bone Miner.*, 5, 271, 1989.
80. **Krukowski, M. and Kahn, A.,** Inductive specificity of mineralized bone matrix in ectopic osteoclast differentation, *Calcif. Tissue Int.*, 34, 474, 1982.
81. **Webber, D., Osdoby, P., Hauschka, P., and Krukowski, M.,** Correlation of an osteoclast antigen and ruffled border on giant cells formed in response to resorbable substrates, *J. Bone Miner. Res.*, 5, 401, 1990.
82. **Groessner-Schreiber, B., Krukowski, M., Hertwick, D., and Osdoby, P.,** Osteoclast formation is related to bone matrix age, *Calcif. Tissue Int.*, 48, 335, 1991.
83. **Groessner-Schreiber, B., Krukowski, M., Lyons, C., and Osdoby, P.,** Osteoclast formation is related to human bone matrix age, *Mech. Ageing Dev.*, 62, 143, 1992.
84. **Popoff, S. and Marks, S.,** Ultrastructure of the giant cell infiltrate of subcutaneously implanted bone particles in rats and mice, *Am. J. Anat.*, 177, 491, 1986.
85. **Walters, L. and Schneider, G.,** A functional and morphological study of cells adjacent to ectopic bone implants in rats, *Am. J. Anat.*, 173, 287, 1985.
86. **Webber, D., Menton, D., and Osdoby, P.,** An *in vivo* model for the study of avian osteoclast recruitment and activity, *Bone Miner.*, 11, 127, 1990.
87. **Yoshida, H., Hayashi, S., Kunisada, T., Ogawa, M., Nishikawa, S., Okamura, H., Suda, T., Shultz, L., and Nishikawa, S.,** The murine mutation osteopetrosis is the coding region of the macrophage colony stimulating factor gene, *Nature (London)*, 345, 442, 1990.
88. **Felix, R., Cecchini, M., Hofstetter, W., Elford, P., Stutzer, A., and Fleisch, H.,** Impairment of macrophage colony-stimulating factor production and lack of resident bone marrow macrophages in the osteopetrotic op/op mouse, *J. Bone Miner. Res.*, 5, 781, 1990.
89. **Hofstetter, W., Wetterwald, A., Cecchini, M., Felix, R., Mueller, C., and Fleisch, H.,** Expression of M-CSF and of its receptor C-FMS in bone, as detected by *in situ* hybridization, *J. Bone Miner. Res.*, 6 (Suppl. 1), 199, 1991.
90. **Mills, B. and Singer, F.,** Nuclear inclusion in Paget's disease of bone, *Science*, 194, 257, 1976.

91. **Mundy, G. and Spiro, J.,** The mechanisms of bone metastasis and bone destruction by tumor cells, in *Bone Metastases,* Weiss, L. and Gilbert, H., Eds., G. K. Hall, Boston, 1981.
92. **Hough, A. and Sokoloff, L.,** Pathology in *Rheumatoid Arthritis,* Utsinger, P., Availer, N., and Ehrlich, L. Eds., Lippincott, New York, 1986, 49.
93. **Mills, B.,** Bone resorbing cells and human clinical conditions, in *The Osteoclast,* Hall, B., Ed., CRC Press, Boca Raton, FL, 1991, chap. 7.
94. **Soriano, P., Montgomery, C., Geske, M., and Bradley, A.,** Targeted disruption of the c-SRC proto-oncogene leads to osteoporosis in mice, *Cell,* 64, 693, 1991.
95. **Baron, R., Chatterjee, D., Horne, W., Levy, J., Louri, A., and Neff, L.,** Osteoclasts express high levels of the c-SRC proto-oncogene, *J. Bone Miner. Res.,* 6 (Suppl. 1), 197, 1991.
96. **Reis, W., Gong, J., and Gunsolley, G.,** The distribution and kinetics of nuclei in rat osteoclasts, *Cell Tissue Res.,* 20, 1, 1987.
97. **Addison, W.,** Enzyme histochemical properties of kitten osteoclasts in bone imprint preparations, *Histochem. J.,* 10, 645, 1978.
98. **Walker, D.,** Enzymatic and electron microscopic analysis of isolated osteoclasts, *Calcif. Tissue Res.,* 9, 296, 1972.
99. **Athanasou, N., Quinn, J., Gatter, K., Mason, D., and McGee, J.,** Osteoclasts contain macrophage and megakaryocyte antigens, *J. Pathol.,* 150, 239, 1986.
100. **Athanasou, N., Hall, P., Ardenne, A., Quinn, J., and McGee, J.,** A monoclonal antibody (anti-L-35) which reacts with human osteoclasts and cells of the mononuclear phagocyte system, *Br. J. Exp. Pathol.,* 69, 309, 1988.
101. **Horton, M., Rimmer, E., Lewis, D., Pringle, J., Fuller, K., and Chambers, T.,** Cell surface characterization of the human osteoclast: phenotypic relationship to other bone marrow cell types, *J. Pathol.,* 144, 281, 1984.
102. **Sminia, T. and Dijkstra, C.,** The origin of osteoclasts: an immunohistochemical study on macrophage and osteoclasts in embryonic rat bone, *Calcif. Tissue Int.,* 39, 263, 1986.
103. **Oursler, M., Bell, L., Clevinger, B., and Osdoby, P.,** Identification of osteoclast specific monoclonal antibodies, *J. Cell Biol.,* 100, 1592, 1985.
104. **Anderson, R., Schraer, H., and Gay, C.,** Ultrastructural immunocytochemical localization of carbonic anhydrase in normal and calcitonin-treated chick osteoclasts, *Anat. Rec.,* 204, 9, 1982.
105. **Vaes, G.,** Lysosomes and the cellular physiology of bone resorption, in *Lysosomes in Biology and Pathology,* Vol. 1, Dingle, J. and Fell, H., Eds., North-Holland, Amsterdam, 1969, 217.
106. **Baron, R., Neff, L., Louvard, D., and Courtoy, P.,** Cell-mediated exracellular acidification and bone resorption. Evidence for a low pH in resorbing lacunae and localization of a 100 kd lysosomal membrane protein at the osteoclast ruffled border, *J. Cell Biol.,* 101, 2210, 1985.
107. **Miller, S.,** The rapid appearance of acid phosphatase activity at the developing ruffled border of parathyroid hormone activated medullary bone osteoclasts, *Calcif. Tissue Int.,* 37, 526, 1985.
108. **Doty, S. and Schofield, B.,** Electron microscopic localization of hydrolytic enzymes in osteoclasts, *Histochem. J.,* 4, 245, 1972.
109. **Baron, N., Neff, L., Brown, W., Courtoy, P., Louvard, D., and Farqhuar, M.,** Polarized secretion of lysosomal enzymes: co-distribution of cation-independent mannose 6-phosphate receptors and lysosomal enzymes along the osteoclast exocytic pathway, *J. Cell Biol.,* 106, 1863, 1991.
110. **Lucht, U.,** Acid phosphatase of osteoclasts demonstrated by electron microscopic histochemistry, *Histochemistry,* 28, 103, 1971.
111. **Minkin, C.,** Bone acid phosphatase: tartrate-resistant acid phosphatase as a marker of osteoclast function, *Calcif. Tissue Int.,* 34, 285, 1982.

112. **Webber, D., Braidman, I., Robertson, W., and Anderson, D.,** The effect of tartrate on bone cell acid phosphatase activity: a quantitative cytochemical study, *J. Bone Miner. Res.*, 4, 809, 1989.
113. **Baron, R., Neff, L., Brown, W., Louvard, D., and Courtoy, P.,** Selective internalization of the apical plasma membrane and rapid redistribution of lysosomal enzymes and mannose 6-phosphate receptors during osteoclast inactivation by calcitonin, *J. Cell Sci.*, 97, 439, 1990.
114. **Baron, R., Neff, L., Roy, C., Boisuert, A., and Caplan, M.,** Evidence for a high and specific concentration of Na^+, K^+, ATPase in the plasma membrane of osteoclasts, *Cell*, 46, 311, 1986.
115. **Akisaka, T. and Gay, C.,** Intracytochemical evidence for a proton pump ATPase in chick osteoclasts, *Cell Tissue Res.*, 245, 507, 1986.
116. **Blair, H., Teitelbaum, S., Ghiselli, R., and Gluck, S.,** Osteoclastic bone resorption by a polarized vacuolar proton pump, *Science*, 245, 855, 1984.
117. **Vaananen, H., Roman, I., Parvinen, E., Sundquist, K., Hentunen, T., Lakkakorpi, P., Tuukkanen, J., and Wallmark, B.,** Localization and characterization of vacuolar ATPase in osteoclasts, *J. Cell Biol.*, 111, 1305, 1990.
118. **Bekker, P. and Gay, C.,** Biochemical characterization of an electrogenic vacuolar proton pump in purified chicken osteoclast plasma membrane vesicles, *J. Bone Miner. Res.*, 5, 569, 1990.
119. **Bekker, P. and Gay, C.,** Characterization of a Ca^{2+} ATPase in osteoclast plasma membranes, *J. Bone Miner. Res.*, 5, 557, 1990.
120. **Stern, P. and Raisz, L.,** Organ cultures of bones, in *Skeletal Research and Experimental Approaches*, Simmons, D. and Kunin, S., Eds., Academic Press, New York, 1979, 21.
121. **Holtrop, M., Raisz, L., and Simmons, H.,** The effects of parathyroid hormone, colchicine and calcitonin on the ultrastructure and activity of osteoclasts in organ culture, *J. Cell Biol.*, 60, 346, 1974.
122. **Ko, J. and Bernard, G.,** Osteoclast formation *in vitro* from bone marrow mononuclear cells in osteoclast-free bone, *Am. J. Anat.*, 161, 415, 1981.
123. **Burger, E., Van Der Meer, J., and Nijweide, P.,** Osteoclast formation from mononuclear phagocytes: role of bone forming cells, *J. Cell Biol.*, 99, 1901, 1984.
124. **Burger, E., Van der Meer, J., Van de Gevel, J., Gribnau, J., Thesingh, C., and Van Furth, R.,** *In vitro* formation of osteoclasts from long-term cultures of bone marrow mononuclear phagocytes, *J. Exp. Med.*, 156, 1604, 1982.
125. **Klein-Nulend, J., Veldhuijzen, P., de Jong, M., and Burger, E.,** Increased bone formation and decreased bone resorption in fetal mouse calvaria as a result of intermittent comprehensive force *in vitro, Bone Miner.*, 2, 441, 1987.
126. **Taylor, L., Tertinegg, I., Okuda, A., and Heersche, J.,** Expression of calcitonin receptors during osteoclast differentiation in mouse metatarsals, *J. Bone Miner. Res.*, 4, 751, 1989.
127. **Braidman, I., Rothwell, C., Webber, D., Crowe, P., and Anderson, D.,** Location of osteoclast precursors in fetal rat calvaria cultured on collagen gels, *J. Bone Miner. Res.*, 5, 287, 1990.
128. **Kunchisa, J. and Heersche, J.,** Osteoclastic bone resorption: *in vitro* analysis of the rate of resorption and migration of individual osteoclasts, *Bone*, 9, 73, 1988.
129. **Gay, C., Ito, M., and Schraer, H.,** Carbonic anhydrase activity in isolated osteoclasts, *Metab. Bone Dis. Relat. Res.*, 5, 33, 1983.
130. **Gay, C.,** Isolation of multinucleated osteoclasts, in *Isolation and Properties of Osteoclasts*, Stern, P. and Hefley, Eds., American Society for Bone and Mineral Research, 1983, 1 (workshop).
131. **Carano, A., Teitelbaum, J., Konsek, J., Schlesinger, P., and Blair, H.,** Biphosphonates directly inhibit the bone resorption activity of isolated avian osteoclasts *in vitro*, *J. Clin. Invest.*, 85, 456, 1990.

132. **Blair, H., Kahn, A., Crouch, E., Jeffreys, J., and Teitelbaum, S.,** Isolated osteoclasts resorb the organic and inorganic components of bone, *J. Cell Biol.*, 102, 1164, 1986.
133. **de Vernejoul, M., Horowitz, M., Denignin, J., Neff, L., and Baron, R.,** Bone resorption by isolated chick osteoclasts in culture is stimulated by murine spleen cell supernatant fluids (osteoclast-activating factor), and inhibited by calcitonin and prostaglandin E2, *J. Bone Miner. Res.*, 3, 69, 1988.
134. **Oursler, M., Collin-Osdoby, P., Anderson, R., Li, L., Webber, D., and Osdoby, P.,** Isolation of avian osteoclasts: improved techniques to preferentially purify viable cells, *J. Bone Miner. Res.*, 6, 375, 1991.
135. **Collin-Osdoby, P., Oursler, M., Webber, D., and Osdoby, P.,** Osteoclast-specific monoclonal antibodies coupled to magnetic beads provide a rapid and efficient method of purifying avian osteoclasts, *J. Bone Miner. Res.*, 6, 1353, 1991.
136. **Osdoby, P.,** personal communication.
137. **Oursler, M., Osdoby, P., Pyfferoen, J., Riggs, B., and Spelsberg, T.,** Avian osteoclasts as estrogen target cells, *Proc. Natl. Acad. Sci. U.S.A.*, 88, 6613, 1991.
138. **Oursler, M., Pyfferoen, J., Osdoby, P., and Spelsberg, T.,** Avian osteoclast gene responses to 17β-estradiol treatment, *J. Bone Miner. Res.*, 6 (Suppl.), 297, 1991.
139. **Eliann, M., Blasle, Z., Bouizar, J., Beilakoff, M., and de Vernejoul, M.,** Influence of blood calcium on calcitonin receptors in isolated chick osteoclasts, *J. Endocrinol.*, 119, 243, 1988.
140. **Rifkin, B., Auszmann, J., Kleckner, A., Vernillo, A., and Fine, A.,** Calcitonin stimulates cAMP accumulation in chicken osteoclasts, *Life Sci.*, 42, 799, 1988.
141. **Pandalai, S. and Gay, C.,** Effects of parathyroid hormone, calcitonin, and dibutryl-cyclic AMP on osteoclast area in cultured chick tibia, *J. Bone Miner. Res.*, 5, 701, 1990.
142. **Nicholson, G., Mosely, J., Secton, P., and Martin, T.,** Chicken osteoclasts do not possess calcitonin receptors, *J. Bone Miner. Res.*, 2, 53, 1987.
143. **Nicholson, G., Livesey, S., Mosely, J., and Martin, T.,** Action of calcitonin, parathyroid hormone, and prostaglandin E_2 on cyclic AMP formation in chick and rat osteoclasts, *J. Cell Biochem.*, 31, 229, 1986.
144. **Dempster, D., Murrills, J., Horbert, W., and Arnett, T.,** Biological activity of chicken calcitonin: effects on neonatal rat and embryonic chick osteoclasts, *J. Bone Miner. Res.*, 2, 443, 1987.
145. **Jones, S., Boyde, A., and Ali, N.,** The resorption of biological and non-biological substrates by cultured avian and mammalian osteoclasts, *Anat. Embryol.*, 170, 247, 1984.
146. **Ali, N., Boyde, A., and Jones, S.,** Motility and resorption: osteoclastic activity *in vitro, Anat. Embryol.*, 170, 51, 1984.
147. **Jones, S., Boyde, A., Ali, N., and Maconnachie, E.,** Variation of the sizes of resorption lacunae made *in vitro, Scanning Electron Microsc.*, 4, 1571, 1986.
148. **Murrills, J., Shane, E., Lindsey, R., and Dempster, D.,** Bone resorption by isolated human osteoclasts *in vitro:* effects of calcitonin, *J. Bone Miner. Res.*, 4, 259, 1989.
149. **Chambers, T., Fuller, K., McSheehy, P., and Prinsler, J.,** The effects of calcium regulating hormones on bone resorption by isolated human osteoclastoma cells, *J. Pathol.*, 145, 297, 1985.
150. **Athanasou, N., Bliss, E., Gatter, C., Heryek, A., Woods, C., and McGee, J.,** An immunological study of giant-cell tumor of bone: evidence for an osteoclast origin of the giant cells, *J. Pathol.*, 147, 153, 1985.
151. **Goldring, S., Dayer, J. M., and Krane, S.,** Regulation of hormone-induced cAMP response to parathyroid hormone and prostaglandin E_2 in cells cultured from human giant cell tumors of bone, *Calcif. Tissue Int.*, 29, 193, 1979.
152. **Yoneda, T., Mundy, G., and Roodman, G.,** Induction of differentiation of the human promyelocytic HL-60 cells into cells with the osteoclastic phenotype, *J. Bone Miner. Res.*, 4, 559A, 1989.

153. **Gattei, V., Bernabei, P., Pinto, A., Bezzini, R., Ringressi, A., Formigli, L., Tanini, A., Attadia, V., and Brandi, M.,** Phorbol ester induced osteoclast differentiation of a novel human leukemic cell line (FLG 29.1), *J. Cell Biol.,* 116, 437, 1992.
154. **Weiss, L. and Fawcett, D.,** Cytochemical observations on chicken monocytes, macrophages, and giant cells in tissue culture, *J. Histochem. Cytochem.,* 1, 47, 1953.
155. **Sutton, J. and Weiss, L.,** Transformation of monocytes in tissue culture into macrophages, epithelioid cells and multinucleated giant cells, *J. Cell Biol.,* 28, 303, 1966.
156. **Ibbotson, K., Roodman, G., McManus, L., and Mundy, G.,** Identification and characterization of osteoclast-like cells and their progeny in cultures of feline marrow mononuclear cells, *J. Cell Biol.,* 99, 471, 1984.
157. **Testa, N., Allen, T., Lajtha, L., Onions, D., and Jarret, O.,** Generation of osteoclasts *in vitro, J. Cell Sci.,* 47, 127, 1981.
158. **Oursler, M. and Osdoby, P.,** Osteoclast development in marrow cultured in calvaria-conditioned media, *Dev. Biol.,* 127, 170, 1988.
159. **Pharoah, M. and Heersche, J.,** 1,25 dihydroxyvitamin D_3 causes an increase in the number of osteoclast-like cells and their progenitors in cultures of feline marrow mononuclear cells, *J. Cell Biol.,* 94, 471, 1984.
160. **Takahashi, N., Yamana, H., Yoshiki, S., Roodman, G., Mundy, G., Jones, S., Boyde, A., and Suda, T.,** Osteoclast-like cell formation and its regulation by osteotropic hormones in mouse bone marrow culture, *Endocrinology,* 122, 1373, 1988.
161. **Roodman, G., Ibbotson, K., MacDonald, B., Kuehl, T., and Mundy, G.,** 1,25 dihydroxyvitamin D_3 causes formation of multinucleated cells with several osteoclast characteristics in cultures of primate marrow, *Proc. Natl. Acad. Sci. U.S.A.,* 82, 8213, 1985.
162. **MacDonald, B., Takahashi, N., McManus, L., Itolahan, J., Mundy, G., and Roodman, G.,** Formation of multinucleated cells that respond to osteotropic hormones in long term human bone marrow cultures, *Endocrinology,* 120, 2326, 1987.
163. **Hattersley, G. and Chambers, T.,** Generation of osteoclasts from hemopoietic cells and a multipotential cell line *in vitro, J. Cell Physiol.,* 140, 478, 1989.
164. **Roodman, G. and Mundy, G.,** Osteoclast ontogeny and function, in *Annual Review of Bone and Mineral,* Vol. 5, Peck, W., Ed., Elsevier, New York, 1989, 127.
165. **Van de Wijngaret, F., Tas, M., and Burger, E.,** Characteristics of osteoclast precursor-like cells grown from mouse bone marrow, *Bone Miner.,* 3, 111, 1987.
166. **Kurihara, N., Civin, C., and Roodman, G.,** Osteotropic factor responsiveness of highly purified populations of early and late precursors for human multinucleated cells expressing the osteoclast phenotype, *J. Bone Miner. Res.,* 6, 257, 1991.
167. **Takahashi, N., Akatsa, T., Udagawa, N., Sasaki, T., Yamaguchi, A., Mosely, J., Martin, T., and Suda, T.,** Osteoblastic cells are involved in osteoclast formation, *Endocrinology,* 123, 2600, 1988.
168. **Udagawa, N., Takahashi, N., Akatsu, T., Sasaki, T., Yamaguchi, A., Kodama, M., Martin, T., and Suda, T.,** The bone marrow-derived stromal cell lines MC3T3-G2/PA6 and ST2 support osteoclast-like cell differentiation in co-culture with mouse spleen cells, *Endocrinology,* 125, 1805, 1985.
169. **Roodman, G.,** Osteoclast differentiation, *Crit. Rev. Oral Biol. Med.,* 2, 389, 1991.
170. **Boyde, A., Ali, N., and Jones, S.,** Computer-aided measurements of resorptive activity of isolated osteoclasts, *Proc. R. Microsc. Soc.,* 18, 357, 1983.
171. **Chambers, T., Revell, P., Fuller, K., and Athanasou, N.,** Resorption of bone by isolated rabbit osteoclasts, *J. Cell Sci.,* 66, 383, 1984.
172. **Jones, S., Boyde, A., and Ali, N.,** The resorption of biological and non-biological substrates by cultured avian and mammalian osteoclasts, *Anat. Embryol.,* 170, 247, 1984.
173. **Boyde, A.,** Confocal optical microscopy, in *Modern Microscopes: Techniques and Applications,* Duek, P. and Michette, A., Eds., Plenum Press, New York, 1990, 185.

174. **Arnett, T. and Dempster, D.,** A comparative study of disaggregated chick and rat osteoclasts *in vitro:* effect of calcitonin and prostaglandins, *Endocrinology,* 120, 602, 1987.
175. **Arnett, T. and Dempster, D.,** Effect of pH on bone resorption by rat osteoclasts *in vitro, Endocrinology,* 119, 119, 1986.
176. **Boyde, A. and Jones, S.,** Pitfalls in pit measurements, *Calcif. Tissue Int.,* 49, 65, 1991.
177. **Shibutani and Heersche, J.,** In an isolated rabbit osteoclast system, decreasing the pH increases osteoclast activity but reduces the increase in osteoclast number during culture, *J. Bone Miner. Res.,* 6 (Suppl. 1), 594, 1991.
178. **Silver, I., Murrills, R., and Etherington, D.,** Microelectrode studies on the acid microenvironment beneath adherent macrophages and osteoclasts, *Exp. Cell. Res.,* 175, 266, 1988.
179. **Ravesloot, J., Ypey, D., Urigheid-Lammers, J., and Nijweide, P.,** Voltage-activated K^+ conductants in freshly isolated embryonic chicken osteoclasts, *Proc. Natl. Acad. Sci. U.S.A.,* 86, 6821, 1989.
180. **Teti, A., Blair, H., Teitelbaum, S., Kahn, A., Koziol, C., Konsek, J., Zambonin-Zallone, A., and Schlesinger, P.,** Cytoplasmic pH regulation and chloride/bicarbonate exchange in avian osteoclasts, *J. Clin. Invest.,* 83, 227, 1989.
181. **Zaidi, M.,** "Calcium receptors" on eukaryotic cells with special reference to the osteoclast, *Biosci. Rep.,* 10, 493, 1990.
182. **Miyauchi, A., Hruska, K., Greenfield, E., Randall, D., Alvarez, J., Barattolo, R., Colucci, S., Zambonin-Zallone, A., Teitelbaum, S., and Teti, A.,** Osteoclast cytosolic calcium, regulated by voltage operated calcium channels and extracellular calcium controls podosome assembly and bone resorption, *J. Cell Biol.,* 111, 2547, 1990.
183. **Zheng, M., Papadimitriou, J., and Nicholson, G.,** A quantitative cytochemical investigation of osteoclasts and multinucleated giant cells, *Histochem. J.,* 23, 180, 1991.
184. **MacIntyre, I., Zardi, M., Towhidul, A., Datta, H., Moonsa, B., Lidbury, P., Hecker, M., and Vane, J.,** Osteoclastic inhibition: an action of nitric oxide not mediated by cyclic GMP, *Proc. Natl. Acad. Sci. U.S.A.,* 88, 2936, 1991.
185. **Ferrier, J., Warda, A., Kanehisa, J., and Heersche, J.,** Electrophysiological responses of osteoclasts to hormones, *J. Cell Physiol.,* 128, 23, 1986.

Chapter 2

ANTIGENIC MARKERS OF OSTEOCLASTS

Michael A. Horton and Miep H. Helfrich

TABLE OF CONTENTS

I. SCOPE OF CHAPTER

Immunologically defined cellular antigens have readily found a place in the analysis of hemopoietic differentiation and definition of subpopulations beyond the scope of morphological and function criteria. Over recent years, the tried and tested techniques of monoclonal antibody technology, biochemical analysis, and, more recently, molecular cloning have been applied to questions of bone cell biology. In this chapter, we review the current literature on the expression of antibody-defined cell surface and intracellular antigens of osteoclasts. Our analysis primarily concerns the definition of the immunophenotype of mammalian (mainly human and rodent) and avian osteoclasts; purposely, we have largely omitted discussion of osteoclast enzyme systems, unless they were initially delineated as "antigenic markers", as they are covered elsewhere. (See Chapters 3, 6, 13, and 16 of this volume.) Our knowledge of the function of the majority of osteoclast antigens is limited, as function has generally not been addressed in a skeletal context; we do, however, speculate on a role in bone biology for some molecules, e.g., CD9, aminopeptidase N (CD13), and the transferrin receptor (CD71). A section on adhesive molecules, mainly integrins, is included as they were among the first "specific" markers of human osteoclasts to be reported[1] and the definition of their functional activity in osteoclasts is well advanced. Finally, we review the application of immunological techniques to the phenotypic analysis of human and other marrow culture systems and speculate about the nature of the "osteoclast-like" multinucleated giant cells seen, particularly those derived from human tissue.

II. ANTIGENIC MARKERS OF HUMAN OSTEOCLASTS

A. HUMAN OSTEOCLAST ANTIGENS DEFINED BY MONOCLONAL ANTIBODIES

Osteoclasts are multinucleated giant cells with the property of resorbing mineralized tissues. They are formed by cell fusion of postmitotic progenitor cells which are ultimately derived from multipotential hemopoietic stem cells. (See Chapters 1, 4, 5, 8, and 21 of this volume.) This property, together with some morphological and functional activities, led to the suggestion that they were macrophages, albeit of a lineage with specialized functions limited to the skeleton, and has led to an impetus over the past few years to examine the relationship between osteoclasts and other hemopoietic cells. Many of these studies have involved examination of the distribution of cell surface and other antigens, largely defined by monoclonal antibodies (mabs), on osteoclasts in bone and mononuclear phagocytes in other tissues. Analysis of this type cannot *prove* definitively that one cell is identical to another, or not. However, the finding of a concensus immunophenotype which is markedly different, particularly if molecules involved in differentiated cell function are absent, suggests a more distant lineage relationship. Shared expression of widely expressed, "panreactive" molecules between two hemopoietic cell types again does not prove identity of cell lineage; it merely supports the view that osteoclasts are hemopoietic. The reviewers view, previously expounded and taken with more recently available functional data, is that osteoclasts and tissue macrophages are not highly related; whether they both arise from a common, *committed* stem cell is still not convincingly proven and is outside the scope of this chapter. (See Chapter 4 of this volume.)

A large number of studies on the expression of the majority of well-defined hemopoietic antigens by human osteoclasts have been described and reviewed through 1988.[2] Only details of positive and important negative findings from these earlier studies will be discussed.

Hemopoietic antigens expressed by human osteoclasts are given in Table 1. Details of negative findings with a wide range of monoclonal antibodies (mabs) to lymphoid, myeloid, and other antigens are given elsewhere.[2-5] More recent data on newer molecules and data as yet largely unconfirmed or only presented in preliminary form are found in the proceedings of source books on the definition of human hemopoietic markers.[6,7]

In general, the antigens expressed by osteoclasts fall into three groups. First, there are those which have a wide tissue distribution, such as ICAM1 (CD54), CD45, and VLA-2 (CD49b), and which are shared by many hemopoietic cell types. Preliminary data from the groups of Parravicini[8] and Athanasou[9] on the expression of molecules such as CD47, CD53, and CD63 would also fall into this class. Second, there are molecules of more limited tissue distribution, such as CD9, CD13, transferrin receptor (CD71), and vitronectin receptor (CD51). These might have particular specific functions in osteoclasts, which may differ from those defined in other hemopoietic cells; their potential functions are discussed elsewhere in this chapter. Third, there are antigens present in both osteoclasts and macrophages. Example include CD68, 5/9, and a number of ill-defined (possibly related) mabs to antigens of macrophages which are shared with osteoclasts and generally have an intracellular distribution. Their functions, enzymatic or other, have yet to be explored in either cell type. Finally, several molecules crucial for the phagocytic and antigen-presenting functions of macrophages, for example HLA-DR (Class II histocompatibility antigen), and leukocyte function antigens CD11b and c are absent from osteoclasts. Taken together, immunological studies support the view that osteoclasts express a distinct phenotype which differs from cells of the mononuclear phagocyte system. The inability of macrophages to resorb bone, express receptors for the regulatory hormone calcitonin, and respond to calcitonin all confirm that osteoclasts and macrophages are at best only distant cousins. This has led to attempts to derive osteoclast-specific mabs for use in defining the cell lineage of osteoclasts vs. other hemopoietic cell types. A number of groups have attempted to accom-

TABLE 1
Immunologically Defined Hemopoietic Antigens Expressed by Human Osteoclasts

Antigen[a]		Antibody (e.g.)	Myeloid cell cross-reactivity[b]	Molecular mass (kDa)	cDNA cloned
	CD9	J2	+ (all)	25	
Aminopeptidase N	CD13	My7	+ + (all)	150	√
Collagen receptor					
α_2 chain	CD49b	Gi14	+ (subpop)	170 (red.)[c]	√
β_1 chain	CD29	A-1A5	+ + (all)	130 (red.)	√
Vitronectin receptor					
α_v chain	CD51	13C2, 23C6	±	125 (red.)	√
β_3 chain	CD61	V1-PL2	± (subpop)	110 (red.)	√
Leukocyte common	CD45	F8-11-13	+ + + (all)	"200"	√
ICAM1	CD54	RR7/7F7	+ + (mono)	85	√
—	CD68	EBM11	+ + + (mono, mac)	110	√
—	—	5/9	+ + (some macs)	?	
Transferrin receptor	CD71	T9	± (macs)	95	√

[a] Antigen definition using CD nomenclature.[7]
[b] ± to + + +, range of cellular reactivity; subpop, subpopulation of myeloid cells reactive; mono, monocyte; mac, macrophage.
[c] red., reduced.

plish this by raising mabs to human and avian osteoclasts. These mabs have indeed proven to be of wide use in the definition of osteoclast lineage, analysis of the differentiation of mature osteoclasts from their precursors, and in the study of osteoclast function. Such data as have been published are discussed below; however, it must be admitted that no entirely osteoclast-specific reagent has been discovered and, realistically, it is unlikely that one will.

1. Possible Functions of Selected Osteoclast Antigens

a. Aminopeptidase N (CD13)

The CD13 molecule was shown to be expressed by human osteoclasts in earlier phenotypic studies[10] and later, although at lower levels, by mouse osteoclasts.[81] CD13 has been shown to be identical to membrane aminopeptidase N (EC3.4.11.2). It has been speculated[11] that this enzyme acts in concert with a further membrane peptidase, neutral endopeptidase or CD10 (EC3.4.24.11), in the regulation, by proteolysis, of the tissue half-life of short peptide mediators of the inflammatory response and of other proteins involved in normal tissue function, such as encephalins and atrial natriuretic factor. CD10 has recently been shown to be present on cells of an osteoblastic phenotype,[12] which also express low levels of CD13; osteoclasts do not express CD10. Clearly, the presence of two membrane-associated peptidases, known to be involved in the downregulation of the cellular response to peptide mediators, may well be of importance in bone. They may act together to modulate the local skeletal response to (as yet unspecified) factors produced either systemically or locally to limit stimulatory or inhibitory effects on osteoclastic resorption or on osteoblastic osteoid synthesis.

b. Transferrin Receptor (CD71)

Iron is an essential cofactor for cell growth. Rapidly proliferating cells, in addition to developing erythrocytes, express readily detectable membrane transferrin receptor. Human osteoclasts, although known to be postmitotic, have immunodetectable transferrin receptor. Its function is not known, but it is tempting to speculate that it may enable the osteoclast to take in significant amounts of iron required for synthesis of the tartrate-resistant isoenzyme of acid phosphatase, an enzyme found at high level in osteoclasts, which contains two atoms of iron per molecule of enzyme.[13]

c. ICAM1 (CD54)

ICAM1 (CD54), the cellular ligand for the β_2 integrins LFA-1 and Mac-1 (CD11a and b) and member of the immunoglobulin gene superfamily (see Reference 14), is expressed by human osteoclasts.[5] Whether the homologous proteins ICAM2 and 3 are present has not been analyzed. Levels appear to be higher in osteoclasts associated with tumors[82] and may represent changes in expression secondary to cytokines released by the tumor.[15] Its role in the function of normal or tumor-associated osteoclasts is not known. It could mediate regulatory interactions between osteoclasts and hemopoietic cells in the marrow cavity, events which have been hypothesized to be important in the regulation of skeletal homeostasis. (See Chapter 8 of this volume.) No experimental data have been published which address the question of the role of ICAM1 in osteoclast function.

d. CD9

This 24-kDa membrane protein is expressed by osteoclasts in addition to a number of other cell types and tissues, including platelets. Antibodies to CD9 have been shown to affect platelet function, such as inducing platelet aggregation and granule release; this is in part due to increases in levels of the platelet integrin fibrinogen receptor gpIIb/IIIa, a member of the β_3 integrin family, which occur both secondarily as a response to secreted adenosine 5′ diphosphate (ADP) and thromboxane[16] and also via a physical interaction of CD9 and gpIIb/IIIa in the platelet membrane.[17] It is not known if CD9 is physically associated with vitronectin receptor in the osteoclast membrane in a way analogous to that observed in platelets. Whether CD9 is somehow involved in the regulation of osteoclast activity by analogous or different mechanisms is clearly important and deserves further investigation.

B. UNCHARACTERIZED ANTIGENS OF HUMAN OSTEOCLASTS

A number of other mabs which define cellular "antigens" have been examined for reactivity with human osteoclasts and are included for completeness. These antibodies were raised against hemopoietic cells other than osteoclasts and, in general, little is known about the tissue distribution and the molecular species they recognize or their relationship to well-defined CD antigens. Kn22[18] was made to cultured human marrow cells and recognizes many cell types besides osteoclasts and cultured marrow giant cells. The YTH series of rat anti-human macrophage mabs[19] are well characterized immunohistochemically. Two mabs, YTH25.7 and YTH85.12, cross-react with antigens of osteoclasts in addition to macrophages; their relationship to the similar CD68 antigen is not known. Two mabs, MB1 and 3MA-134, react with osteoclasts (and other cells) in standard decalcified paraffin sections;[20,21] they may be useful in identifying and quantitating osteoclasts in calcified tissues. Finally, James et al.[22] have raised a series of antibodies to human osteoclastoma tissue, including mabs to CD13, CD61, and CD68-like antigens. Among their panel of mabs which reacted with a range of novel antigens is C35; this mab appears to selectively stain bone osteoclasts and fails to react with macrophage polykarya, including those giant cells appearing in long-term human bone marrow cultures (see Section V). It will be important to define the biochemical nature of this intracellular molecule.

C. ANTIGENIC EXPRESSION IN TUMOR-ASSOCIATED OSTEOCLASTS

The immunophenotype of "osteoclasts" in tumor variants with osteoclast-like giant cells and giant cell tumors of bone has generally been found to be similar to that of osteoclasts isolated directly from human fetal bone or in sections of adult bone (from hip resections or bone biopsies), although extensive analysis has been limited to a few examples.[2,10,23,24] Thus, osteoclasts in giant cell tumors of bone/osteoclastoma have a profile identical to that of most hemopoietic antigens with the exception of ICAM1 (CD54)[82] and HLA-DR,[24] which show a variably increased expression. In two examples of variant breast carcinoma with osteoclast-like giant cells, the osteoclast phenotype (and functional response to calcium-regulating hormones) differed, the giant cells tending to have a phenotype more like that of macrophages.[25,83] Presumptively, these differences reflect secondary modifications of osteoclast phenotype in response to an abnormal or nonosseous environment and/or modification by "factors" produced by the associated tumor cells.

III. IMMUNOPHENOTYPIC ANALYSIS OF RODENT AND AVIAN OSTEOCLASTS

A. ANTIGENS OF RODENT OSTEOCLASTS

There have been few reports on the antigenic profile of rodent osteoclasts.[3,26,27] Earlier studies were limited by the lack of an adequate range of

TABLE 2
Immunophenotype of Rat and Mouse Osteoclasts

Antigen	Expression on osteoclasts: Rat	Mouse
Integrins		
Vitronectin receptor		
α_v chain	+	+
β_3 chain	+	+
Integrin β_1 chain	+	+
Mac-1 [CD11b ($\alpha_m\beta_2$)]	−	−
Myeloid antigens		
MHC Class II (Ia)	−	−
CD45	±	±
Mouse CD13 (ER-BMDM1/1)	na	+
ED1	+	na
CD68/KiM6	+	na
F4/80	na	−
ED2 to ED9	−	na
OX41 to OX44	−	na
Leucosialin (W3/13)	−	na
Mac-2, Mac-3	na	−
ER series of 7 antimacrophage antigens	na	−

Note: +, positive; ±, weak or variable; −, negative; na, not applicable due to species specificity.

mabs against nonlymphoid antigens in both rat and mouse with the exception of the ED1 antigen, a cytoplasmic molecule expressed at high level in both osteoclasts and tissue macrophages.[26] The immunohistochemical profile of ED1 in the rat is reminiscent of the CD68 antigen (*vide supra*), and it is possible that ED1 recognizes the rat homolog of CD68; it is of interest that one anti-CD68 mab, KiM6, cross-reacts with rat osteoclasts and macrophages with pattern identical to that of ED1.[82]

We[81] have recently taken advantage of the current availability of mabs to rodent myeloid antigens to reexamine their phenotype (Table 2). Thus far, no major differences in immunophenotype between rodent and human cells have been discovered. Vitronectin receptor is strongly expressed by rat osteoclasts, although its expression appears to be less in mouse (see Section IV.A.2). Likewise, membrane aminopeptidase N (CD13) defined by mab ER-BMDM1/1[28] is expressed less in mouse than in other species. The generalization that membrane antigens of monocytes, including those of functional importance, are absent from rodent osteoclasts holds true; some cytoplasmic antigens of terminally differentiated macrophages (such as ED1 and CD68) are shared between osteoclasts and macrophages, as in man.

Several studies have been carried out using lectins as probes of membrane glycoproteins of rat osteoclasts. While they do not identify single classes of

TABLE 3
Antigens Expressed by Avian Osteoclasts

Antibody	Antigen/mol wt (kDa)	Osteoclast expression[a]
121F	Membrane SOD/150 kDa	√
75B	40 kDa	√
29C		√ (+ macs)
K2, K5, etc.	—	√
K20	150 kDa	√
K51, K70, etc.	—	√ (+ macs)
K36, etc.	—	√ (polarized distribution)
OC6.9	—	√ (polarized distribution)
OC5.3, OC6.8	—	√ (+ macs)
OC6.1, OC6.3	—	√ (+ macs)[b]
23C6	Vitronectin receptor (CD51)	√
33	Vitronectin receptor β_3 chain (CD61)	√

[a] Mabs raised against and reactive with chicken OCs apart from "OC series" which are anti-quail OC mabs. √, OC expression; macs, expression in other myeloid cells.

[b] Also react with chick cells.

antigens, at least one report[29] has provided important differences between macrophage giant cells and osteoclasts. *Arachis hypogaea* (PNA) lectin bound to osteoclasts, whereas giant cells were unreactive; moreover, the PNA binding was most marked in the ruffled border of osteoclasts in zones of active bone resorption. Clearly, PNA may react with a functionally important, and as yet unidentified, molecule of osteoclasts.

B. ANTIGENS OF AVIAN OSTEOCLASTS

Osdoby's group has raised an extensive panel of mabs to freshly isolated chick osteoclasts[30] (Table 3). Two of these, 121F and 29C, have been analyzed in detail. The 121F mab shows specific membrane staining only of osteoclasts, whereas 29C shared reactivity with both osteoclasts and tissue macrophages. The 121F epitope is upregulated in culture on development of osteoclasts from their mononuclear precursors.[31] Of crucial importance, antigenic expression was dependent upon addition of undefined factors in bone cell conditioned medium; in its absence, multinucleated cells formed, but these were 121F negative and showed membrane structural features of macrophage polykarya rather than true osteoclasts (*vide infra*). The 121F antigen is also expressed in giant cells forming on chick chorioallantoic membrane in response to bone particles and hydroxyapatite, but not to nonresorbable materials such as methacrylate;[32] 121F expression thus correlates with the acquisition of the osteoclastic phenotype. Recently, structural studies have been performed on the 121F antigen which show it to be of 150 kDa size. Partial amino acid sequencing shows that it has close homology to a number of prokaryotic and mitochondrial Mn^{2+}/Fe^{2+} superoxide dismutase (SOD) enzymes.[33] Functional analysis has demonstrated that purified 121F antigen has SOD activity[33] and that 121F mab inhibits bone resorption.[84] A role for SOD in bone has not been previously demonstrated, although an involvement of free radicals in bone resorption has been suggested. (See Chapters 8 and 15 of this volume.)

Relatively little is known about antigens recognized by the other mabs. Antibody 29C recognizes an intracellular epitope strongly expressed by osteoclasts and macrophages.[30] Its relationship to the many similar antibodies raised against human, rat, or avian cells which exhibit a similar staining pattern and the CD68 antigen remains to be established. Mab 75B has recently been shown to be an "osteoclast-specific" 40-kDa cytoskeleton-associated protein.[34]

Three further studies have concerned the definition of avian osteoclast antigens by mab techniques (Table 3). A number of mabs to quail cells were studied by Nijweide and co-workers.[35] Several (e.g., OC5.3, 6.1, 6.3, and 6.8) showed cross-reactivity with marrow and tissue macrophages in addition to osteoclasts. One, OC6.9, was of particular interest as it appeared to localize an antigen primarily found adjacent to the bone surface in osteoclasts that were actively resorbing bone, but was uniformly distributed in quiescent osteoclasts. The molecular structure of functional activity of the OC6.9 and other anti-quail osteoclast antigens has not been reported.

Hentunen et al.[36] produced a panel of 13 anti-chick osteoclast mabs, some of which (e.g., K2 and K5) were restricted in expression to osteoclasts (and a few mononuclear cells), whereas others (e.g., K51 and K70) had a broader reactivity profile. Of note, some mabs (e.g., K36) showed polar staining reminiscent of OC6.9. Several of these mabs have now been screened for effects on osteoclast activity in bone resorption assays.[37] Some (K27, K35, and K41) showed moderate inhibitory activity; a new mab, K20, markedly inhibited bone resorption, which did not appear to be via effects on osteoclast proton pump activity. K20 recognizes a 150-kDa protein by Western blotting;[38] its exact relationship to the 121F antigen and aminopeptidase N (CD13), both similar in molecular size to K20, remains to be definitively investigated.

A further series of mabs to chick osteoclasts have been reported by Chanoki et al.[39] which show apparent osteoclast specificity. No functional or biochemical studies have been published on this potentially interesting group of antibodies.

Immunological analysis of avian osteoclasts with reagents not purposely produced against bone tissue have been limited. Athanasou and co-workers[40] have recently examined the expression of mabs to human myeloid cells by chick marrow-derived giant cells and found a broader reactivity pattern, with CD11b and c being expressed, than observed with freshly isolated human osteoclasts. Whether these results suggest that avian osteoclasts are phenotypically distinct from their mammalian counterparts or whether the phenotype reflects the presence of a heterogeneous population of giant cells in the cultured chick marrow (i.e., contains both macrophage polykarya *and* osteoclasts) remains to be established (*vide infra*). We[82] have failed to see reactivity with chick osteoclasts of a series of mabs (JS1 to 9) raised against AEV-transformed erythroblasts[41] which reacted with a variety of avian hemopoietic cell types.

IV. RECEPTORS MEDIATING OSTEOCLAST ADHESION

A. INTEGRINS

Immunocytochemical evidence for expression of members of the integrin superfamily of adhesion receptors[42] has been found in osteoclasts in all species examined so far. The most extensive studies have been performed with human osteoclasts because integrin-specific antibodies are more widely available in man than in other species.

Integrins are transmembrane glycoproteins consisting of two noncovalently linked subunits (usually a large α- and smaller β-subunit). They function as receptors for a number of protein ligands — either cell bound, matrix bound, or free ligand — the binding of which influences the cytoskeleton and results in changes in cell adhesion and motility.[43]

Integrins are classically divided into three families[42] according to their β-subunits (β_{1-3}), which are shared by a number of different α-subunits. Recently, however, this classification has become inadequate due to the discovery of additional β-subunits (β_{4-8}), some of which share α-chains with β_1 or β_3. Also, alternative associations of previously identified α- and β-chains creating novel receptors have recently been discovered. A review of integrin structure and the classification of all integrins currently known is outside the scope of this chapter; the reader is referred to a recent review by Hemler.[44] Due to the promiscuous behavior of integrin α- and β-subunits and the fact that novel chain associations are still being discovered, it is impossible to predict the molecular structure of an integrin merely by immunocytochemical detection of individual subunits; further biochemical analysis of the molecules is required to elucidate the exact association of the subunits into heterodimeric receptors. However, biochemical analyses of osteoclast integrins have been limited by the absence of osteoclastic cell lines and by the low numbers of osteoclasts that can be obtained from bone. The most complete data so far have been obtained with osteoclastoma osteoclasts.[45]

1. Integrins of Human Osteoclasts

Human osteoclasts, either derived from osteoclastomas or from normal bone tissue, express high levels of β_3 (CD61) and α_v (CD51). In addition, they express β_1 (CD29) and VLA-α_2 (CD49b)[46] (see Table 1). Davies et al.[45] provided biochemical evidence that in osteoclasts, α_v and β_3 are associated in the vitronectin receptor $\alpha_v\beta_3$. No such biochemical evidence is available

TABLE 4
Integrin Receptors of Human Osteoclasts

Detected
α_2
β_1
$\alpha_v\beta_3$ (vitronectin receptor)
Not detected
α_1, $\alpha_{3\text{-}6}$, α_8 (other VLAs)
β_2 and CD11a, b, c (LFAs)
β_4
β_5
$\beta P/\beta_7$
β_8
$gpIIb_\alpha$

for the association of the two other chains due to their widespread distribution, which makes it impossible to selectively analyze osteoclast-expressed receptor in a heterogeneous cell population. However, it is probable that, as in many other cell types, α_2 and β_1 are associated in the collagen/laminin/fibronectin receptor.[44] No other integrins have been detected on human osteoclast so far (Table 4), but novel integrins and subunit associations are still being discovered in other cell types, and it remains possible that additional, possibly ''bone-related'' integrin(s) will be found.

Thus, osteoclasts have a remarkably restricted expression of integrins: they lack common VLA-type integrins, such as the fibronectin receptor $\alpha_5\beta_1$, and do not express members of the β_2 family, unlike tissue macrophages. In contrast, they do express high levels of vitronectin receptor, an integrin with a restricted expression *in vivo,* although it is more generally expressed by cell lines and by cultured cells.[47] Because of the restricted *in vivo* expression of vitronectin receptor, mabs 13C2 and 23C6 that recognize α_v and the $\alpha_v\beta_3$ complex, respectively (see Table 1), were originally thought to be ''osteoclast specific''.[10] Now that the molecular structures of their antigens has been defined and it has become evident that other cell types also express the vitronectin receptor, it seems more appropriate to abandon the term ''osteoclast specific'' and refer to these and other mabs that identify the same molecules by their CD numbers.

2. Integrins of Nonhuman Mammalian Osteoclasts

The integrin repertoire of nonhuman mammalian osteoclasts appears, as far as can be assessed with mabs, polyclonal antisera, and limited biochemical data, to be similar to that of human osteoclasts. Mabs that recognize α_v (13C2) or the $\alpha_v\beta_3$ complex (23C6) in man[1] also react with osteoclasts from all mammalian species tested with the exception of rat and mouse, indicating that they also express the vitronectin receptor. Biochemical evidence for a molecular species of the appropriate size was obtained with 23C6 in immunoprecipitation studies using osteoclast-enriched rabbit cell suspensions.[48] Rat osteoclasts express α_v, β_3, and β_1.[49] Antibodies to detect rat VLA α-chains are not available. Biochemical confirmation for the molecular structure of rat osteoclast integrins is also not available, but functional studies (*vide infra*) suggest that the chains may associate as $\alpha_v\beta_3$ and $\alpha_2\beta_1$.

Polyclonal antisera to human α_v and β_3 that cross-react with other species tend to have lower immunoreactivity with the mouse osteoclasts than with osteoclasts of man or rat. Whether this reflects a genuine lower expression of vitronectin receptor in mouse osteoclasts or is the result of poor cross-reactivity in this species is presently unclear. Attempts to raise mabs specifically reactive with mouse osteoclasts by using these cells as immogen have so far been unsuccessful. Two mabs, H9.2B8 and 8.18E12 (hamster × mouse and rat × mouse interspecific hybridomas, respectively), were recently raised to murine dendritic epidermal T-cell lines.[50] Functional and biochemical studies with these antibodies suggest that they recognize the mouse α_v subunit.[51] Again, reactivity of these mabs with mouse osteoclasts occurs, but is relatively weak.[83] Mouse osteoclasts also express β_1 (Table 2). Antibodies that react with mouse VLA-α chains are not available.

3. Integrins of Avian Osteoclasts

Chick osteoclasts express vitronectin receptor[23,45,52] (Table 3) defined by mabs (such as 23C6) to epitope B of the vitronectin receptor; this can been confirmed biochemically.[52] Likewise, α_v and β_3 anti-human C-terminal peptide sera react with osteoclasts, as does mab 33 to a functional epitope on the human platelet IIIa (β_3) molecule.[53] Chick vitronectin receptor α_v cDNA has recently been cloned[54] and found to be highly homologous to its human counterpart; peptide antisera to chick α_v amino acid sequences also react with chick osteoclasts. Anti-chick β_1 C-terminal peptide antiserum[55] reacts weakly with chick osteoclasts; no information is available on the associated α-chain specificity.

Thus, within the limitations of immunohistochemical analysis and the lack of available species-specific reagents, the integrin phenotype of chick osteoclasts is identical to its mammalian counterpart.

B. NONINTEGRIN ADHESION RECEPTORS IN BONE

A number of homophylic and heterophylic cell surface receptors whose structures differ from those of the integrin superfamily have been described. They are primarily involved in cell-to-cell contact rather than cell-to-extracellular matrix interaction, e.g., in epithelial tissues and between lymphocytes. These receptors include (1) the calcium-dependent (CADS) cadherins and L-CAM (reviewed in Reference 56), (2) the calcium-independent (CIDS) receptors such as N-CAM (reviewed in Reference 57), some of which belong to the immunoglobulin gene superfamily (reviewed in Reference 14), and the two new groups of hemopoietic cell "homing" receptors, (3) lectin-like adhesion molecules (such as the murine MEL14 antigen), and (4) the CD44/Hermes antigen, which has cartilage proteoglycan homology (reviewed in References 15 and 58 to 60).

With the exception of ICAM1 (CD54) (a member of the immunoglobulin gene superfamily), we have failed to detect receptors of the nonintegrin type in an extensive study of osteoclasts from mammalian and avian species.[85] This negative finding does not exclude the possibility that such receptors may be involved in stages of osteoclast development prior to their morphological identification or that novel osteoclast-specific nonintegrin receptors exist for which reagents allowing their detection are not yet available.

C. FUNCTIONAL ANALYSIS OF ADHESION RECEPTORS ON OSTEOCLASTS

The first indication that adhesion receptors play a functional role in osteoclastic bone resorption was obtained when mabs 13C2 and 23C6 were found to inhibit resorption of bone slices by osteoclastoma osteoclasts *in vitro*.[61] At the time, it was unknown which antigen was being affected by these mabs, and it was therefore named OFA: osteoclast functional antigen. It is now well established that OFA is identical or at least highly related to the vitronectin receptor,[45] and additional evidence for blocking of *in vitro* osteoclastic resorption by antibodies to the vitronectin receptor in chick[52] has recently confirmed the role of this integrin in resorption by freshly bone-derived osteoclasts in a second species. The molecular basis for the role of vitronectin receptor in bone resorption is not fully known. It has been hypothesized that the receptor mediates the attachment of osteoclast to the bone surface. When osteoclasts were settled onto a serum-coated surface in the presence of the snake venom echistatin or short peptides, containing an arginine-glycine-aspartic acid (RGD) amino acid sequence which represents the cell-binding sequence in known ligands for the receptor in other cell types (see Reference 47), osteoclasts did not attach, and osteoclasts which had already spread retracted and finally detached from the substrate.[52,62-64] Similar morphological changes were obtained when osteoclasts were treated with mabs to α_v or β_3.[52,63] So, it appears that the reduction in bone resorption observed with the mabs, and also with echistatin and a number of RGD-containing short peptides,[52,62-64] is the result of disruption of osteoclast attachment to the bone surface via interference with the receptor or its ligand, respectively.

The extent to which the vitronectin receptor is involved in the formation of the tight adhesion of osteoclasts to the bone surface observed in clear zones of podosomes remains unclear, as conflicting evidence to this effect has been presented. Zambonin-Zallone et al.[65] reported accumulation of β_3 immunoreactivity in podosomes in osteoclastoma osteoclasts, and Reinholt et al.[56] found selective immunostaining in the clear zones of rat osteoclast with a polyclonal antiserum that recognizes predominantly α_v and β_3. However, Helfrich[86] found no evidence for accumulation of β_3 in rat osteoclast podosomes and Lakkakorpi et al.[64] reported the specific absence of β_3 and α_v from clear zones in rat and chick osteoclasts, respectively. They postulate that absence of vitronectin receptor in clear zones may be the result of physical exclusion of the receptors from the area of closest cell-matrix contact, and does not necessarily preclude a crucial role for this integrin in initial attachment to the bone matrix and in the motility of osteoclasts. Clearly, more information is needed on the exact localization of the vitronectin receptor on the membrane of osteoclasts in different stages of functional activity, as this will help clarify the role this integrin plays in osteoclast function and may resolve some of the apparent contradictions of the above reports. Also, the interesting possibility that vitronectin receptor is involved in the "homing" of osteoclast precursors to developing bone and their fusion (see Reference 47) remains to be addressed.

The role of other adhesion receptors in osteoclast function is more unclear at present. Some circumstantial evidence for a role in osteoclast adhesion of receptors other than vitronectin receptor has been presented in a study of potential ligands for rat osteoclast integrins. A large number of RGD-containing proteins, including vitronectin, fibronectin, and the bone sialoproteins, were found to mediate osteoclast attachment *in vitro* in a β_3 (presumptively vitronectin receptor)-dependent manner. However, adhesion to type I collagen also occurred. This process was not β_3-dependent, although it was inhibited by RGD peptide. Involvement of the integrin collagen receptor $\alpha_2\beta_1$ could not be conclusively demonstrated, and it was suggested that additional adhesion receptors, possibly including nonintegrin receptors, may be involved in this attachment.[49]

Taken together, much evidence has been presented in support of a role of adhesion receptors in osteoclast attachment, and hence bone resorption *in vitro,* and a number of ligands for these receptors have been identified. The receptor-ligand interactions *in vivo* are potentially very complex, involving different receptors and multiple ligands, and they remain to be fully explored.

V. IMMUNOLOGICAL ANALYSIS OF "OSTEOCLASTS" DERIVED FROM BONE MARROW IN CULTURE

A. OSTEOCLASTS IN HUMAN MARROW CULTURE

Are the multinucleated giant cells of human marrow origin phenotypically related to macrophage polykarya or are they *bona fide* osteoclasts?

Osteoclasts are undoubtedly of hemopoietic origin (see Chapters 4 and 5 of this volume), sharing a cell lineage with cells of the mononuclear phagocyte system, but what is their relationship to the morphologically similar macrophage polykarya? Both are likely to appear in cultures of mononuclear cells from bone marrow, but only osteoclasts are capable of resorbing bone and responding appropriately to calciotrophic hormones and factors. Hemopoietic cells of rabbit and murine origin can be induced to generate osteoclasts *in vitro* when stimulated by 1,25(OH)$_2$ vitamin D$_3$ — they show all the features of genuine osteoclasts isolated from skeletal tissue.[67-69] Production of multinucleated giant cells from human marrow has been demonstrated in analogous experiments and under the stimulatory influence of a range of calcium-regulating hormones, cytokines, and hemopoietic growth factors.[70-72] However, obligate criteria for identification of these cells as osteoclasts have not been rigorously applied. Size, multinuclearity and tartrate-resistant acid phosphatase (TRAP) activity, shared between the two cell types, are insufficient even to call the cells "osteoclast-like". Tumor cells and tissue macrophages can form large, multinucleated cells under appropriate conditions, and TRAP expression is not a specific marker of osteoclasts in man, being expressed constitutively by alveolar macrophages and developing upon macrophage activation *in vivo*. While some cells may be osteoclastic in human marrow cultures, the numbers of bone-resorbing, calcitonin receptor-positive cells are extremely low compared with osteoclast from rabbit or mouse marrow.

We, in a multicenter collaborative study,[73] have recently failed to conclusively demonstrate osteoclastic characteristics in the multinucleated cells appearing in human marrow culture despite application of a wide range of cell developmental stimuli. One of the criteria we analyzed was whether the surface antigenic phenotype of the multinucleated giant cells in these cultures was typical of osteoclasts or of macrophage polykarya (Table 5).

Osteoclasts express high levels of CD13 and vitronectin receptor; the cultured giant cells express only low amounts. In the latter case, the staining was of perinuclear distribution, unlike the membrane distribution in osteoclasts. This vitronectin receptor expression is comparable to reports showing upregulation, albeit at low levels, of receptor in cultured monocytes,[74] which is known to be a novel heterodimer, $\alpha_v\beta_{3b}$.[75] The marrow giant cells expressed high levels of myeloid antigens such as CD11b, CD11c, and CD64 (as was also found by Kukita et al.[76]), whereas osteoclasts fail to express these antigens. Preliminary reports have also shown that a novel mab to human osteoclast, C35, which reacts specifically with bone osteoclast, fails to detect the giant cells forming in human marrow cultures.[22] This further supports the contention that cultured giant cells are distinct from osteoclasts. Thus, the phenotype of human marrow culture-derived giant cells is more akin to that of monocytes and tissue macrophages *in vivo* or cultured marrow or blood myeloid cells than that typical of freshly isolated osteoclasts.

As with the lack of other osteoclastic features, the immunophenotype strongly suggests that the giant cells generated from human marrow are not osteoclasts, but are macrophage polykarya.

B. OSTEOCLASTS IN NONHUMAN MARROW CULTURE

Genuine osteoclasts can be derived in culture from mouse hemopoietic tissues (marrow, spleen), isolated multipotential hemopoietic stem cells, and

TABLE 5
Immunophenotype of Human Long-Term Culture Marrow Giant Cells Compared with Freshly Isolated Human Osteoclasts

	Antigenic profile[a]		
	Cultured bone marrow		
CD number (antigen)	Monocytes	Multinucleated cells	Fresh osteoclasts[b]
HLA DR (class II)	+ +	+	−
45 (leukocyte common)	+ +	+ +	+/±
11b (LFA)	+ +/+ + +	+ +	−
11c (LFA)	+ +/+ + +	+/+ +	−
13 (aminopeptidase N)	+/±	±	+ + +
14	Some + +	−	−
15	Some +	−	−
16	Some +	−	−
33	Some +	−	−
34	Some + +	+/±	−
64	+ +	+	−
68	+ +/+ + +	+	+ + +
51 (vitronectin receptor α_v)	Some ±	±	+ + +
61 (vitronectin receptor β_3)	Some ±	±	+ + +

[a] Data from Reference 73. ± to + + +, range of cellular reactivity; −, negative reaction.
[b] In fetal bone.

immortalized hemopoietic cell lines. (See Chapters 4 and 5 of this volume.) To date, there have been no reports on the surface antigens expressed by these culture-derived osteoclasts. Thus, it is not currently known whether their phenotype conforms to the results from studies in man. It is to be expected that some of the antigenic properties of skeletal osteoclast recently defined (see Section III.A) will be valuable to the question of how similar the cultured osteoclast is to its counterpart appearing on culture. Clearly, if there is disparity in immunophenotype in the rodent system where there is incontrovertible evidence for a genuine functional osteoclast phenotype, then the interpretation of results from human marrow culture will have to be reevaluated. For this reason, it is very important to initiate such studies.

Osteoclasts from other species have not been a focus for immunological analysis, despite being used for functional and developmental analysis. Feline marrow readily forms giant cells on culture;[77,78] these giant cells, however, do not resorb mineralized tissue and are negative for vitronectin receptor unless they are cultured in calf rather than horse serum.[87] Feline osteoclasts express vitronectin receptor and seem to have a phenotype similar to that of human osteoclasts;[79] thus, it would be an appropriate experimental system for study of the regulation of integrin expression *in vitro* in relation to acquisition of osteoclastic function during their terminal differention. Similarly, osteoclasts can be induced to develop from rabbit bone marrow on culture with 1,25$(OH)_2$ vitamin D_3[67] and amplified in the presence of bone matrix.[80] The osteoclasts formed express vitronectin receptor together with the ability to resorb bone, whereas cultured macrophage giant cells fail to do so. The recent development of an osteoclast-specific mab[88] to rabbit osteoclasts would make further exploitation of rabbit marrow culture attractive.

VI. CONCLUDING REMARKS

In this chapter, we have described the range and, where possible, functional attributes of the antigens (mainly surface) expressed by osteoclasts in a range of mammalian and avian species. Of particular note is the current work on osteoclast extracellular matrix receptors of the integrin type, especially the vitronectin receptor, which clearly have functional importance in the regulation of osteoclast recognition of, adhesion to, and resorption of mineralized matrix. Likewise, phenotypic analysis has been useful in defining cell lineage relationships between osteoclasts and other bone and bone marrow cell types. Osteoclasts express some antigens (for example, CD9, CD13, and CD54) which may play significant roles in osteoclast function; to date, these possible activities have not been investigated. Analysis of osteoclast-like cells derived in culture has raised important questions as to the true nature of these cells, particularly in the human system. These will be areas of focus for future research in osteoclast cell biology.

ACKNOWLEDGMENTS

We thank Drs. P. Osdoby, I. James, J. Heersche, and K. Väänänen for providing us with unpublished data which have helped in the writing of this review.

REFERENCES

1. **Horton, M. A., Lewis, D., McNulty, K., Pringle, J. A. S., and Chambers, T. J.,** Monoclonal antibodies to osteoclastomas (giant cell bone tumors): definition of osteoclast-specific antigens, *Cancer Res.*, 45, 5663, 1985.
2. **Horton, M. A.,** Osteoclast-specific antigens, in *ISI Atlas of Science: Immunology,* 1988, 1, 35.
3. **Horton, M. A., Rimmer, E. F., Moore, A., and Chambers, T. J.,** On the origin of the osteoclast: the cell surface phenotype of rodent osteoclasts, *Calcif. Tissue Int.*, 37, 46, 1985.

4. **Athanasou, N. A., Quinn, J., and McGee, J. O.-D.,** Immunocytochemical analysis of the human osteoclast: phenotypic relationship to other marrow-derived cells. *Bone Miner.*, 3, 317, 1988.
5. **Athanasou, N. A. and Quinn, J.,** Immunophenotypic differences between osteoclasts and macrophage polykaryons: immunohistological distinction and implications for osteoclast ontogeny and function, *J. Clin. Pathol.*, 43, 997, 1990.
6. **Mc Michael, A. J., Beverley, P. L. C., Cobbold, S., Crumpton, M. J., Gilks, W., Gotch, F. M., Hogg, N., Horton, M. A., Ling, N., MacLennan, I. C. M., Mason, D. Y., Milstein, C., Spiegelhalter, D., and Waldmann, H.,** *Leucocyte Typing III: White Cell Differentiation Antigens,* Oxford University Press, Oxford, 1987, 1-1050.
7. **Knapp, W., Dörken, B., Gilks, W. R., Reiber, E. P., Schmidt, R. E., Stein, H., and von dem Borne, A. E. G. Kr.,** *Leucocyte Typing IV: White Cell Differentiation Antigens,* Oxford University Press, Oxford, 1989, 1-1182.
8. **Parravicini, C. L., Soligo, D., Berti, E., Cattoretti, G., Gaiera, G., and Vago, L.,** Immunohistochemical reactivity of anti-platelet mAb in normal human tissues and bone marrow, in *Leucocyte Typing IV: White Cell Differentiation Antigens,* Knapp, W., Dörken, B., Gilks, W. R., Reiber, E. P., Schmidt, R. E., Stein, H., and von dem Borne, A. E. G., Kr., Eds., Oxford University Press, Oxford, 1989, 981.
9. **Athanasou, N. A., Quinn, J., and McGee, J. O'D.,** Reactivity of Workshop mAb of the myeloid, platelet, and non-lineage (CD45, 45R, CD11a, b, c) series with cells in bone and synovium, im *Leucocyte Typing IV: White Cell Differentiation Antigens,* Knapp, W., Dörker, B., Gilks, W. R., Rieber, E. P., Schmidt, R. E., Stein, H., and von dem Borne, A. E. G. Kr., Eds., Oxford University Press, Oxford, 1989, 921.
10. **Horton, M. A., Rimmer, E. F., Lewis, D., Pringle, J. A. S., Fuller, K., and Chambers, T. J.,** Cell surface characterization of the human osteoclast: phenotypic relationship to other bone marrow cell types, *J. Pathol.*, 144, 281, 1984.
11. **Ashmun, R. A. and Look, A. T.,** Metalloprotease activity of CD13/aminopeptidase N on the surface of human myeloid cells, *Blood,* 75, 462, 1990.
12. **Indig, F. E., Benayahu, D., Fried, A., Wientroub, S., and Blumberg, S.,** Neutral endopeptidease (EC3.4.24.11) is highly expressed on osteoblastic cells and other marrow stromal cell types, *Biochem. Biophys. Res. Commun.*, 172, 620, 1990.
13. **Ketcham, C. M., Roberts, R. M., Simmen, R. C. M., and Nick, H. S.,** Molecular cloning of the type 5, iron-containing, tartrate-resistant acid phosphatase from human placenta, *J. Biol. Chem.*, 264, 557, 1989.
14. **Hunkapiller, T. and Hood, L.,** Diversity of the immunoglobulin gene superfamily, *Adv. Immunol.*, 44, 1, 1989.
15. **Osborn, L.,** Leukocyte adhesion to endothelium in inflammation, *Cell,* 62, 3, 1990.
16. **Hato, T., Ikeda, K., Yasukawa, M., Watanabe, A., and Kobayashi, Y.,** Exposure of platelet fibrinogen receptors by a monoclonal antibody to CD9 antigen, *Blood,* 72, 224, 1988.
17. **Slupsky, J. R., Seehafer, J. G., Tang, S.-S., Masellis-Smith, A., and Shaw, A. R. E.,** Evidence that monoclonal antibodies against CD9 antigen induce specific association between CD9 and the platelet glycoprotein IIb-IIIa complex, *J. Biol. Chem.*, 264, 12289, 1989.
18. **Kukita, T. and Roodman, G. D.,** Development of a monoclonal antibody to osteoclasts formed *in vitro* which recognizes mononuclear osteoclast precursors in the marrow, *Endocrinology,* 125, 630, 1989.
19. **Aqel, N. M., Clark, M., Jones, D. B., Arnot, J., and Waldmann, H.,** Identification of cells of the human mononuclear phagocyte system with rat monoclonal antibodies, *J. Pathol.*, 152, 265, 1987.
20. **Chilosi, M., Giliol, E., Lestani, M., Menestrina, F., and Fiore-Donati, L.,** Immunohistochemical characterization of osteoclasts and osteoclast-like cells with monoclonal antibody MB1 on paraffin-embedded tissues, *J. Pathol.*, 156, 251, 1988.

21. **Athanasou, N. A., Pulford, K. A. F., Quinn, J., Gatter, K. C., Mason, D. Y., and McGee, J. O.,** A monoclonal antimacrophage antibody (3MA-134) which reacts with human osteoclasts in decalcified paraffin embedded bone sections, *Calcif. Tissue Int.* 42 (Suppl.), A1, 1988.
22. **James, I. E., Walsh, S., Dodds, R. A., and Gowen, M.,** The production and characterization of osteoclast-selective monoclonal antibodies which distinguish between multinucleated cells derived from different human tissues, *J. Histochem. Cytochem.*, 39, 905, 1991.
23. **Horton, M. A. and Chambers, T. J.,** Human osteoclast-specific antigens are expressed by osteoclasts in a wide range of non-human species, *Br. J. Exp. Pathol.*, 67, 95, 1986.
24. **Aqel, N. M., Pringle, J. A. S., and Horton, M. A.,** Cellular heterogeneity in giant cell tumor of bone (osteoclastoma): an immunohistological study of 16 cases, *Histopathology,* 13, 675, 1988.
25. **Athanasou, N. A., Wells, C. A., Quinn, J., Ferguson, D. P., Heryet, A., and McGee, J. O.,** The origin and nature of stromal osteoclast-like multinucleated giant cells in breast carcinoma: implication of tumor osteolysis and macrophage biology, *Br. J. Cancer,* 59, 491, 1989.
26. **Sminia, T. and Dijkstra, C. D.,** The origin of osteoclasts: an immunohistochemical study on macrophages and osteoclasts in embryonic rat bone, *Calcif. Tissue Int.*, 39, 263, 1986.
27. **Hume, D. A., Loutit, J. F., and Gordon, S.,** The mononuclear phagocyte system of the mouse defined by immunohistochemical localization of antigen F4/80: macrophages of bone and associated connective tissue, *J. Cell Sci.*, 66, 189, 1984.
28. **Leenen, P. J. M., Melis, M., Kraal, G., Hoogeveen, A. T., and van Ewijk, W.,** Monoclonal antibody ER-BMDM1 recognizes a macrophage and dendritic cell differentiation antigen with aminopeptidase activity, in Phenotypical Analysis of Murine Macrophage Differentiation, Leenen, P. J. M., Ph.D. thesis, University of Rotterdam, 1989, 141.
29. **Illés, T. and Fischer, J.,** Distribution of lectin binding glycoprotein in osteoclasts, *Histochemistry,* 91, 55, 1989.
30. **Oursler, M. J., Bell, L. V., Clevinger, B., and Osdoby, P.,** Identification of osteoclast-specific monoclonal antibodies, *J. Cell Biol.*, 100, 1592, 1985.
31. **Oursler, M. J. and Osdoby, P.,** Osteoclast development in marrow cultured in calvaria-conditioned media, *Dev. Biol.*, 127, 170, 1988.
32. **Webber, D., Osdoby, P., Hauschka, P., and Krukowski, M.,** Correlation of an osteoclast antigen and ruffled border on giant cells formed in response to resorbable substrates, *J. Bone Miner. Res.*, 5, 401, 1990.
33. **Oursler, M. J. and Osdoby, P.,** Characterization of an osteoclast membrane protein related to superoxide dismutase, in *Calcium Regulation and Bone Metabolism,* Cohn, M. J., Glorieux, F. H., and Martin, T. J., Eds., Elsevier, Amsterdam, 1990, 214.
34. **Collin-Osdoby, P., Oursler, M. J., Schmitt, E. M., and Osdoby, P.,** Identification and isolation of a new osteoclast specific surface protein, *J. Bone Miner. Res.*, 4 (Suppl. 1), 205, 1989.
35. **Nijweide, P. J., Vrijheid-Lammers, T., Mulder, R. J., and Blok, J.,** Cell surface antigens on osteoclasts and related cells in the quail studied with monoclonal antibodies, *Histochemistry,* 83, 315, 1985.
36. **Hentunen, T. A., Tuukkanen, J., and Väänänen, H. K.,** Osteoclasts and a small population of peripheral blood cells share common surface antigens, *Calcif. Tissue Int.*, 47, 8, 1990.
37. **Hentunen, T. A. and Väänänen, H. K.,** Inhibition of bone resorption by osteoclast-recognizing monoclonal antibodies, *J. Bone Miner. Res.*, 4 (Suppl. 1), 213, 1989.
38. **Hentunen, T. A., Lakkakorpi, P. T., Rautiala, T., and Väänänen, K.,** Inhibition of bone resorption by a monoclonal antibody which reacts with a 150 kD membrane protein in chicken osteoclasts, *J. Bone Min. Res.*, 6, 1091, 1991.

39. **Chanoki, Y., Shimazaki, M. Y., Kuwahara, H., Kadoya, A., Morikita, I., Nagahama, S., Shimazu, A., Ogawa, Y., and Yagi, T.,** Production and characterization of osteoclast-specific monoclonal antibodies, *Osaka City Med. J.*, 32, 99, 1986.
40. **Athanasou, N. A., Alvarez, J. I., Ross, F. P., and Teitelbaum, S. L.,** Immunohistochemical characterization and bone-resorbing characteristics of avian marrow-derived mononuclear and mutlinuclear cells, *J. Bone Miner. Res.*, 5 (Suppl. 2), 93, 1990.
41. **Schmidt, J. A., Marshall, J. M., Hayman, M. J., Doederlein, G., and Beug, H.,** Monoclonal antibodies to novel erythroid differentiation antigens reveal specific effects of oncogenes on the transformed cell phenotype, *Leukocyte Res.*, 10, 257, 1985.
42. **Hynes, R.,** Integrins: a family of cell surface receptors, *Cell,* 48, 549, 1987.
43. **Burridge, K. and Fath, K.,** Focal contacts: transmembrane links between the extracellular matrix and the cytoskeleton, *BioEssays,* 10, 104, 1989.
44. **Hemler, M. E.,** VLA proteins in the integrin family: structures, functions, and their role on leukocytes, *Annu. Rev. Immunol.*, 8, 365, 1990.
45. **Davies, J., Warwick, J., Totty, N., Philp, R., Helfrich, M., and Horton, M.,** The osteoclast functional antigen implicated in the regulation of bone resorption is biochemically related to the vitronectin receptor, *J. Cell Biol.*, 109, 1817, 1989.
46. **Horton, M. A. and Davies, J.,** Perspectives: adhesion receptors in bone, *J. Bone Miner. Res.*, 4, 803, 1989.
47. **Horton, M.,** Current status review. Vitronectin receptor: tissue specific expression or adaptation to culture, *Int. J. Exp. Pathol.*, 71, 741, 1990.
48. **Davies, J. D.,** The Use of Monoclonal Antibodies to Identify and Characterize Functional Antigens on Osteoclasts, Ph.D. thesis, University of London, 1990.
49. **Helfrich, M. H., Nesbitt, S. A., Dorey, E. L., and Horton, M. A.,** Rat osteoclasts adhere to a wide range of RGD (Arg-Gly-Asp) peptide-containing proteins, including the bone sialoproteins and fibronectin via a β3 integrin, *J. Bone Min. Res.*, 7, 335, 1992.
50. **Maxfield, S. R., Moulder, K., Koning, F., Elbe, S., Stingl, G., Coligan, J. E., Shevach, E. M., and Yokoyama, W. M.,** Murine T cells express a cell surface receptor for multiple extracellular matrix protein. Identification and characterization with monoclonal antibodies, *J. Exp. Med.*, 169, 2173, 1989.
51. **Moulder, K., Roberts, K., Shevach, E. M., and Coligan, J. E.,** The mouse vitronectin receptor is a T cell activation antigen, *J. Exp. Med.*, 173, 343, 1991.
52. **Horton, M. A., Taylor, M. L., Arnett, T. R., and Helfrich, M. H.,** Arg-Gly-Asp (RGD) peptides and the anti-vitronectin receptor antibody 23C6 inhibit cell spreading and dentine resorption by osteoclasts, *Exp. Cell Res.*, 195, 368, 1991.
53. **D'Souza, S. E., Ginsberg, M. H., and Plow, E. F.,** The ligand binding site of the platelet integrin receptor GPIIb-IIIa is proximal to the second calcium binding domain of its alpha subunit, *J. Biol. Chem.*, 265, 3440, 1990.
54. **Bossy, B. and Reichardt, L. F.,** Chicken integrin αv subunit molecular analysis reveals high conservation of structural domains and association with multiple β subunits in embryo fibroblasts, *Biochemistry,* 29, 10191, 1990.
55. **Marcantonio, E. E. and Hynes, R. O.,** Antibodies to the conserved cytoplasmic domain of the integrin β subunit react with proteins in vertebrates, invertebrates, and fungi, *J. Cell Biol.*, 106, 1765, 1988.
56. **Takeichi, M.,** The cadherins: cell-cell adhesion molecules controlling animal morphogenesis, *Development,* 102, 639, 1988.
57. **Rutishauser, U. and Jessell, T. M.,** Cell adhesion molecules in vertebrate neural development, *Physiol. Rev.*, 68, 819, 1988.
58. **Duijvestijn, A. and Hamann, A.,** Mechanisms and regulation of lymphocyte migration, *Immunol. Today,* 10, 23, 1989.
59. **Coombe, D. R. and Rider, C. C.,** Lymphocyte homing receptors cloned — a role for anionic polysaccharides in lymphocyte adhesion, *Immunol. Today,* 10, 289, 1989.
60. **Stoolman, L. M.,** Adhesion molecules controlling lymphocyte migration, *Cell,* 56, 907, 1989.

61. **Chambers, T. J., Fuller, K., Darby, J. A., Pringel, J. A. S., and Horton, M. A.,** Monoclonal antibodies against osteoclasts inhibit bone resorption *in vitro, Bone Miner.,* 1, 127, 1986.
62. **Sato, M., Sardana, M. K., Grasser, W. A., Garsky, V. M., Murray, J. M., and Gould, R. J.,** Echistatin is a potent inhibitor of bone resorption in culture, *J. Cell Biol.,* 111, 1713, 1990.
63. **Horton, M. A., Dorey, E. L., Nesbitt, S., Samanen, J., Ali, F. E., Stadel, J. M., Nichols, A., Greig, R., and Helfrich, M. H.,** An initial structure-function analysis of RGD-peptide analogue modulation of osteoclast function through vitronectin receptor inhibition, submitted.
64. **Lakkakorpi, P. T., Horton, M. A., Helfrich, M. H., Karhukorpi, E.-K., and Väänänen, H. K.,** Vitronectin receptor has a role in bone resorption but does not mediate tight sealing zone attachment of osteoclasts to the bone surface, *J. Cell Biol.,* 115, 1179, 1991.
65. **Zambonin-Zallone, A., Teti, A., Grano, M., Rubinacci, A., Abbadini, M., Gaboli, M., and Marchisio, P. C.,** Immunocytochemical distribution of extracellular matrix receptors in human osteoclasts: a β3 antigen is colocalized with vinculin and talin in the podosomes of osteoclastoma giant cells, *Exp. Cell Res.,* 182, 645, 1989.
66. **Reinholt, F. P., Hultenby, K., Oldberg, Å., and Heinegård, D.,** Osteopontin — a possible anchor of osteoclasts to bone, *Proc. Natl. Acad. Sci. U.S.A.,* 87, 4473, 1990.
67. **Fuller, K. and Chambers, T. J.,** Generation of osteoclasts in cultures of rabbit bone marrow and spleen cells, *J. Cell Physiol.,* 132, 441, 1987.
68. **Hattersley, G. and Chambers, T. J.,** Calcitonin receptors as markers for osteoclastic differentiation: correlation between generation of bone resorptive cells and cells that express calcitonin receptors in mouse bone marrow cultures, *Endocrinology,* 125, 1606, 1989.
69. **Takahashi, N., Akatsu, T., Udagawa, N., Sasaki, T., Moseley, J. M., Martin, T. J., and Suda, T.,** Osteoblastic cells are involved in osteoclast formation, *Endocrinology,* 123, 2600, 1988.
70. **Kurihara, N., Bertolini, D., Suda, T., Akayama, Y., and Rooodman, G. D.,** IL-6 stimulated osteoclast-like multinucleated cell formation in long-term human cultures by inducing IL-1 release, *J. Immunol.,* 144, 4226, 1990.
71. **MacDonald, B. R., Takahashi, N., McManus, L. M., Holohan, J., Mundy, G. R., and Roodman, G. D.,** Formation of multinucleated cells that respond to osteotropic hormones in long term human marrow cultures, *Endocrinology,* 120, 2326, 1987.
72. **MacDonald, B. R., Clark, S., Wang, E. A., Kuehl, T. J., Stanley, E. R., and Roodman, G. D.,** Effects of human recombinant CSF-GM and high purified CSF-1 on the formation of multinucleated cells with osteoclast characteristics in long-term bone marrow cultures, *J. Bone Miner. Res.,* 1, 227, 1986.
73. **Flanagan, A. M., Horton, M. A., Dorey, E. L., Evely, R. S., Moseley, J. M., Firkin, F. C., Chambers, T. J., Helfrich, M. H., and Martin, T. J.,** An assessment of the ability of human bone marrow cultures to generate osteoclasts, *Int. J. Exptl. Pathol.,* in press.
74. **Andreesen, R., Brugger, W., Scheibenbogen, C., Kreutz, M., Lesser, H. G., Rehm, A., and Löhr, G. W.,** Surface phenotype analysis of human monocyte to macrophage maturation, *J. Leukocyte Biol.,* 47, 490, 1990.
75. **Krissansen, G. W., Elliott, M. J., Lucas, C. M., Stomski, F. C., Berndt, M. C., Cheresh, D. A., Lopez, A. F., and Burns, G. F.,** Identification of a novel integrin β subunit expressed on cultured monocytes (macrophages): evidence that one α subunit can associate with multiple β subunits, *J. Biol. Chem.,* 265, 823, 1990.
76. **Kukita, T., McManus, L. M., Miller, M., Civin, C., and Roodman, G. D.,** Osteoclast-like cells formed in long-term human bone marrow cultures express a similar surface phentoype as authentic osteoclasts, *Lab. Invest.,* 60, 532, 1989.
77. **Testa, N. G., Allen, T. D., Lajtha, L. G., Onions, D., and Jarret, O.,** Generation of osteoclasts *in vitro, J. Cell Sci.,* 47, 127, 1981.

78. **Pharoah, M. J. and Heersche, J. N. M.,** 1,25-Dihydroxyvitamin D_3 causes an increase in the number of osteoclast-like cells in cat bone marrow cultures, *Calcif. Tissue Int.*, 37, 276, 1985.
79. **Horton, M. A., Fowler, P., Simpson, A., and Onions, D.,** Monoclonal antibodies to human antigens recognize feline myeloid cells, *Vet. Immunol. Immunopathol.*, 18, 213, 1988.
80. **Fuller, K. and Chambers, T. J.,** Bone matrix stimulates osteoclastic differentiation in cultures of rabbit bone marrow cells, *J. Bone Miner. Res.*, 4, 179, 1989.
81. **Smith, T. A., Leenen, P. J. M., and Horton, M. A.,** unpublished data, 1991.
82. **Horton, M. A.,** unpublished data, 1990.
83. **Horton, M. A. and Helfrich, M. H.,** unpublished data, 1990.
84. **Osdoby, P.,** personal communication, 1991.
85. **Horton, M. A.,** unpublished data, 1991.
86. **Helfrich, M. H.,** unpublished data, 1991.
87. **Horton, M. A.,** unpublished data, 1989.
88. **Heersche, J.,** personal communication, 1990.

Chapter 3

ACID PHOSPHATASES

Göran Andersson, Barbro Ek-Rylander, and Cedric Minkin

TABLE OF CONTENTS

I. OVERVIEW OF ACID PHOSPHATASES

Acid phosphatase (EC3.1.3.2) is represented by a number of enzymes that can be differentiated according to structural and catalytic properties, tissue distribution, and subcellular location[1] (Table 1). Polyacrylamide gel electrophoresis (PAGE) under nondenaturing conditions at acidic pH, reveals five major bands of acid phosphatase activity in tissues and body fluids.[1] The lysosomal (LAP) (bands 1 and 3) and prostatic (PAP) (bands 2 and 4) acid phosphatases belong to the group of acid phosphatases that are sensitive to inhibition by L-tartrate. LAP is confined to lysosomes and is expressed in all nucleated mammalian cells. Cloning of cDNAs for rat[2] and human[3] LAP shows that the mature LAPs consist of 393 amino acids with a molecular mass (unglycosylated) of approximately 45 kDa. Expression of PAP is confined to the secretory epithelia of the prostate gland, from which the enzyme is secreted into semen and blood. PAP, but not LAP, exhibits phosphotyrosyl protein phosphatase activity.[4] Isolation and sequencing of rat[5] and human[5,6] PAP cDNA clones has revealed that the mature PAPs consist of 350 (rat) or 354 (human) amino acids with a deduced molecular weight (unglycosylated) of around 40 kDa. Comparison of the protein coding areas of rat PAP and LAP show an overall sequence similarity of 46%. However, there are three regions of much higher similarity, as well as conserved positions of six cysteine residues and some of the N-glycosylation sites, suggesting an evolutionary relationship between LAP and PAP genes.[3,5]

There are at least two tartrate-resistant acid phosphatases expressed in the cytosol of mammalian cells which do not migrate in the acidic PAGE system. The cytosolic acid phosphatase (CAP) of Mr 17 to 18 kDa is probably expressed in all mammalian cells and has phosphotyrosyl protein phosphatase activity.[7,8] This enzyme may play a role in the regulation of the signal transduction pathway which involves protein tyrosine phosphorylation.[9] The entire protein sequence of CAP from bovine liver has been determined.[10] The enzyme consists of 157 amino acids, has an acetylated NH_2 terminus and a calculated molecular weight of around 18 kDa.

Another low molecular weight CAP (cytosolic acid acylphosphatase, CAAP) of 11 to 12 kDa, which is active against acyl phosphate substrates but not against phosphomonoesters such as *p*-nitrophenylphosphate, consists of 98 amino acids and bears no significant sequence similarity to the Mr 18-kDa form.[11] CAAP has been found in erythrocytes and skeletal muscle, where it may be involved in regulation of calcium transport by its ability to dephosphorylate the phosphoenzyme intermediate of cell membrane Ca^{2+}-ATPase.[12]

TABLE 1
Classification and Properties of Acid Phosphatases

Common name	PAGE mobility	Mol wt	Tartrate response	Function
Lysosomal acid phosphatase (LAP)	Bands 1, 3	45 kDa	Tartrate sensitive	Lysosomal catabolism
Prostatic acid phosphatase (PAP)	Bands 2, 4	40 kDa	Tartrate sensitive	Phosphotyrosyl protein phosphatase
Osteoclast/macrophage acid phosphatase (TRAP)	Band 5	35 kDa	Tartrate resistant	Bone/extracellular matrix degradation
Cytosolic acid phosphatase (CAP)	—	18 kDa	Tartrate resistant	Postreceptor signal transduction
Cytosolic acid acylphosphatase (CAAP)	—	11 kDa	Tartrate resistant	Regulation of Ca^{2+}-ATPase

II. BIOCHEMISTRY, MOLECULAR BIOLOGY, AND HISTOCHEMISTRY OF TYPE 5, TARTRATE-RESISTANT AND PURPLE, ACID PHOSPHATASES

Band 5, tartrate-resistant and purple, acid phosphatases are associated with osteoclasts and have been implicated in the process of osteoclastic bone resorption, which is the scope of this volume. In this chapter, we will concentrate on various aspects of this particular subclass of acid phosphatases, namely, their structure, catalytic properties, cellular and subcellular distribution, regulation, and proposed function(s). The general characteristics of the type 5 acid phosphatases are summarized in Table 2.

Characteristic features of the type 5 acid phosphatases include resistance to the inhibition of catalytic activity by L-tartrate, high isoelectric points, molecular weights of around 35 kDa, and the presence of iron in the form of a dinuclear active site.[13] As an excellent and detailed review on the bioinorganic and biophysical aspects on the structure and function of the catalytic iron center in the purple acid phosphatases has recently appeared,[13] we have only covered these aspects briefly.

A. TYPE 5 ACID PHOSPHATASE IN THE SKELETON

1. Purification, Enzymatic and Structural Properties

From the histochemical observation in 1971 that the acid phosphatase activity of rat osteoclasts could be distinguished from the activity of neighboring bone cells by its resistance to inhibition by high concentrations (>10 m*M*) of sodium tartrate,[14] the isolation of the tartrate-resistant isoenzyme was accomplished in 1977 by Anderson and Toverud.[15] Following extraction of neonatal rat long bones with the nonionic detergent Triton® X-100 in the presence of 0.3 *M* KCl, the separation of tartrate-sensitive (LAP) and tartrate-resistant (TRAP) forms was achieved by utilizing the strong affinity of TRAP to the negatively charged carboxymethyl (CM)-cellulose matrix at an acidic

TABLE 2
General Characteristics of Type 5 Acid Phosphatases

Iron-containing lysosomal glycoproteins
Molecular weights between 30 and 40 kDa
Basic isoelectric points
Acidic pH optima
Resistant to inhibition of enzyme activity by tartrate
Hydrolyze phosphomonoester substrates, including nucleotides and phosphoproteins

pH.[16,17] Further purification of TRAP by gel permeation chromatography,[16] Blue-Sepharose,[18] and concanavalin A-agarose affinity chromatography[19,20] yielded highly purified and enzymatically active preparations. Similar procedures were later used for the purification of TRAP from slightly older (i.e., 3-week-old) rats,[21] from bovine cortical bone,[22,23] and from adult human femoral bone.[24]

In long bones of young rats, the tartrate-resistant form constitutes around 15% of the total acid phosphatase activity and 0.05 to 0.1% of the detergent-extractable protein.[21] The molecular weights of the purified rat, bovine, and human TRAPs are in the range of 33 to 36 kDa.[18-23] The rat enzyme is isolated entirely as a disulfide-linked dimer consisting of subunits of 20 and 16 kDa.[25] The pI of purified rat and human bone TRAP is 9.3 and 8.2, respectively.[19,24] In contrast, the bovine bone TRAP is acidic, having a pI of 4 to 5.[23] Hydrodynamic studies indicate that rat bone TRAP behaves as a perfectly spherical particle and binds low amounts of nonionic detergents,[25] properties uncharacteristic for membrane-integrated proteins. The iron content of the rat bone enzyme has been determined to be 1.3 to 2.1 mol Fe per mol enzyme,[18,19,25] similar to purple acid phosphatases isolated from other tissues.[13] Substrate specificity studies have demonstrated that bone TRAPs are active against nucleotide tri- and diphosphates, *p*-nitrophenylphosphate, α-naphthylphosphate, phenylphosphate, phosphotyrosine, pyrophosphate, the phosphoproteins phosvitin and α-casein, and in the case of the bovine enzyme, phosphotyrosyl histone.[16-23] Enzyme activity is stimulated three- to tenfold by ferrous iron, particularly in the presence of ascorbate as a reducing agent.[18,19,21,23,26,27] A stimulatory effect of ascorbate alone is probably due to reduction of enzyme-bound ferric iron. In some preparations, slight activation can also be achieved by the reducing agents dithiotreitol or β-mercaptoethanol.[18,19,23,24,26] The bovine bone enzyme is slightly activated by manganese ions.[23] Potent inhibitory agents are dithionite, molybdate, and *p*-chloromercuribenzoate, whereas phosphate, arsenate, fluoride, diphosphonates, and meta- and orthovanadate are weaker inhibitors of tartrate-resistant acid *p*-nitrophenylphosphatase activity.[16,18,19,21,23,24] Of cations, zinc, copper, and cobalt are strongly inhibitory.[18,19,21,23] Interestingly, the nucleotide triphosphatase activity of the rat bone enzyme is, in contrast to *p*-nitrophenylphosphatase, noncompetitively inhibited by magnesium and calcium ions.[21]

Based on the structural and enzymatic properties summarized above, it can be concluded that rat and human bone TRAP are similar in terms of catalytic and structural properties, but that the bovine enzyme differs with respect to pI and exhibits phosphotyrosyl protein phosphatase activity. N-terminal amino acid sequence analysis of rat bone TRAP shows a high degree of similarity to TRAP enzymes from bovine spleen, human placenta, human hairy cell leukemia, human giant cell tumor of bone (osteoclastoma), and uteroferrin,[25] indicating, in combination with similarities in enzymatic and physicochemical properties, a close structural relationship between type 5 acid phosphatases expressed in different tissues. Additionally, the presence of approximately two atoms of Fe in rat bone TRAP/tartrate-resistant acid ATPase (TrATPase) lends further support to the hypothesis by Vincent and Averill[13] that purple acid phosphatases and TRAPs constitute one and the same entity.

TRAP appears to be expressed in several tissues besides bone. In an initial attempt to clarify whether tissue- and/or cell-specific forms of TRAP exist (i.e., whether type 5 acid phosphatases constitute a single protein entity or not), a 1397-bp full-length TRAP/TrATPase cDNA clone was isolated from a neonatal rat calvaria λ-gt11 cDNA library.[28] The cDNA clone consists of a 92-bp untranslated 5′-flanking region, an open reading frame of 981 bp, and a 324-bp untranslated 3′-poly A-containing region. The deduced protein sequence of 327 amino acids contains a putative cleavable signal sequence of 21 amino acids. The mature polypeptide of 306 amino acids has a calculated Mr of 34,350 Da, a pI of 9.18, and contains two potential N-glycosylation sites and the lysosomal targeting sequence DKRFQ (Table 3). The sequence is schematically depicted in Figure 1. The deduced protein sequence displays 89 to 94% homology to TRAP enzymes from human placenta, beef spleen, and uteroferrin, and is identical to the N terminus of purified rat bone TRAP/TrATPase.[25,28] Interestingly, an N-terminal amino acid segment is strikingly homologous to the corresponding region in lysosomal and prostatic acid phosphatases, suggesting a possible evolutionary relationship between these acid phosphatases.[28] The cDNA recognized a 1.5-kb mRNA in long bones and calvaria. *In vitro* translation using an mRNA template transcribed from the full-length insert yielded an immunoprecipitated product of 34 kDa.[28] The amino acid sequence and properties of TRAP deduced from the nucleotide sequence of the full-length cDNA corroborate earlier findings from structural studies on the purified enzyme protein, and emphasizes the similarity of TRAP enzymes expressed in different tissues.

2. Cellular and Subcellular Distribution of TRAP in Bone

a. Methodological Considerations

As previously mentioned, acid phosphatase has long been associated with osteoclasts, and with the discovery that the osteoclastic enzyme is resistant to inhibition by tartrate,[14] TRAP has emerged as a useful histochemical tissue

TABLE 3
Properties of Rat Bone Type 5 Acid Phosphatase (ATPase)

Molecular weight	
SDS-PAGE	33,000 Da
Sedimentation equilibrium	31,900 Da
Sequence	34,350 Da
Iron content	
(mol/mol enzyme)	1.7 ± 0.2
Isoelectric point	9.18
Number of amino acids	
With signal peptide	327
Without signal peptide	306
Number of potential N-glycosylation sites	2
Number of cysteine residues	3

Note: Data from References 25 and 28.

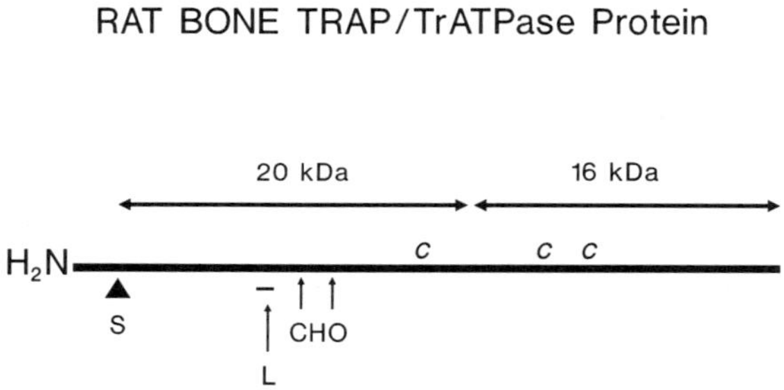

FIGURE 1. Schematic representation of the rat bone TrATPase/TRAP protein sequence. S, signal peptidase cleavage site; L, lysosomal targeting sequence; CHO, potential N-glycosylation sites: C, cysteine residues. (Data derived from Ek-Rylander, B., Bill, P., Norgård, M., Nilsson, S., and Andersson, G. N., *J. Biol. Chem.,* 266, 24684, 1991.)

marker for osteoclasts, e.g., in studies of osteoclast ontogeny.[29,30] TRAP staining can also be used to quantitate osteoclasts in histomorphometric studies using human bone biopsies.[31]

Histochemical techniques for the demonstration of TRAP activity in bone have been developed for both light and electron microscopy.[32,33] For conventional light microscopical analysis, bones from young (<2 to 3 weeks) are preferred, since extensive and prolonged decalcification is deleterious to enzyme activity. Thus, rapid decalcification using ethylene diamine tetracetic acid (EDTA), rather than formic acid, is in most cases mild enough to preserve TRAP activity. However, if inactivated by fixation and/or demineralization, TRAP may be reactivated by preincubation of bone sections in alkaline buffers.[34] The enzyme resists quite stringent fixation conditions, e.g., using 2% glutaraldehyde as the fixative does not appreciably inactivate the enzyme.[29] The visualization of TRAP activity can be accomplished using α-naphthylphosphate as substrate, which can be coupled to the chromogen Fast Garnet added directly to the incubation medium;[35,36] alternatively, naphthyl-ASBI-phosphate can be used as the substrate, forming the insoluble naphthyl-ASBI which can be visualized following postcoupling with either the Fast Garnet chromogen or with pararosaniline.[37,38] Another recently developed method employs adenosine 5′-triphosphate (ATP) as a substrate for the enzyme.[39] As demonstrated by substrate specificity studies, the purified bone TRAP can efficiently hydrolyze the phosphate groups of ATP and adenosine 5′-diphosphate (ADP) at an acidic pH, e.g., the enzyme can also be described as a TrATPase.[21] In this histochemical reaction, liberated phosphate is complexed and made insoluble by incubation with lead nitrate; the precipitate is then visualized by incubation with ammonium sulfide.[39]

b. Enzyme- and Immuno-histochemical Localization of TRAP/TrATPase in Bone

By light microscopical analysis, TRAP activity can be observed in emerging mono- and multinucleated cells in the periosteum of the bone anlage in fetal long bones and calvaria.[30] In neonatal and postnatal bone, the predominant cellular localization of TRAP is in osteoclasts at the lateral endosteal surfaces of metaphyseal long bones and on the marrow cavity ends of bone trabeculae of the metaphysis beneath the epiphyseal plate.[29] Additionally, some TRAP-positive mononuclear cells can be detected in the marrow cavity. It has recently been observed that fewer mononuclear cells are stained using TrATPase than conventional TRAP staining.[39] If the TRAP- and TrATPase-positive cells in fact represent preosteoclasts, in contrast to the TRAP-positive/TrATPase-negative mononuclear cells, TrATPase may be a more selective histochemical marker than TRAP in studies of osteoclast ontogeny.[29,39] Besides a more restricted mononuclear cell expression, TrATPase does not exhibit the general bone surface staining commonly observed with TRAP.[39] This difference facilitates identification of osteoclasts using staining for TrATPase. However, the bone area undergoing resorption that is juxtaposed to the ruffled border of the osteoclast is heavily stained for TrATPase (Figure 2A), suggesting that the enzyme is secreted in a polarized fashion by the osteoclast.

An antiserum to a rat bone TRAP/TrATPase has been used in immunocytochemical studies at the light microscopical[40] and ultrastructural levels.[41] This antiserum reacts with the purified 33-kDa TRAP/TrATPase on Western blots[25] and precipitates a 34-kDa protein transcribed and translated *in vitro* using the TRAP/TrATPase cDNA clone as template.[28] By light microscopy, specific skeletal immunostaining was observed in osteoclasts situated in resorption lacunae and in some adjacent mononuclear cells.[40] Using the same polyclonal antiserum, and employing an immunogold technique on low temperature-embedded bone in an ultrastructural study,[41] intensive immunolabeling was seen on the bone surface facing the ruffled border zone, while lower amounts of antigen were detected on the bone surfaces facing the clear

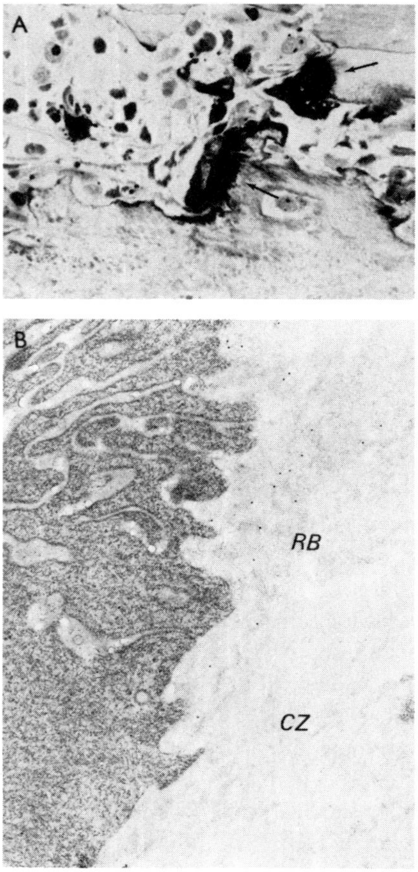

FIGURE 2. Histochemical localization of TRAP/TrATPase in the skeleton of neonatal rats. (A) Enzyme histochemical reaction for TrATPase. Note the accumulation of reaction product in the bone matrix juxtaposed to the ruffled border (arrows). (From Lindunger, A., MacKay, C. A., Ek-Rylander, B., Andersson, G., and Marks, S. C., Jr., *Bone Miner.*, 10, 109, 1990. With permission.) (B) Immunolocalization of TRAP/TrATPase in bone facing the osteoclast. Note the localization of immunogold grains in the bone matrix opposite the ruffled border (RB) compared to the matrix opposing the clear zone (CZ). (Courtesy of Dr. F. P. Reinholt.)

zones as well as deeper into the bone (Figure 2B). Within the osteoclasts, TRAP/TrATPase was detected mainly in vesicular structures interpreted as lysosomes. Immunolabeling was also observed in lysosomes of mononuclear cells in the vicinity of osteoclasts. The morphological appearance of these cells was similar to the recently described endocytic cells.[37] Immunolabeling was absent over osteoblasts and in bone facing the osteoblasts. In contrast, using a different polyclonal TRAP antiserum, as well as a preembedding technique with peroxidase-antiperoxidase and diaminobenzidine detection, Clark et al.[42] found immunoreactivity in intracellular vesicular structures interpreted as lysosomes, but were unable to demonstrate extracellular immunolabeling. A possible explanation for this discrepancy is that the different antisera recognize different proteins or different epitopes on the same protein.

In conclusion, structural features, including the lysosomal targeting sequence domain DKRFQ,[28] as well as direct subcellular localization by organelle fractionation[40] and ultrastructural immunocytochemistry[41] clearly indicate that TRAP is a lysosomal enzyme that appears to be targeted for secretion in actively resorbing osteoclasts.

The use of TRAP as a specific marker for osteoclasts and their putative mononuclear precursor cells has been questioned by the recent demonstration of TRAP activity in osteoblasts,[43,44] osteocytes,[44] and activated macrophages in the bone marrow.[45] At least in the case of osteoblasts, the cellular TRAP reaction is weak compared to that of osteoclasts.[43,44] It remains possible that the general bone surface staining observed with α-naphthylphosphate, but not with ATP, as substrate[35] is in fact generated by another distinct TRAP enzyme. The issue of TRAP in activated macrophages is interesting and may have bearing for studies on osteoclast differentiation, and will therefore be discussed in detail later in this chapter.

c. *Tissue Distribution and Cellular Localization of TRAP/TrATPase mRNA in the Skeleton*

In neonatal rats, TRAP/TrATPase mRNA is highly expressed in skeletal tissues, with much lower ($<10\%$) levels detected in spleen, thymus, liver, skin, brain, kidney, brain, lung, and heart, as assessed by solution hybridization (Table 4). Within bone, *in situ* hybridization showed specific labeling of osteoclasts at endosteal surfaces and bone trabeculae of long bones.[28] Thus, despite the apparent similarity of this osteoclastic TRAP/TrATPase with type 5, tartrate-resistant, and purple acid phosphatases expressed in other mammalian tissues, this gene appears to be preferentially expressed at skeletal sites.

3. Regulation and Function(s) of TRAP/TrATPase in the Skeleton

Osteoclastic bone resorption is a complex process regulated by systemic and local factors.[46] Despite the fact that TRAP has been used as a means of identifying the osteoclast, there is not much useful information on its regulation and function in the skeleton. One reason for the lack of studies on regulation is that systems for studying isolated osteoclasts were not available until recently. Previously published biochemical measurements of TRAP activity in bone extracts or culture media could not distinguish between changes in osteoclast number of alterations in the synthesis, secretion, or activity of the enzyme from individual osteoclasts. Furthermore, until recently, there has not been a sufficient source of authentic bone TRAP for thorough functional studies. In this section, we will review recent studies on TRAP regulation in isolated osteoclasts and our preliminary studies on potential physiological substrates for purified rat bone TRAP.

TABLE 4
Tissue Distribution of TRAP/TrATPase mRNA Expression in the 5-Day-Old Neonatal Rat

Tissue source	Tissue mRNA expression	
	cpm $10^{-3}/\mu g$ DNA	Percent of long bone
Long bone	1.86	100
Calvarium	1.67	90
Skeletal muscle	0.26	14
Liver	0.21	11
Spleen	0.18	10
Placenta	0.16	9
Kidney	0.15	8
Brain	0.15	8
Skin	0.12	7
Thymus	0.11	6
Heart	0.08	4
Lung	0.08	4

Note: Data from Reference 28.

Parathyroid hormone (PTH) is a potent stimulator of bone resorption both *in vivo* and *in vitro.* Following administration of PTH *in vivo,* there is a rapid increase in the number of osteoclasts[47] as well as an expansion of the ruffled border area of each individual osteoclast.[48] PTH also appears to induce the accumulation of TRAP in the area of the bone matrix adjacent to the ruffled border.[48] *In vitro,* PTH enhances TRAP activity in osteoclast-like cells isolated from fetal rat calvaria or neonatal rat long bones only if cocultivated with osteoblast-like cells.[49,50] PTH, in nanomolar concentrations, rapidly (within 1 h) caused a slight increase in the intracellular TRAP activity of individual osteoclasts, whereas after 18 h, intracellular TRAP activity was slightly decreased.[50] Taken together with the *in vivo* observations,[47,48] PTH seems to have weak specific effects on the synthesis and/or secretion of TRAP in osteoclasts.

In contrast to PTH, calcitonin (CT) appears to act directly on the osteoclast to inhibit bone resorption.[46,51] The mechanism of CT action on actively resorbing osteoclasts involves recession and disappearance of the ruffled border,[51] a physical separation from bone, and decreased numbers of nuclei per osteoclast followed by a reduction of osteoclast numbers.[46,47] In isolated neonatal rat osteoclasts cultivated on plastic, CT inhibits the release of TRAP to the medium in a dose-dependent manner.[52,53] The effect was observed after 2 h of CT treatment, and at 6 h, a 50% inhibition of enzyme release was noted.[52] Interestingly, in addition to inhibition of TRAP release, intracellular TRAP activity was also reduced by CT,[52,53] possibly indicating an effect of CT on the synthesis and/or degradation of TRAP.

There is substantial circumstantial evidence that TRAP is involved in the process of bone resorption: (1) high-level expression in osteoclasts, (2) secretion of the enzyme into the bone matrix adjacent to the osteoclast ruffled border, and (3) inhibition of TRAP release and possibly synthesis by CT. However, only one brief report is available in the literature indicating that TRAP may be necessary for osteoclast function in resorption.[54] In this report, osteoclasts were isolated from neonatal rats, cultivated on slices of devitalized bone, and evaluated for their ability to form resorption pits. Antibodies against the bone TRAP homolog uteroferrin produced a quantitative decrease in the number and area of resorption pits. Since TRAP/TrATPase is secreted onto bone by the osteoclast,[41] one explanation for this effect of anti-TRAP antibodies on pit formation is that the role of TRAP in osteoclastic resorption is to act on bone matrix phosphoproteins.

Since others have noted that there is considerable homology between type 5 acid phosphatases and certain phosphoprotein phosphatases,[55] and that purified TRAP/TrATPase can dephosphorylate certain nonbone-derived phosphoproteins, it seemed warranted to examine the action of bone-derived TRAP on bone matrix phosphoproteins. In these studies, we observed that purified rat bone TRAP efficiently dephosphorylated the bone matrix phosphoproteins osteonectin, osteopontin, and bone sialoprotein II.[56]

B. TYPE 5 ACID PHOSPHATASES AT EXTRASKELETAL SITES

1. Spleen

It has been recognized for some time that spleen expresses an iron-containing acid phosphatase.[57] Subsequently, a purple/tartrate-resistant acid phosphatase was purified from bovine[58] and rat[59,60] spleen, and extensively investigated mainly from a biophysical point of view.[13] In bovine and rat spleen, the enzyme constitutes 0.01 to 0.02% of the total protein,[58,59] i.e., a relative abundance of 20% compared to young rat long bone.[21] Compared to rat bone, the relative mRNA expression in rat spleen is around 10%[28] (Table 3), and the relative TRAP activity approximately 25%.[29] The bovine and rat spleen enzymes are similar to the bone enzyme in that they are cationic glycoproteins of Mr 33 to 40 kDa containing two iron atoms per molecule.[58-60] The bovine enzyme is invariably isolated as two "subunits" of Mr 24 and 15 kDa,[58] whereas the rat spleen enzyme can be isolated as the single-chain 33-kDa form.[59] Comparison of the near-complete amino acid sequence of beef spleen TRAP with the sequence of the rat bone enzyme reveals around 89% identity.[28,61] This suggests that the bone and spleen enzymes are identical or at least belong to the same gene family.

Biophysical studies of bovine spleen TRAP show that the oxidized, purple, catalytically inactive form containing two Fe(III) atoms can be converted to the pink, active form containing a Fe(II)-Fe(III) dinuclear unit by treatment with mild reducing agents such as β-mercaptoethanol or ascorbate.[58,62] Stabilization of the bovine TRAP in a high-salt environment results in a more ordered structure, which has permitted a detailed spectroscopical analysis of the interaction of inhibitory anions and substrate analogs with the protein.[63] From these studies, it has been proposed that the noncompetitive inhibitors molybdate and tungstate bind to the di-iron center preferentially at a site other than the active site, whereas the competitive inhibitors phosphate, arsenate, AMP, and phosphotyrosine mainly interact with the active site.[63]

The cellular distribution of purple/tartrate-resistant acid phosphatase in the spleen has been investigated by immunohistochemistry[64] and enzyme histochemistry.[29,64] These studies have localized TRAP protein and activity to lysosomes and siderosomes of ferritin- and hemosiderin-laden phagocytic cells of the red pulp.[64] The mononuclear TRAP-positive cells are present in the marginal zone of the red pulp.[29] Interestingly, when ATP was used as a histochemical TRAP substrate, no reaction could be demonstrated.[29] In addition, very low levels of TrATPase activity were measured biochemically in spleen homogenates.[29] These observations were interpreted to indicate that the conventional TRAP substrates, *p*-nitrophenylphosphate and α-naphthylphosphate, which are commonly used to measure osteoclastic TRAP also measure the TRAP of splenic macrophages. They further indicate that TrATPase may be a more selective marker for cells of the osteoclast lineage.[29]

2. Lung

Extracts of human lung tissue contain significant activities of a Mr 37 kDa tartrate-resistant, molybdate-inhibited acid phosphatase which can be purified by a procedure similar to that for the bone enzyme.[65] The abundance of TRAP mRNA in neonatal rat lung is low, e.g., 5%, compared to bone (Table 4). A tenfold enrichment of TRAP activity over whole lung was noted in isolated human alveolar macrophages.[66] Localization of a purple/tartrate-resistant acid phosphatase to intracellular granules of human and bovine alveolar macrophages has recently been demonstrated by a combined enzyme and immunohistochemical study.[67] Moreover, the immunocytochemical demonstration of TRAP reactivity on the villous surface of the respiratory epithelium[40] suggests that alveolar macrophages, like osteoclasts,[41] secrete the enzyme. Besides alveolar macrophages, TRAP-positive mononuclear cells are encountered in the granulomatous lesions of sarcoidosis and tuberculosis,[68] suggesting that interstitial lung macrophages, when activated by inflammatory stimuli, also express a TRAP.

3. Placenta

The TRAP of human placenta is a Mr 35-kDa protein which, in contrast to the rat bone and spleen enzyme, is not inhibited by molybdate and exhibits negligible activity with ATP as substrate.[69] The TRAP activity per gram of tissue in human placenta at term is approximately equal to that of human spleen,[70] and a similar abundance of TRAP mRNA transcripts in rat placenta and spleen has been noted[28] (Table 4). Human placental TRAP is recognized by a polyclonal antiserum against porcine uteroferrin, indicating the presence of common immunological determinants.[70]

Isolation and sequencing of cDNA clones[70,71] have recently demonstrated that the deduced amino acid sequence of human placenta TRAP is 85% identical to pig uteroferrin,[70] 94% identical to rat bone TRAP/TrATPase,[28] and 82% identical to bovine spleen TRAP.[70] Interestingly, comparison of the amino acid sequence of human placenta TRAP with a partial sequence of human osteoclastoma TRAP, derived from protein sequencing, reveals less than 100% identity.[70,72] If not due to sequencing errors, these data may indicate that a family of highly similar TRAP gene products exists. It may also be of significance that chromosomal assignment of the human TRAP gene has yielded divergent results, i.e., chromosome 15[73] or chromosome 19.[71]

4. Uteroferrin

Uteroferrin is a progesterone-induced, iron-containing, basic 35-kDa glycoprotein with a purple color present in uterine secretions of pigs.[74] Uteroferrin exhibits acid phosphatase activity that resists suppression by tartrate, is inhibited by molybdate, fluoride, arsenate, and phosphate, is activated by mild reducing agents, and readily hydrolyzes phosphate groups from phosphomonoester substrates such as *p*-nitrophenylphosphate, ATP, and certain phosphoproteins.[75,76] Polyclonal anti-uteroferrin antibodies recognize TRAP enzymes of human bone[24] and hairy cell leukemia,[77] a relatively rare B-cell leukemia, suggesting the presence of common structural determinants on these proteins. Isolation and sequencing of uteroferrin cDNA clones[78] have shown that the deduced amino acid sequence of uteroferrin is 82% identical to human placental TRAP[70,71] and 89% identical to rat bone TRAP/TrATPase.[28]

Uteroferrin is synthesized in nonciliated cells of the glandular endometrium, from which the protein is secreted.[79] Uteroferrin synthesized by explanted endometrium contains high-mannose-type N-linked oligosaccharides with mannose-6-phosphate residues covered by a diester-linked N-acetylglucosamine residue.[80] This mannose-P-P-N-acetylglucosamine unit is formed as an intermediate during the biosynthesis of the mannose-6-phosphate residue, and the N-acetylglucosamine-1-P has to be removed by a specific Golgi-associated phosphodiesterase to generate mannose-6-phosphate.[81] Proteins containing this type of oligosaccharide structure are normally sorted to lysosomes due to binding to mannose-6-phosphate-recognizing receptors in the Golgi complex.[81] Thus, the secretion of uteroferrin appears to be due to missorting to the secretory pathway by the inability of the mannose-6-phosphate receptors to bind uteroferrin.

The secreted uteroferrin is extensively taken up by chorionic epithelial cells,[79] and from there passed on to the fetal circulation. In the fetus, uteroferrin is avidly taken up by sinusoidal cells by a process dependent on the high-mannose-type oligosaccharides of uteroferrin,[82] presumably by binding to the mannose/fucose/N-acetylglucosamine-specific lectin present on Kupffer cells and other macrophages.[83] Dissociation of iron from uteroferrin and transfer to transferrin can be reconstituted *in vitro,*[85] and the ability to release one of the two iron atoms in uteroferrin in the presence of reducing agents[84,85] has been proposed to be of physiological significance for the function of uteroferrin in iron transport from the mother to the site of fetal hemopoiesis.[74] However, the significance of the uteroferrin-associated acid phosphatase activity for the iron-transporting function of uteroferrin is entirely unknown.

5. Epidermis

A tartrate-resistant, iron-activated acid phosphatase has been purified from newborn rat epidermis.[86] The enzyme is a basic Mr 32-kDa protein with a substrate specificity similar to that of the rat spleen TRAP, including activity toward certain phosphoproteins. Moreover, polyclonal antibodies against the splenic form cross-react with the epidermal TRAP. The enzyme has been immunolocalized to keratohyalin granules in the granular cell layer and to cornified cells in the lower part of the stratum corneum, possibly indicating a role in the process of cellular degeneration prior to shedding.[87]

C. TYPE 5 ACID PHOSPHATASES IN PATHOLOGICAL TISSUES

1. Gaucher Spleen

Gaucher's disease is an autosomal recessive inborn error of metabolism leading to the accumulation of glucocerebroside in lysosomes of spleen, liver, and bone marrow mononuclear phagocytes due to a severe deficiency of glucocerebrosidase.[88] The affected macrophages, Gaucher cells, express the purple/tartrate-resistant acid phosphatase[89,90] which has been purified from the spleen[91,92] and serum[93] of Gaucher patients. The circulating form of TRAP in Gaucher patients is a desialylated variant, classified as type 5b, in contrast to the normal sialylated form, type 5a.[93] The Gaucher spleen TRAP has been purified to apparent homogeneity and shown to be a basic glycoprotein of Mr 33 to 34 kDa consisting of 20- and 16-kDa "subunits".[92] The enzyme has phosphoseryl protein phosphatase activity and appears to prefer acidic phosphoproteins as substrates.[94] The hydrodynamic properties of the Gaucher spleen TRAP are remarkably similar to the rat bone TRAP/TrATPase,[25,92] suggesting a structural relationship between these proteins. No amino acid sequence information is yet available for the Gaucher spleen TRAP.

2. Hairy Cell Leukemia

Hairy cell leukemia (HCL) is a chronic lymphocytic leukemia of B-cell origin that comprises around 2% of all leukemias.[95] HCL is morphologically characterized by circulating leukemic cells with cytoplasmic projections giving a "hairy" appearance. These cells often express TRAP activity, which is one important diagnostic hallmark.[96] The enlarged spleens of HCL patients are a particularly rich source of human TRAP, which has been purified utilizing CM-cellulose chromatography and immunoaffinity chromatography using anti-uteroferrin antibodies.[77] The human HCL TRAP is a cationic Mr 34-

kDa glycoprotein containing two atoms of iron per molecule, with substrate specificity and response to activators/inhibitors similar to that of TRAP enzymes from bone, spleen, and uteroferrin.[77] Moreover, the N-terminal amino acid sequence of 16 amino acids is identical to the human placental TRAP.[97] However, two distinct peaks of TRAP activity differing in amino acid composition provide circumstantial evidence for more than one TRAP form in HCL spleen.[97] These two forms may be related to the two pools of TRAP activity isolated by Lam et al.[98] which differ in their sensitivity to stimulation by iron.

3. Giant Cell Tumor of Bone (Osteoclastoma)

Giant cell tumor of bone contains a large number of multinucleated cells with the capacity to resorb bone *in vitro*.[99] This tumor tissue has provided a source for the purification of the human osteoclastic TRAP.[72,100] Osteoclastoma TRAP is a basic glycoprotein with an Mr of around 30 kDa, and is composed of 17,5- and 15-kDa ''subunits''.[72] Curiously, osteoclastoma TRAP is not activated by mild reducing agents.[72] Other differences are negligible phosphoprotein phosphatase activity, a higher iron content, and inhibition of enzyme activity by iron and ascorbate,[72] although Lam et al.[98] found a strong enhancement of purified osteoclastoma TRAP activity by iron. Interestingly, the partial amino acid sequence derived from the two subunits of osteoclastoma TRAP is only 94% identical to the corresponding region of human placenta TRAP,[72,100] suggesting that these proteins are structurally similar but possibly nonidentical. Conclusive evidence to this end will require elucidation of the entire amino acid sequence of osteoclastoma TRAP derived from cDNA clones.

III. TARTRATE-RESISTANT ACID PHOSPHATASE AS A MARKER FOR DIFFERENTIATION AND DISEASE

A. TRAP AND MACROPHAGE DIFFERENTIATION/ACTIVATION

As previously indicated, purple/tartrate-resistant acid phosphatase does not appear to be specifically expressed in osteoclasts, but can be detected in low abundance in most extraskeletal tissues (Table 4). It is likely that most tissue macrophages harbor the enzyme. Notable examples of TRAP-positive macrophages are alveolar macrophages[66,67] and splenic phagocytes,[29,64] whereas, for instance, peritoneal macrophages do not constitutively express TRAP.[101-103] Freshly isolated blood monocytes, which are precursors of tissue macrophages, also do not express TRAP.[71,101,104,105] However, following induction of macrophage differentiation by primary culture of monocytes in the presence of autologous serum, these monocyte-derived macrophages do exhibit strong TRAP activity,[71,101,104,105] which is due to a type 5 isoenzyme.[71,105] TRAP expression in the monocyte-derived macrophages is further enhanced by lymphocyte-conditioned media,[101] the phorbol ester TPA,[101] and

1,25-dihydroxyvitamin D_3 (1,25$(OH)_2D_3$).[105] The importance of local mediators as inducers of TRAP expression has been illustrated by the increased number of TRAP-positive bone marrow macrophages in the vicinity of metastatic tumor cells,[45] in the granulomatous lesions of sarcoidosis and tuberculosis,[68] and in activated brain macrophages in AIDS encephalopathy.[106] The reason for the high "constitutive" expression of TRAP in, e.g., alveolar macrophages is unclear, but the capacity of alveolar macrophages to synthesize 1,25$(OH)_2D_3$[107] may be of physiological significance for the induction of TRAP.

Although osteoclasts are no longer believed to originate from the fusion of terminally differentiated macrophages (for a review, see Reference 46), it is interesting to note that a variety of mononuclear phagocytes, including spleen cells, alveolar macrophages, bone marrow cells, thymus cells, and blood monocytes, can be efficiently induced to form TRAP-positive multinucleated cells when cocultivated with bone marrow-derived stromal cells in the presence of 1,25$(OH)_2D_3$ and the synthetic glucocorticoid dexamethasone.[104,108] These multinucleated cells also express other osteoclast characteristics, such as calcitonin receptors and the capacity to excavate bone.[104]

Setting aside the controversy of whether these multinucleated cells are osteoclasts, or whether such mechanisms of osteoclast-like cell formation are operative *in vivo,* this phenomenon of TRAP induction concomitant with multinucleated cell formation may prove to be of importance to elucidate the regulation of TRAP expression.

B. TRAP AND OSTEOCLAST ONTOGENY/DIFFERENTIATION

A compelling body of circumstantial experimental evidence has been accumulated which supports the conclusion that the osteoclast arises from the pluripotent hemopoietic stem cell (see Reference 46 for a review). A reasonable hypothesis of osteoclast ontogeny, suggested by the work of Burger and Nijweide,[109] is presented in Figure 3. In this model, there is an early (multipotent) and a late (committed) progenitor which proliferate and differentiate to give rise to an early (TRAP−) and late (TRAP+) postmitotic precursor (preosteoclast, OC^{PRE}), leading to the formation of a mononuclear osteoclast (OC^{MO}). Multinuclear osteoclasts form when the mononuclear osteoclast fuses either with other mononuclear osteoclasts or with TRAP+ precursors.

Although TRAP is a widely used marker for studies of osteoclast ontogeny, there are problems with its use, especially in the absence of any other phenotypic markers of the osteoclast. Some reports indicate that the criteria of TRAP reactivity and multinuclearity overestimate the formation of osteoclasts.[102,103] This is particularly important when evaluating the results of experimental models where various mononuclear phagocytes are used as a source of osteoclast precursors,[104,108] since splenic macrophages,[29,64] alveolar macrophages,[66,67] and monocytes[71,101,104,105] have been shown to express TRAP and form polykaryons without any apparent relationship to osteoclast for-

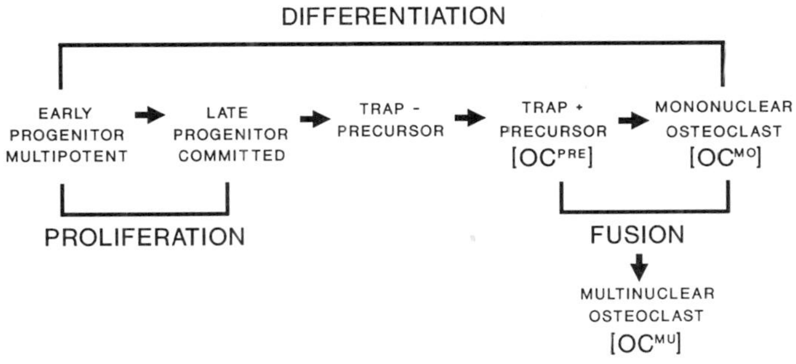

FIGURE 3. A cell lineage model of osteoclast ontogeny. The osteoclast cell lineage is thought to arise from an early multipotent progenitor which forms a late premitotic progenitor which is committed to the osteoclast lineage. This committed progenitor gives rise to an early (TRAP−) and late (TRAP+) postmitotic precursor (preosteoclast, OC^{PRE}) which differentiates to become a mononuclear osteoclast (OC^{MO}). This mononuclear osteoclast either fuses with another mononuclear osteoclast or a preosteoclast to form a multinuclear osteoclast (OC^{MU}). (Adapted from Burger, E. H. and Nijweide, P. J., in *Bone,* Vol. 2, *The Osteoclast,* Hall, B.-K., Ed., CRC Press, Boca Raton, FL, 1991, 31.)

mation. On the other hand, TRAP appears to be a rather selective marker for mononuclear preosteoclasts and osteoclasts in experiments where explanted fetal mouse metatarsal bone[30,110] or calvaria are used.[111] It has been demonstrated that TRAP is expressed rather late in osteoclast ontogeny, i.e., in postmitotic precursors (see Reference 109 for a review) and that TRAP expression in precursors and osteoclasts occurs prior to calcitonin receptor expression.[112,113] Nonetheless, it is clear that there are still difficulties in interpreting the results of experiments on the regulation of osteoclast ontogeny using TRAP as a marker. Major among them is distinguishing when TRAP expression marks a differentiation event (i.e., the differentiation of precursor or osteoclast) from when it indicates increased synthesis or release of TRAP due to regulatory control of an already differentiated precursor or osteoclast.

Recently, we have employed ATP as an alternative histochemical substrate for TRAP, and observed a much more restricted expression of TrATPase in bone marrow mononuclear cells compared to TRAP.[39] In addition, TRAP-positive spleen macrophages do not exhibit TrATPase activity,[29] and the multinucleated TRAP-positive cells recruited by subcutaneous implants of mineralized bone particles[114] show weak TRAP but no TrATPase reactivity.[115] TrATPase is expressed at a late stage in osteoclast differentiation, as demonstrated in the osteopetrotic *op/op* mutation where an elevated number of strongly TrATPase-positive mononuclear cells are clustered in the vicinity of weakly TrATPase-positive vacuolated osteoclasts[27] (Figure 4). Thus, since macrophage TRAP isoenzymes appear to be unable to hydrolyze ATP as a substrate, TrATPase may prove to be a more selective marker for osteoclast ontogeny *in vivo* and *in vitro* than the classical TRAP histochemical method.

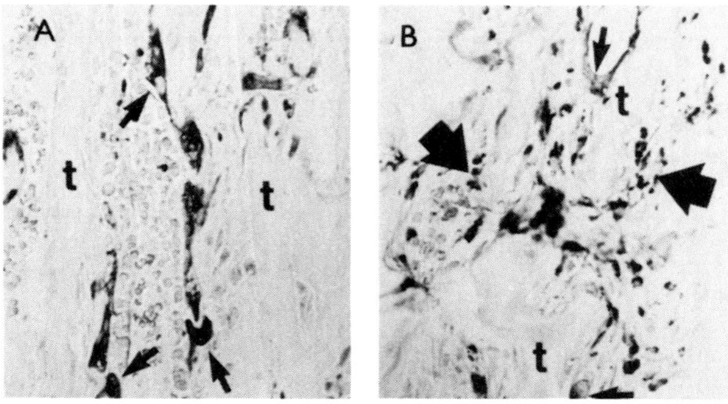

FIGURE 4. Cellular localization of TrATPase activity in tibiae of osteopetrotic mutant (*op*) rats and their normal littermates. Osteoclasts stain intensely for the enzyme (arrows) on trabeculae (t) of normal rats (A). In contrast, osteoclasts (small arrows) are sparsely stained in *op* mutant rats (B), while foci of intensely stained mononuclear cells (large arrows) are typically found. (From Ek-Rylander, B., Marks, S. C., Jr. Hammarström, L. E., and Andersson, G. N., *Bone Miner.*, 5, 309, 1989. With permission.)

Studies are in progress to elucidate the molecular basis for this substrate selectivity by examining whether distinct, but structurally related TRAP isoenzymes are expressed in macrophages and osteoclasts or, alternatively, whether cell-specific modification of a single translation product occurs.

C. TRAP AS A MARKER FOR HEMATOLOGICAL DISEASES

In humans, around 2% of circulating lymphocytes, representing approximately 25% of circulating B-lymphocytes, express histochemical TRAP activity.[116] These TRAP-positive B-cells exhibit cytoplasmic processes similar to those found on the strongly TRAP-positive tumor cells of HCL, a relatively benign form of chronic B-lymphocytic leukemia.[95,96,116] Interestingly, several cases of chronic B-cell leukemias present with an increased bone resorption in combination with an increased number of TRAP-positive mononuclear cells and osteoclasts at endosteal bone surfaces.[117] In B-cell leukemias of various stages of differentiation, TRAP expression increases concomitant with differentiation, and is lost when the B-cells differentiate to plasma cells.[118] When less mature B-cell leukemia cells are induced to differentiate by the phorbol ester TPA, a high incidence of TRAP expression can be observed,[119-123] suggesting that TRAP is expressed only during later stages of B-lymphocyte differentiation. TRAP expression is also observed in subsets of Hodgkin-like B-cell lymphomas[123,124] and in certain cases of monocytic variants of acute myeloid leukemias.[125]

D. TRAP AS A CLINICAL MARKER FOR BONE RESORPTION

Measurements of serum TRAP enzyme activity[126-131] and TRAP immunoreactivity[131-134] have lately been applied as a potential clinical marker for bone resorption in various metabolic bone diseases[127,128,132,134] as well as in patients with metastatic bone tumors.[126,129,130,133] Although significant elevations of TRAP have been consistently observed using either method, several considerations still hamper the use of circulating TRAP as a reliable marker for bone resorption. Enzyme activity measurements lack sufficient specificity due to the absence of specific substrates for the type 5 TRAP of osteoclasts. Further, there is a lack of quantitative inhibition of the prostatic and lysosomal acid phosphatases by tartrate. In addition, the widespread distribution of TRAP in tissue macrophages and certain B-lymphocytes in addition to osteoclasts is the most likely reason for the relatively small difference in serum TRAP between healthy subjects and patients with various resorptive bone diseases. In the immunochemical assays employed, polyclonal antibodies generated against uteroferrin or the hairy cell leukemia TRAP have been used for the detection of the osteoclastic TRAP. If serum TRAP measurements are to be used to accurately monitor changes in the extent of bone resorption, it will be necessary to develop monospecific antibodies reactive against the TRAP expressed only in osteoclasts. Before antibodies specific for the osteoclastic TRAP can be developed, it will be necessary to elucidate and compare the structures and sequences of type 5 TRAP isoenzymes expressed in osteoclasts, B-lymphocytes, and tissue macrophages.

ACKNOWLEDGMENTS

Supported by grants to G. Andersson and B. Ek-Rylander from the Swedish Medical Research Council, Swedish Cancer Society, Karolinska Institutet, and Magnus Bergvalls stiftelse; also supported by National Institutes of Health grants DE04961 and AR40176 awarded to C. Minkin.

REFERENCES

1. **Moss, D. W.,** Multiple forms of acid and alkaline phosphatases: genetics, expression and tissue-specific modification, *Clin. Chim. Acta,* 161, 123, 1986.
2. **Himeno, M., Fujita, H., Noguchi, Y., Kono, A., and Kato, K.,** Isolation and sequencing of a cDNA clone encoding acid phosphatase in rat liver lysosomes, *Biochem. Biophys. Res. Commun.,* 162, 1044, 1989.
3. **Pohlmann, R., Krentler, C., Schmidt, B., Schröder, W., Lorkowski, G., Culley, J., Mersmann, G., Geier, C., Waheed, A., Gottschalk, S., Grzeschik, K.-H., Hasilik, A., and von Figura, K.,** Human lysosomal acid phosphatase: cloning, expression and chromosomal assignment, *EMBO J.,* 7, 2343, 1988.
4. **Lin, M.-F., Li, S. S.-L., Ming Chu, T., and Lee, C.-L.,** Comparison of prostate acid phosphatase with acid phosphatase isoenzymes from the lung and spleen, *J. Clin. Lab. Anal.,* 4, 420, 1990.
5. **Roiko, K., Jänne, O. A., and Vihko, P.,** Primary structure of rat secretory acid phosphatase and comparison to other acid phosphatases, *Gene,* 89, 223, 1990.

6. **Van Etten, R. L., Davidson, R., Stevis, P. E., MacArthur, H., and Moore, D. L.,** Covalent structure, disulfide bonding, and identification of reactive surfaces and active site residues of human prostatic acid phosphatase, *J. Biol. Chem.*, 266, 2313, 1991.
7. **Waheed, A., Laidler, P. M., Wo, Y.-Y. P., and Van Etten, R. L.,** Purification and physicochemical characterization of a human placental acid phosphatase possessing phosphotyrosyl protein phosphatase activity, *Biochemistry*, 27, 4265, 1988.
8. **Chernoff, J. and Li, H.-C.,** A major phosphotyrosyl-protein phosphatase from bovine heart is assoociated with a low-molecular-weight acid phosphatase, *Arch. Biochem. Biophys.*, 240, 135, 1985.
9. **Ramponi, G., Manao, G., Camici, G., Cappugi, G., Ruggiero, M., and Bottaro, D. P.,** The 18kDa cytosolic acid phosphatase from bovine liver has phosphotyrosine phosphatase activity on the autophosphorylated epidermal growth factor receptor, *FEBS Lett.*, 250, 469, 1989.
10. **Camici, G., Manao, G., Cappugi, G., Modesti, A., Stefani, M., and Ramponi, G.,** The complete amino acid sequence of the low molecular weight cytosolic acid phosphatase, *J. Biol. Chem.*, 264, 2560, 1989.
11. **Liguri, G., Camici, G., Manao, G., Cappugi, G., Nassi, P., Modesti, A., and Ramponi, G.,** A new acylphosphatase isoenzyme from human erythrocytes: purification, characterization, and primary structure, *Biochemistry*, 25, 8089, 1986.
12. **Nassi, P., Nediani, C., Liguri, G., Taddei, N., and Ramponi, G.,** Effects of acylphosphatase on the activity of erythrocyte membrane Ca^{2+} pump, *J. Biol. Chem.*, 266, 10867, 1991.
13. **Vincent, J. B. and Averill, B. A.,** An enzyme with a double identity: purple acid phosphatase and tartrate-resistant acid phosphatase, *FASEB J.*, 4, 3009, 1990.
14. **Hammarström, L. E., Hanker, J. S., and Toverud, S. U.,** Cellular differences in acid phosphatase isoenzymes in bone and teeth, *Clin. Orthop.*, 78, 151, 1971.
15. **Anderson, T. R. and Toverud, S. U.,** Chromatographic separation of two acid phosphatases from rat bone, *Calcif. Tissue Res.*, 24, 187, 1977.
16. **Anderson, T. R. and Toverud, S. U.,** Purification and partial characterization of two acid phosphatases from rat bone, *Calcif. Tissue Int.*, 27, 219, 1979.
17. **Anderson, T. R. and Toverud, S. U.,** Further studies on the separation and identification of two phosphatases with acid optima from rat bone, *Calcif. Tissue Int.*, 33, 261, 1981.
18. **Kato, T., Hara, A., Nakayama, T., Sawada, H., Hamatake, M., and Matsumoto, Y.,** Purification and characterization of purple acid phosphatase from rat bone, *Comp. Biochem. Physiol. B*, 83, 813, 1986.
19. **Anderson, T. R. and Toverud, S.,** Purification and characterization of purple acid phosphatase from developing rat bone, *Arch. Biochem. Biophys.*, 247, 131, 1986.
20. **Andersson, G., Lindunger, A., and Ek-Rylander, B.,** Isolation and characterization of skeletal acid ATPase — a new osteoclast marker?, *Connect. Tissue Res.*, 20, 151, 1989.
21. **Andersson, G., Ek-Rylander, B., and Hammarström, L.,** Purification and characterization of a vanadate-sensitive nucleotide tri- and diphosphatase with acid pH optimum from rat bone, *Arch. Biochem. Biophys.*, 228, 431, 1984.
22. **Lau, K.-H. W., Farley, J. R., and Baylink, D. J.,** Phosphotyrosyl-specific protein phosphatase activity of a bovine skeletal acid phosphatase isoenzyme, *J. Biol. Chem.*, 260, 4653, 1985.
23. **Lau, K.-H. W., Freeman, T. K., and Baylink, D. J.,** Purification and characterization of an acid phosphatase that displays phosphotyrosyl-protein phosphatase activity from bovine cortical bone matrix, *J. Biol. Chem.*, 262, 1389, 1987.
24. **Allen, S. H., Nuttleman, P. R., Ketcham, C. M., and Roberts, R. M.,** Purification and characterization of human bone tartrate-resistant acid phosphatase, *J. Bone Miner. Res.*, 4, 47, 1989.

25. **Ek-Rylander, B., Bergman, T., and Andersson, G.,** Characterization of a tartrate-resistant acid phosphatase (ATPase) from rat bone: hydrodynamic properties and N-terminal amino acid sequence, *J. Bone Miner. Res.*, 6, 365, 1991.
26. **Anderson, T. R. and Toverud, S. U.,** Effect of Fe^{2+} and ascorbic acid on acid phosphatases from rat bone, *Calcif. Tissue Int.*, 34, 54, 1982.
27. **Ek-Rylander, B., Marks, S. C., Jr., Hammarström, L. E., and Andersson, G. N.,** Osteoclastic acid ATPase: biochemical and histochemical studies of the osteopetrotic mutations in the rat, *Bone Miner.*, 5, 309, 1989.
28. **Ek-Rylander, B., Bill, P., Norgärd, M., Nilsson, S., and Andersson, G.,** Cloning, sequence and developmental expression of a type 5, tartrate-resistant, acid phosphatase of rat bone, *J. Biol. Chem.*, 266, 24684, 1991.
29. **Lindunger, A., MacKay, C. A., Ek-Rylander, B., Andersson, G., and Marks, S. C., Jr.,** Histochemistry and biochemistry of tartrate-resistant acid phosphatase (TRAP) and tartrate-resistant acid adenosine triphosphatase (TrATPase) in bone, bone marrow and spleen; implications for osteoclast ontogeny, *Bone Miner.*, 10, 109, 1990.
30. **Van der Wijngaert, F. P. and Burger, E. H.,** Demonstration of tartrate-resistant acid phosphatase in un-decalcified, glycolmethacrylate-embedded mouse bone: a possible marker for (pre)osteoclast identification, *J. Histochem. Cytochem.*, 34, 1317, 1986.
31. **Chappard, D., Azema, J., Alexandre, C., and Becker, J.-M.,** Cytomorphometry of osteoclasts, *Med. Lab. Sci.*, 46, 363, 1989.
32. **Cole, A. A. and Walters, L. M.,** Tartrate-resistant acid phosphatase in bone and cartilage following decalcification and cold-embedding in plastic, *J. Histochem. Cytochem.*, 35, 203, 1987.
33. **Akisaka, T., Subita, G. P., Kawaguchi, H., and Shigenaga, Y.,** Different tartrate sensitivity and pH optimum for two isoenzymes of acid phosphatase in osteoclasts. An electron-microscopic enzyme-cytochemical study, *Cell Tissue Res.*, 255, 69, 1989.
34. **Liu, C.-C., Sherrard, D. J., Maloney, N. A., and Howard, G. A.,** Reactivation of inhibited acid phosphatase and its significance in bone histomorphometry, *J. Histochem. Cytochem.*, 35, 1355, 1987.
35. **Minkin, C.,** Bone acid phosphatase: tartrate-resistant acid phosphatase as a marker of osteoclast function, *Calcif. Tissue Int.*, 34, 285, 1982.
36. **Webber, D., Braidman, I. P., Robertson, W. R., and Anderson, D. C.,** The effect of tartrate on bone cell acid phosphatase activity: a quantitative cytochemical study, *J. Bone Miner. Res.*, 4, 809, 1989.
37. **Janckila, A. J., Li, C.-Y., Lam, K.-W., and Yam, L. T.,** The cytochemistry of tartrate-resistant acid phosphatase, *Am. J. Clin. Pathol.*, 70, 45, 1978.
38. **Webber, D. M., Braidman, I. P., Robertson, W. R., and Anderson, D. C.,** A quantitative cytochemical assay for osteoclastic acid phosphatase activity in fetal rat calvaria, *Histochem. J.*, 20, 269, 1988.
39. **Andersson, G. N. and Marks, S. C., Jr.,** Tartrate-resistant acid ATPase as a cytochemical marker for osteoclasts, *J. Histochem. Cytochem.*, 37, 115, 1989.
40. **Andersson, G. N., Ek-Rylander, B., Hammarström, L. E., Lindskog, S., and Toverud, S. U.,** Immunocytochemical localization of a tartrate-resistant and vanadate-sensitive acid nucleotide tri- and diphosphatase, *J. Histochem. Cytochem.*, 34, 293, 1986.
41. **Reinholt, F. P., Mengarelli Widholm, S., Ek-Rylander, B., and Andersson, G.,** Ultrastructural localization of a tartrate-resistant acid ATPase in bone, *J. Bone Miner. Res.*, 5, 1055, 1990.
42. **Clark, S. A., Ambrose, W. W., Anderson, T. R., Terrell, R. S., and Toverud, S. U.,** Ultrastructural localization of tartrate-resistant, purple acid phosphatase in rat osteoclasts by histochemistry and immunocytochemistry, *J. Bone Miner. Res.*, 4, 399, 1989.
43. **Lundy, M. W., Lau, K.-H. W., Blair, H. C., and Baylink, D. J.,** Chick osteoblasts contain fluoride-sensitive acid phosphatase activity, *J. Histochem. Cytochem.*, 36, 1175, 1988.

44. **Bianco, P., Ballanti, P., and Bonucci, E.,** Tartrate-resistant acid phosphatase activity in rat osteoblasts and osteocytes, *Calcif. Tissue Int.*, 43, 167, 1988.
45. **Bianco, P., Costantini, M., Dearden, L. C., and Bonucci, E.,** Expression of tartrate-resistant acid phosphatase in bone marrow macrophages, *Basic Appl. Histochem.*, 31, 433, 1987.
46. **Marks, S. C., Jr., and Popoff, S. N.,** Bone cell biology: the regulation of development, structure, and function in the skeleton, *Am. J. Anat.*, 183, 1, 1988.
47. **Baron, R. and Vignery, A.,** Behavior of osteoclasts during a rapid change in their number induced by high doses of parathyroid hormone or calcitonin in intact rats, *Metab. Bone Dis. Relat. Res.*, 2, 339, 1981.
48. **Miller, S. C.,** The rapid appearance of acid phosphatase activity at the developing ruffled border of parathyroid hormone activated medullary bone osteoclasts, *Calcif. Tissue Int.*, 37, 526, 1985.
49. **Braidman, I. P., St John, J. G., Anderson, D. C., and Robertson, W. R.,** Effects of physiological concentrations of parathyroid hormone on acid phosphatase activity in cultured rat bone cells, *J. Endocrinol.*, 111, 17, 1986.
50. **Rowe, D. J., Kent, G. N., and Nicholson, G. C.,** Effects of parathyroid hormone on intracellular tartrate-resistant acid phosphatase activity in isolated osteoclasts, *Bone Miner.*, 10, S312, 1990.
51. **Kallio, D. M., Garant, P. R., and Minkin, C.,** Ultrastructural effects of calcitonin on osteoclasts in tissue culture, *J. Ultrastruct. Res.*, 39, 205, 1972.
52. **Chambers, T. J., Fuller, K., and Darby, J. A.,** Hormonal regulation of acid phosphatase release by osteoclasts disaggregated from neonatal rat bone, *J. Cell. Physiol.*, 132, 90, 1987.
53. **Yumita, S., Nicholson, G. C., Rowe, D. J., Kent, G. N., and Martin, T. J.,** Biphasic effect of calcitonin on tartrate-resistant acid phosphatase activity in isolated rat osteoclasts, *J. Bone Miner. Res.*, 6, 591, 1991.
54. **Zaidi, M., Moonga, B., Moss, D. W., and MacIntyre, I.,** Inhibition of osteoclastic acid phosphatase abolishes bone resorption, *Biochem. Biophys. Res. Commun.*, 159, 68, 1989.
55. **Vincent, J. B. and Averill, B. A.,** Sequence homology between purple acid phosphatases and phosphoprotein phosphatases, *FEBS Lett.*, 263, 265, 1990.
56. **Ek-Rylander, B., Wendel, M., Heinegärd, D., and Andersson, G.,** Osteoclastic tartrate-resistant acid phosphatase/ATPase can act as a bone matrix phosphoprotein phosphatase, *J. Bone Miner. Res.*, 6 (Suppl. 1), S276, 1991.
57. **Antanaitis, B. C. and Aisen, P.,** Uteroferrin and the purple acid phosphatases, *Adv. Inorg. Biochem.*, 5, 111, 1983.
58. **Davis, J. C., Shieufun, S., and Averill, B. A.,** Kinetics and optical spectroscopic studies on the purple acid phosphatase from beef spleen, *Biochemistry*, 20, 4062, 1981.
59. **Hara, A., Sawada, H., Kato, T., Nakayama, T., Yamamoto, H., and Matsumoto, Y.,** Purification and characterization of a purple acid phosphatase from rat spleen, *J. Biochem.*, 95, 67, 1984.
60. **Hara, A., Sawada, H., Nakayama, T., and Matsumoto, Y.,** Affinity separation of an acid phosphatase from rat tissues and Gaucher spleen with immobilized Cibacron blue, *J. Biochem.*, 94, 721, 1983.
61. **Hunt, D. F., Yates, J. R., III, Shabanowitz, J., Zhu, N.-Z., Zirino, T., Averill, B. A., Daurat-Larroque, S. T., Shewale, J. G., Roberts, R. M., and Brew, K.,** Sequence homology in the metalloproteins: purple acid phosphatase from beef spleen and uteroferrin from porcine uterus, *Biochem. Biophys. Res. Commun.*, 144, 1154, 1987.
62. **Dietrich, M., Munstermann, D., Suerbaum, H., and Witzel, H.,** Purple acid phosphatase from bovine spleen. Interactions at the active site in relation to the reaction mechanism, *Eur. J. Biochem.*, 199, 105, 1991.

63. **Vincent, J. B., Crowder, M. W., and Averill, B. A.,** Spectroscopic and kinetics studies of a high-salt-stabilized form of the purple acid phosphatase from bovine spleen, *Biochemistry,* 30, 3025, 1991.
64. **Schindelmeiser, J., Munstermann, D., and Witzel, H.,** Histochemical investigations on the localization of the purple acid phosphatase in the bovine spleen, *Histochemistry,* 87, 13, 1987.
65. **Efstratiadis, T. and Moss, D. W.,** Tartrate-resistant acid phosphatase of human lung: apparent identity with osteoclastic acid phosphatase, *Enzyme,* 33, 34, 1985.
66. **Efstratiadis, T. and Moss, D. W.,** Tartrate-resistant acid phosphatase in human alveolar macrophages, *Enzyme,* 34, 140, 1985.
67. **Schindelmeiser, J., Schewe, P., Zonka, T., and Munstermann, D.,** Histochemical and immunological demonstration of purple acid phosphatase in human and bovine alveolar macrophages, *Histochemistry,* 92, 81, 1989.
68. **Barth, J., Kreipe, H., Kiemle-Kallee, J., Radzun, H. J., Parwaresch, M. R., and Petermann, W.,** Diminished activity of tartrate resistant acid phosphatase in alveolar macrophages from patients with active sarcoidosis, *Thorax,* 43, 901, 1988.
69. **DiPietro, D. L. and Zengerle, F. S.,** Separation and properties of three acid phosphatases from human placenta, *J. Biol. Chem.,* 242, 3391, 1967.
70. **Ketcham, C. M., Roberts, R. M., Simmen, R. C. M., and Nick, H. S.,** Molecular cloning of the type 5, iron containing, tartrate-resistant acid phosphatase from human placenta, *J. Biol. Chem.,* 264, 287, 1989.
71. **Lord, D. K., Cross, N. C. P., Bevilacqua, M. A., Rider, S. H., Gorman, P. A., Groves, A. V., Moss, D. W., Sheer, D., and Cox, T. M.,** Type 5 acid phosphatase. Sequence, expression and chromosomal localization of a differentiation-associated protein of the human macrophage, *Eur. J. Biochem.,* 189, 287, 1990.
72. **Hayman, A. R., Warburton, M. J., Pringle, J. A. S., Coles, B., and Chambers, T. J.,** Purification and characterization of a tartrate-resistant acid phosphatase from human osteoclastomas, *Biochem. J.,* 261, 601, 1989.
73. **Allen, B. S., Ketcham, C. M., Roberts, R. M., Nick, H. S., and Ostrer, H.,** Localization of the human type 5, tartrate-resistant acid phosphatase gene by *in situ* hybridization, *Genomics,* 4, 597, 1989.
74. **Roberts, R. M. and Bazer, F. W.,** Uteroferrin: a protein in search of a function, *BioEssays,* 1, 8, 1985.
75. **Schlosnagle, D. C., Bazer, F. W., Tsibris, J. C. M., and Roberts, R. M.,** An iron-containing phosphatase induced by progesterone in the uterine fluids of pigs, *J. Biol. Chem.,* 249, 7574, 1974.
76. **Roberts, R. M. and Bazer, F. W.,** Phosphoprotein phosphatase activity of the progesterone-induced purple glycoprotein of the porcine uterus, *Biochem. Biophys. Res. Commun.,* 68, 450, 1976.
77. **Ketcham, C. M., Baumbach, G. A., Bazer, F. W., and Roberts, R. M.,** The type 5, acid phosphatase from spleen of humans with hairy cell leukemia. Purification, properties, immunological characterization, and comparison with porcine uteroferrin, *J. Biol. Chem.,* 260, 5768, 1985.
78. **Simmen, R. C. M., Baumbach, G. A., and Roberts, R. M.,** Molecular cloning and temporal expression during pregnancy of the messenger ribonucleic acid encoding uteroferrin, a progesterone-induced uterine secretory protein, *Mol. Endocrinol.,* 2, 253, 1988.
79. **Raub, T. J., Bazer, F. W., and Roberts, R. M.,** Localization of the iron transport glycoprotein, uteroferrin, in the porcine endometrium and placenta by using immunocolloidal gold, *Anat. Embryol.,* 171, 253, 1985.
80. **Baumbach, G. A., Saunders, P. T. K., Bazer, F. W., and Roberts, R. M.,** Uteroferrin has N-asparagine-linked high-mannose-type oligosaccharides that contain mannose 6-phosphate, *Proc. Natl. Acad. Sci. U.S.A.,* 81, 2985, 1984.
81. **Dahms, N. M., Lobel, P., and Kornfeld, S.,** Mannose 6-phosphate receptors and lysosomal enzyme targeting, *J. Biol. Chem.,* 264, 12115, 1989.

82. **Saunders, P. T. K., Renegar, R. H., Raub, T. J., Baumbach, G. A., Atkinson, P. H., Bazer, F. W., and Roberts, R. M.,** The carbohydrate structure of porcine uteroferrin and the role of the high mannose chains in promoting uptake by the reticuloendothelial cells of the fetal liver, *J. Biol. Chem.,* 260, 3658, 1985.
83. **Lennartz, M. R., Sessions Cole, F., and Stahl, P. D.,** Biosynthesis and processing of the mannose receptor in human macrophages, *J. Biol. Chem.,* 264, 2385, 1989.
84. **Buhi, W. C., Ducsay, C. A., Bazer, F. W., and Roberts, R. M.,** Iron transfer between the purple phosphatase uteroferrin and transferrin and its possible role in iron metabolism of the fetal pig, *J. Biol. Chem.,* 257, 1712, 1982.
85. **Nuttleman, P. R. and Roberts, R. M.,** Transfer of iron from uteroferrin (purple acid phosphatase) to transferrin related to acid phosphatase activity, *J. Biol. Chem.,* 265, 12192, 1990.
86. **Hara, A., Kato, T., Sawada, H., and Fukuyama, K.,** Characterization of Fe^{2+}-activated acid phosphatase in rat epidermis, *Comp. Biochem. Physiol. B,* 82, 269, 1985.
87. **Ohno, J., Fukuyama, K., Hara, A., and Epstein, W. L.,** Immuno- and enzyme-histochemical detection of phosphoprotein phosphatase in rat epidermis, *J. Histochem. Cytochem.,* 37, 629, 1989.
88. **Lee, R. E., Peters, S. P., and Glew, R. H.,** Gaucher's disease: clinical, morphologic, and pathogenetic considerations, *Pathol. Annu.,* 12, 309, 1977.
89. **Troy, K., Cuttner, J., Reilly, M., Grabowski, G., and Desnick, R.,** Tartrate-resistant acid phosphatase staining of monocytes in Gaucher disease, *Am. J. Hematol.,* 19, 237, 1985.
90. **Schindelmeiser, J., Radzun, H. J., and Munstermann, D.,** Tartrate-resistant, purple acid phosphatase in Gaucher cells of the spleen, *Pathol. Res. Pract.,* 187, 209, 1991.
91. **Chambers, J. P., Peters, S. P., Glew, R. H., Lee, R. E., McCafferty, L. R., Mercer, D. W., and Wenger, D. A.,** Multiple forms of acid phosphatase activity in Gaucher's disease, *Metabolism,* 27, 801, 1978.
92. **Robinson, D. B. and Glew, R. H.,** A tartrate-resistant acid phosphatase from Gaucher spleen, *J. Biol. Chem.,* 255, 5864, 1980.
93. **Lam, K.-W., Li, C.-Y., Yam, L. T., and Desnik, R. J.,** Comparison of the tartrate-resistant acid phosphatase in Gaucher's disease and leukemic reticuloendotheliosis, *Clin. Biochem.,* 14, 177, 1981.
94. **Robinson, D. B. and Glew, R. H.,** Substrate specificity of Gaucher spleen phosphoprotein phosphatase, *Arch. Biochem. Biophys.,* 210, 186, 1981.
95. **Bouroncle, B. A.,** Leukemic reticuloendotheliosis, *Blood,* 53, 412, 1979.
96. **Yam, L. T., Janckila, A. J., Li, C.-Y., and Lam, W. K. W.,** Cytochemistry of tartrate-resistant acid phosphatase: 15 year's experience, *Leukemia,* 1, 285, 1987.
97. **Stepan, J. J., Lau, K. H. W., Mohan, S., Kranzlin, M., and Baylink, D. J.,** Purification and N-terminal sequence of two tartrate-resistant acid phosphatase type-5 from the hairy cell leukemia spleen, *Biochem. Biophys. Res. Commun.,* 165, 1027, 1989.
98. **Lam, K.-W., Townsend, D., Garza, A., Li, C. Y., and Yam, L. T.,** Comparison of tartrate resistant acid phosphatase in a giant cell bone tumor and a spleen infiltrated with hairy cells, *Clin. Biochem.,* 23, 335, 1990.
99. **Flanagan, A. M., Tinkler, S. M. B., Horton, M. A., Williams, D. M., and Chambers, T. J.,** The multinucleate cells in giant cell granulomas of the jaw are osteoclasts, *Cancer,* 62, 1139, 1988.
100. **Stepan, J. J., Lau, K. H. W., Mohan, S., Singer, F. R., and Baylink, D. J.,** Purification and N-terminal amino acid sequence of the tartrate-resistant acid phosphatase from human osteoclastoma: evidence for a single structure, *Biochem. Biophys. Res. Commun.,* 168, 792, 1990.
101. **Radzun, H. J., Kreipe, H., and Parwaresch, M. R.,** Tartrate-resistant acid phosphatase as a differentiation marker for the human mononuclear phagocyte system, *Hematol. Oncol.,* 1, 321, 1983.

102. **Hattersley, G. and Chambers, T. J.,** Generation of osteoclastic function in mouse bone marrow cultures: multinuclearity and tartrate-resistant acid phosphatase are unreliable markers for osteoclastic differentiation, *Endocrinology,* 124, 1689, 1989.
103. **Modderman, W. E., Tuinenburg-Bol Raap, A. C., and Nijweide, P. J.,** Tartrate-resistant acid phosphatase is not an exclusive marker for mouse ostoclasts in cell culture, *Bone,* 12, 81, 1991.
104. **Udagawa, N., Takahashi, N., Akatsu, T., Tanaka, H., Sasaki, T., Nichihara, T., Koga, T., Martin, T. J., and Suda, T.,** Origin of osteoclasts: mature monocytes and macrophages are capable of differentiating into osteoclasts under a suitable microenvironment prepared by bone marrow-derived stromal cells, *Proc. Natl. Acad. Sci. U.S.A.,* 87, 7260, 1990.
105. **Snipes, R. G., Lam, K.-W., Dodd, R. C., Gray, T. K., and Cohen, M. S.,** Acid phosphatase activity in mononuclear phagocytes and the U937 cell line: monocyte-derived macrophages express tartrate-resistant acid phosphatase, *Blood,* 67, 729, 1986.
106. **Schindelmeiser, J., Gullotta, F., and Munstermann, D.,** Purple acid phosphatase of human macrophages in AIDS encephalopathy, *Pathol. Res. Pract.,* 185, 184, 1989.
107. **Reichel, H., Koeffler, H. P., and Norman, A. W.,** Synthesis *in vitro* of 1,25-dihydroxyvitamin D_3 and 24,25-dihydroxyvitamin D_3 by interferon-stimulated normal human bone marrow and alveolar macrophages, *J. Biol. Chem.,* 262, 10931, 1987.
108. **Udagawa, N., Takahashi, N., Akatsu, T., Sasaki, T., Yamaguchi, A., Kodama, H., Martin, T. J., and Suda, T.,** The bone marrow-derived stromal cell lines MC3T3-G2/PA6 and ST2 support osteoclast-like cell differentiation in cocultures with mouse spleen cells, *Endocrinology,* 125, 1805, 1989.
109. **Burger, E. H. and Nijweide, P. J.,** Cellular origin and theories of osteoclast differentiation, in *Bone,* Vol. 2 *The Osteoclast,* Hall, B. K., Ed., CRC Press, Boca Raton, FL, 1991, 31.
110. **Minkin, C., St James, S., Tao, H.-H., Yu, X.-H., Pockwinse, S., MacKay, C., and Marks, S. C., Jr.,** Skeletal development and formation of osteoclast-like cells from *in situ* progenitors in fetal mouse metatarsal cultured in chemically defined medium, *Bone Miner.,* 12 141, 1991.
111. **Braidman, I. P., Rothwell, C., Webber, D. M., Crowe, P., and Anderson, D. C.,** Location of osteoclast precursors in fetal rat calvaria cultured on collagen gels, *J. Bone Miner. Res.,* 5, 287, 1990.
112. **Taylor, L. M., Tertinegg, I., Okuda, A., and Heersche, J. N. M.,** Expression of calcitonin receptors during osteoclast differentiation in mouse metatarsals, *J. Bone Miner. Res.,* 4, 751, 1989.
113. **Minkin, C. and Yu, X.-H.,** Calcitonin receptor expression and its regulation by 1α,25-dihydroxyvitamin D_3 during *de novo* osteoclast formation in organ cultures of fetal mouse metatarsal, *Bone Miner.,* 13, 191, 1991.
114. **Glowacki, J., Rey, C., Glimcher, M. J., Cox, K. A., and Lian, J.,** A role for osteocalcin in osteoclast differentiation, *J. Cell. Biochem.,* 45, 292, 1991.
115. **Marks, S. C., Jr. and Chambers, T. J.,** The giant cells recruited by subcutaneous implants of mineralized bone particles and slices in rabbits are not osteoclasts, *J. Bone Miner. Res.,* 6, 395, 1991.
116. **Machii, T. and Kitani, T.,** Similarities between Ig-G bearing lymphocytes and hairy cells: cytologic and cytochemical studies, *Blood,* 64, 166, 1984.
117. **Chappard, D., Rossi, J. F., Bataille, R., and Alexandre, C.,** Osteoclast cytomorphometry and scanning electron microscopy of bone eroded surfaces during leukemic disorders, *Scanning Microsc.,* 4, 323, 1990.
118. **Drexler, H. G., Menon, M., Gaedicke, G., and Minowada, J.,** Expression of FMC7 antigen and tartrate-resistant acid phosphatase isoenzyme in cases of B-lymphoproliferative diseases, *Eur. J. Cancer Clin. Oncol.,* 23, 61, 1987.

119. **Drexler, H. G., Klein, M.,, Bhoopalam, N., Gaedicke, G., and Minowada, J.,** Morphological and isoenzymatic differentiation of B-chronic lymphocytic leukaemia cells induced by phorbolester, *Br. J. Cancer,* 53, 181, 1986.
120. **Ziegler-Heitbrock, H. W. L., Munker, R., Dörken, B., Gaedicke, G., and Thiel, E.,** Induction of features characteristic of hairy cell leukemia in chronic lymphocytic leukemia and prolymphocytic leukemia cells, *Cancer Res.,* 46, 2172, 1986.
121. **Drexler, H. G., Otsuka, K., Gaedicke, G., and Minowada, J.,** Changes in isoenzyme profiles during induction of differentiation in human myelomonocytic leukemia cell lines, *Cancer Res.,* 46, 6078, 1986.
122. **Drexler, H. G., Brenner, M. K., Gignac, S. M., and Hoffbrand, V.,** Expression of tartrate-resistant acid phosphatase in B-CLL treated with phorbol ester or phorbol ester plus calcium ionophore, *Eur. J. Haematol.,* 41, 250, 1988.
123. **Drexler, H. G., Gaedicke, G., and Minowada, J.,** Occurrence of particular isoenzymes in fresh and cultured leukemia-lymphoma cells, *Cancer,* 57, 1776, 1986.
124. **Kettle, P., Morris, T. C. M., Markey, G. M., Alexander, H. D., Curry, R. C., Hayes, D., Cameron, C. H. S., and Toner, P. G.,** Tartrate resistant acid phosphatase positive splenic lymphoma: a relatively benign condition occurring in a time-space cluster?, *J. Clin. Pathol.,* 43, 714, 1990.
125. **Drexler, H. G., Gaedicke, G., Klein, M., Menon, M., and Minowada, J.,** Monocyte-associated acid phosphatase isoenzyme profiles as determined in acute myeloid leukaemia cells, *Scand. J. Haematol.,* 36, 239, 1986.
126. **Tavassoli, M., Rizo, M., and Yam, L. T.,** Evaluation of serum acid phosphatase in cancers with bone metastasis, *Cancer,* 45, 2400, 1980.
127. **Stepan, J. J., Zichova, M., Lachmanova, J., and Pacovsky, V.,** Plasma tartrate resistant acid phosphatase: a marker for osteoclastic bone resorption, *Acta Univ. Carolin. Med.,* 32, 345, 1986.
128. **Lau, K.-H. W., Onishi, T., Wergedal, J. E., Singer, F. R., and Baylink, D. J.,** Characterization and assay of tartrate-resistant acid phosphatase activity in serum: potential use to assess bone resorption, *Clin. Chem.,* 33, 458, 1987.
129. **Coleman, R. E., Whitaker, K. B., Moss, D. W., Mashiter, G., Fogelman, I., and Rubens, R. D.,** Biochemical prediction of response of bone metastases to treatment, *Br. J. Cancer,* 58, 205, 1988.
130. **Desoize, B., Veiler, V., Pourny, C., Comoe, L., and Jardillier, J. C.,** Isoenzymes of alkaline and acid phosphatases as bone metastasis marker in breast cancer patients, *Anticancer Res.,* 9, 1105, 1989.
131. **Lam, K.-W., Dannaher, C., Letchford, S., Eastlund, T., Li, C.-Y., and Yam, L.-T.,** Tartrate-resistant acid phosphatase in serum of cancer patients, *Clin. Chem.,* 30, 457, 1984.
132. **Echeteblu, Z. O., Cox, T. M., and Moss, D. W.,** Antibodies to porcine uteroferrin used in measurement of human tartrate-resistant acid phosphatase, *Clin. Chem.,* 33, 1832, 1987.
133. **Whitaker, K. B., Cox, T. M., and Moss, D. W.,** An immunoassay of human band 5 (''tartrate-resistant'') acid phosphatase that involves use of anti-porcine uteroferrin antibodies, *Clin. Chem.,* 35, 86, 1989.
134. **Kraenzlin, M. E., Lau, K.-H. W., Liang, L., Freeman, T. K., Singer, F. R., Stepan, J., and Baylink, D. J.,** Development of an immunoassay for human serum osteoclastic tartrate-resistant acid phosphatase, *J. Clin. Endocrinol. Metab.,* 71, 442, 1990.

Chapter 4

ONTOGENY OF THE OSTEOCLAST

Peter J. Nijweide and Ricardo de Grooth

TABLE OF CONTENTS

I. INTRODUCTION

There are few cells in the body with such a fascinating appearance and function as the osteoclast. Its size, multinuclearity, self-reliance, capability to resorb a substance so resistant as bone, and ability to stretch its life span by shedding "old" nuclei and taking up new ones have stimulated the interest of many workers in the field of bone research.

Still, in many ways the osteoclast remains a mysterious cell. Although a number of clues have been found and several theories have been proposed, we do not completely understand what determines whether a certain piece of bone is resorbed, what terminates the resorbing activity, or how osteoclast and other cells, notably cells of the osteogenic lineage, interact with each other. Neither do we know how the number of osteoclasts is regulated, by what mechanism the osteoclast is confined to bone tissue, or what causes the osteoclast to be a multinucleated cell with 2 to 100 nuclei.

One of these unsolved, but now partially unraveled questions that has attracted the attention of many investigators is the question of the lineage of the osteoclast. We will not go into the ideas and theories on the ontogeny of the osteoclast that have been developed and proposed since the 1970s, as they have already been reviewed extensively in a number of recent reviews.[1-6] It has been firmly established that the origin of the osteoclast must be sought in the hemopoietic tissues. Here, we will review the evidence for the hemopoietic origin of the osteoclast and summarize and analyze the available information on along which lineage osteoclast differentiation takes place.

In the past decade, the number of publications on the osteoclast have multiplied each year. Methods for the isolation of osteoclasts, although not to purity, have become available. A large number of more or less specific osteoclast markers have been recognized. Several *in vitro* systems for the study of resorption activity by isolated osteoclasts have been presented. Cell culture systems in which osteoclast formation is induced have been proposed. Regulation of osteoclast formation and activity by well-known and less-well-known humoral and local factors have been described. Although most of these subjects will be dealt with extensively in other chapters of this volume, some of them notably the identification of osteoclasts, are addressed in Section II in order to discuss the validity of the arguments for a certain proposed osteoclast lineage. Then, we discuss recent *in vitro* studies that may shed light on the identity of osteoclast precursors and progenitors. We describe the four possible routes which osteoclast differentiation may follow, and review the available evidence for each of these pathways.

In the following section, we discuss the role of some of the experimental conditions on osteoclast formation *in vitro* and analyze in what way these conditions influence the results of *in vitro* lineage studies. We summarize and attempt to integrate the conflicting data on osteoclast differentiation into a model originally proposed for macrophage differentiation.

II. IDENTIFICATION OF OSTEOCLASTS

Osteoclasts are large multinucleated cells capable of bone resorption. Multinuclearity is, however, not unique to the osteoclast, nor is bone resorption. Although osteoclastic bone resorption has a number of features specific to the osteoclast,[7] other cells, in particular mononuclear and multinuclear phagocytes, are also to a certain extent capable of bone dissolution.[8,9] Therefore, bone resorption itself is not so much specific as it is the way by which osteoclasts resorb bone.

When osteoclasts are studied or generated in the absence of bone matrix, identification markers other than osteoclastic bone resorption have to be used.

A. CYTOLOGY

Electron microscopy enables osteoclasts to be recognized by the presence of a unique ruffled border adjacent to the bone surface which is being resorbed,[10] surrounded by an organelle-free area, the clear zone.[11] The large number of mitochondria and lysosomes are also well-known characteristics of the osteoclast cytoplasm,[12] as discussed in greater detail in Chapters 6 and 9 of this volume.

For a long time, the identification of osteoclast precursors has been greatly dependent on the structural similarities between the putative mononuclear precursor and the osteoclast, as seen by transmission electron microscopy (TEM).[12,13] Scott,[13] for instance, used the morphological differences between two types of cells in the periosteum to identify them as osteoblast and osteoclast precursors. The latter she described as rounded, mononuclear cells rich in mitochondria, free ribosomes, and dense granules. However, since TEM is a very laborious technique and several morphological characteristics of osteoclasts are only present when the cell is attached to the bone surface (clear zone, ruffled border), other techniques have been developed.

B. ENZYME HISTOCHEMISTRY

Another characteristic that can be used is the presence of enzymes specific for the osteoclast. Several enzymes have been examined in this respect. Acid phosphatase is such a marker, and is discussed in greater detail in Chapter 3 of this volume. It has been shown that in osteoclasts, this enzyme is characterized by its resistance to suppression by tartrate, whereas the isoenzyme of monocytes and macrophages is tartrate sensitive.[14-18] This established tartrate-resistant acid phosphatase (TRAP) as a convenient cytochemical marker for osteoclasts and their direct precursors in histological sections and bone or bone marrow cell cultures.[19-23] Several recent reports, however, have indicated that the enzyme is not solely specific for osteoclasts. Nonskeletal tissues, e.g., lung[24,25] and spleen,[26,27] as well as osteoblasts,[28,29] have recently been shown to contain TRAP activity. TRAP has also been reported in the blood monocytes of the human,[30] quail, and chicken.[31] Moreover, Hattersly and Chambers[32] and Modderman et al.[33] found that TRAP and multinuclearity are unreliable markers for the osteoclast phenotype in cell cultures, even of mouse origin.

Several authors have suggested that the enzyme tartrate-resistant nucleotide tri- and diphosphatase (TrATPase), an enzyme belonging to the TRAP family, is a more specific osteoclast marker.[34-38] It is postulated that this TrATPase provides the energy necessary for the acidification of the microenvironment of bone resorption.[36] However, although acid ATPase may be a more substrate-specific indicator for osteoclasts than TRAP, it is not exclusively an osteoclastic marker, since other tissues (e.g., liver) and cells (e.g., macrophages) also contain this enzyme.[35]

Despite the problems just mentioned, histochemical staining for TRAP is, especially in combination with other markers, a much-used, useful method for osteoclast identification.

The high level of carbonic anhydrase in the osteoclast also appears to be a helpful marker in the identification of osteoclasts, since carbonic anhydrase has been associated with bone resorption.[39-41] Minkin and Jennings[39] showed that acetozolamide, an inhibitor of carbonic anhydrase, could inhibit bone resorption. Väänänen and Parvinnen[42] used an antiserum against carbonic anhydrase to detect the enzyme in the osteoclasts of rat calvaria. The specificity of carbonic anhydrase as an osteoclast marker is greatly enhanced with the use of such antisera. The function of osteoclastic anhydrase could be to assist in mineral solubilization by utilizing metabolic CO_2 to produce hydrogen ions, which are then actively transported across the membrane of the ruffled border to acidify the site of resorption.[40] This is in keeping with the fact that the pH in the osteoclast resorption cavity is acidic,[43,44] that the membrane of the osteoclast ruffled border contains a protein which cross-reacts with a lysosomal proton pump ATPase,[43] and that a low pH facilitates osteoclast adhesion and/or subsequent excavation of a lacuna on the bone surface.[45] Other authors also reported the presence of an ATP-dependent proton pump in the ruffled border of the osteoclast.[46-48] Antibodies directed against the H^+-translocating ATPase in osteoclasts can be used for immunocytochemical identification of osteoclasts.[48]

C. IMMUNOCYTOCHEMISTRY

As noted above, the lack of an exclusive marker for enzyme histochemical identification of osteoclasts has hindered the solution of the question of the osteoclast origin. One possible approach to this problem is the use of monoclonal antibodies. Immunocytochemistry avoids the problems inherent in enzyme histochemical studies, such as the specificity of the substrate, concentration of tartrate as an inhibitor for acid phosphatase, and pH.

Nijweide et al.[49] have raised a number of monoclonal antibodies (mabs) recognizing predominantly multinuclear osteoclasts in chicken and quail. Although these antibodies proved to be fairly specific for the osteoclast in bone, macrophages showed similar antigens when kept in culture for a few days.

Oursler et al.[50] obtained a number of mabs directed to chicken osteoclasts. Again, however, these antibodies interacted with monocyte- or macrophage-like cells in culture, especially in the presence of bone tissue or conditioned medium of bone tissue.[51,52] Although at that time these results were disappointing in that the mabs did not appear to be specific for osteoclasts in cell culture, the newly found possible close relationship between macrophages and osteoclasts (see later) forces a reevaluation of the results. It has to be reestablished whether the presence of bone tissue can induce such high numbers of osteoclasts in cultured bone marrow of chick.

Currently, the most universally used mab is 23C6, developed by Horton et al.[53] This antibody, raised against the osteoclast-like cells in giant cell bone tumors, recognizes the vitronectin receptor on human osteoclasts, but cross-reacts with osteoclasts of other species (rabbit, chicken, and baboon).[54] The antibody appears to be reasonably specific for osteoclasts, especially when one considers the intensity of the reaction. The antigen involved, however, is also present in a range of other cells, including macrophage polykaryons.[55]

Other monoclonals that have been used to identify osteoclasts are MB1, an antibody that has been shown to be useful in paraffin-embedded surgical specimens,[56] and kn22, which reacts very strongly with freshly isolated bone-derived baboon osteoclasts.[57]

Other antigens that have been described on osteoclasts and other bone marrow-derived cells include the pan leukocyte (CD45) antigen,[58] platelet-associated glycoproteins,[59] CD13, CD15a, CD44, CD54, CD68, and CD71 (transferrin receptor antigen), but these antigens are also expressed on monocytes, macrophages, and macrophage polykaryons.[60]

Instead of using monoclonal antibodies raised against osteoclast-specific markers, mabs can also be used as negative controls in order to differentiate between osteoclasts and phagocytes in mixed cultures. The macrophage-reactive mabs F4/80 and MAC-1 are especially used in this respect.[22]

D. CALCITONIN RECEPTOR

Osteoclastic function and differentiation is, in part, regulated by osteotropic hormones such as parathyroid hormone (PTH), 1,25-Dihydroxyvitamin D_3 (1,25$(OH)_2D_3$), and calcitonin (CT). Of these, CT is thought to act directly on the osteoclast, causing changes in the motility and morphology of the cell.[61,62] Indeed, Nicholson et al.[63] demonstrated specific binding sites for CT on isolated osteoclasts. Therefore, CT receptor positivity and the especially high abundance of these receptors is considered to be one of the most specific markers for the osteoclast phenotype.[64] However, this marker also has to be used with some caution: CT receptors can be downregulated by pretreatment with CT,[65] and not all species have abundant CT receptors expressed on the osteoclast.[62]

E. OSTEOCLAST ACTIVITY

The final proof that a cell population contains osteoclasts or osteoclast precursors is the demonstration of osteoclastic bone resorption, although recently the specificity of even this property has been questioned.[66]

There are, in principle, two techniques. (1) The coculture technique in which an isolated cell population is cocultured with living bone tissue, either a calvarium[67,68] or a long bone,[69-72] is especially suitable for osteoclast induction. Osteoclast formation and bone resorption may be studied with histological, enzyme histochemical, or biochemical methods. (2) The bone slice technique involves seeding existing osteoclasts or precursors onto a piece of devitalized bone. Bone resorption is monitored by evaluating resorption pit volume or area with scanning electron microscopy[73,74] or light microscopy,[62,75] or by measuring ^{45}Ca release.[76]

Another marker for osteoclast activity is the oxygen-derived free radical (ODFR) production by activated osteoclasts. It has been suggested that osteoclasts generate ODFR capable of reducing nitroble tetrazolium (NBT).[77-79]

III. OSTEOCLAST FORMATION *IN VITRO*

Although the search for the origin of the osteoclast may have started with *in vivo* studies,[1-6] notably experiments in which the bone marrow of healthy animals was transplanted into irradiated osteopetrotic animals to cure this disease,[80] most of the recent contributions to the unraveling of the osteoclast lineage came from *in vitro* studies.

A. BONE MARROW CULTURES

Transplantation studies in osteopetrotic animals revealed that the osteoclast progenitors are present in the bone marrow.[1-6,80] Osteoclast formation studies in cell cultures of bone marrow probably started with the work of Testa et al.,[81,82] who cultured bone marrow of adult cats. After 2 weeks of culture, many multinucleated cells formed, possessing some, although not highly discriminative, osteoclastic features. Thereafter, many other investigators used cell cultures to study osteoclast formation. Cat,[81-83] baboon,[84,85] human,[86,87] mouse,[88,89] rabbit,[90] and chicken[52] bone marrow cultures have been used. Progressively more osteoclast markers were applied to identify the multinucleated cells that formed in these cultures; however, it was also found that many of these markers are not purely osteoclast specific. Nevertheless, the evidence that osteoclast formation may take place in bone marrow cultures is now overwhelming. Many, and in some cases the majority, of the multinucleated cells formed during 1 to 3 weeks of culture show several of the markers also present on authentic osteoclast: TRAP activity, CT sensitivity, CT receptors, and staining with mabs such as 23C6[53] and 121F.[50]

B. STUDIES WITH SELECTED HEMOPOIETIC STEM CELL POPULATIONS

Although these bone marrow culture studies corroborated the *in vivo* findings that osteoclast progenitors are present in bone marrow, and were applied to the study of the effects of several hormones and growth factors on osteoclast formation, they gave no indisputable answer to the question of the osteoclast origin. Indeed, neither the osteopetrotic animal *in vivo* studies nor the bone marrow culture studies furnished definite proof that the osteoclast belongs to the hemopoietic lineage.

The first convincing evidence that the osteoclast stems from the pluripotent hemopoietic stem cell came from the work of Scheven et al.[91] Via a multistep procedure, cell populations highly enriched in hemopoietic stem cells were isolated. These stem cell populations formed osteoclasts in cocultures with stripped metatarsal bones devoid of endogenous osteoclast progenitors.[69-72] Similar results were obtained by Schneider and co-workers.[92,93] They prepared stem cell-enriched populations from rat bone marrow and showed that these populations gave rise to osteoclast formation both *in vivo* (transplantation)[92] and *in vitro*.[93] The final proof for the hemopoietic origin of the osteoclast was supplied by Hagenaars et al.[94] Cloned hemopoietic stem cells were shown to give rise to the formation of osteoclasts, granulocytes, macrophages, and erythrocytes. These findings were confirmed by studies showing that hemopoietic stem cell lines[95] can be induced to generate osteoclasts together with primarily granulocytes and macrophages in the bone slice technique[96] and the stripped metatarsal coculture technique.[72,97] In addition, Kurihara et al.[98] obtained blast cell colonies from spleen cell populations isolated from 5-fluorouracil-treated mice and seeded in methylcellulose. These blast cells formed neutrophil, macrophage, eosinophil, mast cell, megakaryocyte, and erythroblast colonies in secondary cultures in the presence of interleukin-3 (IL-3) and erythropoietin. In the presence of IL-3 or granulocyte-macrophage colony-stimulating factor (GM-CSF) and $1,25(OH)_2D_3$, similar blast cell populations, but not macrophage colonies, formed multinucleated, TRAP-positive cells sensitive for CT and capable of bone resorption in cocultures with stripped mouse calvaria. The same authors showed that both macrophages and osteoclast-like cells (TRAP-positive multinuclear cells) could derive from a single cloned blast cell.

C. POSSIBLE ROUTES OF OSTEOCLAST DIFFERENTIATION

Although the hemopoietic origin of the osteoclast is now firmly established, the relationship between the osteoclast lineage and other hemopoietic lineages is still being debated. As the pluripotent stem cell and the multipotent stem cells CFU-S, spleen colony-forming unit, (12 to 14) and CFU-S (8) can differentiate into osteoclasts,[94,99] the osteoclast is considered to be a member of the myeloid branch of the hemopoietic lineage.[99] There are four different routes which the osteoclast lineage may follow (Figure 1). According to

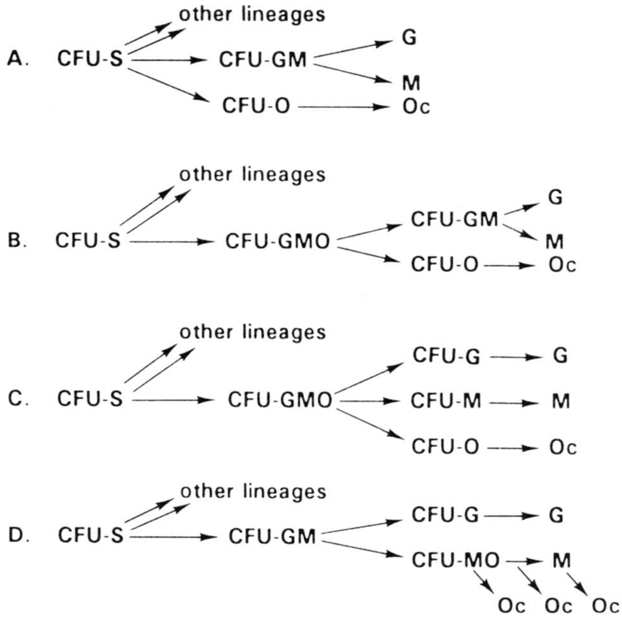

FIGURE 1. Possible routes of osteoclast differentiation. On the basis of the available evidence, four different routes of osteoclast differentiation can be postulated.

pathway A, the osteoclast has its own, independent lineage. In pathway B, the osteoclast shares its committed progenitor with the monocyte-granulocyte line, but the osteoclast line branches off before CFU-GM. It is also possible (pathway C) that granulocytes, monocytes, and osteoclasts arise at the same differentiation stage from a common progenitor. This progenitor is generally called CFU-GM, but should then be renamed CFU-GMO. In the final pathway (D), the osteoclast line branches off from the monocyte line at any point between the CFU-M and the monocyte.

There is circumstantial evidence for all four possibilities in the literature.

D. PROGENITORS, PRECURSORS, AND COLONY-STIMULATING FACTORS

Evidence for any of the four possible osteoclast differentiation pathways is generally obtained in two ways. First, as the osteoclast is now recognized as belonging to the hemopoietic cell lineage, the class of growth factors which has been known to regulate proliferation and differentiation processes in hemopoiesis[100] is also likely to be involved in osteoclastogenesis. These growth factors, the CSFs, have differential effects on the various sublines of the hemopoietic lineage. Although the subline specificity of the various growth factors is by no means absolute, each CSF preferentially stimulates the differentiation of stem or progenitor cells along one or two sublines. Hence, the effects or lack of effects of these various CSFs have been used as indications for a certain differentiation pathway.

The second way to approach the problem is to use cell populations of a certain differentiation stage. The analysis of the progeny of such populations *in vitro* may give information about which mature cell types have a common progenitor. In essence, such an approach is much more direct. It demands, however, a very rigorous purification of the required populations or, even better, the cloning of a required progenitor.

1. Pathway A: A Separate Lineage for the Osteoclast

The evidence for pathway A (Figure 1) is not very strong. This route was proposed by Loutit and Nisbet,[101] but no evidence was offered. Regarding this possibility, Flanagan[102] recently observed that in patients with aplastic anemia, a disease which is characterized by a depletion of erythrocytes, granulocytes, monocytes, and platelets, the osteoclast numbers are not reduced. These results agree with a very early bifurcation of the osteoclast line from the other myeloid lines, even before the progenitor stage common to the granulocytes, erythrocytes, monocytes, and megakaryocytes, the CFU-GEMM. Most of the available evidence, however, points toward a later separation from the mainstream of myeloid differentiation, as we will see in the following section.

2. Pathway B: CFU-GMO Bifurcates into CFU-GM and CFU-O

The evidence for this route primarily comes from negative arguments: A CFU-GMO gave rise to the formation of a CFU-GM and a CFU-O has not yet been identified. GM-CSF has been found by several groups to inhibit or have no effect on osteoclast formation. Schneider and Relfson[93,103] observed no effect of GM-CSF on osteoclast formation and resorption when added to fetal rat long-bone cultures. Hattersley and Chambers[104] and Shinar et al.[105] reported an inhibition of osteoclast-like cell formation (defined by TRAP, CT binding, and resorption) in mouse bone marrow cultures. Lorenzo et al.[106] found no effect or an inhibitory effect on new osteoclast formation in cultures of fetal mouse long bones, depending on the experimental circumstances. The explanation offered by most of these authors for their findings was that GM-CSF diverts common progenitor cells to pathways other than the osteoclast line. It should be mentioned, however, that other workers have found a beneficial effect of GM-CSF on osteoclast formation[85] (see Section III.D.3). Furthermore, it must be stressed that studies on the effects of CSFs have to be interpreted with caution. CSFs may have multiple effects, depending on the experimental conditions and the presence or absence of other growth factors. In the presence of a monocyte stimulatory factor, GM-CSF may very well stimulate monocyte formation and inhibit osteoclast formation from a common progenitor, whereas in the presence of an osteoclast stimulatory factor, it may act the opposite.

The only other evidence for route B comes from a recent paper by Hattersley et al.[107] These authors studied the resorption activity of single colonies obtained from mouse spleen cells grown in methylcellulose, and subsequently seeded onto bone slices together with a stromal cell line (ts 8). No single lineage colony (M, G, etc.) contained cells with the capacity to differentiate into functionally competent osteoclasts (pit formation in bone slices). GM and multilineage colonies had this capability, but only to 16 to 25 and 38 to 60%, respectively, depending on the circumstances. They reasoned that since such a low percentage of the GM colonies gave rise to osteoclast formation, true CFU-GMs are not capable of differentiating into osteoclasts. The osteoclast-producing GM colonies contained, in their opinion, osteoclast progenitors together with CFU-GM. An alternative explanation, however, is that CFU-GM differentiates to osteoclasts much less frequently than to granulocytes or monocytes. Furthermore, their induction system may have a low efficiency. In short, the evidence is inconclusive.

3. Pathway C: Separate Granulocyte, Monocyte, and Osteoclast Lines Originate from One Common Progenitor

Support for this osteoclast differentiation route derives primarily from the work of Kurihara et al.[108,109] who presented direct evidence of an osteoclast progenitor role for what has until now been called CFU-GM. Kurihara et al.[108] cloned and cultured the proliferative fraction of human bone marrow in methylcellulose in the presence of GM-CSF, and somewhat later added $1{,}25(OH)_2D_3$. After a few weeks of culture, they observed four types of colonies: mab 23C6-positive polygonal cells, macrophages, granulocytes, and mixed colonies in which about 30% of the cells reacted with mab 23C6. The polygonal cells rescued from the methylcellulose formed many multinuclear cells in liquid cultures. Virtually all of these stained with mab 23C6 and contracted in the presence of CT. Populations of the mab 23C6-positive polygonal cells seeded onto bone slices caused pit formation. The most important conclusions from this work are that (1) monocytes/macrophages, granulocytes, and osteoclasts share a GM-CSF-sensitive progenitor, the CFU-GM (or, rather, the CFU-GMO), (2) the osteoclast and monocyte lineages separate beyond that progenitor stage, and (3) the more differentiated stages of the monocyte lineage cannot give rise to osteoclast formation.

The earlier work of MacDonald et al.,[85] who found a stimulation of osteoclast-like cell formation by GM-CSF in baboon bone marrow cultures, corroborates the studies of Kurihara et al. Although the multinucleated cells were identified at that time as osteoclast-like by inconclusive markers, the same authors later defined the multinucleated cells in similar cultures with more convincing osteoclast characteristics. On the other hand, several other investigators[103-106] described either a lack of effect or an inhibitory effect of GM-CSF in various culture systems.

Although the evidence is fairly convincing, the final acceptance of pathway C as the osteoclast differentiation line must await further corroboration by other investigators and by results obtained under other circumstances and from other species.

4. Pathway D: Osteoclasts Derive from Monocyte Precursors, Monocytes, or Macrophages

This final possibility is perhaps the most controversial one. Initially, the many similarities between macrophages and osteoclasts suggested a very close relationship between the two cell types. Later, it was found that clear distinctions exist between phagocytes and osteoclasts, and evidence emerged for a more separated origin. Indeed, the majority of the recent papers on the subject deny the possibility that osteoclasts derive from monocytic cells. Burger et al.[69] showed that blood leukocytes or peritoneal cells do not form osteoclasts in cocultures with stripped metatarsal bones. Although inflammatory peritoneal exudate cells were capable of osteoclast formation,[71] and a number of years later Helfrich et al.[23] showed that the leukocyte fraction in blood can give rise to osteoclast formation, the number of osteoclasts formed appeared to be very small compared to the number of leukocytes put into coculture. Schneider and Relfson[103,110] isolated CFU-GM, CFU-G, and CFU-M from rat bone marrow. *In vivo* (transplantation in osteopetrotic rats) as well as *in vitro* (coculture with fetal rat long bones) CFU-GM and CFU-G generated osteoclasts, while CFU-M did not. As both populations (CFU-GM, CFU-G) were contaminated with pluripotent stem cells,[111] the negative results with CFU-M are more informative than the positive results with CFU-GM and CFU-G. Similar negative results on osteoclast formation by CFU-M were obtained by Haltersley et al.,[107] Kurihara et al.,[98] and Kurihara et al.[108] All three groups observed that monocyte colonies in methylcellulose cultures of hemopoietic tissues did not give rise to osteoclast formation. Furthermore, van de Wijngaert et al.[112] observed in mouse bone marrow cultures an inhibition of the formation of TRAP-positive mononuclear cells by M-CSF, concomitant with a stimulation of TRAP-negative mononuclear phagocytes. They did not characterize the TRAP-positive cells any further, but did show that the number of TRAP-positive cells in cell culture correlated with the number of osteoclast nuclei formed when the cells were cocultured with stripped metatarsal bones. A similar inhibition of the formation of osteoclast-like cells in mouse bone marrow cultures was found by Hattersley and Chambers[104] and Shinar et al.[105] In both studies, multinuclear cell formation and bone resorption on devitalized bone slices was inhibited by M-CSF. The multinucleated cells were characterized as osteoclast-like by TRAP positivity and CT binding.

Not quite in line with these findings are the results of Felix et al.[113] and Kodama et al.[114] The *op/op* variant of murine osteopetrosis has a lack of osteoclasts and an impaired resorption.[80] Recently, it has been found that not only the osteoclast, but also the macrophage line is affected,[115] and that the impairment resides in the production of the M-CSF.[115-118] The mutant mice were shown to have a defect in the coding region of the M-CSF gene on chromosome 3.[117] The correlation between this defect and the impaired macrophage and osteoclast formation was further established by experiments showing that transplantation of bone marrow could not cure the osteopetrosis,[119] but that implantation of M-CSF-producing L929 cells partially corrected the deficiency of macrophages.[118] Felix et al.[113] and Kodama et al.[114] injected *op/op* mice with M-CSF. This treatment appeared to restore bone resorption and induced the reappearance of osteoclasts and resident bone marrow macrophages. Finally, Takahashi et al.[120] showed that M-CSF added to the culture medium *in vitro* restored the osteoclast-like cell formation in cocultures of *op/op*-derived osteoblastic cells and wild type-derived spleen cells.

Of course, these observations offer only circumstantial evidence for a close relationship between osteoclasts and macrophages. The effects of M-CSF are perhaps indirect; moreover, they are not restricted to the monocyte subline. In conjunction with IL-3, M-CSF stimulates the proliferation of stem cells.[94,121] M-CSF combined with GM-CSF increases the number and size of GM colonies *in vitro* more than their additive effect.[122] Treatment time and the concentration of M-CSF may be of utmost importance in determining whether stimulation of proliferation of the common monocyte-osteoclast progenitor occurs or the more restricted stimulation of monocyte/macrophage differentiation dominates. Careful studies of dose-response and time-response relationships are clearly needed to resolve this question and find an explanation for the conflicting evidence.

Although these M-CSF and *op/op* experiments may still be reconciled with the CFU-M studies mentioned earlier, the recently published results of Udagawa et al.[123] are clearly opposed. These authors used ST-2 cells, a bone marrow stromal cell line,[124] as a feeder layer for the development of osteoclast-like cells from spleen, bone marrow, and thymus, but not only from these well-known osteoclast-generating tissues. Alveolar macrophages seeded in low density on ST-2 monolayers also gave rise to colonies of TRAP-positive cells, both mononuclear and multinuclear. More than 90% of these cells could bind CT. When alveolar macrophages were cultured together with ST-2 cells on bone slices, resorption pits were formed. Peripheral blood cells gave similar results: when seeded onto ST-2 cell layers, TRAP-positive colonies were formed. The number of colonies increased as the peripheral blood cell population was enriched for monocytes by glass adherence. These results are so conflicting with those of others that confirmation by additional groups must be available before a definitive evaluation can be performed.

IV. ROLE OF EXPERIMENTAL CONDITIONS ON OSTEOCLAST FORMATION *IN VITRO:* 1,25(OH)$_2$D$_3$, CALCIFIED MATRIX, AND OSTEOGENIC CELLS

Almost all studies in which osteoclast formation is investigated in cell cultures agree on three things. Unphysiological concentrations of 1,25(OH)$_2$D$_3$

are obligatory.[125] Usually, concentrations around 10^{-8} *M* are used. These high concentrations are not needed in cocultures of hemopoietic cells and stripped metatarsal bones. Without the addition of $1{,}25(OH)_2D_3$, osteoclasts form readily in cocultures of metatarsal bones and various hemopoietic stem cells or progenitor populations. Clearly, some factor(s) is (are) missing in the hemopoietic cell cultures. As high $1{,}25(OH)_2D_3$ concentrations have a strong differentiation-inducing effect, it is fairly possible that osteogenic tissue, bone or hypertrophic cartilage, produces one or more osteoclast differentiation factors. These factors are, in this notion, not produced or presented in the right way (see below) by isolated primary osteoblast populations or osteogenic/stromal cell lines. The nature of these factors is not yet known, but several candidates have already been proposed in the literature. An in-depth analysis of these putative factors is beyond the scope of this chapter. It is also possible that osteoclast formation in cell culture represents an aberrant, overstimulated osteoclast formation pathway (see Section V).

Second, almost all studies show that cells of the osteogenic lineage, osteoblast-like or stromal derived, must be present in order for osteoclast differentiation to occur. These cells produce some of the factors that induce osteoclast differentiation. Some, but not all of them, are presented in the correct way, since high $1{,}25(OH)_2D_3$ concentrations are needed in conjunction with the presence of osteogenic cells for complete osteoclast differentiation. One of the few exceptions to this general rule is the study of Kurihara and co-workers,[98] which reported osteoclast-like cell formation from hemopoietic blast cell colonies in the absence of stromal or osteoblast-like cells. Osteoclasts were characterized by multinuclearity, TRAP staining, and CT sensitivity. Final determination of the osteoclast nature of the multinuclear cells obtained in this study was done in a resorption assay using living, stripped calvaria. So, perhaps also in this study bone cells were needed when complete osteoclast differentiation was shown. The authors did not, however, report how their osteoclast-like cells behaved on slices of devitalized bone.

The third component needed for osteoclast differentiation studies is, of course, the calcified matrix. In the first place, calcified matrix is needed to show osteoclastic bone resorption as the ultimate marker of the osteoclastic nature. Second, bone matrix may have a beneficial influence on terminal osteoclast differentiation. Chambers' group has shown in many papers that bone slices alone or in combination with osteogenic cells do not induce osteoclast formation from progenitor cells *in vitro*. $1{,}25(OH)_2D_3$ is obligatory. The calcified matrix used in these and other studies is, however, often dentin. Furthermore, the bone and dentin slices used in the slice resorption assay are extensively washed and extracted with water or ethanol. These procedures could easily cause the removal of osteoclast differentiation factors from the bone matrix. Even so, Fuller and Chambers[126] showed that osteoclast-like cells (23C6 positive, resorption pit formation) do form in larger numbers when bone marrow is seeded onto bone slices than onto glass cover slips. In

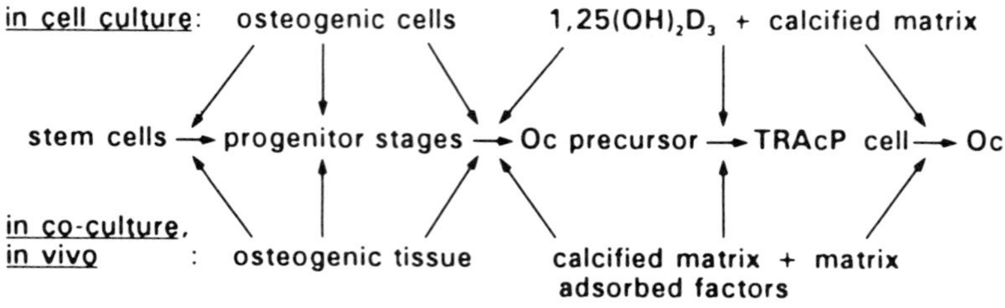

FIGURE 2. Induction of osteoclast differentiation. Induction of early and late stages of osteoclast differentiation in cell cultures (hemopoietic together with osteogenic cells) and under *in vivo* or coculture (hemopoietic cells with stripped fetal bones) conditions.

addition, Krukowski and Kahn[127] have shown that primarily bone particles seeded onto the chorioallantoic membrane of chick embryos induce osteoclast formation. Glowacki and co-workers[128] reported similar findings in studies in which particles of various origin were implanted in subcutaneous pockets in rats. Only cells adjacent to bone particles elaborated osteoclastic characteristics. Löwik et al.[129] demonstrated that formation of TRAP-positive cells in coculture of stripped metatarsal rudiments and bone marrow depended heavily on the contact between calcified matrix and osteoclast precursor cells. Bacterial filters placed between the bone and osteoclast progenitor source completely inhibited TRAP-positive cell formation. The fact that devitalized long-bone rudiments, stripped of their periosteum, do not induce osteoclast formation in cocultures with hemopoietic cells[71,130] is not contradictory. In this case, stimulation of the early osteoclast differentiation steps is missing. These results allow the following conclusion (Figure 2). Osteoclast formation *in vivo* is confined to bone tissue. Only bone tissue appears to create the right microenvironment for osteoclast differentiation, although very low numbers of osteoclast progenitors are ubiquitous in the circulation.[23] This microenvironment is created by osteogenic cells, and the bone matrix offers the right substrate for (probably) the late steps of osteoclast differentiation. The factor(s) responsible for this calcified matrix effect may be the matrix itself and/or factors produced by osteogenic cells and adsorbed to the matrix surface. It is possible that these adsorbed factors have to be replaced by high $1,25(OH)_2D_3$ concentrations in the cell culture studies.

V. CONCLUSIONS

As reviewed above, the origin of the osteoclast from the pluripotent, hemopoietic stem cell is now generally accepted. Although it is not completely excluded that the osteoclast may accept the nuclei of cells of other lineages,[131,132] we think it is very unlikely. Similarly, the great majority of studies agree that the osteoclast line is a member of the myeloid branch of the hemopoietic lineage. Furthermore, most of the available evidence sustains the notion that the committed macrophage-granulocyte progenitor or a differentiation stage just before that progenitor is the committed progenitor of the osteoclast. At this point, the various opinions diverge. There is reasonable evidence for all three possible routes (B, C, and D) of Figure 1. Moreover, some of the evidence for one option appears to exclude the possibility of one of the other options. At the moment, it is difficult to decide which is the true *in vivo* route of osteoclast differentiation. The term *in vivo* is stressed here because most of the data on osteoclast differentiation are obtained from *in vitro* experiments. Some of the contradictions in the results obtained by different groups may be caused by differences of *in vitro* circumstances. *In vitro* circumstances may cause progenitors to differentiate along pathways not normally used *in vivo,* or used only under pathological of high-stress conditions. Perhaps osteoclastologists want too much in too short a time. The osteoclast differentiation line is perhaps not the unidirectional one-way route we would like it to be, especially not under extreme conditions such as *in vitro* circumstances.

Macrophage development has been intensively studied for a much longer time by numerous investigators, and is not fully understood. Conventionally, the monocyte-granulocyte lines bifurcate at the CFU-GM stage. Promyelocytes and even such mature granulocytic stages as metamyelocytes and band forms can be induced, however, to change into monocytic cells by γ-interferon.[133] Sterry and Czarnetzki[134] demonstrated mast cell differentiation *in vitro* from elicited rat peritoneal macrophages. The mast cell lineage is supposed to separate from the granulocyte-monocyte line at the multipotent stem cell level. There are more such examples showing that the different hemopoietic cell lines are less strictly separated than the conventional scheme of the hemopoietic lineage suggests.

There is no general agreement in the macrophage development studies on how the diversity of macrophage phenotypes arises. Differences in maturation stage, in local environment, in the origin and differentiation route, and in separate macrophage populations with self-renewal capacities in the different tissues have all been put forward to explain the diversity in phenotype and function of the macrophage. To integrate these various views, Leenen[135] has recently proposed the flexible lineage hypothesis:

1. In the steady state, different macrophage sublineages differentiate from a common progenitor along different pathways. Monocyte belonging to a certain subline show a preferential development into a particular macrophage type and are preferentially homing in a particular tissue. This implies that the blood monocyte population is heterogenous with regard to its differentiation potential.
2. Under stress conditions, the regular differentiation pathways can be overruled by extreme environmental conditions. This means that members of a subline can be diverted from their regular differentiation route to the end stage of another subline. The stress conditions causing such diversion may consist of high concentrations of cytokines or combi-

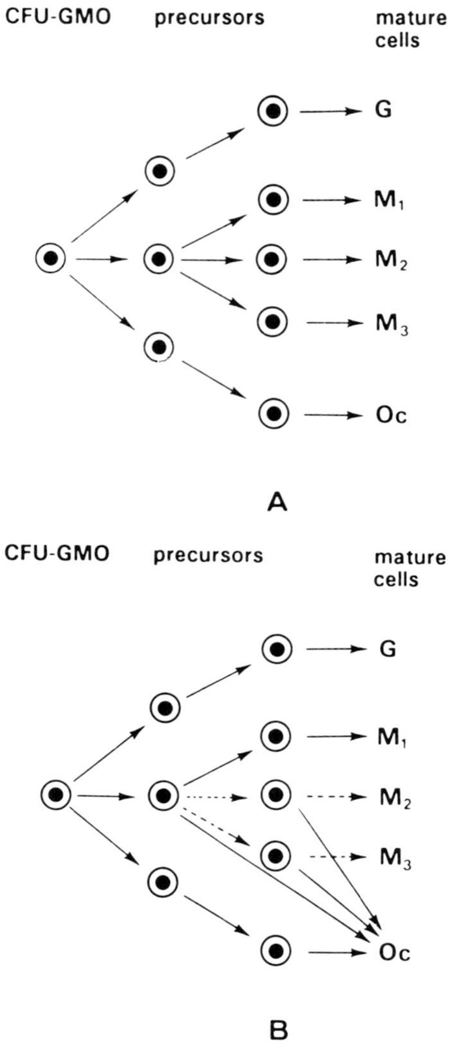

FIGURE 3. Osteoclast formation in steady-state and stress conditions. (A) Granulocytes (G), various macrophage lineages ($M_{1\text{-}3}$), and osteoclasts (Oc) stem from a common committed progenitor, CFU-GMO, each via its own preferential pathway. (B) Under stress conditions, some of the regular macrophage differentiation pathways are overruled, leading to enhanced osteoclast formation, based on the flexible lineage hypothesis for macrophages by Leenen.[135]

> nations of cytokines which a particular subline does not see under steady-state conditions. These cytokines may then induce the appearance of receptors for other cytokines and homing receptors normally not present on the cell membrane of cells of that particular subline.

The conflicting data on osteoclast differentiation may very well fit this hypothesis (Figure 3). The extremely high concentrations of $1,25(OH)_2D_3$

needed for osteoclast formation in cell cultures and the high percentage of osteoclast-like mononuclear and multinuclear cells (compared to macrophages) found under these conditions suggest that aberrant, stress-induced differentiation pathways are involved.

This may mean that the search for the true osteoclast lineage with cell culture methods is perhaps somewhat superfluous. It may also mean that coculture of selected cell populations with stripped metatarsal bones approaches the physiological situation better. *In vivo* experiments in which selected, labeled cell populations are injected into animals and osteoclast formation is monitored may, however, give the best answers to the question of the physiological, steady-state osteoclast differentiation pathway. It is, however, also possible that the aberrant routes of osteoclast formation from macrophage progenitors/precursors also have some significance in *in vivo* situations where bone resorption is stimulated above normal levels.

REFERENCES

1. **Loutit, J. F. and Nisbet, N. W.,** The origin of the osteoclast, *Immunobiology,* 161, 193, 1982.
2. **Marks, S. C.,** The origin of the osteoclast, *J. Oral Pathol.,* 12, 226, 1983.
3. **Chambers, T. J.,** The pathobiology of the osteoclast, *J. Clin. Pathol.,* 38, 241, 1985.
4. **Nijweide, P. J., Burger, E. H., and Feyen, J. H.,** Cells of bone: proliferation, differentiation, and hormonal regulation, *Physiol. Rev.,* 66, 855, 1986.
5. **Mundy, G. R. and Roodman, G. D.,** Osteoclast ontogeny and function, in *Bone and Mineral Research,* Vol. 5, Peck, W. A., Ed., Elsevier, Amsterdam, 1987, 209.
6. **Vaes, G.,** Cellular biology and biochemical mechanism of bone resorption, *Clin. Orthop. Relat. Res.,* 231, 239, 1988.
7. **Marks, S. C. and Popoff, S. N.,** Ultrastructural biology and pathology of the osteoclast, in *Ultrastructure of Skeletal Tissues,* Bonucci, E. and Motta, P. M., Kluwer Academic Publishers, Boston, 1990, 239.
8. **Mundy, G. R., Altman, A. J., Gondek, M. D., and Bandelin, G. J.,** Direct resorption of bone by human monocytes, *Science,* 196, 1109, 1977.
9. **Kahn, A. J., Stewart, C. C., and Teitelbaum, S. L.,** Contact-mediated bone resorption by human monocytes *in vitro, Science,* 199, 988, 1978.
10. **Kallio, D. M., Garant, P. R., and Minkin, C.,** Evidence of coated membranes in the ruffled border of the osteoclast, *J. Ultrastruct. Res.,* 37, 169, 1971.
11. **Malkani, K., Luxembourger, M. M., and Rebel, A.,** Cytoplasmic modifications at the contact zone of osteoclasts and calcified tissue in the diaphyseal growing plate of fetal quinea-pig tibia, *Calcif. Tissue Res.,* 11, 258, 1973.
12. **Rifkin, B. R., Brand, J. S., Cushing, J. E., Coleman, S. J., and Sanavi, F.,** Fine structure of fetal rat calvarium: provisional identification of preosteoclasts, *Calcif. Tissue Int.,* 31, 21, 1980.
13. **Scott, B. L.,** The occurrence of specific cytoplasmic granules in the osteoclast, *J. Ultrastruct. Res.,* 19, 417, 1967.
14. **Wergedal, J. E. and Baylink, D. J.,** Distribution of acid and alkaline phosphatase activity in undemineralized sections of the rat tibial diaphysis, *J. Histochem. Cytochem.,* 17, 799, 1969.

15. **Hammarström, L. E., Hanker, J. S., and Toverud, S. U.,** Cellular differences in acid phosphatase as a marker of osteoclast function, *Calcif. Tissue Int.*, 34, 285, 1982.
16. **Chilosi, M., Pizzolo, G., Janossi, G., Bofill, M., and Fiore-Donati, L.,** Enzyme histochemical analysis on cryostat sections of human bone marrow, *J. Clin. Pathol.*, 35, 1220, 1982.
17. **Webber, D., Braidman, I. P., Robertson, W. R., and Anderson, D. C.,** A quantitative cytochemical assay for acid phosphatase in osteoclasts in the fetal rat calvaria, *J. Bone Miner. Res.*, 1, 789, 1986.
18. **Braidman, I. P., St John, J. G., Anderson, D. C., and Robertson, W. R.,** Effects of physiological concentrations of parathyroid hormone on acid phosphatase activity in cultured rat bone cells, *J. Endocrinol.*, 111, 17, 1986.
19. **Minkin, C.,** Bone acid phosphatase: tartrate-resistant acid phosphatase as a marker of osteoclast function, *Calcif. Tissue Int.*, 34, 285, 1982.
20. **van de Wijngaert, F. P. and Burger, E. H.,** Demonstration of tartrate-resistant acid phosphatase in undecalcified, glycomethacrylate-embedded mouse bone: a possible marker for (pre) osteoclast identification, *J. Histochem. Cytochem.*, 34, 131, 7, 1986.
21. **Scheven, B. A. A., Kawilarang-De Haas, E. W. M., Wassenaar, A. M., and Nijweide, P. J.,** Differentiation kinetics of osteoclasts in the periosteum of embryonic bones *in vivo* and *in vitro, Anat. Rec.*, 214, 418, 1986.
22. **van de Wijngaert, F. P., Tas, M. C., and Burger, E. H.,** Characteristics of osteoclast precursor-like cells grown from mouse bone marrow, *Bone Miner.*, 3, 111, 1987.
23. **Helfrich, M. H., Mieremet, R. H. P., and Thesingh, C. W.,** Osteoclast formation *in vitro* from progenitor cells present in the adult mouse circulation, *J. Bone Miner. Res.*, 4, 325, 1989.
24. **Efstratiadis, T. and Moss, D. W.,** Tartrate resistant acid phosphatase of human lung. Apparent identity with osteoclastic acid phosphatase, *Enzyme*, 33, 34, 1985.
25. **Efstratiadis, T. and Moss, D. W.,** Tartrate resistant acid phosphatase in human alveolar macrophages, *Enzyme*, 34, 140, 1985.
26. **Hara, A., Sawada, H., Kato, T., Nakayama, T., Yamamoto, H., and Matsumoto, Y.,** Purification and characterization of a purple acid phosphatase from rat spleen, *J. Biochem.*, 95, 67, 1984.
27. **Schindelmeiser, J., Munstermann, D., and Witzel, H.,** Histochemical investigations on the localization of the purple acid phosphatase in the bovine spleen, *Histochemistry*, 87, 13, 1987.
28. **Wergedal, J. E., Lau, K. H. W., and Baylink, D. J.,** Fluoride and bovine bone extract influence cell proliferation and phosphatase activities in human bone cell cultures, *Clin. Orthop. Relat. Res.*, 233, 274, 1988.
29. **Bianco, P., Ballanti, P., and Bonucci, E.,** Tartrate-resistant acid phosphatase activity in rat osteoblasts and osteocytes, *Calcif. Tissue Int.*, 43, 167, 1988.
30. **Helfrich, M. H., Thesingh, C. W., Mieremet, R. H. P., and van Iperen-van Gent, A. S.,** Osteoclast generation from human fetal bone marrow in coculture with murine fetal long bones, *Cell Tissue Res.*, 249, 125, 1987.
31. **Mulder, R. J. P., Vrijheid-Lammers, T., and Nijweide, P. J.,** Osteoclast marker expression in a monocyte enriched leucocyte population of egg laying birds, *Calcif. Tissue Int.*, 41 (Suppl. 2), 99, 1987.
32. **Hattersley, G. and Chambers, T. J.,** Generation of osteoclastic function in mouse bone marrow cultures: multinuclearity and tartrate-resistant acid phosphatase are unreliable markers for osteoclastic differentiation, *Endocrinology*, 124, 1689, 1989.
33. **Modderman, W. E., Tuinenburg-Bol Raap, A. C., and Nijweide, P. J.,** Tartrate-resistant acid phosphatase is not an exclusive marker for mouse osteoclasts in cell culture, *Bone*, 12, 81, 1991.
34. **Andersson, G. N., Ek-Rylander, B., and Hammarström, L. E.,** Purification and characterization of a vanadate-sensitive nucleotide tri- and diphosphatase with acid pH optimum from rat bone, *Arch. Biochem. Biophys.*, 228, 431, 1984.

35. **Andersson, G. N., Ek-Rylander, B., Hammarström, L. E., Lindskog, S., and Toverud, S. U.,** Immunocytochemical localization of a tartrate-resistant and vanadate-sensitive acid nucleotide tri- and diphosphatase, *J. Histochem. Cytochem.*, 34, 293, 1986.
36. **Andersson, G. N. and Marks, S. C., Jr.,** Tartrate-resistant acid ATPase as a cytochemical marker for osteoclasts, *J. Histochem. Cytochem.*, 37, 115, 1989.
37. **Ek-Rylander, B., Marks, S. C., Jr., Hammarström, L. E., and Andersson, G. N.,** Osteoclastic acid ATPase: biochemical and histochemical studies of the osteopetrotic mutations in the rat, *Bone Miner.*, 5, 309, 1989.
38. **Lindunger, A., MacKay, C. A., Ek-Rylander, B., Andersson, G. N., and Marks, S. C. Jr.,** Histochemistry and biochemistry of tartrate-resistant acid phosphatase (TRAP) and tartrate-resistant acid adenosine triphosphatase (TrATPase) in bone, bone marrow and spleen: implications for osteoclast ontogeny, *Bone Miner.*, 10, 109, 1990.
39. **Minkin, C. and Jennings, J. M.,** Carbonic anhydrase and bone remodelling: sulfonamide inhibition on bone resorption in organ cultures, *Science*, 176, 1031, 1972.
40. **Hall, G. and Kenny, A. D.,** Role of carbonic anhydrase in bone resorption induced by prostaglandin E_2 *in vitro, Pharmacology*, 30, 339, 1985.
41. **Gay, C. V., Ito, M. B., and Schraer, H.,** Carbonic anhydrase activity in isolated osteoclasts, *Metab. Bone Dis. Relat. Res.*, 5, 33, 1983.
42. **Väänänen, H. K., and Parvinnen, E. K.,** High active isoenzyme of carbonic anhydrase in rat calvaria osteoclasts, *Histochemistry*, 78, 481, 1983.
43. **Baron, R., Neff, L., Louvard, D., and Courtoy, P.,** Acidification and bone resorption: immunocytochemical localization of a lysosomal membrane protein at the ruffled border of osteoclasts, *Calcif. Tissue Int.*, 36, 451, 1984.
44. **Fallon, M. D.,** Alterations in the pH of osteoclast resorbing fluid reflects changes in bone degradative activity, *Calcif. Tissue Int.*, 36, 458, 1984.
45. **Walsh, C. A., Dawson, W. E., Birch, M. A., and Gallagher, J. A.,** The effects of extracellular pH on bone resorption by avian osteoclasts *in vitro, J. Bone Miner. Res.*, 5, 1273, 1990.
46. **Akisaka, T. and Gay, G. V.,** Ultracytochemical evidence for a proton-pump adenosine triphosphate in chick osteoclasts, *Cell Tissue Res.*, 245, 507, 1986.
47. **Baron, R., Neff, L., Louvard, D., and Courtoy, P. J.,** Cell-mediated extracellular acidification and bone resorption: evidence for a low pH in resorbing lacunae and localization of a 100-kD lysosomal membrane protein of the osteoclast ruffled border, *J. Cell Biol.*, 101, 2210, 1985.
48. **Väänänen, H. K., Karhukorpi, E. K., Sundquist, K., Wallmark, B., Roininen, I., Hentunen, T., Tuukkanen, J., and Lakkakorpi, P.,** Evidence for the presence of a proton pump of the vacuolar H^+-ATPase type in the ruffled borders of osteoclasts, *J. Cell Biol.*, 111, 1305, 1990.
49. **Nijweide, P. J., Vrijheid-Lammers, T., Mulder, R. J. P., and Blok, J.,** Cell surface antigens on osteoclasts and related cells in the quail studied with monoclonal antibodies, *Histochemistry*, 83, 315, 1985.
50. **Oursler, M. J., Bell, L. V., Clevinger, B., and Osdoby, P.,** Identification of osteoclast-specific monoclonal antibodies, *J. Cell Biol.*, 100, 1592, 1985.
51. **Osdoby, P., Oursler, M. J., and Anderson, F.,** The osteoblast and osteoclast cytodifferentiation, in *Progress in Developmental Biology*, Part B, Alan R. Liss, New York, 1986, 409.
52. **Oursler, M. J. and Osdoby, P.,** Osteoclast development in marrow cultured in calvaria-conditioned media, *Dev. Biol.*, 127, 170, 1988.
53. **Horton, M. A., Lewis, D., McNulty, K., Pringle, J. A. S., and Chambers, T. J.,** Monoclonal antibodies to osteoblastomas (giant cell bone tumors): definition of osteoclast specific cellular antigens, *Cancer Res.*, 45, 5663, 1985.
54. **Horton, M. A. and Chambers, T. J.,** Human osteoclast-specific antigens are expressed by osteoclasts in a wide range of non-human species, *Br. J. Exp. Pathol.*, 67, 95, 1986.

55. **Athanasou, N. A., Quinn, J., Horton, M. A., and McGee, J. O. D.,** New sites of cellular vitronectin receptor immunoreactivity detected with osteoclast-reacting monoclonal antibodies 13 C2 and 23 C6, *Bone Miner.*, 8, 7, 1990.
56. **Chilosi, M., Gilioli, E., Lestani, M., Menestrina, F., and Fiore-Donati, L.,** Immunohistochemical characterization of osteoclasts and osteoclast-like cells with monoclonal antibody MB1 on paraffin-embedded tissues, *J. Pathol.*, 156, 251, 1988.
57. **Kukita, T. and Roodman, G. D.,** Development of a monoclonal antibody to osteoclasts formed *in vitro* which recognizes mononuclear osteoclast precursors in the marrow, *Endocrinology*, 125, 630, 1989.
58. **Athanasou, N. A., Quinn, J., and McGee, J. O. D.,** Leucocyte common antigen is present on osteoclasts, *J. Pathol.*, 153, 121, 1987.
59. **Athanasou, N. A., Quinn, J., Gatter, K. C., Mason, D. Y., and McGee, J. O. D.,** Osteoclasts contain macrophage and megakaryocyte antigens, *J. Pathol.*, 150, 239, 1986.
60. **Athanasou, N. A. and Quinn, J.,** Immunophenotypic differences between osteoclasts and macrophage polykaryons: immunohistological distinction and implications for osteoclast ontogeny and function, *J. Clin. Pathol.*, 43, 997, 1990.
61. **Chambers, T. J. and Magnus, C. J.,** Calcitonin alters behavior of isolated osteoclasts, *J. Pathol.*, 136, 27, 1982.
62. **Arnett, T. R. and Dempster, D. W.,** A comparative study of disaggregated chick and rat osteoclasts *in vitro:* effects of calcitonin and prostaglandins, *Endocrinology,* 120, 602, 1987.
63. **Nicholson, G. C., Moseley, J. M., Sexton, P. M., Mendelsohn, F. A. O., and Martin, T. J.,** Abundant calcitonin receptors in isolated rat osteoclasts, *J. Clin. Invest.*, 78, 355, 1986.
64. **Hattersly, G. and Chambers, T. J.,** Calcitonin receptors as markers for osteoclastic differentiation: correlation between generation of bone-resorptive cells and cells that express calcitonin receptors in mouse bone marrow cultures, *Endocrinology,* 125, 1606, 1989.
65. **Bouilar, Z., Rostene, W. H., and Milhaud, C.,** Down-regulation of rat kidney calcitonin receptors by salmon calcitonin infusion evidenced by autoradiography, *Proc. Natl. Acad. Sci. U.S.A.*, 84, 5125, 1987.
66. **Athanasou, N. A., Wells, C. A., Quinn, J., Ferguson, D. P., Heryet, A., and McGee, J. O. D.,** The origin and nature of stromal osteoclast-like multinucleated giant cells in breast carcinoma: implications for tumor osteolysis and macrophage biology, *Br. J. Cancer,* 59, 491, 1989.
67. **Ko, J. S. and Bernard, G. W.,** Osteoclast formation *in vitro* from bone marrow mononuclear cells in osteoclast free bone, *Am. J. Anat.*, 161, 415, 1981.
68. **Braidman, I. P., Rothwell, C., Webber, D. M., Crowe, P., and Anderson, D. C.,** Location of osteoclast precursors in fetal calvaria cultured on collagen gels, *J. Bone Miner. Res.*, 5, 287, 1990.
69. **Burger, E. H., van der Meer, J. W. M., van de Gevel, J. S., Gribnau, J. C., Thesingh, C. W., and van Furth, R.,** *In vitro* formation of osteoclasts from long term culture of bone marrow mononuclear phagocytes, *J. Exp. Med.*, 156, 1604, 1982.
70. **Thesingh, C. W. and Burger, E. H.,** The role of mesenchyme in embryonic long bones as early deposition site for osteoclast progenitor cells, *Dev. Biol.*, 95, 429, 1983.
71. **Burger, E. H., van der Meer, J. W. M., and Nijweide, P. J.,** Osteoclast formation from mononuclear phagocytes: role of bone forming cells, *J. Cell Biol.*, 99, 1901, 1984.
72. **Hagenaars, C. E., van der Kraan, A. A. M., Kawilarang-de Haas, E. W. M., Spooncer, E., Dexter, T. M., and Nijweide, P. J.,** Interleukine-3 dependent hemopoietic stem cell lines are capable of osteoclast formation *in vitro:* a model system for the study on osteoclast formation, in *Calcium Regulation and Bone Metabolism,* Cohn, D. V., Glorieux, F. H., and Martin, T. J., Eds., Elsevier, Amsterdam, 1990, 280.
73. **Chambers, T. J., Revell, P. A., Fuller, K., and Athanasou, N. A.,** Resorption of bone by isolated rabbit osteoclasts, *J. Cell Sci.*, 66, 383, 1984.

74. **Boyde, A., Ali, N. N., and Jones, S.,** Resorption of dentine by isolated osteoclasts *in vitro, Br. Dent. J.*, 156, 216, 1984.
75. **Kanehisa, J. and Heersche, J. N. M.,** Osteoclastic bone resorption: *in vitro* analysis of the rate of resorption and migration of individual osteoclasts, *Bone,* 9, 73, 1988.
76. **de Vernejoul, M. C., Horowitz, M., Demignon, J., Neff, L., and Baron, R.,** Bone resorption by isolated chick osteoclasts in culture is stimulated by murine spleen cell supernatant fluids (osteoclast activating factor) and inhibited by calcitonin and prostaglandin E_2, *J. Bone Miner. Res.*, 3, 69, 1988.
77. **Garrett, I. R., Boyce, B. F., and Mundy, G. R.,** Osteoclastic bone resorption *in vitro* is dependent on oxygen-derived radical production, *J. Bone Miner. Res.*, 2 (Suppl. 1), 1A, 1987.
78. **Fallon, M. D., Silverton, S. F., Smith, P., Moskal, T., Constantinescu, C., Feldman, R., Golub, E., and Shapiro, I.,** The oxidative metabolism of isolated osteoclasts is regulated by calcitropic agents, *J. Bone Miner. Res.*, 1 (Suppl. 1), 1A, 1986.
79. **Key, L. L., Jr., Ries, W. L., Taylor, R. G., Hays, B. D., and Pitzer, B. L.,** Oxygen derived free radicals in osteoclasts: the specificity and localization of nitroblue tetrazolium reaction, *Bone,* 11, 115, 1990.
80. **Marks, S. C.,** Congenital osteopetrotic mutations as probes of the origin, structure and function of osteoclasts, *Clin. Orthop. Relat. Res.*, 189, 239, 1984.
81. **Testa, N. G., Allen, T. D., Lajtha, L. G., Onions, D., and Jarret, O.,** Generation of osteoclasts *in vitro, J. Cell Sci.*, 47, 127, 1981.
82. **Allen, T. D., Testa, N. G., Suda, T., Schor, S. L., Onions, D., Jarret, O., and Boyde, A.,** The production of putative osteoclasts in tissue culture-ultrastructure, formation and behavior, *Scanning Electron Microsc.*, 111, 347, 1981.
83. **Ibbotson, K. J., Roodman, G. D., McManus, L. M., and Mundy, G. R.,** Identification and characterization of osteoclast-like cells and their progenitors in cultures of feline marrow mononuclear cells, *J. Cell Biol.*, 99, 471, 1984.
84. **Roodman, G. D., Ibbotson, K. J., MacDonald, B. R., Kuehl, T. J., and Mundy, G. R.,** 1,25-dihydroxyvitamin D_3 causes formation of multinucleated cells with several osteoclastic characteristic in cultures of primate marrow, *Proc. Natl. Acad. Sci. U.S.A.*, 82, 8213, 1985.
85. **MacDonald, B. R., Mundy, G. R., Clark, S., Wang, E. A., Kuehl, T. J., Stanley, E. R., and Roodman, G. D.,** Effects of human recombinant CSF-GM and highly purified CSF-1 on the formation of multinucleated cells with osteoclast characteristics in long term bone marrow cultures, *J. Bone Miner Res.*, 1, 227, 1986.
86. **Takahashi, N., MacDonald, B. R., Hon, J., Winkler, M. E., Derynck, B., Mundy, G. R., and Roodman, G. D.,** Recombinant human transforming growth factor-alpha stimulates the formation of osteoclast-like cells in long term human marrow cultures, *J. Clin. Invest.*, 78, 894, 1986.
87. **Takahashi, N., Kukita, T., MacDonald, B. R., Bird, A., Mundy, G. R., McManus, L. M., Milles, M., Boyde, A., Jones, S. J., and Roodman, G. D.,** Osteoclast-like cells form in long-term human bone marrow but not in peripheral blood cultures, *J. Clin. Invest.*, 83, 543, 1989.
88. **Takahashi, N., Yamana, H., Yoshiki, S., Roodman, G. D., Mundy, G. R., Jones, S. J., Boyde, A., and Suda, T.,** Osteoclast-like cell formation and its regulation by osteotropic hormones in mouse bone marrow cultures, *Endocrinology,* 122, 1373, 1988.
89. **Shinar, D. M., Sato, M., and Rodan, G. A.,** The effect of hemopoietic growth factors on the generation of osteoclast-like cells in mouse bone marrow cultures, *Endocrinology,* 126, 1728, 1990.
90. **Fuller, K. and Chambers, T. J.,** Generation of osteoclasts in cultures of rabbit bone marrow and spleen cells, *J. Cell Physiol.*, 132, 441, 1987.
91. **Scheven, B. A. A., Visser, J. W. M., and Nijweide, P. J.,** *In vitro* osteoclast generation from different bone marrow fractions, including a highly enriched haematopoietic stem cell population, *Nature (London),* 321, 79, 1986.

92. **Schneider, G. B., Relfson, M., and Nicolas, J.,** Pluripotent hemopoietic stem cells give rise to osteoclasts, *Am. J. Anat.,* 177, 505, 1986.
93. **Schneider, G. B. and Relfson, M.,** Pluripotent hemopoietic stem cells give rise to osteoclasts *in vitro:* effects of rGM-CSF, *Bone Miner.,* 5, 129, 1989.
94. **Hagenaars, C. E., van der Kraan, A. A. M., Kawilarang-de Haas, E. W. M., Visser, J. W. M., and Nijweide, P. J.,** Osteoclast formation from cloned pluripotent hemopoietic stem cells, *Bone Miner.,* 6, 179, 1989.
95. **Spooncer, E., Heyworth, C. M., Dunn, A., and Dexter, T. M.,** Self-renewal and differentiation of interleukin-3-dependent multipotent stem cells are modulated by stromal cells and serum factors, *Differentiation,* 31, 111, 1986.
96. **Hattersley, G. and Chambers, T. J.,** Generation of osteoclasts from hemopoietic cells and a multipotential cell line *in vitro, J. Cell Physiol.,* 140, 478, 1989.
97. **Hagenaars, C. E., Kawilarang-de Haas, E. W. M., van de Kraan, A. A. M., Spooncer, E., Dexter, T. M., and Nijweide, P. J.,** Interleukin-3 dependent hemopoietic stem cell lines are capable of osteoclast formation *in vitro, J. Bone Miner. Res.,* 6, 947, 1991.
98. **Kurihara, N., Suda, T., Miura, Y., Nakauchi, H., Kodama, H., Hiura, K., Hakeda, Y., and Kumegawa, M.,** Generation of osteoclasts from isolated hematopoietic progenitor cells, *Blood,* 74, 1295, 1989.
99. **Nijweide, P. J., Hagenaars, C. E., Modderman, W. E., and Mulder, R. J. P.,** The cell lineage of the osteoclast, in *Calcium Regulation and Bone Metabolism,* Cohn, D. V., Glorieux, F. H., and Martin, T. J., Eds., Elsevier, Amsterdam, 1990, 415.
100. **Metcalf, D.,** *The Haemopoietic Colony Stimulating Factors,* Elsevier, Amsterdam, 1984.
101. **Loutit, J. F. and Nisbet, N. W.,** The origin of the osteoclast,, *Immunobiology,* 161, 193, 1982.
102. **Flanagan, A. M.,** The osteoclast, which derives from a haemopoietic stem cell, is not depleted in aplastic anaemia, *J. Pathol.,* 164, 261, 1990.
103. **Schneider, G. B. and Relfson, M.,** A bone marrow fraction enriched for granulocyte-macrophage progenitors gives rise to osteoclasts *in vitro, Bone,* 9, 303, 1988.
104. **Hattersley, G. and Chambers, T. J.,** Effects of interleukin 3 and of granulocyte-macrophage and macrophage colony stimulating factors on osteoclast differentiation from mouse hemopoietic tissue, *J. Cell. Physiol.,* 142, 201, 1990.
105. **Shinar, D. M., Sato, M., and Rodan, G. A.,** The effect of hemopoietic growth factors on the generation of osteoclast-like cells in mouse bone marrow cultures, *Endocrinology,* 126, 1728, 1990.
106. **Lorenzo, J. A., Sousa, S. L., Fonseca, J. M., Hock, J. M., and Medlock, E. S.,** Colony-stimulating factors regulate the development of multinucleated osteoclasts from recently replicated cells *in vitro, J. Clin. Invest.,* 80, 160, 1987.
107. **Hattersley, G., Kerby, J. A., and Chambers, T. J.,** Identification of osteoclast precursors in multilineage hemopoietic colonies, *Endocrinology,* 128, 259, 1991.
108. **Kurihara, N., Chenu, C., Miller, M., Civin, C., and Roodman, G. D.,** Identification of committed mononuclear precursors for osteoclast-like cells formed in long term human marrow cultures, *Endocrinology,* 126, 2733, 1990.
109. **Kurihara, N., Gluck, S., and Roodman, G. D.,** Sequential expression of phenotype markers for osteoclasts during differentiation of precursors for multinucleated cells formed in long term human marrow cultures, *Endocrinology,* 127, 3215, 1990.
110. **Schneider, G. B. and Relfson, M.,** The effects of transplantation of granulocyte-macrophage-progenitors on bone resorption in osteopetrotic rats, *J. Bone Miner. Res.,* 3, 225, 1988.
111. **Goldschneider, I., Metcalf, D., Battye, F., and Mandel, T.,** Analysis of rat hemopoietic cells on the fluorescence-activated cell sorter. I. Isolation of pluripotent hemopoietic stem cells and granulocyte-macrophage progenitor cells, *J. Exp. Med.,* 152, 419, 1980.
112. **van de Wijngaert, F. P., Tas, M. C., van der Meer, J. W. M., and Burger, E. H.,** Growth of osteoclast precursor-like cells from whole mouse bone marrow: inhibitory effect of CSF-1, *Bone Miner.,* 3, 97, 1987.

113. **Felix, R., Cecchini, M. G., and Fleisch, H.,** Macrophage colony stimulating factor restores *in vivo* bone resorption in the op/op osteopetrotic mouse, *Endocrinology,* 127, 2592, 1990.
114. **Kodama, H., Yamasaki, A., Nose, M., Niida, S., Ohgame, Y., Abe, M., Kumegawa, M., and Suda, T.,** Congenital osteoclast deficiency in osteopetrotic (op/op) mice is cured by injections of macrophage colony-stimulating factor, *J. Exp. Med.,* 173, 269, 1991.
115. **Wiktor-Jedrzejczak, W., Ahmed, A., Szczylik, C., and Skelly, P. R.,** Hematological characterization of congenital osteoporosis in op/op mouse. Possible mechanism for abnormal macrophage differentiation, *J. Exp. Med.,* 156, 1516, 1982.
116. **Felix, R., Cecchini, M. G., Hofstetter, W., Elford, P. R., Stutzer, A., and Fleisch, H.,** Impairment of macrophage colony-stimulating factor production and lack of resident bone marrow macrophages in the osteopetrotic op/op mouse, *J. Bone Miner. Res.,* 5, 781, 1990.
117. **Yoshida, H., Hayashi, S. I., Kunisada, T., Ogawa, M., Nishikawa, S., Okamura, H., Suda, T., Schultz, L. D., and Nishikawa, S. I.,** The murine mutation osteopetrosis is in the coding region of the macrophage colony stimulating factor gene, *Nature (London),* 345, 442, 1990.
118. **Wiktor-Jedrzejczak, W., Bartocci, A., Ferrante, A. W., Ahmed-Ansari, A., Sell, K. W., Pollard, J. W., and Stanley, E. R.,** Total absence of colony-stimulating factor 1 in the macrophage-deficient osteopetrotic (op/op) mouse, *Proc. Natl. Acad. Sci. U.S.A.,* 87, 4828, 1990.
119. **Marks, S. C., Seifert, M. F., and McGuire, J. L.,** Congenitally osteopetrotic (op/op) mice are not cured by transplants of spleen or bone marrow cells from normal littermates, *Metab. Bone Dis. Relat. Res.,* 5, 183, 1984.
120. **Takahashi, N., Udagawa, N., Akatsu, T., Tanaka, H., Isogai, Y., and Suda, T.,** Deficiency of osteoclasts in osteopetrotic mice is due to a defect in the local microenvironment provided by osteoblastic cells, *Endocrinology,* 128, 1792, 1991.
121. **Bartelmez, S. H., Bradley, T. R., Bertoncello, I., Mochizuki, D. Y., Tushinski, R. J., Stanley, E. R., Hapel, A. J., Young, I. G., Kriegler, A. B., and Hodgson, G. S.,** Interleukin-1 plus interleukin-3 plus colony stimulating factor 1 are essential for clonal proliferation of primitive myeloid bone marrow cells, *Exp. Hematol.,* 17, 240, 1989.
122. **Williams, D. E., Straneva, J. E., Cooper, S., Shadduck, R. K., Waheed, A., Gillis, S., Urdal, D., and Broxmeyer, H. E.,** Interactions between purified murine colony-stimulating factors (natural CSF-1, recombinant GM-CSF, and recombinant IL-3) on the *in vitro* proliferation of purified murine granulocyte-macrophage progenitor cells, *Exp. Hematol.,* 15, 1007, 1987.
123. **Udagawa, N., Takahashi, N., Akatsu, T., Tanaka, H., Sasaki, T., Nishihara, T., Koga, T., Martin, T. J., and Suda, T.,** Origin of osteoclasts: mature monocytes and macrophages are capable of differentiating into osteoclasts under a suitable microenvironment prepared by bone marrow-derived stromal cells, *Proc. Natl. Acad. Sci. U.S.A.,* 87, 7260, 1990.
124. **Ogawa, M., Nishikawa, S., Ikuta, K., Yamamura, F., Naito, M., Takahashi, K., and Nishikawa, S.,** B cell ontogeny in murine embryo studied by a culture system with the monolayer of a stromal cell clone, ST2: B cell progenitor develops first in the embryonal body rather than in the yolk sac, *EMBO J.,* 7, 1337, 1988.
125. **Fuller, K. and Chambers, T. J.,** Generation of osteoclasts in cultures of rabbit bone marrow and spleen cells, *J. Cell. Physiol.,* 132, 441, 1987.
126. **Fuller, K. and Chambers, T. J.,** Bone matrix stimulates osteoclastic differentiation in cultures of rabbit bone marrow cells, *J. Bone Miner. Res.,* 4, 79, 1989.
127. **Krukowski, M. and Kahn, A. J.,** Inductive specificity of mineralized bone matrix in ectopic osteoclast differentiation, *Calcif. Tissue Int.,* 34, 474, 1982.

128. **Glowacki, J., Jasty, M., and Goldring, S.,** Comparison of multinucleated cells elicited in rats by particulate bone, polyethylene, or poly methyl-methacrylate, *J. Bone Miner. Res.,* 1, 327, 1986.
129. **Löwik, C. W. G. M., van der Pluijm, G., van der Wee-Pals, L. J. A., Bloys van Treslong-de Groot, H., and Bijvoet, O. L. M.,** Migration and phenotypic transformation of osteoclast precursors into mature osteoclasts: the effects of a bisphosphonate, *J. Bone Miner. Res.,* 3, 185, 1988.
130. **Boonekamp, P. M., van der Wee-Pals, L. J. A., van Wijk-van Lennep, M. M. L., Thesingh, C. W., and Bijvoet, O. L. M.,** Two modes of action of bisphosphonates on osteoclastic resorption of mineralized matrix, *Bone Miner.,* 1, 27, 1986.
131. **Hanaoka, H.,** The origin of the osteoclast, *Clin. Orthop. Relat. Res.,* 145, 252, 1979.
132. **Yabe, H. and Hanaoka, H.,** Investigation of the origin of the osteoclast by use of transplantation on chick chorio-allantoic membrane, *Clin. Orthop. Relat. Res.,* 197, 255, 1985.
133. **Perussia, B., Dayton, E. T., Fanning, V., Thiagarajan, P., Hoxie, J., and Trinchieri, G.,** Immune interferon and leukocyte-conditioned medium induce normal and leukemic myeloid cells to differentiate along the monocytic pathway, *J. Exp. Med.,* 158, 2058, 1983.
134. **Sterry, W. and Czarnetzki, B. M.,** *In vitro* differentiation of rat peritoneal macrophages into mast cells: an enzyme cytochemical study, *Blood,* 44, 211, 1982.
135. **Leenen, P. J. M.,** Phenotypical Analysis of Murine Macrophage Differentiation, Ph.D. thesis, Erasmus University of Rotterdam, The Netherlands, 1989 (ISBN 90-9003179-0).

Chapter 5

REGULATION OF OSTEOCLAST DEVELOPMENT AND FUNCTION

Timothy J. Chambers

TABLE OF CONTENTS

I. INTRODUCTION

The morphogenesis and restructuring of bone depends upon the integrated activity of the cells that form bone (osteoblasts) and the cells that resorb bone (osteoclasts). Osteoclasts are essentially wandering cells, and the complex and dynamic patterns of resorption that enable morphogenesis and restructuring are likely to be achieved with the assistance of instructions from local mesenchymal cells. A major component of this control consists of the regulation of osteoclast formation from local precursors, and ultimately from hemopoietic stem cells. In this chapter, the mechanisms by which osteoclasts are formed are discussed, followed by a discussion of the mechanisms by which the resorptive activity of mature osteoclasts may be induced, localized, and modulated on bone surfaces.

II. REGULATION OF OSTEOCLAST FORMATION

The life span of osteoclasts *in vivo* appears to be up to 2 weeks, with a half-life of around 6 to 10 d.[1-3] Osteoclasts thus need to be continuously replenished for bone resorption to continue. Moreover, the rate of bone resorption *in vivo* correlates with the number of multinuclear cells present, e.g., increased numbers are seen in response to parathyroid hormone, and calcitonin administration causes a fall in osteoclast numbers (see Reference 4). The relationship between number and activity, and the short half-life imply that, as for other relatively short-lived hemopoietic progeny, regulation of formation represents an important mechanism by which osteoclastic function can be regulated.

A. ORIGIN OF OSTEOCLASTS

It has been established for 15 years by parabiosis experiments, quail-chick chimeras, and bone marrow transplantation experiments that osteoclasts can be supplied to bone via the circulatory system, from a precursor which derives ultimately from hemopoietic tissue (see References 5 to 7 for reviews). In view of the evidence that not only hemopoietic cells, but also elements of the stroma of hemopoietic tissue can be transplanted and can reach bone through the bloodstream, the osteoclasts may have derived from hemopoietic cells, cells of the bone marrow stroma, or from some other unidentified precursor that may be present in hemopoietic tissue (see References 7 and 8). Experiments in which limb rudiments were cocultured with hemopoietic cells showed that the cells that led to osteoclast formation in the rudiment were present in greatest number in the fraction that also contained the greatest number of hemopoietic stem cells.[9] That the hemopoietic stem cell was indeed responsible for osteoclast formation in the rudiment was demonstrated when osteoclast formation was induced by coculture of rudiments with a single hemopoietic stem cell.[10] The same conclusion was reached in experiments in which a hemopoietic cell line, and colonies derived from hemopoietic cells, were induced to form osteoclasts when incubated on bone marrow stromal cells.[11,12] Since osteoclast formation was induced in the hemopoietic cell line even when nonvital stromal cells were used, these experiments also demonstrated that stromal cells induce osteoclasts in hemopoietic cells, rather than the converse.

Circumstantial evidence and phenotypic similarities quickly established the mononuclear phagocyte lineage as the favorite candidate for osteoclast precursor.[13] However, comparison of the properties of osteoclasts with those of mononuclear phagocytes also reveals significant distinctions between the two cell types. Not only do osteoclasts lack many characteristics shared by all mononuclear phagocytes, but they also possess properties lacking in mononuclear phagocytes, most notably calcitonin receptors (CTRs) and the ability to excavate bone. Our conclusion in a recent review,[7] in which the evidence available at that time is more fully discussed, favored an origin for osteoclasts distinct from the mononuclear phagocyte series, on the basis that the phenotypic differences seemed sufficiently extensive that it would be inefficient for osteoclasts to develop from a cell that had already invested in a significant degree of differentiation down the mononuclear phagocyte lineage. Moreover, attempts to induce osteoclastic differentiation from macrophages had failed. In the present review, we will discuss only more recent data concerning the induction and lineage relationships of osteoclasts.

B. ROLE OF STROMAL CELLS IN OSTEOCLAST DIFFERENTIATION

Experimental systems have recently been developed in which osteoclasts can be induced to differentiate *in vitro* from hemopoietic tissue of rabbits[14] and mice.[15] Induction of osteoclastic differentiation in other species, particularly man, remains controversial since, in the absence of substantial excavation of bone in these cultures,[16] other, less specific markers of osteoclastic differentiation are difficult to interpret.

Osteoclastic differentiation from hemopoietic precursors is a characteristic of cultures of bone marrow but not hemopoietic spleen cells, and occurs if bone marrow stromal cells and hemopoietic cells are in contact, even if the stromal cells are devitalized.[11,14,15,17,18]

A similar contact-dependent interaction appears to occur during the formation of other types of hemopoietic cells in liquid culture.[19-21] For other lineages, colony-stimulating factors (CSFs) have been identified which are necessary for the survival, proliferation, and differentiation of hemopoietic cells (see Reference 22). CSFs have been difficult to identify in the fluid phase of liquid bone marrow cultures, despite hemopoiesis in such cultures, but at least some can be eluted from the bone marrow stromal cells.[23,24] Thus, the explanation for the proliferation of hemopoietic cells despite an apparent lack of the CSFs presumed to be necessary for the process may be that the cytokines are bound to the matrix/surface of the stromal cells that support hemopoiesis, where they are available, through contact interactions, to the hemopoietic cells. We have attempted to similarly extract the stromal osteoclast-forming activity (SOFA) from bone marrow stromal cells and stromal cell lines by elution or heparitinase treatment. Although CSFs for other lineages (including macrophages) can be demonstrated in the eluates, such fractions are unable to induce osteoclast formation in hemopoietic spleen cell cultures.[184] This suggests that SOFA may be more irreversibly bound to the stromal cells, perhaps via membrane insertion, analogous to the c-*kit* ligand (see References 25 to 28) that supports proliferation of hemopoietic stem cells.

The nature of the osteoclast-inducing stromal cell,that is present in bone marrow but not in spleen, is unknown. It has been suggested that it is osteoblastic,[15] but cell lines of known osteoblastic phenotype are unable to induce osteoclasts.[17,18] Moreover, the stromal cell lines we have developed that do induce osteoclast formation do not stimulate bone resorption when cocultured with mature osteoclasts in the presence of parathyroid hormone.[184] This suggests that osteoclast induction and osteoclast stimulation may be through different cells and/or different agents. The differentiation potential of at least one of the stromal cell lines that induces osteoclasts (ST-2 cells)[17] is known to include that of osteoblasts, but such cells can differentiate into other phenotypes, including cartilage, fat, and muscle,[29-31] and thus the phenotype of the ST-2 cells during osteoclast induction is uncertain.

C. ROLE OF 1,25(OH)$_2$ VITAMIN D$_3$ AND PROSTAGLANDINS IN OSTEOCLAST FORMATION

Prostaglandin (PG) production appears to be necessary for osteoclast formation. Osteoclast formation in liquid cultures of bone marrow cells has been shown to be stimulated by 1,25(OH)$_2$D$_3$,[14] and not only does 1,25(OH)$_2$D$_3$ stimulate PGE$_2$ production in hemopoietic cultures, but osteoclast-like cell formation in response to 1,25(OH)$_2$D$_3$ is inhibited by indomethacin.[32] PGE$_2$ itself induces osteoclast formation.[33,34] This dependence on PG for osteoclast formation is surprising: if the osteoclast-inductive stromal cell is of the osteoblastic lineage, this suggests that osteoclast formation occurs on or adjacent to bone surfaces, yet PGs are known to directly inhibit resorption by mature osteoclasts disaggregated from rat and rabbit[35-37] and also inhibit resorption for several hours in organ culture.[38,39] A role for PG in osteoclast formation is, however, consistent with evidence from longer-term organ culture experiments, in which PG production mediates increased bone resorption by EGF, PDGF, TGF-α and -β, TNF, and IL-1.[40-44]

We have found that although some osteoclast-inductive stromal cell types produce PGE$_2$ in response to 1,25(OH)$_2$D$_3$, others do not, either alone or in combination with hemopoietic cells. On the other hand, both bone marrow and spleen cells produced amounts of PGE$_2$ in response to 1,25(OH)$_2$D$_3$ that were sufficient to account for osteoclast formation.[45]

We also found that stromal cells were able to support osteoclast formation from spleen cells in the presence of $1,25(OH)_2D_3$ or PGE_2 even if the stromal cells were devitalized. Incubation of stromal cells in these agents prior to devitalization did not avoid the requirement for the presence of PGE_2 or $1,25(OH)_2D_3$ during subsequent incubation with spleen cells. This suggests that induction of PGE_2 production in hemopoietic tissue by $1,25(OH)_2D_3$ causes the generation of precursors that can be induced to become osteoclasts on contact with osteoclast-inductive stromal cells.

An action by PGE_2 to alter hemopoietic precursors in such a way that they become competent to become osteoclastic in an appropriate microenvironment is consistent with the known ability of PGE_2 to influence macrophage colony formation in semisolid medium. Because the osteoclast-inductive cells are not required to produce, induce, or respond to PGs, the PGE_2-dependent component of osteoclast formation may normally occur, in intact tissue, in a separate microenvironment from that in which resorptive function is induced, by a process that does not depend on the local presence of PGE_2. *In vivo,* hematogenous cells competent to become osteoclasts only develop recognizable osteoclastic characteristics when they arrive in bone tissue.[5] Thus, it may be that PGE_2 is required, in the hemopoietic microenvironment, to provide precursors, which mature in the distinct microenvironment of bone surfaces by a PG-independent process. This model allows the possibility that on bone surfaces PGs are antiresorptive, in keeping with their ability to inhibit the function of mature osteoclasts[46] and induce bone formation,[47,48] while the action of PGs to generate osteoclast precursors in the hemopoietic microenvironment may account for the otherwise paradoxical PG-dependent stimulation of bone resorption in organ culture, in which both hemopoietic and resorptive microenvironments are present, by agents such as EGF and PDGF.

D. OSTEOCLAST LINEAGE ANALYSIS ON OSTEOCLAST-INDUCTIVE STROMAL CELLS

It was concluded in a previous section that while the mononuclear phagocyte system appears to be a likely candidate for provision of osteoclast precursors, osteoclasts and macrophages show distinctive phenotypes, and attempts to induce osteoclastic characteristics in mononuclear phagocytes had failed. The demonstration that osteoclast formation requires contact between precursors and osteoclast-inductive stromal cells raises the possibility that osteoclastic differentiation might require that mononuclear phagocytes be incubated on stromal cells.

This experiment has recently been performed.[49] Alveolar macrophages, peritoneal macrophages, and monocytes formed macrophage colonies, 80% of which were reported to contain CTR-positive cells, when incubated on ST-2-cell monolayers. Excavation of bone was demonstrated. Unpublished data were also cited in which macrophage colonies, induced in semisolid medium in M-CSF, similarly showed osteoclastic differentiation when incubated on ST-2 cells. We, however, have performed similar experiments (unpublished) using ST-2 cells, and also a bone marrow stromal cell line we have developed expressly for its capacity to induce osteoclasts (ts8). We have been unable to induce osteoclastic differentiation in alveolar or peritoneal macrophages, or in macrophage colonies formed in semisolid medium in the presence of M-CSF and transferred to stromal cell monolayers. We have also incubated single colony-forming cells (derived from blast cell colonies) on the stromal monolayers and found that osteoclast formation was not a feature of macrophage colonies, but of multilineage colonies.[50] This discrepancy is difficult to explain. Our results are consistent with those of Scheven et al.,[9] who found that osteoclast-forming cells copurified with hemopoietic stem cells, and with transplantation experiments in which macrophages and macrophage colony-forming cells were unable to cure the osteoclast deficiency of a form of osteopetrosis that can be cured by bone marrow transplantation.[51,52]

E. ROLE OF M-CSF IN OSTEOCLAST FORMATION

Recent data derived from experiments with the *op/op* (osteopetrotic) mouse model have established a role for M-CSF in osteoclast formation.

Osteopetrosis in the *op/op* mouse is due to an osteoclast deficiency[53] that cannot be cured by transplantation of normal bone marrow cells.[54,55] It has been known for some time that macrophage numbers are also low, probably due to deficient M-CSF production,[54] and it has recently been shown that this is absent[56,57] due to a mutation in the M-CSF gene in which a stop codon is inserted early in the coding region.[58] Osteoclast formation has been induced in *op/op* mice by M-CSF administration *in vivo*.[59,60]

The mononuclear phagocyte system, the main target cell system for M-CSF,[61] is also severely deficient in *op/op* mice, suggesting that this lineage may be directly or indirectly responsible for osteoclast formation. However, the relationship between the bone disorder, the macrophage deficiency, and M-CSF was not established from the *in vivo* work. In two animals, Kodama et al.[60] found cure of osteopetrosis without monocyte recovery. However, monocyte numbers may not be a reliable guide to macrophage production, since monocytes will be depleted from blood by deficient target tissues. The half-life of M-CSF is measured in hours *in vivo*,[62] but despite this, Felix et al.[59] found that very large quantities of M-CSF were required for cure *in vivo*. Wiktor-Jedrzejczak et al.[57] found that intraperitoneal implantation of L-cells in diffusion chambers induced many peritoneal macrophages and restored circulating levels of M-CSF to near normal, but did not cure the osteopetrosis. The data from *in vivo* experiments thus do not distinguish between a physiological role for M-CSF in the induction of osteoclasts and other possibilities, such as an action by M-CSF as a weak agonist for an alternative product of the M-CSF gene, which is known to undergo alternative splicing,[63] or the correction of some indirect disturbance cause by M-CSF deficiency, e.g., *op/*

op animals have increased levels of GM-CSF,[57] which suppresses osteoclast formation.[64]

We therefore assessed the ability of M-CSF to induce osteoclast and macrophage formation in *op/op* hemopoietic cells *in vitro*.[65] We found that M-CSF restored macrophage formation to cultures of *op/op* spleen cells. However, as in cultures of normal spleen cells, no osteoclasts were formed. If *op/op* spleen cells were cocultured with osteoclast-inductive (calvarial) stromal cells from *op/op* mice (as a source of SOFA), then again neither osteoclasts nor macrophages were detected, but both cell types were induced by the addition of M-CSF. M-CSF thus restored the behavior of these culture systems toward normal. However, osteoclast-induction occurred at concentrations approximately an order of magnitude higher than those that induced macrophages. Moreover, while macrophage numbers could be increased to higher-than-normal levels, osteoclast formation was always suboptimal, compared with parallel cultures from normal animals. Thus, although M-CSF was essential for osteoclast formation, it favored macrophage formation and did not restore the behavior of the cultures to normal.

The observation that tenfold higher concentrations of M-CSF were required to generate osteoclasts than to form macrophages, and the fact that osteoclast formation remained suboptimal compared to macrophage formation over the entire dose-response range (0.5 to 500 ng/ml), suggest that osteoclast-induction by M-CSF cannot be explained simply as the provision by M-CSF of mononuclear phagocytic precursors for osteoclast induction by stromal cells. This view is supported by the failure of either GM-CSF or IL-3 to induce osteoclast formation in the *op/op* system, a distinction more absolute than previously reported differences between the actions of M-CSF and other cytokines on the mononuclear phagocyte system (see, e.g., Reference 66).

The ratio of osteoclasts to macrophages was approximately 20-fold lower throughout the dose-response range in *op/op* cultures compared to similar cultures of normal cells incubated without exogenous M-CSF. Why was osteoclast formation so suboptimal in *op/op* cultures even with the addition of M-CSF? Our observation that M-CSF induces osteoclasts in *op/op* cocultures of calvaria with spleen cells but not in cultures of spleen cells, despite formation of similar numbers of macrophages in spleen cell cultures, shows that SOFA does not represent an alternative product of the M-CSF gene, or an alternative presentation of the same product (e.g., membrane bound): if it did, we would expect M-CSF to induce osteoclasts not only from cocultures, but also from spleen cells alone. Suboptimal osteoclast formation could be due to reduced SOFA in *op/op* cultures, but if SOFA is unrelated to M-CSF, there is no clear reason why this should be so. Last, since M-CSF exists as both soluble and membrane forms, the explanation may be that the membrane-bound form favors osteoclast formation in the presence of SOFA (although this form also induces macrophages in the absence of an additional, osteoclast-inductive signal[67]), while the soluble form favors macrophage formation

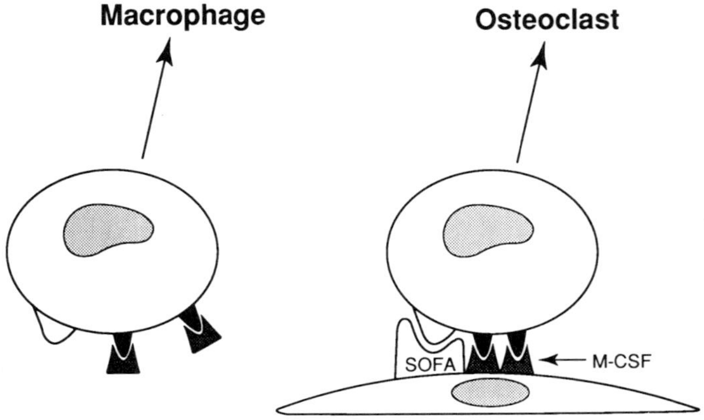

FIGURE 1. A model for osteoclast formation from hemopoietic precursors. Bone marrow stromal cells, through stromal osteoclast-forming activity (SOFA), induce osteoclastic differentiation in hemopoietic precursors through a contact-dependent interaction. Both SOFA and M-CSF are required for osteoclast formation. If M-CSF is presented as a soluble form, then, without SOFA, only macrophages are induced; and even with SOFA, the proportion of macrophages to osteoclasts is 20-fold greater than normal.[65] This suggests that M-CSF may normally be presented for osteoclast formation as a membrane-bound form, and that this form increases the likelihood (or rate, or sensitivity) of osteoclast formation. Without SOFA, both soluble and membrane-bound M-CSFs induce macrophage formation. Lineage analysis experiments suggest that hemopoietic precursors are only sensitive to osteoclast induction at a very early stage, from multilineage colonies. This is consistent with evidence from the *op/op* mouse, discussed in the text, that M-CSF does not enable osteoclast formation through the provision of precursors, a process that would imply an origin of ostoclasts from the mononuclear phagocyte series.

(but can nevertheless allow osteoclast formation in the presence of SOFA) (Figure 1).

This model might also explain why, while soluble M-CSF induces a dose-dependent increase in both macrophages and osteoclasts in *op/op* cultures, even small additions of M-CSF to cultures of normal hemopoietic cells increase macrophage numbers, but strongly suppress osteoclast formation:[64,68] addition of soluble M-CSF to normal cultures might alter the proportion of soluble vs. membrane-bound M-CSF in the cultures and divert shared precursors to macrophage formation. An alternative explanation might be that cultures of normal bone marrow contain preexisting macrophages, and these might, in response to M-CSF, produce an inhibitor of osteoclast differentiation. GM-CSF and γ-IFN, which also increase macrophage number and/or activity, similarly inhibit osteoclast formation.[64] We are currently investigating whether these three agents suppress osteoclast formation through a direct or indirect effect on osteoclast precursors.

III. REGULATION OF OSTEOCLAST FUNCTION

When osteoclast progenitors first arrive in bone during development, they are known to be still proliferative and highly radiosensitive. They are at this stage CTR- and tartrate-resistant acid phosphatase (TRAP)-negative, but differentiate into postmitotic TRAP-negative cells, which then become TRAP positive and, simultaneously or soon afterwards, CTR positive (see References 5 and 6 for reviews).[69] It is known that osteoclast can also be supplied from hematogenous precursors in later life if osteoclasts are abnormally deficient, as occurs in osteopetrosis or after irradiation. It is not known, however, whether the hematogenous supply provides osteoclasts in normal adult bone physiology. It may be that hemopoietic tissue supplies a radiosensitive stem cell to bone surface during development that becomes dedicated to osteoclast production and sustains osteoclast number independent of both hemopoiesis and continued vascular supply. Alternatively, there may be continued recruitment of osteoclastic precursors through the vascular system, as occurs in the embryo, throughout life. In either case, a population of osteoclast precursors is likely to exist on bone surfaces that is capable of proliferation, but which is unidentified since it consists of TRAP-negative, CTR-negative mononuclear cells: the earliest precursor recognizable by morphological criteria is already TRAP positive.[70] Thus, initiation of bone resorption is likely to involve both recruitment of osteoclastic precursors and induction of bone resorption in the mature cell so formed, on bone surfaces.

A. OSTEOCLAST RECRUITMENT

As discussed more fully elsewhere,[71] for the osteoclast, essentially a wandering cell derived from hematogenous precursors, to achieve the complex and dynamic patterns of localization that enable skeletal morphogenesis and restructuring suggests that its localization is assisted by resident bone cells (osteocytes, bone-lining cells, and other cells of the osteoblastic lineage). Thus, osteoclastic precursors, whether already present on bone surfaces or emigrating from the bloodstream, are likely to be guided to resorptive sites and induced to mature by osteoblastic cells. Constituents of bone matrix, such as osteocalcin or type 1 collagen peptides, are chemotactic for monocytes.[72-74] Osteocalcin-deficient bone particles implanted subcutaneously are also less efficient than normal bone particles in eliciting the recruitment of multinuclear cells that resorb these particles.[75] However, the osteoclastic nature of the multinucleate giant cells that surround these particles is controversial. Although some report osteoclastic characteristics,[76-79] these observations have not been repeatable by some groups of investigators.[76,80-83] (Also see Chapter 9 of this volume.)

B. INDUCTION OF BONE RESORPTION

Once induced to a site appropriate for resorption, osteoblastic cells (most likely bone-lining cells) may initiate resorptive behavior in osteoclasts through mineral exposure. Implantation of inorganic materials such as hydroxyapatite crystals into tissues evokes a macrophage foreign body response that results in sequestration or dissolution of the implanted material.[84] Presumably, therefore, bone has evolved some means of protecting bone mineral from such phagocytic attack. This protective mechanism is unlikely to be cellular, for when bone-lining cells die (as in osteonecrosis), bone does not evoke a foreign body response,[85,86] but may well be the unmineralized layer of organic material that is observed lining bone surfaces.[87-89] Then, when bone resorption is required, bone-lining cells may remove the organic protective material and induce exposure of mineral to contact with the specialized digestive cells of bone, the osteoclasts.[71]

We tested this model using devitalized slices of cortical bone. Because these slices are sections of bone, mineral is exposed on the cut surface. We found that such slices, lacking the unmineralized organic covering present on native surfaces, are spontaneously resorbed by osteoclasts *in vitro*. However, if bone mineral was removed, resorption did not occur.[90]

The model was also tested using uncut bone surfaces. Surface cells were removed from calvaria of young adult rats. In young adult rats, this bone surface is actively forming bone, so the surface consists of osteoid. Osteoclasts were then incubated on the endocranial surface. We found that when osteoclasts were sedimented onto such specimens, the bone was not resorbed, but was resorbed if the osteoid was first removed by collagenase.[91,92] That osteoclasts are capable of destroying both inorganic and organic components of bone[90] indicates that contact with bone mineral induces osteoclasts to resorptive activity, while unmineralized organic material does not. We also found, consistent with the model, that osteoblastic cells are able to remove the surface layer of unmineralized organic material *in vitro* in response to PTH and expose underlying bone mineral, and that bone so modified has an increased susceptibility to osteoclastic resorption, which is abrogated by demineralization.[91]

The model is supported by the known ability of osteoblastic cells to produce collagenase and tissue plasminogen activator *in vitro,* and secrete increased amounts of these neutral proteases and decreased amounts of protease inhibitor in the presence of PTH and other bone-resorbing hormones.[44,93-97] Collagenase production has not been detected in human bone cells *in vitro*.[98] However, human bone cell systems do not reproduce *in vivo* behavior *in vitro,* as judged by bone formation and osteoclast formation, as convincingly as rodent systems. Even so, human cells were shown to produce the neutral proteases gelatinase and tissue plasminogen activator *in vitro*.

Collagenase has also been identified in rodent bone-lining cells *in situ* in circumstances of increased bone resorption, and synthesis in bone is increased by exposure to resorptive stimuli.[99-101] This may represent a role as accessory cells in bone dissolution (viz., accessory cells remove collagen, osteoclasts digest mineral[99]). However, even if (as is likely) collagenase is absent from osteoclasts,[102,103] it is not required by them for collagen digestion, since osteoclasts resorb both organic and inorganic components of bone.[104] Acid protease (cathepsin) release correlates with ^{45}Ca release in organ culture, and cysteine protease (including cathepsin) inhibitors inhibit bone resorption.[105] Thus, osteoclasts are likely to digest organic matrix through acid hydrolase secretion and there seems little need to invoke an additional, collagenase-secreting cell as an accessory cell in bone dissolution.

Alternatively, osteoblastic neutral protease release may be a mechanism by which osteoclasts are induced to resorptive activity. According to this model, there may be distinct processes: osteoblastic neutral protease digestion of the organic layer, preceding and enabling acid hydrolytic resorption of bone. Reduction of the latter process, due to acetazolamide, CT, bisphosphonate, or in osteopetrosis, does not affect collagenase release and uncalcified collagen digestion in response to PTH,[100,101,106,107] but there is failure to degrade mineralized tissue. On the other hand, digestion of unmineralized collagen in bone organ cultures is inhibited by a synthetic collagenase inhibitor.[101,108] Thus, neutral protease secretion accounts for digestion of unmineralized collagen and occurs when osteoclasts are inactive, but cannot by itself account for bone resorption. Nevertheless, osteoclastic resorption seems dependent on this neutral protease secretion: collagenase and products of collagen digestion are detected before ^{45}Ca release;[99] inhibition of collagenase by anticollagenase antibody,[94] cartilage-derived anticollagenase factor,[109] or a synthetic collagenase inhibitor[108] inhibits PTH stimulation of ^{45}Ca release in organ culture. Inhibition is not complete, but increases with time;[109] the explanation may be that osteoclasts already in contact with mineral continue to resorb, but there is inhibition of the induction of osteoclasts to new resorptive sites in response to the hormone.

The model is based on the assumption that bone surfaces are lined by a layer of unmineralized material. This is indisputedly present, as osteoid, on bone-formative surfaces; the general consensus is that unmineralized material persists, after bone formation and mineralization has been completed, on resting surfaces.[87,89,110] However, this view has been challenged on the basis that resting bone surfaces appear similar whether or not they had been treated to remove organic material.[111,112] We therefore undertook a scanning electron microscopy (SEM) study to assess the prevalence of unmineralized organic material on the endosteal surface of adult human rib, in which 85% of the bone surface was resting by light microscopy.[113] Cells were removed from the bone surface and a total of 1200 SEM photographs taken. Organic material was then removed from the specimen and the same surfaces were reinspected.

In only three photographs could we identify areas in which the bone surface was not altered by removal of organic material from the specimen, a proportion consistent with the model, in which we would anticipate that such areas would occasionally exist. Moreover, it is known that superficial to the surface collagen fibers, which are easily recognizable after dehydration for SEM, is a layer of glycosaminoglycans (GAGs),[89,110,114] constituting the most superficial component of the endosteal membrane, immediately subjacent to the bone-lining cells.[115,116] GAGs decrease dramatically in size with dehydration, and have not been noted in (dehydrated) SEM observations of the bone surface. That collagen fibers are clearly seen on the surface despite submersion below these GAGs shows that the GAGs, at least on the bone surface, are not mineralized.

It is also noteworthy that resting bone surfaces show an electron-dense layer of organic material, the *lamina limitans,* a structure which is associated with cessation of mineralization.[117,118] This structure is not seen on mineralizing surfaces,[118] and may represent deposition of material that terminates calcification. Alternatively, the organic surface of resting bone may have a special constitution which is less prone to mineralization than is osteoid, and may be laid down by osteoblasts at the termination of bone formation. Such an alternative is consistent with the existence of the endosteal membrane that has been described, between the *lamina limitans* and cells on the bone surface, which differs ultrastructurally from osteoid.[115,116,119] Both possibilities suggest that the nature of the material secreted by osteoblasts may change immediately before bone formation ceases. It would clearly be of interest to identify the associated cell products, both for an understanding of the mechanism by which calcification is inhibited and as markers for the stimuli that determine cessation of bone formation by osteoblasts.

Because the composition of bone-forming surfaces differs from that of resting surfaces, the conclusion that osteoclasts do not resorb osteoid cannot necessarily be extended to resting surfaces. For example, it is conceivable that such surfaces may, either in general or in the particular areas appropriate for resorption, possess adhesion or other molecules that are not found on osteoid that induce bone-resorptive (acid hydrolytic) or neutral protease (e.g., tissue plasminogen activator, which has been shown to be present on osteoclasts[120]) activity in osteoclasts. Either of these may enable osteoclastic access to underlying mineral. Nevertheless, the hypothesis that neutral protease release by osteoblastic cells in response to bone-resorbing hormones occurs in order to localize and induce osteoclastic resorption remains consistent with the available data.

C. MODULATION OF RESORPTIVE ACTIVITY

Much of our current information on the regulation of bone resorption derives from studies using organ culture. However, while this is an effective way to identify the direct effect of a systemic hormone on bone, the many cellular interactions possible in organ culture make analysis of the mechanism of action very difficult. We therefore developed methods whereby osteoclasts could be disaggregated from other bone cells and dispersed at sufficiently low density that indirect actions are minimized. The hormonal responsiveness of such populations could then be compared in the presence and absence of other bone cells.

Calcitonin (CT) inhibits osteoclastic function through a direct effect on osteoclasts. In time-lapse video recordings, osteoclasts, widely dispersed from other cells, respond to the addition of CT with cessation of cytoplasmic motility within 1 min, followed by gradual retraction of pseudopods.[121] This effect is highly sensitive, prolonged, occurs at concentrations of CT as low as 3×10^{-12} *M*,[122] and is specific, with no apparent effect on macrophages or macrophage polykaryons. CT also inhibits bone resorption by isolated mammalian osteoclasts.[35,36] A direct effect on osteoclasts is consistent with the demonstration of CTR on osteoclasts.[123-125] These CT effects on osteoclasts, as in other target cells, are probably mediated by cyclic AMP, since CT causes a dose-dependent rise in cyclic AMP in osteoclasts.[126,127] Moreover, dibutryl cyclic AMP and forskolin, which increases cyclic AMP by stimulating adenyl cyclase, both induce changes in the motility, shape, TRAP secretion, and bone resorption of isolated osteoclasts identical to those induced by CT.[128-131] Increased intracellular calcium by CT may also be involved in signal transduction.[132] CT gene-related peptide and amylin have similar actions on osteoclasts at 1000 and 30 times higher concentrations, respectively,[133,134] probably as partial agonists of the CT receptor. It has been suggested that CT has two distinct effects on osteoclastic behavior (inhibition of motility and cell retraction) that may be mediated by different receptors.[132]

Chicken osteoclasts seem to differ from mammalian osteoclasts, with bone resorption not being affected by CT[36,129] and no CTRs demonstrable.[135] If, however, chickens are fed a low-calcium diet before osteoclast isolation, CT receptors and responsiveness, including inhibition of bone resorption, can be demonstrated, albeit at high CT concentrations.[127,136-138]

Parathyroid hormone (PTH) stimulates bone resorption in organ culture, but has no effect on isolated osteoclasts in any of the assays in which osteoclasts respond to CT. PTH has no effect on osteoclastic motility or cell spreading,[139] enzyme release,[131] cyclic AMP production,[126] or bone resorption.[35,137,140] This suggests that the preparations of osteoclasts are deficient in some cell type that mediates the PTH responsiveness of bone. This deficiency can be restored by incubation of osteoclasts with osteoblastic cell lines or primary cultures of osteoblastic cells.[140,141] These observations are consistent with the well-documented PTH receptors and responsiveness of osteoblastic cells (see References 71 and 142), and the failure in most,[143-145] but not all,[146] reports to detect receptors on osteoclasts.

When osteoblastic cells are incubated with bone-resorbing hormones, they may release, into culture supernatants, osteoclast resorption-stimulating activity (ORSA) which directly stimulates resorption by osteoclasts disaggregated from bone.[140,147-149] However, the activity is not always detectable in culture supernatants.[139,150,151] Furthermore, supernatants have to be used at concentrations of 50% or above, with further dilution leading to loss of activity.[184] The low levels of ORSA release suggested by these observations, especially when coupled with the tedious and poorly productive nature of the bone slice assay, have proved a major impediment to its identification.

Detection of hemopoietic cell growth factors by liquid cultures of bone marrow cells may represent an analogous situation. Under circumstances where supernatants of bone marrow cell cultures have no detectable growth factor activity, the cytokine can be shown to be bound to the GAGs of the extracellular matrix (ECM) produced in culture by the bone marrow stromal cells.[23,24] Only when stromal binding sites are saturated does cytokine become detectable in the culture supernatant.[152-154] GAGs have a high affinity for a number of other growth factors which possess a variety of biological actions.[155-158] The retention of ORSA by GAGs or other ECM components produced by osteoblasts[159,160] would explain why stimulation of bone resorption occurs in cocultures of osteoclasts and osteoblasts but is often not detected when isolated osteoclasts are incubated with supernatant medium from hormone-stimulated osteoblastic cultures.

Cytokines may be bound to the cells or matrix that line bone surfaces through lectin-like interactions (see Reference 161), or may be an integral cell membrane protein (see, e.g., References 162 to 166). These ionic, lectin-like, membrane-inserted interactions represent a hierarchy in which there is increasing dependence upon close contact for the interaction to occur, and an increasing potential for control over cytokine localization. For lectin-like binding, either a cytokine or an ECM binding site may be the lectin, and the degree of specificity in such interactions may enable a cell or group of cells that emits such a factor to establish a gradient of the factor on the surface of emiting and adjacent cells. The need for a binding partner would limit the gradient to those cells in a sheet that express the partner, with binding proportional to factor concentration and partner density. Such a system may enable complex domains and gradients of ORSA to be established on the bone surface, and would seem to be well suited to the regulation of osteoclastic migration and resorption.

An alternative approach to ORSA identification is to test agents known to stimulate bone resorption in organ culture for a direct effect on osteoclasts. This could be especially productive, since several such cytokines have been found to be produced by osteoblasts, and often in increased amounts in response to resorbing hormones: any agent that is produced by osteoblastic cells, in response to bone-resorbing hormones, that stimulates resorption in organ culture would be a candidate as an osteoblast-derived mediator of osteoclastic stimulation. There are many such agents, including IL-1,[167-169] TGF-β,[170-172] PDGF,[173] GM-CSF,[174,175] G-CSF,[175] M-CSF,[176,177] and DIF/LIF.[178,179] The very number of agents that fulfill the above criteria represents a caveat which extends to yet unidentified osteoblastic products which similarly cause bone resorption in organ culture.[180-183] The number of agents produced by osteoblastic cells to regulate osteoclasts is likely to be small.

Conversely, we have been surprised by the failure of the long list of cytokines we have tried to have any discernible effect on bone resorption by isolated osteoclasts. There was a theoretical basis for the suggestion that systemic hormones such as PTH were unlikely to act directly on osteoclasts,[71] but no basis for extending this prediction to agents that did not act systemically: on the contrary, a direct-acting, osteoblast-generated local messenger is part of the model. Many cytokines are multifunctional, produced by an acting on many types of cells, and owe specificity as much to localization of production as to structure. For some, at least, the osteoclast has been shown to lack receptors, but it is difficult to think of another cell type with a comparable lack of cytokine responsiveness, and surprising that so many hormone-responsive, osteoblast-derived cytokines that stimulate bone resorption in intact bone fail to stimulate isolated osteoclasts. Perhaps due to its role in the dissolution of substrate that accumulates so many cytokines. (see, e.g., Reference 156), by adsorption onto hydroxyapatite during bone formation, regulation is best effected through a membrane-bound or -inserted form of cytokine, a possibility also favored by the essentially contact-dependent mechanism by which PTH and other agents stimulate osteoclasts.

REFERENCES

1. **Jaworski, Z. F. G., Duck, B., and Sekaly, G.,** Kinetics of osteoclasts and their nuclei in evolving secondary Haversian systems, *J. Anat.,* 133, 397, 1981.
2. **Marks, S. C. J. and Schneider, G. B.,** Transformation of osteoclast phenotype in rats cured of congenital osteopetrosis, *J. Morphol.,* 174, 141, 1982.
3. **Loutit, J. F. and Townsend, K. M. S.,** Longevity of osteoclasts in radiation chimeras of osteopetrotic beige and normal mice, *Br. J. Exp. Pathol.,* 63, 221, 1982.
4. **Baron, R. and Vignery, A.,** Behavior of osteoclasts during a rapid change in their number induced by high doses of parathyroid hormone or calcitonin in intact rats, *Metab. Bone Dis. Relat. Res.,* 2, 339, 1981.
5. **Nijweide, P. J., Burger, E. H., and Feyen, J. H. M.,** Cells of bone: proliferation, differentiation, and hormonal regulation, *Physiol. Rev.,* 66, 855, 1986.
6. **Marks, S. C. and Popoff, S. N.,** Bone cell biology: the regulation of development, structure and function in the skeleton, *Am. J. Anat.,* 183, 1, 1988.
7. **Chambers, T. J.,** The origin of the osteoclast, in *Bone and Mineral Research Annual,* Vol. 6, Peck, W., Ed., Elsevier, Amsterdam, 1989, 1.
8. **Loutit, J. F. and Nisbet, N. W.,** The origin of the osteoclast, *Immunobiology,* 161, 193, 1982.
9. **Scheven, B. A. A., Visser, J. W. M., and Nijweide, P. J.,** *In vitro* osteoclast generation from different bone marrow fractions, including a highly enriched haemopoietic stem cell population, *Nature, (London),* 321, 79, 1986.
10. **Hagenaars, C. E., Van der Kraan, A. A. M., Kawilarang-de Haas, E. W. M., Visser, J. W. M., and Nijweide, P. J.,** Osteoclast formation from cloned pluripotent haemopoietic stem cells, *Bone Miner.,* 6, 187, 1989.
11. **Hattersley, G. and Chambers, T. J.,** Generation of osteoclasts from hemopoietic cells and a multipotential cell line *in vitro, J. Cell. Physiol.,* 140, 478, 1989.

12. **Hattersley, G., Kerby, J. A., and Chambers, T. J.,** Identification of osteoclast precursors in multilineage hemopoietic colonies, *Endocrinology,* 128, 259, 1991.
13. **Loutit, J. F. and Nisbet, N. W.,** Resorption of bone, *Lancet,* 2, 26, 1979.
14. **Fuller, K. and Chambers, T. J.,** Generation of osteoclasts in cultures of rabbit bone marrow and spleen cells, *J. Cell. Physiol.,* 132, 441, 1987.
15. **Takahashi, N., Yamama, H., Yoshiki, S., Roodman, D. G., Mundy, G. R., Jones, S. J., Boyde, A, and Suda, T.,** Osteoclast-like formation and its regulation by osteotropic hormones in mouse marrow cultures, *Endocrinology,* 122, 1373, 1988.
16. **Takahashi, N., Kukita, T., MacDonald, B. R., Bird, A., Mundy, G. R., McManus, L. M., Miller, M., Boyde, A., Jones, S. J., and Roodman, G. D.,** Osteoclast-like cells form in long-term human bone marrow but not in peripheral blood, *J. Clin. Invest.,* 84, 543, 1989.
17. **Udagawa, N., Takahashi, N., Akatsu, T., Sasaki, T., Yamaguchi, A., Kodama, H., Martin, T. J., and Suda, T.,** The bone marrow-derived stromal cell lines MC3T3-G2/PA6 and ST2 support osteoclast-like cell differentiation in co-cultures with mouse spleen cells, *Endocrinology,* 125, 1805, 1989.
18. **Hattersley, G. and Chambers, T. J.,** The role of bone marrow stroma in induction of osteoclastic differentiation, in *Calcium Regulation and Bone Metabolism: Basic and Clinical Aspects,* Vol. 10, Cohn, D. V., Glorieux, F. H., and Martin, T. J., Eds., Elsevier, Amsterdam, 1990, 439.
19. **Dexter, T. M., Allen, T. D., and Lajtha, L. G.,** Conditions controlling the proliferation of haemopoietic stem cells *in vitro, J. Cell Physiol.,* 91, 335, 1977.
20. **Roberts, R. A., Spooncer, E., Parkinson, E. K., Lord, B. I., Allen, T. D., and Dexter, T. M.,** Metabolically inactive 3T3 cells can substitute for marrow stromal cells to promote the proliferation and development of multipotent haemopoietic stem cells, *J. Cell. Physiol.,* 132, 203, 1987.
21. **Kodama, H., Izuka, M., Tomiyama, T., Yoshida, K., Seki, M., Suda, T., and Nishikawa, S.-I.,** Response of newly established mouse myeloid leukemic cell lines to MC3T3-G2/PA6 preadipocytes and hematopoietic factors, *Blood,* 1, 49, 1991.
22. **Metcalf, D.,** *The Molecular Control of Blood Cells,* Harvard University Press, Cambridge, 1988.
23. **Gordon, M. Y., Riley, G. P., Watt, S. M., and Greaves, M. F.,** Compartmentalization of a haemopoietic growth factor (GM-CSF) by glycosaminoglycans in the bone marrow microenvironment, *Nature (London),* 326, 403, 1987.
24. **Roberts, R., Gallagher, J., Spooner, E., Allen, T. D., Bloomfield, F., and Dexter, T. M.,** Heparan sulphate bound growth factors: a mechanism for stromal cell mediated haemopoiesis, *Nature (London),* 332, 376, 1988.
25. **Flanagan, J. G. and Leder, P.,** The *kit* ligand: a cell surface molecule altered in steel mutant fibroblasts, *Cell,* 63, 185, 1990.
26. **Zsebo, K. M., Wypych, J., McNiece, I. K., Lu, H. S., Smith, K. A., Karkare, S. B., Sachdev, R. K., Yuschenkoff, V. N., Birkett, N. C., Wiliams, L. R., Satyagai, V. N., Tung, W., Bosselman, R. A., Mendiaz, E. A., and Langley, K. E.,** Identification, purification and biological characterization of hematopoietic stem cell factor from buffalo rat liver-conditioned medium, *Cell,* 63, 195, 1990.
27. **Anderson, D. M., Lyman, S. D., Baird, A., Wignall, J. M., Eiseman, J., Rauch, C., March, C. J., Boswell, H. S., Gimpel, S. D., Cosman, D., and Williams, D. E.,** Molecular cloning of mast cell growth factor, a hematopoietin that is active in both membrane bound and soluble forms, *Cell,* 63, 235, 1990.
28. **Huang, E., Nocka, K., Beier, D. R., Chu, T.-Y., Buck, J., Lahm, H.-W., Wellner, D., Leder, P., and Besmer, P.,** The hematopoietic growth factor KL is encoded by the *SI* locus and is the ligand of the c-*kit* receptor, the gene product of the *W* locus, *Cell,* 63, 225, 1990.

29. **Owen, M.,** Lineage of osteogenic cells and their relationship to the stromal system, in *Bone and Mineral Research,* Vol. 3, Peck, W. A., Ed., Elsevier, Amsterdam, 1985, 1.
30. **Grigoriadis, A. E., Heersche, J. N. M., and Aubin, J. E.,** Differentiation of muscle, fat, cartilage, and bone from progenitor cells present in a bone-derived clonal cell population: effect of dexamethasone, *J. Cell Biol.,* 106, 2139, 1988.
31. **Ashton, B. A., Allen, T. D., Howlett, C. R., Eaglesom, C. C., Hattori, A., and Owen, M.,** Formation of bone and cartilage by marrow stromal cells in diffusion chambers *in vivo, Clin. Orthop.,* 151, 294, 1980.
32. **Shinar, D. M. and Rodan, G. A.,** Biphasic effects of transforming growth factor β on the production of osteoclast-like cells in mouse bone marrow cultures: the role of prostaglandins in the generation of these cells, *Endocrinology,* 126, 3153, 1990.
33. **Akatsu, T., Takahashi, N., Debari, K., Morita, I., Murota, S., Nagata, N., Takatani, O., and Suda, T.,** Prostaglandins promote osteoclast like cell formation by a mechanism involving cyclic adenosine 3′,5′-monophosphate in mouse bone marrow cell cultures, *J. Bone Miner. Res.,* 4, 29, 1989.
34. **Collins, D. A. and Chambers, T. J.,** Effect of prostaglandins E_1, E_2 and $F_{2\alpha}$ on osteoclast formation in mouse bone marrow cultures, *J. Bone Miner. Res.,* 6, 157, 1991.
35. **Chambers, T. J., McSheehy, P. M. J., Thomson, B. M., and Fuller, K.,** The effect of calcium-regulating hormones and prostaglandins on bone resorption by osteoclasts disaggregated from neonatal rabbit bones, *Endocrinology,* 116, 234, 1985.
36. **Arnett, T. R. and Dempster, D. W.,** A comparative study of disaggregated chick and rat osteoclasts *in vitro:* effects of calcitonin and prostaglandins, *Endocrinology,* 120, 602, 1987.
37. **Fuller, K. and Chambers, T. J.,** Effect of arachidonic acid metabolites on bone resorption by isolated rat osteoclasts, *J. Bone Miner. Res.,* 4, 209, 1989.
38. **Conaway, H. H., Diez, L. F., and Raisz, L. G.,** Effects of prostacyclin and prostaglandin E_1 (PGE_1) on bone resorption in the presence and absence of parathyroid hormone, *Calcif. Tissue Int.,* 38, 130, 1986.
39. **Lerner, U. H., Ransjo, M., and Ljunngren, O.,** Prostaglandin E_2 causes a transient inhibition of mineral mobilization, matrix degradation, and lysosomal enzyme release from mouse calvarial bones *in vitro, Calcif. Tissue Int.,* 40, 323, 1987.
40. **Tashjian, A. H., Jr. and Levine, L.,** Epidermal growth factor stimulates prostaglandin production and bone resorption in cultured mouse calvariae, *Biochem. Biophys. Res. Commun.,* 85, 966, 1978.
41. **Tashjian, A. H., Jr., Hohmann, E. L., Antoniades, H. N., and Levine, L.,** Platelet-derived growth factor stimulates bone resorption via a prostaglandin-mediated mechanism, *Endocrinology,* 111, 118, 1982.
42. **Tashjian, A. H., Jr., Voelkel, E. F., Lazzaro, M., Singer, F. R., Roberts, A. B., Derynck, R., Winkler, M. E., and Levine, L.,** α and β human transforming growth factors stimulate prostaglandin production and bone resorption in cultured mouse calvaria, *Proc. Natl. Acad. Sci. U.S.A.,* 82, 4535, 1985.
43. **Tashjian, A. H. Jr., Voelkel, E. F., Lazzaro, M., Goad, D., Bosma, T., and Levine, L.,** Tumor necrosis factor-α (cachectin) stimulates bone resorption in mouse calvaria via a prostaglandin mediated mechanism, *Endocrinology,* 120, 2029, 1987.
44. **Heath, J. K., Atkinson, S. J., Meikle, M. C., and Reynolds, J. J.,** Mouse osteoblasts synthesize collagenase in response to bone resorbing agents, *Biochim. Biophys. Acta,* 802, 151, 1984.
45. **Collins, D. A. and Chambers, T. J.,** Prostaglandin E_2 promotes osteoclast formation in murine hematopoietic cultures through an action on hematopoietic cells, *J. Bone Miner. Res.,* 7, 555, 1992.
46. **Chambers, T. J.,** The pathobiology of the osteoclast, *J. Clin. Pathol.,* 38, 241, 1985.
47. **Raisz, L. G. and Martin, T. J.,** Prostaglandins in bone and mineral metabolism, in *Bone and Mineral Research Annual,* Vol. 2, Peck, W. A., Ed., Elsevier, Amsterdam, 1984, 286.

48. **Norrdin, R. W., Jee, W. S. S., and High, W. B.,** Review: the role of prostaglandins in bone *in vivo, Prostagl. Leuk. Essent. Fatty Acids,* 41, 139, 1990.
49. **Udagawa, N., Takahashi, N., Akatsu, T., Tanaka, H., Sasaki, T., Nishihara, T., Koga, T., Martin, T. J., and Suda, T.,** Origin of osteoclasts: mature monocytes and macrophages are capable of differentiating into osteoclasts under a suitable microenvironment prepared by bone marrow-derived stromal cells, *Proc. Natl. Acad. Sci. U.S.A.,* 87, 7260, 1990.
50. **Kerby, J. A., Hattersley, G., Collins, D. A., and Chambers, T. J.,** Derivation of osteoclasts from hemopoietic colony-forming cells in culture, *J. Bone Miner. Res.,* 7, 353, 1992.
51. **Schneider, G. B. and Byrnes, J. E.,** The cellular specificity of the cure for neonatal osteopetrosis in the ia rat, *Exp. Cell Biol.,* 51, 44, 1983.
52. **Schneider, G. B. and Relfson, M.,** The effects of transplantation of granulocyte-macrophage progenitors on bone resorption in osteopetrotic rats, *J. Bone Miner. Res.,* 3, 255, 1988.
53. **Marks, S. C. and Lane, P. W.,** Osteopetrosis, a new recessive skeletal mutation on chromosome 12 of the mouse, *J. Hered.,* 67, 11, 1976.
54. **Wiktor-Jedrzejczak, W., Ahmed, A., Szczylik, C., and Skelly, R. R.,** Hematological characterization of congential osteopetrosis in *op/op* mouse, *J. Exp. Med.,* 156, 1516, 1982.
55. **Marks, S. C., Seifert, M. F., and McGuire, J. L.,** Congenitally osteopetrotic (*op/op*) mice are not cured by transplants of spleen or bone marrow cells from normal littermates, *Metab. Bone Dis. Relat. Res.,* 5, 183, 1984.
56. **Felix, R., Cecchini, M. G., Hofstetter, W., Elford, P. R., Stutzer, A., and Fleisch, H.,** Impairment of macrophage colony-stimulating factor production and lack of resident bone marrow macrophages in the osteopetrotic *op/op* mouse, *J. Bone Miner. Res.,* 5, 781, 1990.
57. **Wiktor-Jedrzejczak, W., Bartocci, A., Ferrante, A. W., Jr., Ahmed-Ansari, A., Sell, K. W., Pollard, J. W., and Stanley, E. R.,** Total absence of colony-stimulating factor 1 in the macrophage-deficient osteopetrotic (*op/op*) mouse, *Proc. Natl. Acad. Sci. U.S.A.,* 87, 4828, 1990.
58. **Yoshida, H., Hayashi, S.-I., Kunisada, T., Ogawa, M., Nishikawa, S., Okamura, H., Sudo, T., Shultz, L. D., and Nishikawa, S.-I.,** The murine mutation osteopetrosis is in the coding region of the macrophage colony stimulating factor gene, *Nature (London),* 345, 442, 1990.
59. **Felix, R., Cecchini, M. G., and Fleisch, H.,** Macrophage colony stimulating factor restores *in vivo* bone resorption in the *op/op* osteopetrotic mouse, *Endocrinology,* 127, 2592, 1990.
60. **Kodama, H., Yamasaki, A., Nose, M., Niida, S., Ohgame, Y., Abe, M., Kumegawa, M., and Suda, T.,** Congenital osteoclast deficiency in osteopetrotic (*op/op*) mice is cured by injections of macrophage colony-stimulating factor, *J. Exp. Med.,* 173, 269, 1991.
61. **Byrne, P. V., Guilbert, L. J., and Stanley, E. R.,** Distribution of cells bearing receptors for a colony-stimulating factor (CSF-1) in murine tissues, *J. Cell Biol.,* 91, 848, 1981.
62. **Stoudemire, J., Metzger, M., and Bree, T. G.,** Pharmacokinetics of recombinant human macrophage colony-stimulating factor (rhM-CSF) in primates and rodents, paper presented at Annu. Meet. American Society for Cancer Research, 1989, A2139.
63. **Ladner, M. B., Martin, G. A., Noble, J. A., Nikoloff, D. M., Tal, R., Kawasaki, E. S., and White, T. J.,** Human CSF-1: gene structure and alternative splicing of mRNA precursors, *EMBO J.,* 6, 2693, 1987.
64. **Hattersley, G. and Chambers, T. J.,** Effects of interleukin 3 and of granulocyte-macrophage and macrophage colony stimulating factors on osteoclast differentiation from mouse hemopoietic tissue, *J. Cell. Physiol.,* 142, 201, 1990.

65. **Hattersley, G., Owens, J., Flanagan, A. M., and Chambers, T. J.,** Macrophage colony stimulating factor (M-CSF) is essential for osteoclast formation *in vitro, Biochem. Biophys. Res., Commun., BBRC,* 177, 526, 1991.
66. **Witsell, A. L. and Schook, L. B.,** Macrophage heterogeneity occurs through a developmental mechanism, *Proc. Natl. Acad. Sci. U.S.A.,* 88, 1963, 1991.
67. **Stein, J., Borzillo, G. V., and Rettenmier, C. W.,** Direct stimulation of cells expressing receptors for macrophage colony-stimulating factor (CSF-1) by a plasma membrane-bound precursor of human CSF-1, *Blood,* 76, 1308, 1990.
68. **Shinar, D. M., Sato, M., and Rodan, G. A.,** The effect of hemopoietic growth factors on the generation of osteoclast-like cells in mouse bone marrow cultures, *Endocrinology,* 126, 1728, 1990.
69. **Taylor, L. M., Tertinegg, I., Okuda, A., and Heersche, J. N. M.,** Expression of calcitonin receptors during osteoclast differentiation in mouse metatarsals, *J. Bone Miner. Res.,* 4, 751, 1989.
70. **Scott, B. L.,** Thymidine-^{3}H electron microscope radioautography of osteogenic cells in the fetal rat, *J. Cell Biol.,* 35, 115, 1967.
71. **Chambers, T. J.,** The cellular basis of bone resorption, *Clin. Orthop. Relat. Res.,* 151, 283, 1980.
72. **Malone, J. D., Teitelbaum, S., Griffin, G. L., Senior, R. M., and Kahn, A. J.,** Recruitment of osteoclast precursors by purified bone matrix constituents, *J. Cell Biol.,* 92, 227, 1982.
73. **Mundy, G. R. and Poser, J. W.,** Chemotactic activity of the γ-carboxy-glutamic acid containing protein in bone, *Calcif. Tissue Int.,* 35, 164, 1983.
74. **Mundy, G. R., Varani, J., Orr, W., Gondek, M. D., and Ward, P. A.,** Resorbing bone in chemotactic for monocytes, *Nature (London),* 275, 132, 1978.
75. **Lian, J. B., Tissinari, M., and Glowacki, J.,** Resorption of implanted bone prepared from normal and warfarin-treated rats, *J. Clin. Invest.,* 73, 1223, 1984.
76. **Peignoux-Deville, J., Bordat, C., and Vidal, B.,** Demonstration of bone resorbing cells in elasmobranchs: comparison with osteoclasts, *Tissue Cell,* 21, 925, 1989.
77. **Glowacki, J. and Cox, K. A.,** Osteoclast features of cells that resorb bone implants in rats, *Calcif. Tissue Int.,* 39, 97, 1986.
78. **Glowacki, J., Jasty, M., and Goldring, S.,** Comparison of multinucleated cells elicited in rats by particulate bone, polyethylene, or polymethylmethacrylate, *J. Bone Miner. Res.,* 1, 327, 1986.
79. **Goldring, S. R., Roelke, M., and Glowacki, J.,** Multinucleated cells elicited in response to implatns of devitalized bone particles possess receptors for calcitonin, *J. Bone Miner. Res.,* 1, 117, 1988.
80. **Popoff, S. N. and Marks, S. C. J.,** Ultrastructure of the giant cell infiltrate of subcutaneously implanted bone particles in rats and mice, *Am. J. Anat.,* 177, 491, 1986.
81. **Walters, L. M. and Schneider, G. B.,** Cellular response to ectopically implanted silk sutures and osteopetrotic bone, *Cell Tissue Res.,* 248, 79, 1987.
82. **Walters, L. M. and Schneider, G. B.,** Tartrate-resistant acid phosphatase activity in tibial osteoclasts and cells elicited by ectopic bone and suture implants in normal and osteopetrotic rats, *Bone Miner.,* 4, 49, 1988.
83. **Bagi, C. M. and Miller, S. C.,** Osteoclast features of cells that resorb demineralized and mineral-containing bone implants in rats, *Scanning Microsc.,* 3, 963, 1989.
84. **Chambers, T. J.,** Multinucleate giant cells, *J. Pathol.,* 126, 125, 1978.
85. **Ham, A. W. and Harris, W. R.,** Repair and transplantation of bone, in *The Biochemistry and Physiology of Bone,* Vol. 1, Bourne, G. H., Ed., Academic Press, New York, 1982, 337.
86. **Glimcher, M. J. and Kenzora, J. E.,** The biology of osteonecrosis of the human femoral head and its clinical implications, *Clin. Orthop.,* 138, 284, 1979.
87. **Raina, V.,** Normal osteoid tissue, *J. Clin. Pathol.,* 25, 229, 1972.

88. **Fornasier, V. L.,** Transmission electron microscopy studies of osteoid maturation, *Metab. Bone Dis. Relat. Res.,* 25, 103, 1980.
89. **Vanderwiel, C. J.,** An ultrastructural study of the components which make up the resting surface of bone, *Metab. Bone Dis. Relat. Res.,* 2S, 109, 1980.
90. **Chambers, T. J., Thomson, B. M., and Fuller, K.,** Effect of substrate composition on bone resorption by rabbit osteoclasts, *J. Cell Sci.,* 70, 61, 1984.
91. **Chambers, T. J. and Fuller, K.,** Bone cells predispose bone surfaces to resorption by exposure of mineral to osteoclastic contact, *J. Cell Sci.,* 76, 155, 1985.
92. **Chambers, T. J., Darby, J. A., and Fuller, K.,** Mammalian collagenase predisposes bone surfaces to osteoclastic resorption, *Cell Tissue Res.,* 241, 671, 1985.
93. **Hamilton, J. A., Lingelback, S. R., Partridge, N. C., and Martin, T. J.,** Stimulation of plasminogen activator in osteoblast-like cells by bone-resorbing hormones, *Biochem. Biophys. Res. Commun.,* 122, 230, 1984.
94. **Sakamoto, S. and Sakamoto, M.,** Osteoblast collagenase: collagenase synthesis by clonally-derived mouse osteogenic cells, *Biochem. Int.,* 9, 51, 1984.
95. **Thomson, B. M., Atkinson, S. J., Reynolds, J. J., and Meickle, M. C.,** Degradation of type I collagen fibres by mouse osteoblasts is stimulated by 1,25 dihydroxyvitamin D_3 and inhibited by human recombinant TIMP, *Biochem. Biophys. Res. Commun.,* 148, 596, 1987.
96. **Shen, V., Kohler, G., Jeffrey, J. J., and Peck, W. A.,** Bone-resorbing agents promote and interferon-γ inhibits bone cell collagenase production, *J. Bone Miner. Res.,* 3, 657, 1988.
97. **Pfeilschifter, J., Erdmann, J., Schmidt, W., Naumann, A., Minne, H. W., and Ziegler, R.,** Differential regulation of plasminogen activator and plasminogen activator inhibitor by osteotropic factors in primary cultures of mature osteoblasts and osteoblast precursors, *Endocrinology,* 126, 703, 1990.
98. **Rifas, L., Halstead, L. R., Peck, W. A., Avioli, L. V., and Welgus, H. C.,** Human osteoblasts *in vitro* secrete tissue inhibitor of metalloproteinases and gelatinase but not interstitial collagenase as major cellular products, *J. Clin. Invest.,* 84, 686, 1989.
99. **Sakamoto, S. and Sakamoto, M.,** On the possibility that bone matrix collagen is removed prior to bone mineral during cell-mediated bone resorption, in *Current Advances in Skeletogenesis,* Ornoy, A., Harell, A., and Sela, J., Eds., Elsevier, Amsterdam, 1985, 65.
100. **Eeckhout, Y., Delaisse, J. M., and Vaes, G.,** Direct extraction and assay of bone tissue collagenase and its relation to parathyroid-hormone-induced bone resorption, *Biochem. J.,* 239, 793, 1986.
101. **Delaisse, J. M., Eeckhout, Y., and Vaes, L. G.,** Bone-resorbing agents affect the production and distribution of procollagenase as well as the activity of collagens in bone tissue, *Endocrionology,* 123, 264, 1988.
102. **Blair, H. C., Kahn, A. J., Crouch, E. C., Jeffrey, J. J., and Teitelbaum, S. L.,** Isolated osteoclasts resorb the organic and inorganic components of bone, *J. Cell Biol.,* 102, 1164, 1986.
103. **Sakamoto, S. and Sakamoto, M.,** Bone collagenase, osteoblasts, and cell mediated bone resorption, in *Bone and Mineral Research Annual,* Vol. 4, Peck, W. A., Ed., Elsevier, Amsterdam, 1986, 49.
104. **Chambers, T. J., Revell, P. A., Fuller, K., and Athanasou, N. A.,** Resorption of bone by isolated rabbit osteoclasts, *J. Cell Sci.,* 66, 383, 1984.
105. **Delaisse, J.-M., Eeckhout, Y., and Vaes, G.,** *In vivo* and *in vitro* evidence for the involvement of cysteine proteinases in bone resorption, *Biochem. Biophys. Res. Commun.,* 125, 441, 1984.
106. **Jilka, R. L. and Cohn, D. V.,** A collagenolytic response to parathormone, 1,25 dihydroxycholecalciferol D_3 and prostaglandin E_2 in bone of osteoporotic (mi/mi) mice, *Endocrinology,* 112, 945, 1983.

107. **Jilka, R. L. and Hamilton, J. W.,** Evidence for two pathways for stimulation of collagenolysis in bone, *Calcif. Tissue Int.*, 37, 300, 1985.
108. **Delaisse, J. M., Eeckhout, Y., Sear, C., Galloway, A., McCullagh, K.,, and Vaes, G.,** A new synthetic inhibitor of mammalian tissue collagenase inhibits bone resorption in culture, *Biochem. Biophys. Res. Commun.*, 133, 483, 1985.
109. **Horton, J. E., Wezeman, F. H., and Kuettner, K. E.,** Inhibition of bone resorption *in vitro* by a cartilage-derived anticollagenase factor, *Science,* 199, 1342, 1978.
110. **Fornasier, V. L.,** Osteoid: an ultrastructural study, *Hum. Pathol.*, 8, 243, 1977.
111. **Boyde, A. and Hobdell, M. H.,** Scanning electron microscopy of lamellar bone, *Z. Zellforsch.*, 93, 213, 1969.
112. **Boyde, A.,** Scanning electron microscopy studies of bone, in *The Biochemistry and Physiology of Bone,* Vol. 1, Bourne, G. H., Ed., Academic Press, New York, 1972, 259.
113. **Chow, J. and Chambers, T. J.,** An assessment of the prevalence of organic material on bone surfaces, *Calcif. Tissue Int.*, 50, 118, 1992.
114. **Miller, S. C., de Saint-George, L., Bowman, B. M., and Jee, W. S. S.,** Bone lining cells: structure and function, *Scanning Microsc.*, 3, 953, 1989.
115. **Miller, S. C. and Jee, W. S. S.,** The bone lining cell: a distinct phenotype?, *Calcif. Tissue Int.*, 41, 1, 1987.
116. **Miller, S. C., Bowman, B. M., Smith, J. M., and Jee, W. S. S.,** Characterization of endosteal bone-lining cells from fatty marrow bone sites in adult beagle, *Anat. Rec.*, 198, 163, 1980.
117. **Scherft, J. P.,** The lamina limitations of the organic matrix of calcified cartilage and bone, *J. Ultrastruct. Res.*, 38, 318, 1972.
118. **Luk, S. C., Nopajaroonsri, C., and Simon, G. T.,** The ultrastructure of endosteum: a topographic study in young adult rabbits, *J. Ultrastruct. Res.*, 46, 165, 1974.
119. **Parfitt, A. M.,** The cellular basis of bone remodeling: the quantum concept reexamined in light of recent advances in the cell biology of bone, *Calcif. Tissue Int.*, 36 (Suppl.), 37, 1984.
120. **Grills, B. L., Gallagher, J. A., Allan, E. H., Yumita, S., and Martin, T. J.,** Identification of plasminogen activator in osteoclasts, *J. Bone Miner. Res.*, 5, 499, 1990.
121. **Chambers, T. J. and Magnus, C. J.,** Calcitonin alters behavior of isolated osteoclasts, *J. Pathol.*, 136, 27, 1982.
122. **Chambers, T. J. and Moore, A.,** The sensitivity of isolated osteoclasts to morphological transformation by calcitonin, *J. Clin. Endocrinol. Metab.*, 57, 819, 1983.
123. **Rao, L. G., Heersche, J. N. M., Marchu, L. L., and Sturtridge, W.,** Immunohistochemical demonstration of calcitonin binding to specific cell types in fixed rat bone tissue, *Endocrinology,* 108, 1982, 1981.
124. **Warshawsky, J., Goltzman, D., Rouleau, M. F., and Bergeron, J. J. M.,** Direct *in vivo* demonstration by autoradiography of specific binding sites for calcitonin in skeletal and renal tissues of the rat, *J. Cell Biol.*, 85, 682, 1980.
125. **Nicholson, G. C., Moseley, J. M., Sexton, P. M., Mendelsohn, F. A. O., and Martin, T. J.,** Abundant calcitonin receptors in isolated rat osteoclasts, *J. Clin. Invest.*, 78, 355, 1986.
126. **Nicholson, G. C., Livesey, S. A., Moseley, J. M., and Martin, T. J.,** Actions of calcitonin, parathyroid hormone, and prostaglandin E_2 on cyclic AMP formation in chicken and rat osteoclasts, *J. Cell Biochem.*, 31, 229, 1986.
127. **Rifkin, B. R., Auszmann, J. M., Kleckner, A. P., Vernillo, A. T., and Fine, A. S.,** Calcitonin stimulates cAMP accumulation in chicken osteoclasts, *Life Sci.*, 42, 799, 1988.
128. **Nicholson, G. C., Yumita, S., Moseley, J. M., Yates, A. J. P., and Martin, T. J.,** Forskolin augments the effects of calcitonin on cytoplasmic spreading of isolated rat osteoclasts and plasma calcium levels in the rat, *J. Bone Miner. Res.*, 3, 181, 1988.

129. **Murrills, R. J. and Dempster, D. W.,** The effects of stimulators of intracellular cyclic AMP on rat and chick osteoclasts *in vitro:* validation of a simplified light microscope assay of bone resorption, *Bone,* 11, 333, 1990.
130. **Ransjö, M., Lerner, U. H., and Heersche, J. N. M.,** Calcitonin-like effects of forskolin and choleratoxin on surface area motility of isolated rabbit osteoclasts, *J. Bone Miner. Res.,* 3, 611, 1988.
131. **Chambers, T. J., Fuller, K., and Darby, J. A.,** Hormonal regulation of acid phosphatase release by osteoclasts disaggregated from neonatal rat bone, *J. Cell. Physiol.,* 132, 90, 1987.
132. **Zaidi, M., Datta, H. M., Moonga, B. S., and MacIntyre, I.,** Evidence that the action of calcitonin on the osteoclast is mediated by two G proteins acting via separate post-receptor pathways, *J. Endocrinol.,* 125, 473, 1990.
133. **Zaidi, M., Datta, H. K., Bevis, P. J. R., Wimalawansa, S. J., and MacIntyre, I.,** Amylin-amide: a new bone conserving peptide from the pancreas, *Exp. Physiol.,* 75, 529, 1992.
134. **Zaidi, M., Fuller, K., Bevis, P. J. R., GainesDas, R. E., Chambers, T. J., and MacIntyre, I.,** Calcitonin gene-related peptide inhibits osteoclastic bone resorption: a comparative study, *Calcif. Tissue Int.,* 40, 149, 1987.
135. **Nicholson, G. C., Moseley, J. M., Sexton, P. M., and Martin, T. J.,** Chicken osteoclasts do not possess calcitonin receptors, *J. Bone Miner. Res.,* 2, 53, 1987.
136. **Hunter, S. J., Schraer, H., and Gay, C. V.,** Characterization of isolated and cultured chick osteoclasts: the effects of acetazolamide, calcitonin, and parathyroid hormone on acid production, *J. Bone Miner. Res.,* 3, 297, 1988.
137. **de Vernejoule, M.-C., Horowitz, M., Dernignon, J., Neff, L., and Baron, R.,** Bone resorption by isolated chick osteoclasts in culture is stimulated by murine spleen cell supernatant fluids (osteoclast-activating factor) and inhibited by calcitonin and prostaglandin E_2, *J. Bone Miner. Res.,* 3, 69, 1988.
138. **Eliam, M. C., Baslé, M., Bouizar, Z., Bielakoff, J., Moukhtar, M., and de Vernejoul, M. C.,** Influence of blood calcium on calcitonin receptors in isolated chick osteoclasts, *J. Endocrinol.,* 119, 243, 1988.
139. **Chambers, T. J., Athanasou, N. A., and Fuller, K.,** The effect of parathyroid hormone and calcitonin on the cytoplasmic spreading of isolated osteoclasts, *J. Endocrinol.,* 102, 281, 1984.
140. **Nicholson, G. C., Kent, G. N., Rowe, D. J., Cronin, B. J., Evely, R. S., and Moseley, J. M.,** PTH acts indirectly on isolated rat osteoclasts to increase resorption pit number, *J. Bone Miner. Res.,* 4 (Suppl. 1), 210, 1989.
141. **McSheehy, P. M. J. and Chambers, T. J.,** Osteoblastic cells mediate osteoclastic responsiveness to parathyroid hormone, *Endocrinology,* 118, 824, 1986.
142. **Rodan, G. A. and Martin, T. J.,** The role of osteoblasts in hormonal control of bone resorption, *Calcif. Tissue Int.,* 33, 349, 1981.
143. **O'Grady, R. L. and Cameron, D. A.,** Demonstration of binding sites of parathyroid hormone in bone cells, in *Endocrinology 1971: Proceedings of the 3rd Internatioanl Symposium,* Heinemann, London, 1971, 374.
144. **Silve, C. M., Hradek, G. T., Jones, A. L., and Arnaud, C. D.,** Parathyroid hormone receptor in intact embryonic chicken bone: characterization and cellular localization, *J. Cell Biol.,* 94, 379, 1982.
145. **Rouleau, M. F., Warshawsky, H., and Goltzman, D.,** Parathyroid hormone binding *in vivo* to renal, hepatic and skeletal tissues of the rat using a radioautographic technique, *Endocrinology,* 118, 919, 1986.
146. **Rao, L. G., Murray, T. M., and Heersche, T. N. M.,** Immunohistochemical demonstration of parathyroid hormone binding to specific cell types in fixed rat bone tissue, *Endocrinology,* 113, 805, 1983.

147. **McSheehy, P. M. J. and Chambers, T. J.,** Osteoblast-like cells in the presence of parathyroid hormone release soluble factor that stimulates osteoclastic bone resorption, *Endocrinology,* 119, 1654, 1986.
148. **McSheehy, P. M. J. and Chambers, T. J.,** 1,25-dihydroxyvitamin D_3 stimulates rat osteoblastic cells to release a soluble factor that increases osteoclastic bone resorption, *J. Clin. Invest.,* 80, 425, 1987.
149. **Thomson, B. M., Mundy, G. R., and Chambers, T. J.,** Tumor necrosis factors α and β induce osteoblastic cells to stimulate osteoclastic bone resorption, *J. Immunol.,* 138, 775, 1987.
150. **Chambers, T. J.,** Osteoblasts release osteoclasts from calcitonin-induced quiescence, *J. Cell Sci.,* 57, 247, 1982.
151. **Thomson, B. M., Saklatvala, J., and Chambers, T. J.,** Osteoblasts mediate interleukin 1 stimulation of bone resorption by rat osteoclasts, *J. Exp. Med.,* 164, 104, 1986.
152. **Gualtieri, R. J., Shadduck, R. K., Baker, D. G., and Queensberry, P. J.,** Hematopoietic regulatory factors produced in long-term murine bone marrow cultures and the effect of *in vitro* irradiation, *Blood,* 64, 516, 1984.
153. **Heard, J.-M., Fichelson, S., and Varet, B.,** Role of colony-stimulating activity in murine long-term bone marrow cultures: evidence for its production and consumption by the adherent cells, *Blood,* 59, 761, 1982.
154. **Lipschitz, D. A., Udupa, K. B., Taylor, J. M., Shadduck, R. K., and Waheed, A.,** Role of colony-stimulating factor in myelopoiesis in murine long-term bone marrow cultures, *Blood,* 69, 1211, 1987.
155. **Gospodarowicz, D., Cheng, J., Lui G.-M., Baird, A., and Böhlent, P.,** Isolation of brain fibroblast growth factor by heparin-Sepharose affinity chromatography: identity with pituitary fibroblast growth factor, *Proc. Natl. Acad. Sci. U.S.A.,* 81, 6963, 1984.
156. **Hauschka, P. V., Mavrakos, A. E., Iafrati, M. D., Doleman, S. E., and Klagsbrun, M.,** Growth factors in bone matrix. Isolation of multiple types by affinity chromatography on heparin-Sepharose, *J. Biol. Chem.,* 261, 12665, 1986.
157. **Maciag, I., Mehlman, T., Friesel, R., and Schreiber, A. B.,** Heparin binds endothelial cell growth factor, the principal endothelial cell mitogen in bovine brain, *Science,* 225, 932, 1984.
158. **Schreiber, A. B., Kenney, J., Kowalski, W. J., Friesel, R., Mehlman, T., and Maciag, T.,** Interaction of endothelial cell growth factor with heparin: characterization by receptor and antibody recognition, *Proc. Natl. Acad. Sci. U.S.A.,* 82, 6138, 1985.
159. **Ecarot-Charrier, B. and Broekhuyse, H.,** Proteoglycans synthesized by cultured mouse osteoblasts, *J. Biol. Chem.,* 262, 5345, 1987.
160. **Rodan, G. A. and Rodan, S. B.,** Expression of the osteoblastic phenotype, in *Advances in Bone and Mineral Research Annual,* Vol. 2, Peck, W. A., Ed., Elsevier, Amsterdam, 1984, 244.
161. **Springer, T. A. and Lasky, L. A.,** Sticky sugars for selectins, *Nature (London),* 349, 196, 1991.
162. **Nishihara, T., Ishihara, Y., Noguchi, T., and Koga, T.,** Membrane Il-1 induces bone resorption in organ culture, *J. Immunol.,* 143, 1881, 1989.
163. **Wong, S. T., Winchell, L. F., McCune, B. K., Earp, H. S., Teixido, J., Massague, J., Herman, B., and Lee, D. C.,** The TGF-α precursor expressed on the cell surface binds to the EGF receptor on adjacent cells, leading to signal transduction, *Cell,* 56, 495, 1989.
164. **Brachmann, R., Lindquist, P. B., Nagashima, M., Kohr, W., Lipari, T., Napier, M., and Derynck, R.,** Transmembrane TGF-α precursors activate EGF/EGFα receptors, *Cell,* 56, 691, 1989.
165. **Witte, O. N.,** Steel locus defines new multipotent growth factor, *Cell,* 63, 5, 1990.
166. **Lee, B. A. and Donoghue, D. J.,** Membrane-anchored form *v-SIS*/PDGF-B induces mitogenesis without detectable PDGF receptor autophosphorylation, *J. Cell Biol.,* 113, 361, 1991.

167. **Hanazawa, S., Ohmori, Y., Amano, S., Miyoshi, T., Kumegawa, M., and Kitano, S.,** Spontaneous production of interleukin-1-like cytokine from a mouse osteoblastic cell line, *Biochem. Biophys. Res. Commun.*, 131, 774, 1985.
168. **Hanazawa, S., Amano, S., Nakada, K., Ohmori, Y., Miyoshi, T., Hirose, K., and Kitano, S.,** Biological characterization of interleukin-1-like cytokine produced by cultured bone cells from newborn mouse calvaria, *Calcif. Tissue Int.*, 41, 31, 1987.
169. **Lorenzo, J. A., Sousa, S. L., Van Den Brink-Webb, S. E., and Korn, J. H.,** Production of both interleukin-1α and β by newborn mouse calvarial cultures, *J. Bone Miner. Res.*, 5, 77, 1990.
170. **Centrella, M. and Canalis, E.,** Transforming and non-transforming growth factors are present in the medium conditioned by fetal rat calvaria, *Proc. Natl. Acad. Sci. U.S.A.*, 82, 7335, 1985.
171. **Robey, P. G., Young, M. F., Flanders, K. C., Roche, N. S., Kondaiah, P., Reddi, A., Termine, J. D., Sporn, M. B., and Roberts, A. N.,** Osteoblasts synthesize and respond to transforming growth factor-type β (TGFβ), *J. Cell Biol.*, 105, 457, 1987.
172. **Canalis, E., McCarthy, T., and Centrella, M.,** Isolation of growth factors from adult bovine bone, *Calcif. Tissue Int.*, 43, 346, 1988.
173. **Valentin-Opran, A., Delgado, R., Valente, T., Mundy, G. R., and Graves, D. T.,** Autocrine production of platelet-derived growth factor (PDGF)-like peptides by cultured normal human bone cells, *J. Bone Miner. Res.*, 2 (Suppl. 1) (Abstr.), 254, 1987.
174. **Horowitz, M. C., Coleman, D. L., Ryaby, J. T., and Einhorn, T. A.,** Osteotropic agents induce the differential secretion of granulocyte-macrophage colony-stimulating factor by the osteoblast cell line MC3T3-E1, *J. Bone Miner. Res.*, 4, 911, 1989.
175. **Felix, R., Elford, P. R., Stoerckle, M., Cecchini, M., Wetterwald, A., Trechsel, U., Fleisch, H., and Stradler, B. M.,** Production of haemopoietic growth factors by bone tissue and bone cells in culture, *J. Bone Miner. Res.*, 3, 27, 1988.
176. **Sato, K., Fujii, Y., Asano, S., Ohtsuki, T., Kawakami, M., Kasono, K., Tsushima, T., and Shizume, K.,** Recombinant human interleukin 1α and β stimulate mouse osteoblast-like cells (MC3T3-E1) to produce macrophage-colony stimulating activity and prostaglandin E_2, *Biochem. Biophys. Res. Commun.*, 141, 285, 1986.
177. **Elford, P. R., Felix, R., Cecchini, M., Trechsel, U., and Fleisch, H.,** Murine osteoblast-like cells and the osteogenic cell MC3T3-E1 release a macrophage colony-stimulating activity in culture, *Calcif. Tissue Int.*, 41, 151, 1987.
178. **Shiina-Ishimi, Y., Abe, E., Tanaka, H., and Suda, T.,** Synthesis of colony-stimulating factor (CSF) and differentiation-inducing factor (D-factor) by osteoblastic cell, clone MC3T3-E1, *Biochem. Biophys. Res. Commun.*, 134, 400, 1986.
179. **Abe, E., Ishimi, Y., Takahashi, N., Akatsu, T., Ozawa, H., Yamana, H., Yoshiki, S., and Suda, T.,** A differentiation-inducing factor produced by the osteoblastic cell line MC3T3-E1 stimulates bone resorption by promoting osteoclast formation, *J. Bone Miner. Res.*, 3, 635, 1988.
180. **Perry, H. M., Skogen, W., Chappel, J. C., Wilner, G. D., Kahn, A. J., and Teitelbaum, S. L.,** Conditioned medium from osteoblast-like cells mediates parathyroid hormone induced bone resorption, *Calcif. Tissue Int.*, 40, 298, 1987.
181. **Perry, H. M., Skogen, W., Chappel, J., Kahn, A. J., Wilner, G., and Teitelbaum, S. L.,** Partial characterization of a parathyroid hormone-stimulated resorption factor(s) from osteoblast-like cells, *Endocrinology*, 125, 2075, 1989.
182. **Meghji, S., Sandy, J. R., Scutt, A. M., Harvey, W., and Harris, M.,** Heterogeneity of bone resorbing factors produced by unstimulated murine osteoblasts *in vitro* and in response to stimulation by parathyroid hormone and mononuclear cell factors, *Arch. Oral Biol.*, 33, 773, 1988.
183. **Morris, C. A., Mitnick, M. E., Weir, E. C., Horowitz, M., Kreider, B. L., and Insogna, K. L.,** The parathyroid hormone-related protein stimulates human osteoblast-like cells to secrete a 9,000 dalton bone-resorbing protein, *Endocrinology*, 126, 1783, 1990.
184. **Chambers, T. J.,** unpublished observations.

Chapter 6

OSTEOCLAST ULTRASTRUCTURE AND ENZYME HISTOCHEMISTRY: FUNCTIONAL IMPLICATIONS

Carol V. Gay

TABLE OF CONTENTS

I. INTRODUCTION

Morphological studies have provided considerable insight into the function of bone and bone cells. Indeed, large, multinucleated cells on bone surfaces were first described by Robin[1] in 1849, and were named "osteoklasten" by Kölliker[2] in 1873. These early anatomists set forth the correct interpretation that osteoclasts resorb bone. However, deeper insight into the mechanism of osteoclastic resorption had to wait more than a century for the development of other powerful methods, particularly cell isolation and culture procedures and the application of biochemical and molecular biology techniques.

The use of histochemical and immunocytochemical methods at the ultrastructural level to demonstrate intracellular sites of metabolically and structurally important molecules has considerable value in understanding cell function. Particularly helpful are discoveries of asymmetric distributions of molecules. On the other hand, there are major frustrations in utilizing localization methods due to the relative insensitivity and time consumption associated with such techniques.

II. ULTRASTRUCTURE OF THE OSTEOCLAST

Osteoclasts exist in many shapes and sizes. When actively resorbing bone, an osteoclast may be flattened onto a bone surface or may be highly contorted when it seals itself over the end of a spicule identified for destruction. The cross-sectional diameter of osteoclasts varies from around 20 to several hundred microns. Typically, there are six to eight nuclei, but there can be lesser and greater amounts. Ribosomes, in small clusters and free, are readily apparent, suggesting protein synthetic capacity for renewal of intracellular protein. The rough endoplasmic reticulum is scant, relative to the entire cytoplasmic volume, and perinuclear Golgi are few and small, indicating that synthesis of protein for export is not a major function of osteoclasts. However, as discussed later, a number of proteolytic enzymes are secreted into the resorption lacuna and are essential for digestion of the organic matrix phase. (See also Chapter 13 of this volume.)

Mitochondria are notably abundant, and it is estimated that osteoclasts contain more mitochondria per unit of cytoplasmic volume than any other cell in the body.[3] The striking number of this organelle in osteoclasts provides important evidence that these cells function at a high level of aerobic metabolism. There are numerous granules in osteoclast mitochondria, and, by use of anhydrous preparation methods and X-ray microanalysis, these granules have been shown to be mineral deposits of calcium and phosphorus.[4] Gay and colleagues also found Ca/P-rich granules in mitochondria of osteoclasts[148] and chondrocytes.[5] Yamamoto et al.[6] suggest that the high levels of cytochrome oxidase in osteoclast mitochondria are coincident with those mitochondria that contain numerous Ca/P-rich granules. The concept that mitochondria serve as storage sites to buffer intracellular surges of ionic calcium was developed by Lehninger,[7] and cells that handle large amounts of calcium, such as the avian shell gland, have mitochondria that take up and release calcium more efficiently and to a greater extent than mitochondria from other sources.[8] Osteoclast mitochondria have not been studied in this manner, but now that osteoclast isolation methods are available, such a study is now possible. (See Chapter 1 of this volume.)

The ruffled border is an organelle that has received considerable attention. The ruffled border forms the cellular side of the resorption lacuna and is completely surrounded by the clear zone, the structure by which the cell adheres to the bone surface. Proof of this has required three-dimensional analysis of serial sections (see Reference 3 for a review). First named ''striated border'' by Kölliker,[2] it has had a number of descriptors; the term ''ruffled border'' was given by Scott and Pease[9] in the first published electron micrographs of osteoclasts. Like the entire cell, the ruffled border seems to conform to many shapes of bone surfaces. The ruffled border is a series of cytoplasmic extensions which are often finger-like, but may also be folded sheets (see Figure 8 in Reference 3). The fingers and folds are widely variable in thickness and length, but basically fall into two categories, finger-like processes and plate-like processes, as shown in a three-dimensional analysis by Domon and Wakita.[10] The extracellular channels and spaces between the fingers and folds are also highly variable. The ruffled border can be extensive, and thereby provides an enormous surface area for the exchange of materials between the cell and extracellular space. The fine structure of the ruffled border membrane is distinct from the remainder of the plasma membrane. Kallio et al.[11] reported that the cytoplasmic face of the ruffled border contains evenly spaced (20 to 25 nm) perpendicular bristles. It was speculated at the time that the bristles might be sites of enzymatic activity related to bone resorption. As discussed below, it is now known that the ruffled border uniquely contains a number of enzymes involved in the resorptive process.

Holtrop and colleagues,[12-14] using serial section analysis, showed that the ruffled border size varies according to cell activity; parathyroid hormone and vitamin D_3, known stimulators of bone resorption, stimulate an increase in osteoclast ruffled border area in sections of rat bone. Inactive cells have little or no ruffled border.[15,16] Osteoclasts have been shown to vacillate between active and inactive states,[17] and are activated by parathyroid hormone.[18] Careful analysis of ultrastructural images indicates that active osteoclasts are attached to bone surfaces and possess ruffled borders, whereas inactive cells are detached and lack ruffled borders.[14,16] Apparently, detached osteoclasts migrate between the perivascular space and bone surface to new resorption sites.[16] Detached osteoclasts migrate at a rate of 30 to 250 μm/h and do not resorb bone.[19] Large osteoclasts may have more than one ruffled border which resorb at different rates.[19] Figures 1 and 2 show osteoclasts between the

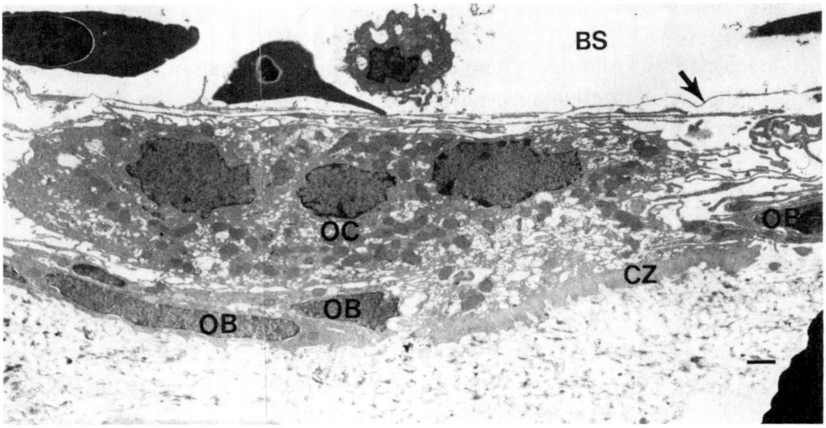

FIGURE 1. Collage of electron micrographs of an osteoclast (OC) attached to bone and flanked by osteoblasts (OB). This view shows the clear zone (CZ) adhering to extracellular matrix. Ruffled border is not in view; it is out of the plane of section. Also apparent is a layer of endothelial cells (arrow) which form a blood sinusoid (BS) and define the bone cell compartment. Bar = 1 μm. (Original magnification × 6200.) (Courtesy of Dr. Osamu Fukushima.)

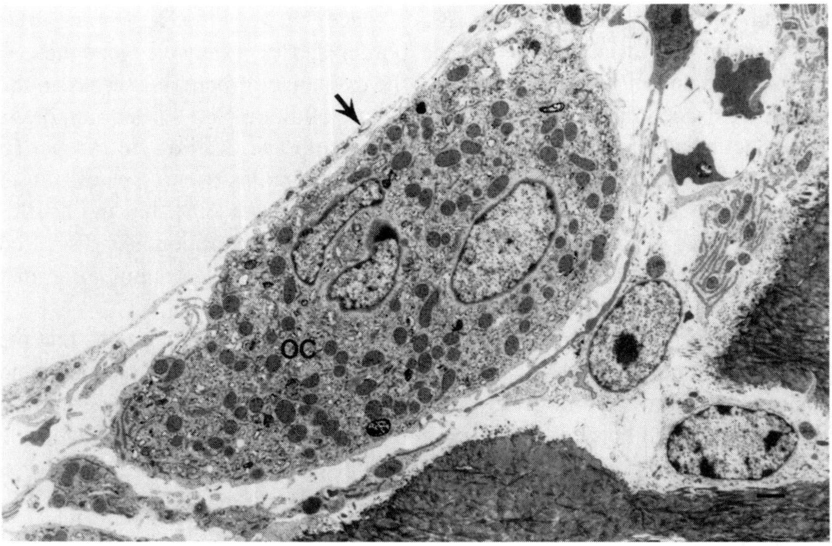

FIGURE 2. Electron micrograph of an osteoclast (OC) which, on the basis of serial sectioning, lacks a ruffled border and is not attached to bone. The boundary between the blood sinusoid and bone compartment (arrow) is apparent. Bar = 1 μm. (Original magnification × 6200.) (Courtesy of Dr. Osamu Fukushima.)

perivascular space and bone surface; one has active morphology, the other inactive. Additional electron micrographs of osteoclasts are provided in Chapter 9 of this volume.

The ultrastructure of the resorption lacuna shows a frayed bone surface. Close to the cell, there are few collagen fibers. Moving deeper into the lacuna, one sees more fibers and an increasing amount of associated mineral.[3] These observations are in agreement with the view that resorption occurs as a result of secretion of hydrogen ions and proteolytic enzymes into the resorption lacuna and results in a resorption gradient.

Osteoclasts are phagocytic, and bits of resorbed bone have been reported between the infoldings of the ruffled border (see Reference 3). Whether these particles are taken up by the cell is still a matter of debate.[20] As bone is resorbed, the fate of the osteocytes appears to be phagocytosis by osteoclasts.[21,22]

The clear zone always surrounds the ruffled border.[3] The term "clear zone" has come into common usage because the zone looks clear at low magnifications; however, many names have been given to this structure.[3] At higher magnification, a few small vesicles are seen in the clear zone, but none of the other cell organelles are present. In well-preserved preparations, numerous thin filaments are present. The filaments are undoubtedly actin filaments, since they are 5 to 7 nm in diameter and stain specifically with heavy meromyosin.[23] In addition, immunofluorescent antibodies specific to F-actin have been shown to bind to filaments in osteoclasts.[24-28]

In an ultrastructural study of the distribution of microtubules in osteoclasts, Sakuraoka[29] found that the ruffled border is devoid of microtubules, but microtubules are present in central areas and as a "plasmalemmal undercoat" of the vascular side of the osteoclast. When inactivated by calcitonin, the ruffled border disappeared and the entire plasma membrane was undercoated with microtubules. Actin filaments were also present in the undercoat. The role of the cytoskeleton in osteoclasts is discussed in Chapter 12 of this volume.

The means by which osteoclasts adhere to the underlying bone surface are presented in Chapters 2, 9, and 12 of this volume. Whatever the sealant may be, it would appear that, since osteoclasts can move along bone surfaces in amoeboid fashion, a well-controlled unsealing and resealing process must occur. The tightness of the seal might be questioned, since the extracellular marker, horseradish peroxidase, is seen in resorption lacunae as early as 5 min after injection.[30] However, the idea of a tight seal is upheld by the observations that both pH (~4.5 to 5.0)[31,32] and Ca^{2+} concentration (26.7 mM)[32] in resorption lacunae are substantially different from those in nearby interstitial fluid.

What is the fate of the ruffled border membrane when an osteoclast becomes inactive? Miller[17] has shown that osteoclasts cycle between active and inactive states daily in Japanese quail medullary bone during the egg-laying cycle. When egg shell calcification is complete and bone resorption has temporarily ceased, the osteoclasts are devoid of ruffled border and extensive membrane interdigitations appear along the sides and apex of the cells. This phenomena has not been reported for other species.

Another reservoir of ruffled border membrane appears to be vacuoles within the cell. The vacuoles are so plentiful that, by light microscopy, the cytoplasm of osteoclasts is described as "foamy". The vacuoles vary from 0.3 to 5.0 μm in diameter, with the larger ones being closest to the ruffled border (see Reference 3). Many vacuoles have a similar membrane thickness (~9 nm)[33,34] and possess a bristle coat,[11,33,35,36] as does the ruffled border,[11] features that distinguish them from the remainder of the plasma membrane. We found both ruffled border and vesicles in osteoclasts of chicken tibial metaphysis to be enriched in an alkaline *p*-nitrophenyl phosphatase (alkaline *p*-NPPase).[16] Inactive osteoclasts which were detached from bone lacked a ruffled border and contained numerous alkaline *p*-NPPase-stained vesicles. Cells that had an extensive ruffled border had markedly fewer alkaline *p*-NPPase-stained vesicles.

Many of the vacuoles and vesicles present in osteoclasts are involved in either endocytosis or exocytosis. Lucht[33] provided ultrastructural evidence that revealed one morphological type of vacuole as being endocytotic and another type as being exocytotic. In addition, Pierce and Lindskog[37] identified two distinct structures involved in endocytosis: clathrin-coated vesicles on the lateral and dorsal osteoclast surfaces involved in receptor-mediated endocytosis of nutrients, and nonclathrin-associated vesicles for endocytosis of bone components at the ruffled border.

A number of small, dense bodies of unknown function are seen in osteoclast cytoplasm. Codification of these is provided by Holtrop.[3]

Osteoclasts are in direct contact with several other types of cells. Evidence that osteoblasts influence osteoclasts, either by direct contact or by secretion of a messenger molecule, has been building for more than a decade. Recently, we observed increased amounts of adenylate cyclase histochemical reaction product at sites of osteoclast-osteoblast contact,[38] a finding which suggests the involvement of direct contact in cell regulation. Soskolne[39] reported that 85% of osteoclasts in young rabbit femur made direct contact with neighboring endothelial cells, a finding recently confirmed in rats, mice, and human.[40] Osteoclasts also contact pericytes,[40] fibroblasts,[41] macrophages,[41] and mononuclear phagocytes.[42] The fusion of preosteoclasts to each other and to existing osteoclasts also involves direct cell contact, as discussed in Chapters 4 and 21 of this volume. The act of fusion has been captured by electron microscopy,[16] but it is not commonly observed.

Scanning electron microscopy (SEM) provides a means of viewing the osteoclast surface.[27,43-47] A notable feature is that isolated osteoclasts possess numerous microvilli-like cell processes that are about 1 μm in diameter.[45] In an *in situ* SEM study of osteoclasts on the endocortical surface of mouse femurs, osteoclasts were polymorphic, lacked microvilli-like cell processes, and appeared to be connected to each other through long pseudopod extensions.[43]

The observations of osteoclasts contacting each other, as well as other types of cells,[38-43] suggest that osteoclasts may function as a syncytium.

Other reviews that discuss osteoclast structure include References 3 and 48 to 53.

III. LOCALIZATION OF ENZYMES

A. CARBONIC ANHYDRASE

1. Intracellular Distribution

Carbonic anhydrase has been shown to be present in osteoclast cytoplasm of chicks,[54,55] rats,[56-58] mice,[59] and humans[60] by autoradiographic, histochemical, and immunocytochemical means. The first sighting of carbonic anhydrase in osteoclasts[61] in tissue sections was in 1960; however, the method used was later shown to be unreliable. Carbonic anhydrase is present throughout the cytoplasm and is also associated with the inner surface of the ruffled border.[54,62] The role of cytosolic carbonic anhydrase in osteoclast cytoplasm is not well understood, although it may serve to facilitate diffusion of CO_2 from the cell interior to its surface, as has been shown to occur in solution.[63,64] On the basis of autoradiographic grain counts of specifically bound ^{3}H-acetazolamide to carbonic anhydrase, the level of carbonic anhydrase in osteoclasts appears to be similar to that found in renal proximal tubules and in pancreas,[55] tissues known to utilize carbonic anhydrase in the elaboration of acidic and bicarbonate-rich fluids, respectively. This suggested that carbonic anhydrase plays an important role in osteoclast function. Membrane-associated carbonic anhydrase has also been observed in a multinucleated giant-cell tumor of bone[65] and in the villus-cavity cell of chorioallantoic membrane of fertilized chicken eggs.[66] Villus-cavity cells are believed to resorb egg shells by secretion of H^+, and thereby provide calcium for the developing embryo.

2. Isoforms of Carbonic Anhydrase

The first immunolocalization of carbonic anhydrase in osteoclasts utilized an antibody against the high-activity form of carbonic anhydrase, CA II, isolated from avian red blood cells.[67] CA II reactivity and the absence of CA I and III in mammalian osteoclasts has also been reported.[58,60]

In humans, a CA II deficiency syndrome has been characterized.[68] Afflicted individuals have a diminished capacity to resorb bone, along with other clinical manifestations. Notably, there is an overabundance of bone in the interior shaft of the long bones.

3. Carbonic Anhydrase Inhibition

A number of specific inhibitors of carbonic anhydrase are known, and considerable effort has been placed on elucidating the effects of carbonic anhydrase inhibition on bone resorption both *in vitro* and *in vivo*. Acetazolamide blocks the hypercalcemic response to parathyroid hormone and dibutyryl cyclic AMP in nephrectomized rats[69,70] and partially blocks the progression of disuse atrophy in rats.[71] The effect can be shown in intact rats if renal-mediated metabolic acidosis caused by acetazolamide is neutralized.[72] Prevention of metabolic acidosis appears to be of critical importance in achieving the desired effect of reducing bone resorption with carbonic anhydrase inhibitors.

The addition of acetazolamide to bone cultures inhibits PTH-induced bone resorption.[73-76] The more potent CA inhibitors, methazolamide,[77] ethoxolamide,[77] and 5-hydroxythiophene sulfonamide,[78,79] are effective at lower doses. The inactive analog of acetazolamide, CL 13,850, is ineffective.[77] Contributions over the past 20 years or so have made it eminently clear that carbonic anhydrase has an important role in osteoclastic bone resorption. The exact mechanism of action and ways of controlling osteoclasts through carbonic anhydrase inhibition remain challenges for the future. However, a likely role is provision of hydrogen ions to a proton-pumping ATPase in the ruffled border. The effects of carbonic anhydrase inhibition on osteoclasts are also discussed in Chapters 14 and 17 of this volume.

4. Regulation of Carbonic Anhydrase

Parathyroid hormone (PTH) is a well-known stimulator of bone resorption, and a number of studies have shown that carbonic anhydrase inhibition will block this effect *in vivo*,[70] in organ cultures,[73,74,80,81] and in isolated osteoclasts.[79,82] PTH potentiated the association of carbonic anhydrase with the ruffled border membrane.[62] In light of studies that reveal specific binding of PTH to osteoclasts,[83-86] it is appropriate to consider the mechanism of PTH regulation of carbonic anhydrase. Silverton and colleagues[87] have shown activation of carbonic anhydrase by PTH in isolated osteoclasts. The effect is believed to be cyclic AMP-mediated, since administration of dibutyryl cyclic AMP (cAMP) to rats elicits hypercalcemia that mimics the PTH response; both the cAMP and PTH responses are blocked by carbonic anhydrase inhibition.[69,70] A PTH-induced transient increase in acidification by isolated osteoclasts is blocked by acetazolamide[80,82] and is mimicked by dibutyryl cAMP administration.[149] In cultures of mouse calvaria treated with dibutyryl cAMP, carbonic anhydrase activity and bone resorption have been found to increase.[81] However, the activity increase may be due to increased numbers of osteoclasts during the 96-h treatment period, since acid phosphatase activity also increased.

Additional evidence for cAMP-mediated activation by phosphorylation of carbonic anhydrase in osteoclasts has been provided by Dietsch.[88] He showed that carbonic anhydrase from chicken epiphyses was activated by prior injection of PTH or by experimentally lowering plasma calcium with ethylene glycoltetra acetic acid (EGTA) infusion or vitamin D deficiency.

The activated carbonic anhydrase was shown, by gel isoelectric focusing, to be phosphorylated. Further, the study revealed that protein kinase activity increased after PTH injection on a time scale that places the kinase activation subsequent to a cAMP response and preceding carbonic anhydrase activation.

Calcitonin, a well-known inhibitor of osteoclast function, has no effect on carbonic anhydrase activity in isolated osteoclasts.[89] However, the association of carbonic anhydrase with the cytoplasmic face of the ruffled border is lost following calcitonin treatment.[54,62] Further, acidification of isolated osteoclasts is blocked to the same extent by 10^{-7} *M* acetazolamide or 10^{-9} *M* 5-hydroxythiophene sulfonamide as by 0.31 mU/ml (2.2×10^{-11} *M*) calcitonin.[79,82] The evidence suggests that calcitonin regulates carbonic anhydrase function by physical orientation of the molecule rather than through a biochemical activation pathway.

Involvement of vitamin D in the regulation of carbonic anhydrase-supported bone resorption has been known for some time. Pierce and colleagues[90] showed that 1,25-dihydroxyvitamin D_3-induced hypercalcemia in nephrectomized rats was blocked by sulfonamide (acetazolamide, methazolamide, and ethoxolamide) inhibition of carbonic anhydrase. Hall and Kenny[91] have shown that ^{45}Ca release from cultured mouse calvaria stimulated by the 1,25-vitamin D metabolite is blocked by acetazolamide. In these studies, the effect of vitamin D_3 on calcium levels in both plasma and culture media required a period of 12, 48, or 96 h, which is in line with evidence that vitamin D metabolites act at the level of the nucleus. Significantly, the 1,25-vitamin D_3 metabolite has been shown to regulate expression of CA II in bone marrow enriched in osteoclast precursor cells.[92]

In summary, three levels of regulation of carbonic anhydrase involvement in osteoclastic bone resorption have been identified: (1) PTH appears to activate the enzyme by a cAMP-mediated phosphorylation, (2) calcitonin is inhibitory and causes a shift in position of the enzyme, and (3) vitamin D_3 regulates the expression of carbonic anhydrase.

B. PROTON-PUMPING ATPases

As evidence mounted that indicated a role for carbonic anhydrase in bone resorption, investigators, including ourselves, began to consider the possibility that carbonic anhydrase in osteoclasts might provide hydrogen ions to a proton pump in the ruffled border membrane in a manner similar to other acidifying tissues. This possibility became a particularly relevant consideration in light of the finding that carbonic anhydrase is deployed along the cytoplasmic face of the ruffled border[54,62] and is therefore appropriately positioned to provide hydrogen ions to a proton pump. Initially, two laboratories sought to demonstrate a proton-generating ATPase in osteoclasts. Baron and colleagues[93] used an antibody to gastric H^+,K^+-ATPase and, in well-controlled immunolocalizations, revealed such an enzyme along the ruffled border. Akisaka and Gay[94] developed a histochemical procedure to demonstrate the presence of an H^+-ATPase in osteoclast ruffled border membranes. Reaction-product development was blocked by the proton-pump inhibitors duramycin, tri-*n*-butyl tin, and *p*-chloromercuribenzoic acid.

The histoimmunological studies implicated the presence of two types of proton pumps in osteoclast ruffled membranes; the immunolocalization suggested the gastric type and, on the basis of inhibitors, the histochemical localization revealed a vacuolar-type ATPase. Functional studies on plasma membrane vesicles isolated from osteoclasts have demonstrated the presence of both types of proton pumps. The vacuolar-type proton pump in osteoclasts[95-97] has been well characterized and is described in Chapter 13 of this volume. Biochemical evidence for the gastric-type pump has also been reported. Omeprazole, a gastric H^+,K^+-ATPase inhibitor, decreased basal, prostaglandin E_2-stimulated, and PTH-stimulated ^{45}Ca release from cultured, prelabeled neonatal mouse calvaria.[98] Anderson et al.[99] showed that omeprazole caused a 17% decrease in osteoclast acidity in a cultured osteoclast system. In addition, the vesicle preparation of Bekker and Gay[96] revealed both a strong vacuolar-type pump and a less apparent nolinium bromide-sensitive pump that could be of the gastric type.

C. Na^+,K^+-ATPase

Structural evidence for the presence of this enzyme in osteoclasts has been provided by histochemistry and immunocytochemistry. Using the histochemical method, which is based on enzyme activity, we found ouabain-inhibitable enzyme in the ruffled border membrane.[100] Enzyme activity was undetectable following calcitonin injection prior to tissue sampling, which indicates sodium pump regulation by calcitonin. Immunocytochemistry revealed antigenically identifiable Na^+,K^+-ATPase subunits around the entire osteoclast plasma membrane.[101] Together, these observations raise the interesting possibility that the sodium pump may be active when at the ruffled border and that inactive enzyme reserves are held in other regions of plasma membrane. It is also possible that the enzyme is active in all parts of the osteoclast plasma membrane, but that the greatest activity is at the ruffled border where activity is sufficiently high to be detected by the relatively insensitive histochemical method.

The well-established role of Na^+,K^+-ATPase is to maintain low Na^+ and high K^+ in cells; maintenance of a sodium-potassium gradient creates membrane potential. Anderson et al.[99] have shown that osteoclast acidity is dependent on the maintenance of Na^+-K^+ gradients. PTH-stimulated resorption of bone prelabeled with ^{45}Ca is inhibited by ouabain, a specific inhibitor of Na^+,K^+-ATPase.[102] Ion channels and membrane potential are discussed in Chapter 11 of this volume.

D. Ca^{2+}-ATPase

In osteoclasts, we found specific Ca^{2+}-ATPase staining in the plasma membrane on the side of the cell facing the marrow; ruffled borders were devoid of stain.[103] Subsequently, we have prepared osteoclast plasma membrane vesicles and demonstrated levels of Ca-ATPase activity that, at 19.8 μM/mg protein per h,[104] is two- to 25-fold greater than that reported for other tissues (see Reference 105). Using a calcium-specific electrode and fura-2 fluorescence, we have also shown outwardly directed Ca-pumping in plasma membrane vesicles.[150]

The role of Ca^{2+}-ATPase in osteoclasts is likely to be maintenance of low levels of cytosolic Ca^{2+}, as has been documented in many other cells.[106] Transcellular flux from resorption lacuna to the circulation does not seem likely, since the ruffled border and clear-zone regions are enriched in actin filaments.[3,23-28] Actin filament stability is well established to be dependent on low calcium levels.

E. CYTOCHROME OXIDASE

One of the most notable features of osteoclasts is their abundant mitochondria. Cytochrome oxidase staining is so pronounced that, along with cell size, it can be employed as a marker of osteoclasts even at the light microscopic level.[16,107-111] Several studies document the ultrastructural distribution and staining intensity of cytochrome oxidase in osteoclast mitochondria.[6,16,112-114]

It is well accepted that the staining intensity of cytochrome oxidase reflects the level of oxidative metabolism in cells.[115-117] In osteoclasts, staining intensity in mitochondria has been shown to vary among cells in the same section of bone, and the suggestion has been made that differences may provide a means of distinguishing active and inactive osteoclasts.[6]

Cytochrome oxidase staining intensity has been used to demonstrate the activity level of osteoclasts during experimental tooth movement. When mechanical force was applied, cytochrome oxidase activity increased and when calcitonin treatment was subsequently administered, staining intensity was reduced.[113,114]

F. ADENYLATE CYCLASE

The action of adenylate cyclase is well established as generation of the widely distributed second-messenger cAMP from ATP. Adenylate cyclase then plays a central role in the implementation of cellular response to numerous hormones; in bone, these include parathyroid hormone, calcitonin, and prostaglandin E_2. (See Chapters 5 and 8 in this volume.)

Since it is obvious that osteoclasts contain adenylate cyclase, what can histochemical localization reveal? Our first search for adenylate cyclase[118] in ultrathin sections of bone revealed a distribution of reaction product on the plasma membrane of osteoblasts, preosteoblasts, and forming osteocytes at sites of contact between the cells, but osteoclasts and mature osteocytes were notably devoid of reaction product.

We repeated the adenylate cyclase localization study on bone from birds maintained on a low (0.3%)-calcium diet,[38] following the study of Eliam et al.[119] These investigators found that osteoclasts isolated from calcium-deficient chicks responded to calcitonin by a doubling of cAMP, as well as an increase in calcitonin receptors; no response was observed in osteoclasts from calcium-sufficient chicks. We found demonstrable adenylate cyclase activity on the nonruffled border plasma membrane of osteoclasts from calcium-deficient chicks. Specificity was shown by alloxan and dideoxyadenosine inhibition. Adenylate cyclase activity was notably strong in regions of membrane which were in contact with neighboring osteoblasts. The role of cAMP generation at cell-cell contact sites might implicate a mechanism for concerted regulation of the cells. Chambers[120] reported contact-dependent stimulation of osteoclasts by osteoblasts; potentially, this is mediated by cAMP production.

The finding of histochemically detectable adenylate cyclase in osteoclasts of calcium-deficient chickens, but not in normo-calcium controls, suggests that adenylate cyclase activity levels may change in relation to metabolic demands. This may be an additional regulatory mechanism in the complex process of osteoclastic bone resorption.

Histochemical detection of adenylate cyclase requires careful attention to substrate, buffer, pH range, and capture reagent.[121,122]

G. TARTRATE-RESISTANT (TRAP) AND TARTRATE-SENSITIVE ACID PHOSPHATASE (TSAP)

The resistance of acid phosphatase in osteoclasts to tartrate was first shown histochemically.[123] Later, a biochemical study revealed two acid phosphatases in bone, one sensitive to tartrate and the other resistant.[124]

Histochemical studies of TRAP in osteoclasts often employ a colored azo-dye to trap released phosphate, and therefore are not useful at the ultrastructural level. In order to evaluate the distribution of TRAP by electron microscopy, a method that produces an electron-dense reaction product, such as a lead precipitate or an immunocytochemical stain, is required. Evaluation of the distribution of TRAP in osteoclasts at the ultrastructural level by lead salt methods[125-127] and by immunocytochemistry[126,128,129] revealed TRAP in intracellular vesicular structures which were probably lysosomes. Also, most studies also show TRAP in the matrix beneath the osteoclast. This is in agreement with the view that TRAP is secreted by osteoclasts and accumulates in the matrix beneath osteoclasts. Polarized secretion by osteoclast lysosomes has been shown by Baron and colleagues.[130,131]

TRAP was also reported in the matrix, but it was most intense in lysosomes and Golgi,[127] in contrast to the distribution of TRAP. Chapter 3 of this volume is devoted to TRAP in osteoclasts and provides more thorough coverage of this topic.

Table 1 provides a comprehensive view of the distribution of enzymes found in osteoclasts by ultrastructural localization methods.

TABLE 1
Enzyme Localization in Osteoclasts as Revealed by Ultrastructural Histochemistry and Immunocytochemistry

Enzyme	Ref.
Ruffled border	
Carbonic anhydrase	54, 62
H^+-ATPase	93, 94, 97
Na^+,K^+-ATPase	100, 101, 131
Acid glycerophosphatase	132
Trimetaphosphatase	133
p-Nitrophenyl phosphatase	125
Aryl sulfatase	35
Nonruffled border plasma membrane	
Ca^{2+}-ATPase	16, 103
Na^+,K^+-ATPase	101
Adenylate cyclase	38
Neutral *p*-nitrophenyl phosphatase	16
Lactate dehydrogenase	134
Mitochondria[a]	
Cytochrome oxidase	6, 16, 112, 113, 114
Ca^{2+}-ATPase	103
H^+-ATPase	94
Lactate dehydrogenase	134
Lysosomes	
Acid phosphatase	16, 135
Acid glycerophosphatase	132
Tartrate-resistant acid phosphatase	125, 127, 128, 129
Aryl sulfatase	132
Glycerophosphatase	132
Trimetaphosphatase	132
Thiamine pyrophosphatase	132
Acid *p*-nitrophenyl phosphatase	132
Phosphoprotein phosphatase	35
Cathepsin B[b]	136
Alkaline *p*-nitrophenyl phosphatase	16
Neutral pyrophosphatase	137
Golgi	
Tartrate-sensitive acid phosphatase	16, 126
Neutral phosphatases	16, 35
Lactate dehydrogenase	134
Secreted to underlying bone surface	
Tartrate-sensitive acid phosphatase	16
Tartrate-resistant acid phosphatase	127, 128, 129
Carbonic anhydrase	54
Neutral phosphatase	16
Collagenase	138
Secretory vesicle membrane	
Alkaline *p*-nitrophenyl phosphatase	16
Carbonic anhydrase	54
Vacuoles	
Lactate dehydrogenase	134, 139

TABLE 1 (continued)
Enzyme Localization in Osteoclasts as Revealed by Ultrastructural Histochemistry and Immunocytochemistry

Enzyme	Ref.
Neutral phosphatases	35
Tartrate-resistant acid phosphatase	126
Endoplasmic reticulum	
p-Nitrophenyl phosphatase	125
Neutral phosphatases	35
Phagocytic vesicles	
Microperoxidase and other exogenous markers	36, 140, 141
Acid glycerophosphatase[c]	132
Aryl sulfatase	35, 130
β-Glucuronidase	130
Cytosol	
Carbonic anhydrase	54, 142, 143
Enzymes not detected in osteoclasts by localization methods	
Collagenase[d]	138, 144
Stromelysin	145
Guanylate cyclase	146
Nonspecific esterase[e]	35
Nicotinamide adenine dinucleotide phosphatase	132

[a] Many other enzymes localized by light microscopy.
[b] Only light microscopic level available.
[c] Diminishes as cell matures.
[d] Has been found biochemically.[147]
[e] Found under some conditions by light microscopic methods (see Reference 132).

IV. SUMMARY

In this chapter, osteoclast ultrastructure has been reviewed and, where possible, structure has been related to function. The second part of the chapter reviewed recent enzyme localization studies that have provided leads for understanding function. These include carbonic anhydrase, H^+-ATPase, Na^+,K^+-ATPase, Ca^{2+}-ATPase, adenylate cyclase, tartrate-sensitive acid phosphatase and TRAP. A codification of enzyme localization at the electron microscopic level was also provided.

REFERENCES

1. **Robin, C. H.,** Sur l'existence de deux espèces nouvelles d'éléments anatomiques qui se trouvent dans le canal medullaire des os, *C. R. Seances Soc. Biol. Paris,* 1, 149, 1849.
2. **Kölliker, A.,** *Die normale resorption des Knochengewehes und ihre bedentung für die entstehung der typischen Knochenformen,* Vogel, Leipzig, 1873.
3. **Holtrop, M. E.,** Light and electronmicroscopic structure of osteoclasts, in *Bone,* Vol. 2, *The Osteoclast,* Hall, B. K., Ed., CRC Press, Boca Raton, FL, 1991, 1.
4. **Lucht, U. and Maunsbach, A. B.,** Effects of parathyroid hormone on osteoclasts *in vivo.* An ultrastructural and histochemical study, *Z. Zellforsch.,* 141, 529, 1973.
5. **Hargest, T. E., Gay, C. V., Schraer, H., and Wasserman, A. J.,** Vertical distribution of elements in cells and matrix of epiphyseal growth plate cartilage determined by quantitative electron probe analysis, *J. Histochem. Cytochem.,* 33, 275, 1985.
6. **Yamamoto, T., Nagai, H., and Gay, C. V.,** An ultracytochemical study of cytochrome oxidase activity in rat metaphyseal bone cells, *Acta Histochem. Cytochem.,* 24, 411, 1991.
7. **Lehninger, A. L.,** Mitochondria and calcium ion transport, *Biochem. J.,* 119, 129, 1970.
8. **Hohman, W. and Schraer, H.,** The intracellular distribution of calcium in the mucosa of the avian shell gland, *J. Cell Biol.,* 30, 317, 1966.
9. **Scott, B. L. and Pease, D. C.,** Electron microscopy of the epiphyseal apparatus, *Anat. Rec.,* 126, 465, 1956.
10. **Domon, T. and Wakita, M.,** Electron microscope study of osteoclasts with special reference to the three-dimensional structure of the ruffled border, *Arch. Histol. Jpn. (Niigata Jpn.),* 49, 593, 1986.
11. **Kallio, D. M., Garant, P. R., and Minkin, C.,** Evidence of coated membranes in the ruffled border of osteoclasts, *J. Ultrastruct. Res.,* 37, 169, 1971.
12. **Holtrop, M. E. and Raisz, L. G.,** Comaprison of the effects of 1,25-dihydroxy-cholecalciferol, prostaglandin E_2, and osteoclast-activating factor with parathyroid hormone on the ultrastructure of osteoclasts in cultured long bones of fetal rats. *Calcif. Tissue Int.,* 29, 201, 1979.
13. **Holtrop, M. E., Cox, K. A., Clark, M. B., Holick, M. F., and Anast, C. S.,** 1,25-Dihydroxycholecalciferol stimulates osteoclasts in rat bones in the absence of parathyroid hormone, *Endocrinology,* 108, 2293, 1981.
14. **King, G. J., Holtrop, M. E., and Raisz, L. G.,** The relation of ultrastructural changes in osteoclasts to resorption in bone cultures stimulated with parathyroid hormone, *Metab. Bone Dis. Relat. Res.,* 1, 67, 1978.
15. **Lucht, U.,** Osteoclasts and their relationship to bone as studied by electron microscopy, *Z. Zellforsch.,* 135, 211, 1972.
16. **Fukushima, O., Bekker, P. J., and Gay, C. V.,** Characterization of the functional stages of osteoclasts by enzyme histochemistry and electron microscopy, *Anat. Rec.,* 231, 298, 1991.
17. **Miller, S. C.,** Osteoclast cell-surface changes during the egg-laying cycle in Japanese quail, *J. Cell Biol.,* 75, 104, 1977.
18. **Miller, S. C.,** Rapid activation of the medullary bone osteoclast cell surface by parathyroid hormone, *J. Cell Biol.,* 76, 615, 1978.
19. **Kanehisa, J. and Heersche, J. N. M.,** Osteoclastic bone resorption: *in vitro* analysis of the rate of resorption and migration of individual osteoclasts, *Bone,* 9, 73, 1988.
20. **Kahn, A. J. and Partridge, N. C.,** Bone resorption *in vivo,* in *Bone,* Vol. 2, *The Osteoclast,* Hall, B. K., Ed., CRC Press, Boca Raton, FL, 1991, 119.
21. **Soskolne, W. A.,** Phagocytosis of osteocytes by osteoclasts in femora of two week-old rabbits, *Cell Tissue Res.,* 195, 557, 1978.
22. **Elmardi, A. S., Katchburian, M. V., and Katchburian, E.,** Electron microscopy of developing calvaria reveals images that suggest that osteoclasts engulf and destroy osteocytes during bone resorption, *Calcif. Tissue Int.,* 46, 239, 1990.

23. **King, G. J. and Holtrop, M. E.,** Actin-like filaments in bone cells of cultured mouse calvaria as demonstrated by binding to heavy meromyosin, *J. Cell Biol.,* 66, 445, 1975.
24. **Marchisio, P. C., Cirillo, D., Naldini, L., Primavera, M. V., Teti, A., and Zambonin-Zallone, A.,** Cell-substratum interaction of cultured avian osteoclasts is mediated by specific adhesion structures, *J. Cell Biol.,* 99, 1696, 1984.
25. **Zambonin-Zallone, A., Teti, A., Carano, A., and Marchisio, P. C.,** The distribution of podosomes in osteoclasts cultured on bone laminae: effect of retinol, *J. Bone Miner. Res.,* 3, 517, 1988.
26. **Hunter, S. J., Schraer, H., and Gay, C. V.,** Characterization of the cytoskeleton of isolated chick osteoclasts: effect of calcitonin, *J. Histochem. Cytochem.,* 37, 1529, 1989.
27. **Yoshida, S., Domon, T., and Wakita, M.,** Studies of the clear zone of osteoclasts: immunohistological aspects of its form and distribution, *Arch. Histol. Cytol.,* 52, 513, 1989.
28. **Lakkakorpi, P., Tuukkanen, J., Hentunen, T., Järvelin, R., and Väänänen, K.,** Organization of osteoclast microfilaments during the attachment to bone surface *in vitro, J. Bone Miner. Res.,* 6, 817, 1989.
29. **Sakuraoka, T.,** An ultrastructural study of the distribution of microtubules in osteoclasts, *Jpn. J. Oral Biol.,* 26, 1245, 1984.
30. **Lucht, U.,** Absorption of peroxidase by osteoclasts as studied by electron microscope histochemistry, *Histochemie,* 29, 274, 1972.
31. **Fallon, M. D.,** Bone resorbing fluid from osteoclasts is acidic — an *in vitro* micropuncture study, in *Endocrine Control of Bone and Calcium Metabolism,* Cohn, D. V., Fujita, T., Potts, J. T., and Talmage, R. V., Eds., Elsevier, Amsterdam, 1984, 144.
32. **Silver, I. A., Murrills, R. J., and Etherington, D. J.,** Microelectrode studies on the acid microenvironment beneath adherent macrophages and osteoclasts, *Exp. Cell Res.,* 175, 266, 1988.
33. **Lucht, U.,** Cytoplasmic vacuoles and bodies of the osteoclast, *Z. Zellforsch.,* 135, 229, 1972.
34. **Thyberg, J., Nilsson, S., and Friberg, U.,** Electron microscopic and enzyme cytochemical studies on the guinea pig metaphysis with special reference to the lysosomal system of different cell types, *Cell Tissue Res.,* 156, 273, 1975.
35. **Doty, S. B. and Schofield, B. H.,** Electron microscopic localization of hydrolytic enzymes in osteoclasts, *Histochem. J.,* 4, 245, 1972.
36. **Thyberg, J.,** Electron microscopic studies on the uptake of exogenous marker particles by different cell types in guinea pig metaphysis, *Cell Tissue Res.,* 156, 301, 1975.
37. **Pierce, A. and Lindskog, S.,** Coated pits and vesicles in the osteoclast, *J. Submicrosc. Cytol. Pathol.,* 20, 161, 1988.
38. **Fukushima, O., Yamamoto, T., and Gay, C. V.,** Ultrastructural localization of adenylate cyclase activity in chicken osteoclasts, *J. Histochem. Cytochem.,* 39, 1207, 1991.
39. **Soskolne, W. A.,** The osteoclast-endothelium interface during bone resorption in the femurs of young rabbits, *Cell Tissue Res.,* 203, 487, 1979.
40. **Cox, K. A., Wilcon, S., Dethlefsen S., and Glowacki, J.,** Detailed transmission electron microscopy and immunohistological characteristics of vessels of rat, mouse, and human growth plates, *J. Bone Miner. Res.,* 3, S183, 1988.
41. **Rifkin, B. R. and Heijl, L.,** The occurrence of mononuclear cells at sites of osteoclastic bone resorption in experimental periodontitis, *J. Periodontol.,* 50, 636, 1979.
42. **Tran Van, P., Vignery, A., and Baron, R.,** An electron-microscopic study of the bone-remodeling sequence in the rat, *Cell Tissue Res.,* 225, 283, 1982.
43. **de Saint-Georges, L., Miller, S. C., Bowman, B. M., and Jee, W. S. S.,** Ultrastructural features of osteoclasts *in situ, Scanning Microsc.,* 3, 1201, 1989.
44. **Boyde, A., Ali, N. N., and Jones, S. J.,** Optical and scanning electron microscopy in the single osteoclast resorption assay, *Scanning Electron Microscopy III,* AMF O'Hare, Chicago, 1259, 1985.

45. **Ubios, A. M., Martin, C. I., and Cabrini, R. L.,** Isolated osteoclasts: scanning electron microscopy, *Naturwissenschaften,* 69, 298, 1982.
46. **Abe, K., Kanno, T., and Schneider, G. B.,** Surface structures and osteoclasts of mouse parietal bones: a light and scanning electron microscopic study, *Arch. Histol. Jpn. (Niigata Jpn.),* 46, 663, 1983.
47. **Helfrich, M. H. and Mieremet, R. H. P.,** A morphologic study of osteoclasts isolated from osteopetrotic microphthalmic (mi/mi) mouse and human fetal long bones using an instrument permitting combination of light and scanning electron microscopy, *Bone,* 9, 113, 1988.
48. **Göthlin, G. and Ericsson, J. L. E.,** The osteoclast, *Clin. Orthop.,* 120, 201, 1976.
49. **Ham, A. H. and Cormack, D. H.,** *Histology,* 8th ed., J. P. Lippincott, Philadelphia, 1979, 407.
50. **Holtrop, M. E. and King, G. J.,** The ultrastructure of the osteoclast and its functional implications, *Clin. Orthop.,* 123, 177, 1977.
51. **Vaes, G.,** Collagenase, lysosomes and osteoclastic bone resorption, in *Collagenase in Normal and Pathological Connective Tissues,* Woolley, D. E. and Evanson, J. M., Eds., John Wiley & Sons, New York, 1980, 185.
52. **Lucht, U.,** Osteoclasts, ultrastructure and functions, in *The Reticuloendothelial System,* Vol. 1, Carr, I. and Daems, W. T., Eds., Plenum Press, New York, 1980, 705.
53. **Marks, S. C. and Popoff, S. N.,** Bone cell biology: the regulation of development, structure and function in the skeleton, *Am. J. Anat.,* 183, 1, 1988.
54. **Anderson, R. E., Schraer, H., and Gay, C. V.,** Ultrastructural immunocytochemical localization of carbonic anhydrase in normal and calcitonin-treated chick osteoclasts, *Anat. Rec.,* 204, 9, 1982.
55. **Gay, C. V. and Mueller, W. J.,** Carbonic anhydrase and osteoclasts: localization by labeled inhibitor autoradiography, *Science,* 183, 432, 1974.
56. **Marie, P. J. and Hott, M.,** Histomorphometric identification of carbonic anhydrase in fetal rat bone embedded in glycolmethacrylate, *J. Histochem. Cytochem.,* 35, 245, 1987.
57. **Sundquist, K. T., Leppilampi, M., Jarvelin, K., Kumpulainen, T., and Väänänen, H. K.,** Carbonic anhydrase isoenzymes in isolated rat peripheral monocytes, tissue macrophages, and osteoclasts, *Bone,* 8, 33, 1987.
58. **Väänänen, H. K. and Parvinen, E.-K.,** High active isoenzyme of carbonic anhydrase in rat calvaria osteoclasts, immunohistochemical study, *Histochemistry,* 78, 481, 1983.
59. **Jilka, R. L., Rogers, J. I., Khalifah, R. G., and Väänänen, H. K.,** Carbonic anhydrase isoenzymes of osteoclasts and erythrocytes of osteopetrotic microphthalmic mice, *Bone,* 6, 445, 1985.
60. **Väänänen, H. K.,** Immunohistochemical localization of carbonic anhydrase isoenzymes I and II in human bone, cartilage and giant cell tumor, *Histochemistry,* 81, 485, 1984.
61. **Simasaki, M. and Yagi, T.,** Histochemistry of carbonic anhydrase with special reference to the osteoclast, *Dent. Bull. Osaka Univ.,* 1, 89, 1960.
62. **Cao, H. and Gay, C. V.,** Effects of parathyroid hormone and calcitonin on carbonic anhydrase location in osteoclasts of cultured embryonic chick bone, *Experientia,* 41, 1472, 1985.
63. **Enns, T.,** Facilitation by carbonic anhydrase of carbon dioxide transport, *Science,* 155, 44, 1967.
64. **Shultz, J. S.,** Facilitation of CO_2 through layers with a spatial distribution of carbonic anhydrase, in *Biophysics and Physiology of Carbon Dioxide,* Bauer, C., Gros, G., and Bartels, H., Eds., Springer-Verlag, New York, 1980, 15.
65. **Kuwahara, H., Shimazaki, M., Kadoya, Y., Chanoki, Y., and Mitsuhashi, T.,** Ultrastructural immunocytochemical localization of carbonic anhydrase in a benign giant cell tumor of bone, *Osaka City Med. J.,* 32, 89, 1986.
66. **Anderson, R. E., Gay, C. V., and Schraer, H.,** Ultrastructural localization of carbonic anhydrase in the chorioallantoic membrane by immunocytochemistry, *J. Histochem. Cytochem.,* 29, 1121, 1981.

67. **Gay, C. V., Faleski, E. J., Schraer, H., and Schraer, R.,** Localization of carbonic anhydrase in avian gastric mucosa, shell gland and bone by immunohistochemistry, *J. Histochem. Cytochem.*, 22, 819, 1974.
68. **Sly, W. S.,** Carbonic anhydrase II deficiency syndrome, in *The Carbonic Anhydrases,* Dodgson, S., Tashian, R. E., Gros, G., and Carter, N. D., Eds., Plenum Press, New York, 1991, 183.
69. **Waite, L. C. and Kenny, A. D.,** Acetazolamide and calcium metabolism in the rat, in *Calcitonin 1969,* Taylor, S. and Foster, G. V., Eds., William Heinemann Medical Books, London, 1970, 442.
70. **Waite, L. C., Volkert, W. A., and Kenny, A. D.,** Inhibition of bone resorption by acetazolamide in the rat, *Endocrinology,* 87, 1129, 1970.
71. **Conaway, H. H., Waite, L. C., and Kenny, A. D.,** Immobilization and bone mass in rats: effects of parathyroidectomy and acetazolamide, *Calcif. Tissue Res.*, 11, 323, 1973.
72. **Lineberry, M. D. and Waite, L. C.,** Acidosis inhibits the hypocalcemic effect of acetazolamide, *J. Pharmacol. Exp. Ther.*, 211, 452, 1979.
73. **Minkin, C. and Jennings, J. M.,** Carbonic anhydrase and bone remodeling: sulfonamide inhibition of bone resorption in organ culture, *Science,* 176, 1031, 1972.
74. **Mahgoub, A. and Stern, P. H.,** Carbon dioxide and the effect of parathyroid hormone on bone *in vitro, Am. J. Physiol.*, 226, 1272, 1974.
75. **Pierce, W. M. and Waite, L. C.,** Acetazolamide inhibition of bone resorption: lack of effect of phosphate release from bone *in vitro, Horm. Metab. Res.*, 13, 591, 1981.
76. **Hall, G. E. and Kenny, A. D.,** Parathyroid hormone increases carbonic anhydrase activity in cultured mouse calvaria, *Calcif. Tissue Int.*, 35, 682, 1983.
77. **Waite, L. C.,** Carbonic anhydrase inhibitors, parathyroid hormone and calcium metabolism, *Endocrinology,* 91, 1160, 1972.
78. **Raisz, L. G., Simmons, H. A., Thompson, W. J., Shepard, K. L., Anderson, P. S., and Rodan, G. A.,** Effects of a potent carbonic anhydrase inhibitor on bone resorption in organ culture, *Endocrinology,* 122, 1083, 1988.
79. **Hunter, S. J., Rosen, C. J., and Gay, C. V.,** *In vitro* resorptive activity of isolated chick osteoclasts: effects of carbonic anhydrase inhibition, *J. Bone Miner. Res.*, 6, 61, 1991.
80. **Anderson, R. E., Jee, W., and Woodbury, D. M.,** Stimulation of carbonic anhydrase in osteoclasts by parathyroid hormone, *Calcif. Tissue Int.*, 37, 646, 1985.
81. **Hall, G. E. and Kenny, A. D.,** Bone resorption induced by parathyroid hormone and dibutyryl cyclic AMP: role of carbonic anhydrase, *J. Pharmacol. Exp. Ther.*, 238, 778, 1986.
82. **Hunter, S. J., Schraer, H., and Gay, C. V.,** Characterization of isolated and cultured chick osteoclasts: the effects of acetazolamide, calcitonin and parathyroid hormone on acid production, *J. Bone Miner. Res.*, 3, 297, 1988.
83. **Duong, L. T., Glasser, W., DeHaven, P. A., and Sato, M.,** Parathyroid hormone receptors identified on avian and rat osteoclasts, *J. Bone Miner. Res.*, 5, S203, 1990.
84. **Agarwala, N. and Gay, C. V.,** Specific binding of parathyroid hormone to living osteoclasts, *J. Bone Miner. Res.*, in press.
85. **Teti, A., Rizzoli, R., and Zambonin-Zallone, A.,** Parathyroid hormone binding to cultured avian osteoclasts, *Biochem. Biophys. Res. Commun.*, 174, 1217, 1991.
86. **Rao, L. G., Murray, T. M., and Heersche, J. N. M.,** Immunohistochemical demonstration of parathyroid hormone binding to specific cell types in fixed rat bone tissue, *Endocrinology,* 113, 805, 1983.
87. **Silverton, S. F., Dodgson, S. J., Fallon, M. D., and Forster, R. E.,** Carbonic anhydrase activity of chick osteoclasts is increased by parathyroid hormone, *Am. J. Physiol.*, 253, E670, 1987.
88. **Dietsch, P.,** Parathyroid hormone, carbonic anhydrase and calcium homeostasis, in *Generalized Bone Diseases,* Kuhlencordt, F., Dietsch, P., Keck, E., and Kruse, H.-P., Eds., Springer-Verlag, Berlin, 1987, 279.

89. **Ito, M. B., Schraer, H., and Gay, C. V.,** The effects of calcitonin, parathyroid hormone and prostaglandin E_2 on cyclic AMP levels of isolated osteoclasts, *Comp. Biochem. Physiol. A,* 81, 653, 1985.
90. **Pierce, W. M., Lineberry, M. D., and Waite, L. C.,** Effect of sulfonamides on the hypercalcemic response to vitamin D, *Horm. Metab. Res.,* 14, 670, 1982.
91. **Hall, G. E. and Kenny, A. D.,** Role of carbonic anhydrase in bone resorption induced by 1,25 dihydroxyvitamin D_3 *in vitro, Calcif. Tissue Int.,* 37, 134, 1985.
92. **Billecocq, A., Emanuel, J. R., Levenson, R., and Baron, R.,** 1α,25-dihydroxyvitamin D_3 regulates the expression of carbonic anhydrase II in nonerythryoid avian bone marrow cells, *Proc. Natl. Acad. Sci. U.S.A.,* 87, 6470, 1990.
93. **Baron, R., Neff, L., Louvard, D., and Courtoy, P. J.,** Cell-mediated extracellular acidification and bone resorption: evidence for a low pH in resorbing lacunae and localization of a 100-kD lysosomal membrane protein at the osteoclast ruffled border, *J. Cell Biol.,* 101, 2210, 1985.
94. **Akisaka, T. and Gay, C. V.,** Ultracytochemical evidence for a proton-pump adenosine triphosphatase in chick osteoclasts, *Cell Tissue Res.,* 245, 507, 1986.
95. **Blair, H. C., Teitelbaum, S. L., Ghiselli, R., and Gluck, S.,** Osteoclastic bone resorption by a polarized vacuolar proton pump, *Science,* 245, 855, 1989.
96. **Bekker, P. J. and Gay, C. V.,** Biochemical characterization of an electrogenic vacuolar proton pump in purified chicken osteoclast plasma membrane vesicles, *J. Bone Miner. Res.,* 5, 569, 1990.
97. **Väänänen, H. K., Karhukorpi, E. K., Sundquist, K., Wallmark, B., Roininen, I., Hentunen, T., Tuukkanen, J., and Lakkakorpi, P.,** Evidence for the presence of a proton pump of the vacuolar proton ATPase type in the ruffled borders of osteoclasts, *J. Cell Biol.,* 111, 1305, 1990.
98. **Tuukkanen, J. and Väänänen, H. K.,** Omeprazole, a specific inhibitor of H^+,K^+-ATPase, inhibits bone resorption *in vitro, Calcif. Tissue Int.,* 38, 123, 1986.
99. **Anderson, R. E., Woodbury, D. M., and Jee, W. S. S.,** Humoral and ionic regulation of osteoclast acidity, *Calcif. Tissue Int.,* 39, 252, 1986.
100. **Akisaka, T. and Gay, C. V.,** An ultracytochemical investigation of ouabain-sensitive *p*-nitrophenylphosphatase in chick osteoclasts, *Cell Tissue Res.,* 244, 57, 1986.
101. **Baron, R., Neff, L., Roy, C., Boisvert, A., and Caplan, M.,** Evidence for a high and specific concentration of (Na^+,K^+) ATPase in the plasma membrane of the osteoclast, *Cell,* 46, 311, 1986.
102. **Krieger, N. S. and Tashjian, A. H.,** Inhibition by ouabain of parathyroid hormone-stimulated bone resorption, *J. Pharmacol. Exp. Ther.,* 217, 586, 1981.
103. **Akisaka, T., Yamamoto, T., and Gay, C. V.,** Ultracytochemical investigation of calcium-activated adenosine triphosphatase (Ca^{++}-ATPase) in chick tibia, *J. Bone Miner. Res.,* 3, 19, 1988.
104. **Bekker, P. J. and Gay, C. V.,** Characterization of a Ca^{2+}-ATPase in osteoclast plasma membrane, *J. Bone Miner. Res.,* 5, 557, 1990.
105. **Rega, A. F. and Garrahan, P. J., Eds.,** *The Ca^{++} Pump of Plasma Membranes,* CRC Press, Boca Raton, FL, 1986, 56.
106. **Schatzmann, H. J.,** Calcium extrusion across the plasma membrane by the calcium-pump and the Ca^{++}-Na^+ exchange system, in *Calcium and Cell Physiology,* Marmé, D., Ed., Springer-Verlag, Berlin, 1985, 18.
107. **Burstone, M. S.,** Histochemical demonstration of cytochrome oxidase activity in osteoclasts, *J. Histochem. Cytochem.,* 8, 225, 1960.
108. **Raekallio, J., Kovács, M., and Mäkinen, P.-L.,** The appearance of oxidoreductases in healing fractures, *Acta Pathol. Microbiol. Scand. Sect. A,* 78, 658, 1970.
109. **Bely, M. and Tanka, D.,** Cytochrome oxidase activity in bone tissue, *Acta Morphol. Acad. Sci. Hung.,* 34, 261, 1982.

110. **Balogh, K.,** Further observations on oxidative enzyme activities in decalcified bone and teeth, *J. Histochem. Cytochem.*, 12, 485, 1964.
111. **Addison, W. C.,** Enzyme histochemical properties of kitten osteoclasts in bone imprint preparations, *Histochem. J.*, 10, 645, 1978.
112. **Noda, K., Kazama, T., Nakamura, Y., and Hirashita, A.,** The localization of cytochrome C oxidase activity in osteoclasts, *J. Electron Microsc.*, 34, 222, 1985.
113. **Noda, K.,** Localization of cytochrome C oxidase activity in osteoclasts treated with calcitonin during experimental tooth movement, *Acta Histochem. Cytochem.*, 21, 301, 1988.
114. **Noda, K. and Kuwahara, Y.,** Cytochrome C oxidase activity in preosteoclasts induced by experimental tooth movement, *J. Electron Microsc.*, 39, 97, 1990.
115. **Seligman, A. M., Karnovsky, M. J., Wasservrug, H. L., and Hanker, J. S.,** Nondroplet ultrastructural demonstration of cytochrome oxidase activity with a polymerizing osmiophillic reagent, diaminobenzidine (DAB), *J. Cell Biol.*, 38, 1, 1968.
116. **Anderson, W. A., Bara, G., and Seligman, A. M.,** The ultrastructural localization of cytochrome oxidase via cytochrome C, *J. Histochem. Cytochem.*, 23, 13, 1975.
117. **Perotti, M. E., Anderson, W. A., and Swift, H.,** Quantitative cytochemistry of the diaminobenzidine cytochrome oxidase reaction product in mitochondria of cardiac muscle and pancreas, *J. Histochem. Cytochem.*, 31, 351, 1983.
118. **Yamamoto, T. and Gay, C. V.,** Ultrastructural localization of adenylate cyclase in chicken bone cells, *J. Histochem. Cytochem.*, 37, 1705, 1989.
119. **Eliam, M. C., Baslé, M., Bouizar, Z., Bielakoff, J., Moukhtar, M., and de Vernejoul, M. C.,** Influence of blood calcium on calcitonin receptors in isolated chick osteoclasts, *J. Endocrinol.*, 119, 243, 1988.
120. **Chambers, T. J.,** Osteoblasts release osteoclasts from calcitonin-induced quiescence, *J. Cell Sci.*, 57, 247, 1982.
121. **Cutler, L. S.,** Cytochemical methods for the localization of adenylate cyclase. A review and evaluation of the efficacy of the procedures, *J. Histochem. Cytochem.*, 31, 85, 1983.
122. **Poeggel, G., Luppa, H., Bernstein, H.-G., and Weiss, J.,** Histochemistry of adenylate cyclase, *Int. Rev. Cytol.*, 89, 35, 1984.
123. **Hammarström, L. E., Hanker, J. S., and Toverud, S. U.,** Cellular differences in acid phosphatase isoenzymes in bone and teeth, *Clin. Orthop.*, 78, 151, 1971.
124. **Anderson, T. R. and Toverud, S. U.,** Chromatographic separation of two acid phosphatases from rat bone, *Calcif. Tissue Res.*, 24, 187, 1977.
125. **Akisaka, T., Subita, G. P., Kawaguchi, H., and Shigenaga, Y.,** Different tartrate sensitivity and pH optimum for two isoenzymes of acid phosphatase in osteoclasts. An electron-microscope enzyme-cytochemical study, *Cell Tissue Res.*, 255, 69, 1989.
126. **Clark, S. A., Ambrose, W. W., Anderson, T. R., Terrell, R. S., and Toverud, S. U.,** Ultrastructural localization of tartrate-resistant, purple acid phosphatase in rat osteoclasts by histochemistry and immunocytochemistry, *J. Bone Miner. Res.*, 4, 399, 1989.
127. **Fukushima, O., Bekker, P. J., and Gay, C. V.,** Ultrastructural localization of tartrate-resistant acid phosphatase (purple acid phosphatase) activity in chicken cartilage and bone, *Am. J. Anat.*, 191, 228, 1991.
128. **Andersson, G. N., Ek-Rylander, B., Hammarström, L. E., Lindskog, S., and Toverud, S. U.,** Immunocytochemical localization of a tartrate-resistant and vanadate-sensitive acid nucleotide tri- and diphosphatase, *J. Histochem. Cytochem.*, 34, 293, 1986.
129. **Reinholt, F. P., Silwa, M. W., Ek-Rylander, B., and Andersson, G.,** Ultrastructural localization of a tartrate-resistant acid ATPase in bone, *J. Bone Miner. Res.*, 5, 1055, 1990.

130. **Baron, R., Neff, L., Brown, W., Courtoy, P. J., Louvard, D., and Farquhar, M. G.,** Polarized secretion of lysosomal enzymes: co-distribution of cation-independent mannose-6-phosphate receptors and lysosomal enzymes along the osteoclast exocytic pathway, *J. Cell Biol.,* 106, 1863, 1988.
131. **Baron, R., Neff, L., Brown, W., Louvard, D., and Courtoy, P. J.,** Selective internalization of the apical plasma membrane and rapid redistribution of lysosomal enzymes and mannose-6-phosphate receptors during osteoclast inactivation by calcitonin, *J. Cell Sci.,* 97, 439, 1990.
132. **Doty, S. B.,** Histochemistry and enzymology of osteoclasts, in *Bone,* Vol. 2, *The Osteoclast,* Hall, B. K., Ed., CRC Press, Boca Raton, FL, 1991, 61.
133. **Doty, S. B. and Schofield, B. H.,** Ultrahistochemistry of calcified tissues, in *Methods of Calcified Tissue Preparation,* Dickson, G. R., Ed., Elsevier, New York, 1984, 149.
134. **Touw, J. A., Hemrika-Wagner, A. M., and Vermeiden, J. P. W.,** An electron microscopic, enzyme cytochemical study on the localization of lactate dehydrogenase (LDH) in osteoclasts and peritoneal macrophages of the rat and its implication for the process of bone resorption and the origin of osteoclasts, *Cell Tissue Res.,* 209, 111, 1980.
135. **Doty, S. B., Schofield, B. H., and Robinson, R. A.,** The electron microscopic identification of acid phosphatase and adenosinetriphosphatase in bone cells following parathyroid extract or thyrocalcitonin administration, in *Parathyroid Hormone and Thyrocalcitonin (Calcitonin). Proceedings of the Third Parathyroid Conference,* Talmage, R. V., Bélanger, L. F., and Clark, I., Eds., Excerpta Medica, Amsterdam, 1968, 169.
136. **Van Noorden, C. J. F., Vogels, I. M. C., and Smith, R. E.,** Localization and cytophotometric analysis of cathepsin B activity in unfixed and undecalcified cryostat sections of whole rat knee joints, *J. Histochem. Cytochem.,* 37, 617, 1989.
137. **Goshi, N. and Fukushima, O.,** The histochemical properties of the neutral pyrophosphatase activity in lysosomes of osteoclasts, *Acta Histochem. Cytochem.,* 16, 31, 1983.
138. **Baron, R., Eeckhout, Y., Neff, L., Francois-Gillet, C., Hennet, P., Delaisse, J. M., and Vaes, G.,** Affinity purified antibodies reveal the presence of (pro) collagenase in the subosteoclastic bone resorbing compartment, *J. Bone Miner. Res.,* 5, S203, 1990.
139. **Coleman, R. A., Ramp, W. K., Toverud, S. U., and Hanker, J. S.,** Electron microscopic localization of lactate dehydrogenase in osteoclasts of chick embryo tibia, *Histochem. J.,* 8, 543, 1976.
140. **Tanaka, T. and Tanaka, M.,** Cytological and functional studies of preosteoclasts and osteoclasts in the alveolar bones from neonatal rats using microperoxidase as a tracer, *Calcif. Tissue Int.,* 42, 267, 1988.
141. **Sasaki, T., Yamaguchi, A., Higashi, S., and Yoshiki, S.,** Uptake of horseradish peroxidase by bone cells during endochondral bone development, *Cell Tissue Res.,* 239, 547, 1985.
142. **Kadoya, Y., Nagahama, S., Kuwahara, H., Shimazaki, M., Shimazu, A., Ogawa, Y., and Yagi, T.,** Isolation of avian carbonic anhydrase and its immunohistochemical demonstration in medullary bone, *Osaka City Med. J.,* 33, 111, 1987.
143. **Väänänen, H. K. and Parvinen, E.,** Localization of carbonic anhydrase isozymes in calcified tissues, in *The Carbonic Anhydrases,* Dodgson, S., Tashian, R E., Gros, G., and Carter, N. D., Eds., Plenum Press, New York, 1991, 351.
144. **Blair, H. C., Dean, D. D., Howell, D. S., Teitelbaum, S. L., and Jeffrey, J. J.,** Hypertrophic chondrocytes produce immunoreactive collagenase *in vivo, Connect. Tissue Res.,* 23, 65, 1989.
145. **Case, J. P., Sano, H., Lafyatis, R., Remmers, E. F., Kumkumian, G. K., and Wilder, R. L.,** Transin/stromelysin expression in the synovium of rats with experimental erosive arthritis. *In situ* localizaion and kinetics of expression of the transformation-associated metalloproteinase in euthymic and athymic Lewis rats, *J. Clin. Invest.,* 84, 1731, 1989.

146. **Fukushima, O. and Gay, C. V.,** Ultrastructural localization of guanylate cyclase in bone cells, *J. Histochem. Cytochem.*, 39, 529, 1991.
147. **Delaisse, J.-M.,** personal communication.
148. **Gay, C. V. and Yamamoto, T.,** unpublished observation.
149. **Gay, C. V.,** unpublished observation.
150. **Bekker, P. J. and Gay, C. V.,** Demonstration of calmodulin-sensitive calcium translocation by isolated osteoclast plasma membrane vesicles, *Calcif. Tissue Int.*, in press.

Chapter 7

SYSTEMIC FACTORS REGULATING OSTEOCLAST FUNCTION

Johan N. M. Heersche

TABLE OF CONTENTS

I. INTRODUCTION

A. OSTEOCLASTIC ACTIVITY

Osteoclasts are mononuclear or multinucleated cells that stain positive for tartrate-resistant acid phosphatase[1] and are able to resorb bone. When resorbing bone, they display specialized cell structures: an attachment zone, called the clear zone or sealing zone,[2] which contains microfilaments and which surrounds and seals an area of extensive membrane ruffling. Current evidence indicates that osteoclasts, once attached to mineralized bone matrix, can initiate the resorptive process and that demineralization occurs between the ruffled border and the bone surface by acidification of this extracellular compartment.[3,4]

It is not clear how the sites to be resorbed are selected, what triggers osteoclasts to start resorbing, and what determines the depth and the lateral extension of the resorption lacuna. It is evident, however, that in the course of physiological remodeling all three aspects of resorptive activity must be regulated. With regard to the selection of resorptive sites, osteoblast-like cells are likely to play a role by regulating access of osteoclasts to the bone surface, as was suggested by Jones and Boyde[5] and Rodan and Martin.[6] According to this view, all osteoclasts are potentially active, but are prevented from actually resorbing bone by cells from the osteoblast lineage that form a continuous protective layer over the bone surface. Since in the adult (human) about 80% of the trabecular bone surface and about 95% of the intracortical surface is covered by inactive osteoblasts or lining cells,[7] the lining cells seem to be the most likely candidates for this regulatory function. The properties of lining cells, however, are virtually unknown: their function is not clear and the factors to which they respond are not known.

Degradation of nonmineralized osteoid is regulated differently from resorption of mineralized bone, and probably involves cells other than osteoclasts.[8-12] In areas where the bone surface is covered by osteoid, degradation of nonmineralized collagen may play an important role in the initiation of bone resorption. This is strongly supported by the observations of Delaisse et al.,[13] who found that a synthetic inhibitor of mammalian tissue collagenase, CL1, almost completely inhibited the formation of resorption lacunae in parathyroid hormone-treated cultured fetal mouse calvaria. Thus, collagenase-mediated degradation of unmineralized collagen appeared to be a prerequisite for the initiation of osteoclast-mediated resorption. That the cells responsible for degradation of osteoid in the resorption process are the osteoblasts and/or the lining cells is suggested by the observations of Sakamoto et al.,[14] who found that osteoblasts, but not osteoclasts, are the source of procollagenase or collagenase in bone tissue.

When isolated osteoclasts are cultured on devitalized slices of mineralized cortical bone, resorption can be studied without the influence of the cell layer normally covering bone surfaces *in vivo* and without the added complication of the presence of a nonmineralized layer of osteoid and/or an unmineralizable connective tissue component.[15,16] Using such a system, we observed[17] that osteoclasts move along the bone surface with the characteristic ruffed advancing edge that is also observed when these cells migrate on glass or plastic surfaces. The osteoclasts appeared to move randomly over the bone surface, not unidirectionally. However, exposed osteocytic lacunae and exposed canals originally containing blood vessels seemed to be preferred sites for osteoclasts to start resorbing. In contrast, osteoclasts very rarely appeared to start resorption on preexcavated sites. Two phases of osteoclast motility could be distinguished: a rapid migratory phase, during which osteoclasts do not resorb, and a resorptive phase. During the resorptive phase, the cells are far less motile, display a characteristic dome-shaped bulge on their superior surface in scanning electron microscopy (SEM) observations, and display adhesion and microfilament distribution patterns that differ from those of rapidly migrating osteoclasts.[18] The migratory phase and the resorptive phase are distinct, and follow each other. We further observed that the same osteoclast can simultaneously resorb two different sites at different rates,[17] suggesting that the rate of resorption is not governed solely by the "state of activity" of the osteoclast, but that other factors play a role in determining the actual amount of bone removed at an individual resorption site. With regard to the rate of resorption, considerable differences were found not only between different osteoclasts, but also between rates measured at different times for the same osteoclast.

Other issues to be considered in a discussion of the regulation of osteoclastic activity by systemic factors are direct vs. indirect stimulation of osteoclastic activity by these factors, the possibility that changes in the responsiveness of osteoclasts to extracellular signals occur by modulation of the number of receptors for these signals, and the fact that bone remodeling is not uniform, but differs between cancellous and cortical bone and may also differ from one cancellous bone site to another. In addition, differences between osteoclasts of different species and genera have to be considered. These aspects will be discussed in the following sections.

B. OSTEOCLAST RECRUITMENT

It is evident that factors that stimulate or inhibit bone resorption could act by changing either the activity or the number of resorbing osteoclasts, or both. It is well established that osteoclasts are cells from the hemopoietic lineage. Early evidence that osteoblasts and osteoclasts might differentiate from two types of precursor cells was provided by Scott,[18a] who observed two types of proliferating cells in the vicinity of blood vessels in the primary spongiosa of rat embryos. She correctly identified cells with some of the morphological characteristics of osteoblasts as the osteoblast progenitors, and mononuclear cells with some of the characteristics of osteoclasts as the osteoclast precursors. Later experiments by Walker[19] made it virtually certain that osteoclasts were cells from the hemopoietic lineage: he transplanted spleen cells or bone marrow cells from normal mice into irradiated osteopetrotic littermates and found that this resulted in the formation of normal osteoclasts. This was confirmed, among others, by Scheven et al.,[20] who cocultured purified bone marrow fractions with living osteoclast-free embryonic mouse long-bone rudiments and found that the fraction containing hemopoietic stem cells gave rise to osteoclasts. The same authors demonstrated that, at least in embryonic mouse long bones, proliferating osteoclast progenitors are seeded via the bloodstream in the mesenchyme surrounding the bone rudiments, where they then proliferate, differentiate, and fuse into multinucleated bone-resorbing osteoclasts.[21]

II. THE MAJOR SYSTEMIC REGULATORS OF OSTEOCLAST ACTIVITY AND OSTEOCLAST NUMBERS: PARATHYROID HORMONE, CALCITONIN, AND 1,25-DIHYDROXYVITAMIN D_3

A. PARATHYROID HORMONE

A prolonged increase in circulating concentrations of parathyroid hormone (PTH) is associated with increased bone turnover, i.e., increased osteoclastic bone resorption and increased osteoblastic activity.[22,23] The bone abnormalities associated with hyperparathyroidism are most striking in cortical bone: histological evaluation of iliac crest biopsies shows that the area occupied by cancellous bone increases at the expense of cortical bone and that increased osteoclastic activity is particularly evident at the endosteal cortical surfaces. Whether this is because the osteoclast populations are different at different anatomical sites, or whether these differences are caused by differences in the populations of osteoblast-like cells or stromal cells in those locations, is not clear.

There exists general agreement, however, that the effects of PTH on bone *in vivo* include stimulation of osteoclastic activity together with an increase in osteoclast numbers. That this is the result of a direct effect of PTH on bone tissue was first shown in intracerebral grafts of bone together with parathyroid tissue.[24] Based on these experiments, Gaillard[25] cocultured human, chick, or mouse parathyroid tissue with parietal bone fragments of mouse embryos and observed a clear stimulation of osteoclastic bone resorption *in vitro* in bone cocultured with parathyroid tissue. Later experiments by Gaillard and by several other investigators, using purified PTH and cultures of embryonic long bones or calvaria from either mouse or rat,[26-28] have confirmed these initial observations and resulted in the long-held view that PTH directly stimulated osteoclast formation and osteoclast activity. Whether PTH indeed stimulates osteoclasts directly is not definitively resolved, but appears less likely now than it did before.

In vitro experiments also consistently show that PTH inhibits osteoblastic activity. This is strikingly different from observations *in vivo* in hyperparathyroid patients, where an increase in osteoblastic activity is seen. The question is whether this stimulatory effect of PTH on osteoblasts *in vivo* is direct or indirect.

A large number of studies using either osteoblast-like populations derived from bone tissue by collagenase digestion or clonal osteoblast-like cell lines derived from normal bone or osteosarcomas, and studies with isolated osteoclast-containing cell populations, have provided new insight into what agents affect osteoclastic and osteoblastic activity directly or indirectly. It is now clear that PTH acts directly on osteoblasts via cAMP and/or phosphatidyl inositol-dependent, receptor-mediated events,[29-31] and that its effect is generally inhibitory with regard to matrix deposition. Studies with isolated osteoclast populations, on the other hand, suggest that PTH does not directly stimulate osteoclastic activity,[32] but stimulates osteoclastic bone resorption indirectly via osteoblast-like cells.[33] In addition to this, other stimulators of bone resorption i.e., (IL-1), tumor necrosis factor (TNF)-α and -β and $1,25(OH)_2D_3$ also stimulated the bone resorptive activity of isolated osteoclasts in the presence, but not in the absence of osteoblast-like cell populations.[34,35]

Since the first report on the existence of a putative osteoclast-activating factor released by osteoblasts after stimulation with PTH, several laboratories have attempted to isolate these factors. However, such factors do not always appear to be present in PTH-stimulated cocultures of osteoblast-like cells and osteoclasts.[117] A further complication is that a large variety of cultured cells and tissues produce bone-resorbing factors *in vitro*.[36-40] For rat osteosarcoma cells, Rodan et al.[39] have shown that prostaglandin E_2 is, in all likelihood, the resorption-stimulating factor, whereas the bone resorptive activity present in conditioned media from fibroblast-like cells derived from alveolar mucosa, gingiva, and periodontal ligament, and from the epithelial cell rests of Malassez[40] was based on the presence of factors other than prostaglandins.

One of the reasons that studying direct or indirect effects on osteoclastic activity is so difficult is that the system used, i.e., dispersed osteoclast-containing cell populations cultured on devitalized bone slices, is much more complicated than was thought initially. First, survival and resorptive activity of isolated osteoclasts appears to depend on the procedures used to isolate them. In our hands, large multinucleated osteoclasts containing more than ten nuclei predominate when cells are isolated using manual agitation of bone fragments, while smaller osteoclasts predominate when using a shaker to agitate the bone fragments.[117] We do not yet know whether such populations of osteoclasts respond differently to agonists. Species differences may also play a role: osteoclasts isolated from newborn rabbit long bones appear to survive for several days in bicarbonate-buffered culture media containing α-minimal essential medium (MEM) and 15% bovine fetal serum, while the majority of osteoclasts isolated from newborn rats survive less than 24 h under similar culture conditions.[41,117] It has also been mentioned that osteoclasts cultured in phosphate-free Eagles MEM and 10% fetal calf serum resorb more extensively than those cultured in other culture media.[32] The possibility that osteoclasts which resorb bone at a maximal rate cannot be stimulated further seems real and should be investigated. It is of interest in this regard that no further stimulation of bone resorption can be achieved by adding PTH or $1,25(OH)_2D_3$ when resorption is maximally stimulated (by unknown factors in the serum or the albumin added to the culture media) in organ cultures of mouse and rat calvaria or rat long bones.[117,118]

Another variable of isolated osteoclast preparations is the varying number of mesenchymal cells and cells of hemopoietic lineage that are present in addition to the osteoclasts, and the degree to which these cells may modify the responsiveness of the system. Relevant to this, it was found that a 24-h incubation with PTH had no stimulatory effect when rabbit osteoclasts were isolated and cultured as described by Chambers et al.,[32] whereas PTH significantly stimulated bone resorption by rat osteoclasts cultured in bicarbonate buffered medium 199 (pH 6.9) with 10% fetal calf serum (FCS).[42] Although the PTH response in the latter experiments was attributed to osteoblast-like cells present in the cultures, other explanations would still seem possible. Of interest here is that observations by Arnett and Dempster[43] and Walsh et al.[44] document that the pH of the culture medium dramatically affects osteoclast activity. Our own observations[45] that pH not only affects osteoclast activity, but also osteoclast survival, and osteoclast formation from progenitor cells add one more factor to be considered.

In order to respond directly to PTH, osteoclasts would have to possess receptors for this hormone. That this is so is not certain. We have shown, using paraffin sections of long bones of 2- to 3-d-old rats, that all identifiable osteoclasts stained intensely when such sections were incubated with high concentrations of bovine PTH 1-84 and then with antiserum to PTH.[46] Staining was also observed in osteocytes and osteoblasts. An inherent shortcoming of the immunohistochemical technique, however, is that it is impossible to distinguish between specific binding of the hormone to a receptor and nonspecific binding of hormone to cell constituents, so this experiment was not conclusive. Specific binding of iodinated or biotinylated bovine PTH 1-84 to chicken osteoclasts was reported by Teti et al.,[47] Agarwala and Gay,[48] and Duong et al.,[49] but others have been unable to demonstrate specific binding of PTH 1-34 osteoclasts in chicken long bones[50] and in rats infused with iodinated PTH 1-34.[51]

In summary, the experimental evidence with regard to osteoclasts responding indirectly to PTH appears to be inconclusive: responsiveness to PTH in isolated osteoclast systems appears to depend on the culture conditions, and the possibility that specific PTH receptors are present on osteoclasts has not been eliminated.

B. CALCITONIN

After the discovery by Copp et al.[52] of the new hypocalcemic hormone calcitonin (CT), Milhaud et al.[53] were the first to demonstrate that its hypocalcemic effect was caused by decreased calcium mobilization from bone. Bone culture studies[54,55] showed conclusively that CT inhibited bone demineralization. Very little controversy exists regarding its mechanism of action: it has been demonstrated, both *in vivo* and *in vitro,* that osteoclasts have receptors for CT[56] and that CT directly acts on osteoclasts.[57,58]

Raisz and co-workers were the first to observe that the inhibition of bone resorption by CT in cultured long bones of newborn rats was transient under certain conditions. They demonstrated[59-61] that the inhibition of PTH-induced bone resorption by CT lasted from 48 to 96 h, depended on the CT concentration used, was not due to inactivation of CT, was dependent on the calcium concentration of the medium, and occurred with salmon, human, and rat CT. This phenomenon was described as the ''escape phenomenon''. Tashjian et al.,[62] in an extensive and detailed reexamination of the ''escape phenomenon'', essentially confirmed their observations. However, the Tashjian et al. results differed from those obtained by Raisz et al. in that escape was not prevented by decreasing the calcium concentration of the culture medium, did not occur unless resorption was stimulated by PTH or PGE_2, and was not reproducibly observed in culture media containing albumin instead of serum.

A possible explanation for the observed differences could be that rat and mouse tissues respond differently. This was investigated by Heersche,[63] who showed that when porcine CT was used in concentrations of 5 or 50 MRC mU/ml, escape occurred with mouse calvaria, but not with rat calvaria. Escape also occurred in cultures where resorption was not stimulated by PTH. However, experiments with much lower CT concentrations (0.1 and 0.01 MRC mU/ml) showed that at these CT concentrations, escape occurred not only in mouse, but also in rat calvaria. These results suggest that escape is dependent on the concentration of CT used, and that at a given concentration of CT, escape may or may not occur, depending on the species of origin of the tissue. In a study on the time course of escape from CT-induced inhibition of motility and resorption of disaggregated osteoclasts, Kanehisa[58] demonstrated that of 23 osteoclasts observed directly, nine resumed motility and resorptive activity after initial inhibition by CT. The resumption of cytoplasmic spreading occurred after approximately 4 h, while resumption of resorptive activity in these same osteoclasts occurred approximately 23 h after the addition of CT. Whether escape in these osteoclasts was dose dependent and could be overcome by increasing the CT concentration was not investigated. What these results do show, however, is that there is heterogeneity with regard to responsiveness of isolated osteoclasts to CT in terms of cytoplasmic motility and resorptive activity. This heterogeneity may be related to the number of CT receptors on these cells, which in turn could depend on the age of the individual osteoclasts.

Osteoclasts derived from neonatal rats respond to CT with an increase in intracellular cAMP, as detected biochemically and immunohistochemically, and with cytoplasmic contraction.[41] Chick osteoclasts, on the other hand, did not appear to respond to CT with an increase in cAMP content and did not bind iodinated CT or respond to CT with cytoplasmic contraction, thus suggesting that chicken osteoclasts have no CT receptors and do not respond to CT.[41] Since chicken osteoclasts clearly resorb bone, both *in vivo* and *in vitro*, this might indicate that bone resorption in birds is regulated in a different way than in mammals.

Prolonged administration of CT *in vivo* also results in an apparent loss of responsiveness to this hormone. No controversy exists with regard to this loss of responsiveness when low to intermediate doses of CT are used.[64-66] The cause of this phenomenon, however, is not so clear. Removal of the CT-releasing mini pumps at day 7 in the experiments of Lausson et al.[66] resulted in an immediate (0 to 2 h after removal) increase of serum Ca, thus suggesting that the apparent loss of responsiveness was not caused by a loss of response to CT, but by compensatory mechanisms. In another series of experiments in rats, Glajchen et al.[67] administered high doses of CT for periods of up to 42 d and found that at all times during the treatment period, ionized plasma calcium levels were slightly decreased. Measurement of serum PTH levels in these animals also revealed that infusion of CT had resulted in secondary hyperparathyroidism, which appeared to have ameliorated the CT-induced hypocalcemia. This confirmed the initial suggestion by Bastian et al.[68] that the development of tolerance to CT might be due, in part, to an increase in endogenous PTH secretion. In humans, loss of responsiveness to CT during prolonged administration in patients with Paget's disease or various osteoporotic syndromes also occurs. This has been attributed in some of the patients to formation of neutralizing antibodies,[69] but in others no detectable antibody formation was found.

In summary, the *in vivo* observations indicate that loss of responsiveness to CT may have three components: (1) loss of CT action through development of antibodies after several months of administration, (2) loss of response due to development of secondary hyperparathyroidism, and (3) a dose-dependent escape from the direct inhibitory action on osteoclasts through mechansims that are not clear.

C. 1,25-DIHYDROXYVITAMIN D_3

1,25-dihydroxyvitamin D_3, $1,25(OH)_2D_3$, is generally considered to be the most important metabolite of vitamin D with regard to the regulation of bone metabolism. It is a potent inducer of bone resorption in mouse and rat bone tissue *in vitro*[70,71] and stimulates bone resorption *in vivo* in both intact and parathyroidectomized animals,[72,73] but at doses that are generally higher than physiological. In short-term organ cultures of bone, the major effect of $1,25(OH)_2D_3$ appears to be an increase in osteoclastic activity,[71] without an increase in osteoclast number. In dispersed osteoclast systems, no direct effects of 1,25$(OH)_2D_3$ on bone resorption have been found, but in cocultures of osteoblast-like cells and osteoclast-containing cell populations, 1,25$(OH)_2D_3$ stimulates resorption.[74] Thus, the experimental evidence for or against a direct action of 1,25$(OH)_2D_3$ on osteoclasts is similar to that for PTH. Receptors for 1,25$(OH)_2D_3$ are present in virtually every cell type studied, including osteoblasts and precursors of osteoclasts, but such receptors have not been demonstrated in osteoclasts.[75]

The increased bone resorption in response to administration of 1,25$(OH)_2D_3$ *in vivo* is clearly associated with an increase in the number of osteoclasts.[76] It is difficult, however, to explore direct effects of vitamin D metabolites on bone tissue in intact animals, because vitamin D deficiency results in hypocalcemia and hyperparathyroidism,[77] while vitamin D administration results in hypercalcemia and hypersecretion of CT.[78] Since both PTH and CT affect osteoclast proliferation and function, it is difficult to ascribe the effects of vitamin D deficiency or hypervitaminosis D to the direct effects of 1,25$(OH)_2D_3$ or other vitamin D metabolites. In thyroparathyroidectomized animals, however, 1,25$(OH)_2D_3$ also stimulates resorption,[73] although the effects of 1,25$(OH)_2D_3$ on osteoclastic resorption appear to be dependent on the calcium content of the diet: in rats fed a 0.05% calcium diet, 1,25$(OH)_2D_3$ caused an increase in osteoclastic activity, whereas when fed a 1% calcium diet, no effects on osteoclastic activity were observed,[79] thus suggesting that the expression of 1,25$(OH)_2D_3$ effects on osteoclast activity is dependent on the calcium requirement of the organism.

D. EFFECTS OF PTH, CALCITONIN, AND 1,25$(OH)_2D_3$ ON OSTEOCLAST DIFFERENTIATION

The effects of PTH and 1,25$(OH)_2D_3$ on bone tissue *in vitro* and *in vivo* consistently include increases in osteoclast numbers. In a mouse bone marrow culture system in which osteoclast-like multinucleated cells are formed after 8 d of culture in the presence of PTH or 1,25$(OH)_2D_3$, Takahashi et al.[80] showed conclusively that both hormones stimulated formation of tartrate-resistant acid phosphatase (TRAP)-positive mononuclear and multinucleated cells. The observation that most TRAP-positive colonies were found adjacent to alkaline phosphatase-positive stromal cell colonies suggests that a specific type of stromal cell, possibly of the osteoblast lineage, may be involved in the stimulatory action of these two hormones. CT had no effect on the number of mononuclear TRAP-positive colonies in PTH- or 1,25$(OH)_2D_3$-treated cultures, but inhibited the formation of multinucleated osteoclast-like cells in such cultures, in keeping with the observation that receptors for CT become detectable in cells of the osteoclast lineage at approximately the same stage as when these cells become TRAP-positive.[81]

It is of note that without added 1,25$(OH)_2D_3$ or PTH, no osteoclast-like cells and only a few mononuclear TRAP-positive colonies develop in the mouse bone marrow culture system, but that functional osteoclasts do develop in cocultures of cartilaginous bone rudiments devoid of osteoclasts and bone marrow cells without the addition of either PTH or $1,25(OH)_2D_3$.[20] Thus, although required in some culture systems, high concentrations of $1,25(OH)_2D_3$ or PTH do not seem to be required for osteoclast formation or differentiation in all culture systems.

III. EFFECTS OF STEROID HORMONES, PARTICULARLY ESTROGEN AND CORTICOSTEROIDS

Estrogen deficiency causes osteoporosis as does chronic treatment with high doses of corticosteroids (for review see References 82 and 83). There is little doubt, therefore, that both hormones are systemic regulators of bone metabolism.

A closer look at estrogen deficiency in humans and rats reveals that the major effect of estrogen deficiency in terms of osteoclastic activity is an increase in the rate of bone remodeling, although there is evidence that the changes in bone metabolism cannot be accounted for by an increase in the frequency of remodeling activation alone.[84] It is also of interest that bone loss in estrogen-deficient rats is not uniform: the amount of cortical bone does not decrease significantly, while the amount of cancellous bone in the metaphysis of the long bones decreases dramatically.[85-87] Estrogen treatment of estrogen-deficient postmenopausal women does not change the average depth of the osteoclastic resorption lacunae,[83] which suggests that the resorptive activity of individual osteoclasts is not affected by estrogen.

Attempts to demonstrate the effects of estrogen on bone resorption in organ cultures or in isolated dispersed osteoclasts cultured on devitalized bone slices have generally been unsuccessful.[88,89] Recently, however, slight inhibition by 17β-estradiol of PTH-stimulated resorption was demonstrated in cultured mouse calvaria under carefully chosen culture conditions.[90] Estrogen appears to have no effect on the development of TRAP-positive mononuclear or multinucleated cells in mouse bone marrow cells cocultured with RCJ 3.1 C 5.18 cells.[119]

The information available to date thus suggests that estrogen deficiency *in vivo* increases osteoclast numbers, that this increase occurs primarily on endosteal cancellous bone surfaces, and that estrogen treatment reverses this effect. The mechanisms appear to involve factors or cells that are not present in bone marrow cultures or in bone tissue culture systems as currently used.

With regard to corticosteroids, it appears that a clear distinction has to be made between the effects of pharmacological doses, as used in the treatment of asthma and other inflammatory diseases, and the physiological effects of corticosteroids. High doses of corticosteroids *in vivo* increase bone resorption,[91,92] but also elevate serum PTH concentrations. Interestingly, parathyroidectomy abolishes the glucocorticoid-induced increase in osteoclastic activity,[93] thus suggesting that this increase is caused by an increase in the production and secretion of PTH. The observation that corticosteroid-induced bone loss in humans is mainly loss of trabecular bone[91,94] suggests, however, that factors other than increased PTH secretion also play a role in corticosteroid-induced bone loss.

In vitro studies using bone explants generally support the view that glucocorticoids reduce osteoclastic resorption[88,95] and decrease the number of osteoclasts developing from precursor cells.[95] Tobias and Chambers[96] recently evaluated the effects of glucocorticoids on the bone resorptive activity of disaggregated rat osteoclasts cultured on devitalized bone slices, and concluded that glucocorticoids decreased resorptive activity by impairing osteoclast viability. This was somewhat surprising in view of the outcome of a previous series of investigations in cat bone marrow cultures,[97,99] where it was observed that hydrocortisone and dexamethasone decreased the number of multinucleated osteoclast-like cells generated over a 4-week culture period, but did not affect the survival of multinucleated cells once they were formed.[99] This raises the possibility that the effects of hydrocortisone and dexamethasone observed by Tobias and Chambers[96] might have been caused by poor viability of the osteoclasts under the culture conditions used. The observation that, in this particular series of experiments, all osteoclastic resorptive activity had ceased after 6 h of culture would seem to support this view.

That the composition of the culture medium can dramatically affect the responses to hormones in various culture systems is well known. An example of this is the observation that, in the absence of serum, epidermal growth factor (EGF) stimulated the proliferation of pituitary GH_4C_1 cells, while in the presence of serum, EGF inhibited proliferation of these cells.[98] Further analysis of these experiments revealed that the inhibition of cell growth in the presence of serum was mediated by an EGF-induced decrease in thyroid hormone receptors, and caused by the resulting decrease in responsiveness to thyroid hormones present in the serum component of the medium.

IV. INTERACTIONS OF LOCAL AND SYSTEMIC FACTORS IN THE CONTROL OF BONE RESORPTION

Glucocorticoids in particular are modulators of both responses to and receptor levels of hormones and growth factors. For example, cortisol has been shown to enhance PTH-induced cAMP accumulation in newborn rat calvaria and fetal rat long bones and in cultured rat bone cells,[100-102] and to cause an increase in PTH receptors in rat osteoblastic osteosarcoma cells.[103] In several other tissues and cell types, glucocorticoids similary change responsiveness to a variety of hormones. Although similar effects have not yet been demonstrated in osteoclasts or osteoclast precursors, it seems likely that such mechanisms exist there as well. However, many effects of systemic factors such as PTH or 1,25$(OH)_2D_3$ could be mediated by osteoblast-like cells or stromal cells. If so, those effects are likely modulated by glucocorticoids.

As pointed out in Chapter 8 of this volume, a large number of locally secreted factors that stimulate bone resorption have now been described. Examples are prostaglandins of the E series (PGEs),[104] IL-1,[105] TNF-α and lymphotoxin (TNF-β),[106] epidermal growth factor (EGF),[107] platelet-derived growth factor (PDGF),[108] and transforming growth factor-α and -β (TGF-α, TGF-β).[109] Several of these local factors affect and regulate each others' production, and systemic hormones can also directly affect the synthesis of several of these cytokines. For example, lymphocytes from estrogen/progesterone-treated postmenopausal women produce less IL-1 than those obtained from nontreated patients,[110] the number of receptors for PTH in ROS 17/2 cells is upregulated by hydrocortisone and downregulated by PTH itself and by 1,25$(OH)_2D_3$,[103,111] and the number of receptors for 1,25$(OH)_2D_3$ is upregulated by glucocorticoids,[112] retinoic acid,[113] and by 1,25$(OH)_2D_3$ itself.[114] This type of interaction will result in synergistic or antagonistic interactions, and it is important to note that such interactions have indeed been demonstrated in the regulation of bone resorption. For example, IL-1, TNF, and lymphotoxin potentiate each others' effects on bone resorption *in vitro*,[115] and IL-1 effects on bone resorption are also potentiated by PTH.[116]

Thus, systemic and local regulators of bone resorption complement each other and cannot really be considered separately. This makes excellent physiological sense, since it is hard otherwise to develop a rationale for the observation that foci of bone formation and bone resorption coexist in close proximity to one another, presumably with similar concentrations of systemically circulating regulatory factors in their environmment. The obvious challenge is to determine the composition of the specific cellular environment, i.e., interaction between cells, between cells and extracellular matrix, and between cells, extracellular matrix, and soluble systemic and local factors, that is responsible for the fine tuning of the cyclical renewal of bone occurring by the coordinated processes of resorption and formation of bone matrix.

REFERENCES

1. **Minkin, C.,** Bone acid phosphatase: tartrate resistant acid phosphatase as a marker of osteoclast function, *Calcif. Tissue Int.*, 34, 285, 1982.
2. **Schenk, R., Spiro, D., and Wiener, J.,** Cartilage resorption in tibial epiphyseal plate of growing rats, *J. Cell Biol.*, 34, 275, 1967.
3. **Vaes, G.,** On the mechanisms of bone resorption: the action of parathyroid hormone on the excretion and synthesis of lysosomal enzymes and on the extracellular release of acid by bone cells, *J. Cell Biol.*, 39, 676, 1968.

4. **Baron, R., Neff, L., Louvard, D., and Courtoy, P. J.,** Cell mediated extracellular acidification and bone resorption: evidence for a low pH in resorbing lacunae and localization of a 100-kDa lysosomal membrane protein at the osteoclast ruffled border, *J. Cell Biol.,* 101, 2210, 1985.
5. **Jones, S. J. and Boyde, A.,** Experimental study of changes in osteoblastic shape induced by calcitonin and parathyroid extract in an organ culture system, *Cell Tissue Res.,* 169, 449, 1976.
6. **Rodan, G. A. and Martin, T. J.,** Role of osteoblasts in hormonal control of bone resorption. A hypothesis, *Calcif. Tissue Int.,* 33, 349, 1981.
7. **Parfitt, A. M.,** The physiological and clinical significance of bone histomorphometric data, in *Bone Histomorphometry. Techniques and Interpretations,* Recker, R., Ed., CRC Press, Boca Raton, FL, 1983, 143.
8. **Heersche, J. N. M.,** The effect of thyrocalcitonin and parathyroid hormone on bone metabolism in tissue culture: the effect on collagen breakdown in relation to bone demineralization, *Proc. K. Ned. Akad. Wet. Ser. C,* 72, 286, 1969.
9. **Heersche, J. N. M.,** The effect of thyrocalcitonin and parathyroid hormone of bone metabolism in tissue culture: the effect of parathyroid hormone and thyrocalcitonin on the process of bone demineralization, *Proc. K. Ned. Akad. Wet. Ser. C,* 72, 594, 1969.
10. **Brand, J. S. and Raisz, L. G.,** Effects of thyrocalcitonin and phosphate ion on the parathyroid hormone stimulated resorption of bone, *Endocrinology,* 90, 479, 1972.
11. **Jilka, R. L. and Cohn, D. V.,** A collagenolytic response to parathormone, 1,25 dihydroxycholecalciferol D_3, and prostaglandin E_2 in bone of osteopetrotic (mi/mi) mice, *Endocrinology,* 112, 945, 1983.
12. **Heersche, J. N. M. and Kanehisa, J.,** The role of osteoblasts in the initiation of osteoclastic resorption and the degradation of the organic bone matrix, in *The Biological Mechanisms of Tooth Eruption and Root Resorption,* Davidovitch, Z., Ed., EBSCO Media, Birmingham, AL, 1988, 101.
13. **Delaisse, J. M., Eeckhout, Y., Sear, C., Galloway, A., McCullagh, K., and Vaes, G.,** A new synthetic inhibitor of mammalian tissue collagenase inhibits bone resorption in culture, *Biochem. Biophys. Res. Commun.,* 133, 483, 1985.
14. **Sakamoto, S., Sakamoto, M., Goldhaber, P., and Glimcher, J.,** Localization of tissue collagenase in bone by indirect immunofluorescent antibody technique, in *Endocrinology of Calcium Metabolism,* Copp, D. H. and Talmage, R. V., Eds., Excerpta Medica, Amsterdam, 1978, 378.
15. **Chambers, T. J., Revell, P. J., Fuller, K., and Athanasou, N. A.,** Resorption of bone by isolated rabbit osteoclasts, *J. Cell Sci.,* 66, 383, 1984.
16. **Boyde, A., Ali, H. N., and Jones, S. J.,** Resorption of dentine by isolated osteoclasts *in vitro, Br. Dent. J.,* 156, 216, 1984.
17. **Kanehisa, J. and Heersche, J. N. M.,** Osteoclastic bone resorption: *in vitro* analysis of the rate of resorption and migration of individual osteoclasts, *Bone,* 9, 73, 1988.
18. **Turksen, K., Kanehisa, J., Opas, M., Heersche, J. N. M., and Aubin, J. E.,** Adhesion patterns and cytoskeleton of rabbit osteoclasts on bone slices and glass, *J. Bone Miner. Res.,* 3, 389, 1988.

18a. **Scott, B. L.,** Thymidine-3H electron microscope radioautography of osteogenic cells in the fetal rat, *J. Cell Biol.,* 35, 115, 1967.

19. **Walker, D. G.,** Control of bone resorption by hematopoietic tissue. The induction and reversal of congenital osteopetrosis in mice through use of bone marrow and splenic transplants, *J. Exp. Med.,* 142, 651, 1975.
20. **Scheven, B. A. A., Visser, J. W. M., and Nyweide, P. J.,** *In vitro* osteoclast generation from different bone marrow fractions, including a highly enriched haematopoietic stem cell population, *Nature (London),* 321, 79, 1986.

21. **Scheven, B. A. A., Kawilarang-De Haas, E. W. M., Wassemaar, E. M., and Nyweide, P. J.,** Differentiation kinetics of osteoclasts in the periosteum of embryonic bones *in vivo* and *in vitro, Anat. Rec.,* 214, 418, 1986.
22. **Meunier, P., Vignon, G., Bernard, J., Edouard, C., and Courpron, P.,** Quantitative bone histology as applied to the diagnosis of hyperparathyroid states, in *Clinical Aspects of Metabolic Bone Disease,* Frame, B., Parfitt, A. M., and Duncan, H., Eds., Exerpta Medica, Amsterdam, 1973, 215.
23. **Eriksen, E. F., Mosekilde, L., and Melsen, F.,** Trabecular bone remodelling and balance in primary hyperparathyroidism, *Bone,* 7, 213, 1986.
24. **Barnicot, N. A.,** The local action of the parathyroid and other tissues on bone in intracerebral grafts, *J. Anat,* 82, 233, 1948.
25. **Gaillard, P. J.,** Parathyroid gland tissue and bone *in vitro, J. Exp. Cell Res.,* 38 (Suppl.), 154, 1955.
26. **Gaillard, P. J.,** The influence of parathormone on the explanted radius of albino mouse embryos, *Proc. K. Ned. Akad. Wet. Ser. C,* 64, 119, 1961.
27. **Kroon, K. B.,** Effect of parathyroid extract on osteogenic tissue, *Acta Morphol. Neerl. Scand.,* 2(1), 38, 1959.
28. **Raisz, L. G., Brand, J. S., Au, W. Y. W., and Niemann, I.,** Interactions of parathyroid hormone and thyrocalcitonin on bone resorption in tissue culture, *Excerpta Med. Sect. 1,* 159, 370, 1968.
29. **Chase, L. R. and Aurbach, G. D.,** Renal adenyl cyclase: anatomically separate sites for parathyroid hormone and vasopressin, *Science,* 159, 545, 1968.
30. **Hruska, K. A., Moskowitz, D., Esbrit, P., Civitelli, R., Westbook, S., and Huskey, M.** Stimulation of inositol trisphosphate and diacylglycerol production in renal tubular cells by parathyroid hormone, *J. Clin. Invest.,* 79, 230, 1987.
31. **Abou-Samra, A.-B., Jueppner, H., Westerberg, D., Potts, J. T., Jr., and Segre, G. V.,** Parathyroid hormone causes translocation of protein kinase-C from cytosol to membranes in rat osteosarcoma cells, *Endocrinology,* 124(3), 1107, 1989.
32. **Chambers, T. J., McSheehy, P. M., Thomson, B. M., and Fuller, K.,** The effect of calcium-regulatory hormones and prostaglandins on bone resorption by osteoclasts disaggregated from neonatal rabbit bones, *Endocrinology,* 116, 234, 1985.
33. **McSheehy, P. M. J. and Chambers, T. J.,** Osteoblastic cells mediate osteoclastic responsiveness to parathyroid hormone, *Endocrinology,* 118, 824, 1986.
34. **Thomson, B. M., Mundy, G. R., and Chambers, T. J.,** Tumor necrosis factors α and β induce osteoblastic cells to stimulate osteoclastic bone resorption, *J. Immunol.,* 138(3), 775, 1987.
35. **Thomson, B. M., Saklatvala, J., and Chambers, T. J.,** Osteoblasts mediate interleukin 1 responsiveness of bone resorption by rat osteoclasts, *J. Exp. Med.,* 164, 104, 1986.
36. **Jacobson, N. and Goldhaber, P.,** Bone resorption induced by epithelial and connective tissue from human gingiva *in vitro, J. Dent. Res.,* 51, 1682, 1972.
37. **Mundy, G. R., Rick, M. E., Turcotte, R., and Kowalky, N. M.,** Pathogenesis of hypercalcemia in lymphosarcoma cell leukemia. Role of an osteoclast activating factor-like substance and a mechanism of action for glucocorticoid therapy, *Am. J. Med.,* 65, 600, 1978.
38. **Josse, R. G., Heersche, J. N. M., Jez, D. M., Mundy, G., and Murray, T. M.,** Observations on the mechanisms of bone resorption induced by multiple myeloma marrow culture fluid and partially purified osteoclast activating factor, *J. Clin., Invest.,* 7, 472, 1981.
39. **Rodan, S. B., Rodan, G. A., Simmons, H. A., Walenga, R. W., Feinstein, M. B., and Raisz, L. G.,** Bone resorptive activity in conditioned medium from rat osteosarcoma cell line, in *Prostaglandins and Cancer,* Powles, T. J., Brockman, R. S., Honn, K. V., and Ramwell, P., Eds., Alan R. Liss, New York, 1982, 573.

40. **Birek, C., Heersche, J. N. M., Jez, D., and Brunette, D. M.,** Secretion of a bone resorbing factor by epithelial cells cultured from porcine rests of Malassez, *J. Periodont. Res.*, 18, 75, 1983.
41. **Nicholson, G. C., Livesey, S. A., Moseley, J. M. and Martin, T. J.,** Actions of calcitonin, parathyroid hormone, and prostaglandin E2 on cyclic AMP formation in chickken and rat osteoclasts, *J. Cell Biochem.*, 31, 229, 1986.
42. **Murrills, R. J., Stein, L. S., Fey, C. P., and Dempster, D. W.,** The effects of parathyroid hormone (PTH) and PTH-related peptide on osteoclast resorption of bone slices *in vitro*: an analysis of pit size and the resorption focus, *Endocrinology*, 127, 2648, 1990.
43. **Arnett, T. R. and Dempster, D. W.,** Effect of pH on bone resorption by rat osteoclasts *in vitro, Endocrinology*, 119, 119, 1986.
44. **Walsh, C. A., Dawson, W. E., Birch, M. A., and Gallagher, J. A.,** The effects of extracellular pH on bone resorption by avian osteoclasts *in vitro, J. Bone Miner. Res.*, 5, 1243, 1991.
45. **Shibutani, T. and Heersche, J. N. M.,** In an isolated rabbit osteoclast system decreasing the pH increases osteoclast activity but reduces the increase in osteoclast number during culture, *J. Bone Miner. Res.*, 6 (Suppl. 1), 94, 1991.
46. **Rao, L. G., Murray, T. M., and Heersche, J. N. M.,** Immunohistochemical demonstration of parathyroid hormone binding to specific cell types in fixed rat bone tissue, *Endocrinology*, 113, 805, 1983.
47. **Teti, A., Rizzoli, R., and Zambonin-Zallone, A.,** Parathyroid hormone binding to cultured avian osteoclasts, *Biochem. Biophys. Res. Commun.*, 174, 1217, 1991.
48. **Agarwala, N. and Gay, C. V.,** Specific binding of parathyroid hormone to living osteoclasts, *J. Bone Miner. Res.*, in press.
49. **Duong, L. T., Grasser, W., DeHaven, P. A., and Sato, M.,** Parathyroid hormone receptors identified on avian and rat osteoclasts, *J. Bone Miner. Res.*, 5, S203, 1990.
50. **Silve, C. M., Hradek, G. T., Jones, A. L., and Arnaud, C. D.,** Parathyroid hormone receptor in intact embryonic chicken bone: characterization and cellular localization, *J. Cell Biol.*, 94, 379, 1982.
51. **Rouleau, M. F., Warshawsky, H., and Goltzman, D.,** Parathyroid hormone binding *in vivo* to renal, hepatic, and skeletal tissues of the rat using a radioautographic approach, *Endocrinology*, 118, 919, 1986.
52. **Copp, D. H., Davidson, A. G. F., and Cheney, B.,** Evidence for a new parathyoroid hormone which lowers blood calcium, *Proc. Can. Fed. Biol. Soc.*, 4, 17, 1961.
53. **Milhaud, G., Perault, A. M., and Moukhtar, M. S.,** Etude du mechanisme de l'action hypocalcemiante de la thyrocalcitonine, *C. R. Acad. Sci.*, 261, 813, 1965.
54. **Aliapoulios, M. A., Goldhaber, P., and Munson, P. L.,** Thyrocalcitonin inhibition of bone resorption induced by parathyroid hormone in tissue culture, *Science*, 151, 330, 1966.
55. **Friedman, J. and Raisz, L. G.,** Thyrocalcitonin: inhibitor of bone resorption in tissue culture, *Science*, 150, 1165, 1965.
56. **Nicholson, G. C., Moseley, J. M., Sexton, P. M., Mendelsohn, F. A. O., and Martin, T. J.,** Abundant calcitonin receptors in isolated osteoclasts, *J. Clin. Invest.*, 78, 355, 1986.
57. **Chambers, T. J., Athanasou, N. A., and Fuller, K.,** Effect of parathyroid hormone and calcitonin on the cytoplasmic spreading of isolated osteoclasts, *J. Endocrinol.*, 102, 281, 1984.
58. **Kanehisa, J.,** Time course of "escape" from calcitonin-induced inhibition of motility and resorption of disaggregated osteoclasts, *Bone*, 10, 125, 1989.
59. **Friedman, J., Au, W. Y. W., and Raisz, L. G.,** Responses of fetal rat bone to thyrocalcitonin in tissue culture, *Endocrinology*, 82, 149, 1968.

60. **Raisz, L. G., Wener, J. A., Trummel, C. L., Feinblatt, J. D., and Au, W. Y. W.,** Induction, inhibition and escape as phenomena of bone resorption, in *Calcium, Parathyroid Hormone and the Calcitonins, Proc. Fourth Parathyroid Conf.*, Excerpta Medica, Amsterdam, 1972, 446.
61. **Wener, J. A., Gorton, S. J., and Raisz, L. G.,** Escape from inhibition of resorption in cultures of fetal bone treated with calcitonin and parathyroid hormone, *Endocrinology,* 90, 752, 1972.
62. **Tashjian, A. M., Wright, D. R., Ivey, J. L., and Pont, A.,** Calcitonin binding sites in bone: relationships to biological response and escape, *Recent Progr. Horm. Res.*, 34, 285, 1978.
63. **Heersche, J. N. M.,** Metabolism of bone in tissue culture: difference between the mouse and rat in sensitivity to parathyroid hormone and thyrocalcitonin and the effect of imidazole, *Proc. K. Ned. Akad. Wet. Ser. C,* 72, 578, 1969.
64. **Messer, H. H. and Copp, D. H.,** Changes in response to calcitonin following prolonged administration to intact rats, *Proc. Soc. Exp. Biol. Med.*, 146, 643, 1974.
65. **Obie, J. F. and Cooper, C. W.,** Loss of calcemic effects of calcitonin and parathyroid hormone infused continuously into rats using the alzet osmotic mini-pumps, *J. Pharmacol. Exp. Ther.*, 209, 422, 1979.
66. **Lausson, S., Tracqui, P., Toubiana, L., Michaud, G., Perault-Staub, A. M., and Staub, J. F.,** Regulation of plasma calcium and phosphate in calcitonin-infused rats, *Am. J. Physiol.*, 259, E370, 1990.
67. **Glajchen, N., Thomas, S., Jowell, P., Epstein, S., Ismail, F., and Fallon, M.,** The effect of high dose salmon calcitonin on bone mineral metabolism in the normal rat, *Calcif. Tissue Int.*, 46, 28, 1990.
68. **Bastian, J. W., Luna, P. D., and Aldred, J. P.,** Calcitonia: considerazonia sui dosaggi e altri aspetti, in *Calcitonia: Nuove Aquisizioni e Prospettive,* DeBastiani, G., Pecile, A., Pietrogrande, V., and Sirtori, C., Eds., Fondazione Carlo Erba, Milano, 1979.
69. **Singer, F. R., Fredericks, R. S., and Minkin, C.,** Salmon calcitonin therapy for Paget's disease of bone. The problem of acquired clinical resistance, *Arthritis Rheum.*, 10, 1148, 1980.
70. **Raisz, L. G., Trummel, C. L., Holick, M. F., and DeLuca, J. F.,** 1,25 dihydroxycholecalciferol: a potent stimulator of bone resorption in tissue culture, *Science,* 175, 768, 1972.
71. **Holtrop, M. E. and Raisz, L. G.,** Comparison of the effects of 1,25 dihydroxycholecalciferol, prostaglandin E_2 and osteoclast activating factor with parathyroid hormone on the ultrastructure of osteoclasts in cultural long bone of fetal rats, *Calcif. Tissue Int.*, 29, 201, 1979.
72. **Tanaka, Y. and DeLuca, H. F.**, Bone mineral mobilization of 1,25 $(OH)_2D_3$, a metabolite of vitamin D, *Arch. Biochem. Biophys.*, 146, 574, 1971.
73. **Reynolds, J. J., Pavlovitch, H., and Balsan, S.,** 1,25 dihydroxycholecalciferol increases bone resorption in thyroparathyroidectomized mice, *Calcif. Tissue Res.*, 21, 207, 1976.
74. **McSheehy, P. M. J. and Chambers, T. J.,** 1,25-dihydroxyvitamin D_3 stimulates rat osteoblastic cells to release a soluble factor that increases osteoclastic resorption, *J. Clin. Invest.*, 80, 425, 1987.
75. **Merke, J., Klaus, G., Hugel, U., Waldherr, R., and Ritz, E.,** No 1,25-dihydroxyvitamin D_3 receptors on osteoclasts of calcium-deficient chicken despite demonstrable receptors on circulating monocytes, *J. Clin. Invest.*, 77, 312, 1986.
76. **Tinkler, S. M. B., Williams, D. M., and Johnson, N. W.,** Osteoclast formation in response to intraperitoneal injection of 1α-hydroxycholecalciferol in mice, *J. Anat.*, 133, 91, 1981.

77. **Thomas, M. L. and Forte, L. R.,** Sex differences during the development of vitamin D deficiency in the rat: serum parathyroid hormone, calcitonin, calcium, and phosphorus, *Endocrinology,* 109, 1528, 1981.
78. **Capen, C. C. and Young, D. M.,** Fine structural alterations in thyroid parafollicular cells of cows in response to experimental hypercalcemia induced by vitamin D, *Am. J. Pathol.,* 57, 365, 1969.
79. **Boyce, R. W. and Weisbrode, S. E.,** Effect of dietary calcium on the response of bone to 1,25 $(OH)_2D_3$, *Lab. Invest.,* 48, 683, 1983.
80. **Takahashi, N., Yamana, H., Yoshiki, S., Roodman, G. D., Mundy, G. R., Jones, S. J., Boyde, A., and Suda, T.,** Osteoclast-like cell formation and its regulation by osteotrophic hormones in mouse bone marrow cultures, *Endocrinology,* 122, 1373, 1988.
81. **Taylor, L. M., Tertinegg, I., Okuda, A., and Heersche, J. N. M.,** Expression of calcitonin receptors during osteoclast differentiation in mouse metatarsals, *J. Bone Miner. Res.,* 4, 751, 1989.
82. **Hahn, T. J.,** Steroid hormones and the skeleton, in *Metabolic Bone Disease: Cellular and Tissue Mechanisms,* Tam, C. S., Heersche, J. N. M., and Murray, T. M., Eds., CRC Press, Boca Raton, FL 1989, 223.
83. **Eriksen, E. F. and Mosekilde, L.,** Estrogens and bone, in *Bone and Mineral Research 7,* Heersche, J. N. M. and Kanis, J. A., Eds., Elsevier, New York, 1990, 273.
84. **Cohen-Solal, M. E., Shih, M. S., and Parfitt, A. M.,** Erosion depth and bone balance in overt osteoporosis, *J. Bone Miner. Res.,* 6(Suppl 1), 225, 1991.
85. **Liu, C. C. and Kalu, D. N.,** Human parathyroid hormone (1-34) prevents bone loss and augments bone formation in sexually mature ovariectomized rats, *J. Bone Miner. Res.,* 5, 973, 1990.
86. **Turner, T. T., Vandersteenhoven, J. J., and Bell, N. H.,** The effects of ovariectomy and 17β-estradiol on cortical bone histomorphometry in growing rats, *J. Bone Miner. Res.,* 2, 115, 1984.
87. **Jee, W. S. S., Mori, S, Li, S. J., and Chan, S.,** Prostaglandin E_2 enhances cortical bone mass and activates intracortical bone remodeling in intact and ovariectomized female rats, *Bone,* 11, 253, 1990.
88. **Stern, P. H.,** Inhibition by steroids of parathyroid hormone-induced ^{45}Ca release from embryonic rat bone *in vitro, J. Pharmacol. Exp. Ther.,* 168, 211, 1969.
89. **Atkins, D., Zanelli, J. M., Peacock, M., and Nordin, B. E. C.,** The effect of oestrogens on the response of bone to parathyroid hormone *in vitro, J. Endocrinol.,* 54, 107, 1972.
90. **Pilbeam, C. C., Klein-Nulend, J., and Raisz, L. G.,** Inhibition by 17β-estradiol of PTH stimulated resorption and prostaglandin production in cultured neonatal mouse calvariae, *Biochem. Biophys. Res. Commun.,* 163, 1319, 1989.
91. **Hahn, T. J., Boisseau, V. C., and Avioli, L. V.,** Effect of chronic corticosteroid administration on diaphyseal and metaphyseal bone mass, *J. Clin. Endocrinol. Metab.,* 39, 274, 1974.
92. **Dykman, T. R., Gluck, O. S., Murphy, W. A., Hahn, T. J., and Hahn, B. H.,** Evaluation of factors associated with glucorticoid-induced osteopenia in patients with rheumatic diseases, *Arthritis Rheum.,* 28, 361, 1985.
93. **Jee, W. S. S., Park, H. Z., Roberts, W. E., et a.,** Corticosteroids and bone, *Am. J. Anat.,* 129, 477, 1970.
94. **Adinoff, A. D. and Hollister, J. R.,** Steroid-induced fractures and bone loss in patients with asthma, *N. Engl. J. Med.,* 309, 265, 1983.
95. **Liskova-Kiar, M.,** Mode of action of cortisol on bone resorption in fetal rat fibula cultured *in vitro, Am. J. Anat.,* 156, 63, 1979.
96. **Tobias, J. and Chambers, T. J.,** Glucocorticoids impair bone resorptive activity and viability of osteoclasts disaggregated from neonatal rat long bones, *Endocrinology,* 125, 1290, 1989.

97. **Suda, T., Testa, N. G., Allen, T. D., Onions, D., and Jarett, O.,** Effect of hydrocortisone on osteoclasts generated in cat bone marrow cultures, *Calcif. Tissue Int.*, 35, 82, 1983.
98. **Kaji, H. and Hinkle, P. M.,** Epidermal growth factor decreases thyroid hormone receptors and attenuates thyroid hormone responses in GH_4C_1 cells, *Endocrinology*, 120, 537, 1987.
99. **Pharoah, M. J. and Heersche, J. N. M.,** Dexamethasone inhibits formation of osteoclast-like cells in bone marrow cultures, *J. Dent. Res.*, 65, 1006, 1986.
100. **Ng, B., Hekkelman, J. W., and Heersche, J. N. M.,** The effect of cortisol on the adenosine 3′,5′-monophosphate response to parathyroid hormone of bone *in vitro, Endocrinology*, 104, 1130, 1979.
101. **Hahn, T. J. and Halstead, L. R.,** Cortisol enhancement of PTH-stimulated cyclic AMP accumulation in cultured fetal rat long bone rudiments, *Calcif. Tissue Int.*, 29, 173, 1979.
102. **Chen, T. L. and Feldman, D.,** Glucocorticoid receptors and actions in subpopulations of cultured rat bone cells: mechanism of dexamethasone potentiation of parathyroid hormone-stimulated cyclic AMP production, *J. Clin. Invest.*, 63, 750, 1979.
103. **Yamamoto, I., Potts, J. T., Jr., and Segre, G. V.,** Glucocorticoids increase parathyroid hormone receptors in rat osteoblastic osteosarcoma cells (ROS 17/2), *J. Bone Miner. Res.*, 3, 707, 1988.
104. **Klein, D. C. and Raisz, L. G.,** Prostaglandins: stimulation of bone resorption in tissue culture, *Endocrinology*, 86, 1436, 1970.
105. **Gowen, M., Wood, D. D., Ihrie, E. G., McGuire, M. K. B., and Russell, R. G. G.,** An interleukin 1-like factor stimulates bone resorption *in vitro, Nature (London)*, 306, 378, 1983.
106. **Bertolini, D. R., Nedwin, G. E., Bringman, T. S., Smith, D. D., and Mundy, G. R.,** Stimulation of bone resorption and inhibition of bone formation *in vitro* by human tumor necrosis factors, *Nature (London)*, 319, 516, 1986.
107. **Tashjian, A. H., Jr. and Levine, L.,** Epidermal growth factor stimulates prostaglandin production and bone resorption in cultured mouse calvaria, *Biochem. Biophys. Res. Commun.*, 85, 966, 1978.
108. **Tashjian, A. H., Jr., Hohmann, E. L., Antoniades, H. N., and Levine, L.,** Platelet-derived growth factor stimulates bone resorption via a prostaglandin mediated mechanism, *Endocrinology*, 111, 118, 1982.
109. **Tashjian, A. H., Jr., Voelkel, E. F., Lazzaro, M., Singer, F. R., Roberts, A. B., Derynck, R., Winkler, M. E., and Levine, L.,** α and β human transforming growth factors stimulate prostaglandin production and bone resorption in cultured mouse calvaria, *Proc. Natl. Acad. Sci. U.S.A.*, 82, 4535, 1985.
110. **Pacifici, R., Rifas, L., McCracken, R., Vered, I., McMurtry, C., Acioli, L. V., and Peck, W. A.,** Ovarian steroid treatment blocks a postmenopausal increase in blood monocyte interleukin 1 release, *Proc. Natl. Sci. U.S.A.*, 86, 2398, 1989.
111. **Yamamoto, I.,** Regulation of receptors for parathyroid hormone in rat osteosarcoma cells, *Jpn. J. Bone Miner. Res.*, 3, 38, 1985.
112. **Chen, T. L., Cone, C. ., Morey-Holton, E., and Feldman, D.,** Glucocorticoid regulation of 1,25 $(OH)_2D_3$ vitamin D_3 receptors in cultured mouse bone cells, *J. Biol. Chem.*, 257, 13563, 1982.
113. **Petkovich, P. M., Heersche, J. N. M., Tinker, D. O., and Jones, G.,** Retinoic acid stimulates 1,25-dihydroxyvitamin D_3 binding in rat osteosarcoma cells, *J. Biol. Chem.*, 259, 8274, 1984.
114. **Costa, E. M., Hirst, M. A., and Feldman, D.,** Regulation of 1,25-dihydroxyvitamin D_3 receptors by vitamin D analogs in cultured mammalian cells, *Endocrinology*, 117, 2204, 1985.

115. **Stashenko, P., Dewhirst, F. E., Peros, W. J., Kent, R. L., and Ago, J. M.,** Synergistic interactions between interleukin 1, tumor necrosis factor and lymphotoxin in bone resorption, *J. Immunol.*, 138, 1464, 1987.
116. **Dewhirst, F. E., Ago, J. M., Peros, W. J., and Stashenko, P.,** Synergism between parathyroid hormone and interleukin 1 in stimulating bone resorption in organ culture, *J. Bone Miner. Res.*, 2, 127, 1987.
117. **Heersche, J. N. M. et al.,** unpublished observation.
118. **Raisz, L. G. et al.,** personal communication.
119. **Shibutani, T. and Heersche, J. N. M.,** unpublished observations.

Chapter 8

LOCAL FACTORS REGULATING OSTEOCLAST FUNCTION

Gregory R. Mundy

TABLE OF CONTENTS

I. INTRODUCTION

In recent years, it has become apparent that osteoclast formation and activation is controlled not just by systemic hormones such as parathyroid hormone and 1,25-dihydroxyvitamin D, but also by a number of potent osteotropic factors which are likely produced in the bone cell microenvironment. In most of the *in vitro* and *in vivo* systems in which they have been investigated, these cytokines are equally as potent as, and somemtimes more potent than, the systemic hormones. Since it appears likely that normal bone remodeling is under the control of locally produced factors, there has been great interest in linking these cytokines not just to abnormal osteoclast function in disease states where they are likely to be overproduced, but also in normal physiological bone remodeling. Currently, there is no direct evidence that they are involved in normal bone remodeling or in mechanisms of disease, although circumstantial evidence is very strong that they are likely mediators of bone destruction associated with chronic inflammatory conditions where immune cells which produce these cytokines are adjacent to areas of bone destruction, or in malignant diseases such as myeloma or metastatic solid tumors where osteolysis occurs adjacent to tumor cell deposits. Recently, it has been shown by one group that peripheral blood monocytes produce excessive cytokines constitutively in patients with osteoporosis, and that this excess production of cytokines may be suppressed by estrogen therapy in postmenopausal women. Workers involved in these studies suggest that since cells in the peripheral blood may be representative of similar cells in the marrow cavity adjacent to bone-resorbing margins, excess cytokine production is a likely mechanism for increased bone resorption occurring during periods of estrogen withdrawal.

The effects of known cytokines on osteoclast formation and activity are discussed in this review.

II. INTERLEUKIN-1

Interleukin-1 (IL-1) was the first bone-resorbing cytokine to be identified definitively.[1-3] Gowen and co-workers[1] first showed that natural human purified IL-1 stimulated bone resorption in organ cultures of neonatal mouse calvaria. Later, it was found that purified porcine IL-1 and recombinant human IL-1 also stimulated bone resorption. There are two types of IL-1: IL-1α and IL-1β. Both of these proteins share considerable amino acid sequence homology and bind to the same receptor. IL-1β is the predominant cytokine in normal activated human leukocyte culture supernatants, although IL-1α seems to be preferentially produced by some tumors associated with increased production of IL-1 activity. Both of the IL-1 molecules have equivalent effects on bone resorption both *in vitro* and *in vivo*.

It appears likely that the predominant form of bone-resorbing activity present in activated normal human leukocyte culture supernatants is IL-1β. Dewhirst et al.[4] purified IL-1β from activated human leukocyte culture supernatants using a fetal rat long-bone assay. Lorenzo et al.[5] then showed that the bone-resorbing activity, formerly referred to as OAF, which was present in activated leukocyte culture supernatants, could be inhibited by neutralizing antibodies to IL-1.

IL-1 probably acts at all stages in osteoclast formation, differentiation, and activation. Pfeilschifter et al.[6] examined the effects of IL-1 in a long-term human bone marrow culture system in which cells with the osteoclast phenotype form. In this system, IL-1 stimulated proliferation of mononuclear progenitor cells as well as fusion of these mononuclear cells to form multinucleated cells. When IL-1 is added to isolated osteoclasts, the osteoclasts form resorption lacunae on calcified matrices when an exogenous population of osteoblasts or possibly stromal cells is present.[7] However, it is also possible that in addition to this effect, IL-1 acts on isolated osteoclasts directly. Some of the actions of IL-1 on osteoclast formation may involve other cytokines such as tumor necrosis factor and IL-6.[8]

It is now apparent that IL-6 may have a very important effect in IL-1-mediated bone resorption. The effects of IL-1 on bone resorption in organ culture systems are partially impaired by neutralizing antibodies to IL-6.[8] Moreover, the effects of IL-1 are enhanced by two orders of magnitude when IL-6 is present in small concentrations in the media.[9] Since IL-1 can stimulate IL-6 production by bone cells or stromal cells, it appears likely that interactions between these two factors may be extremely important not only in pathological bone resorption, where they are both produced in conjunction, but also possibly in normal bone remodeling.

Recently, it has been found that there are two types of receptors for IL-1, an 80-kDa receptor which is present on T-lymphocytes and fibroblasts and a 40-kDa receptor which is present on pre-B-lymphocytes, bone marrow granulocytes, and macrophages.[10-13] Antibodies to the 80-kDa receptor are available which inhibit the binding of IL-1α or IL-1β to this receptor. We have used these antibodies and found that the effects of IL-1 on bone resorption in organ culture are completely inhibited. Moreover, we have found that these antibodies block not only the effects of IL-1, but also the effects of tumor necrosis factor on osteoclastic bone resorption in these organ culture systems.[9] These data further confirm the complex interactions between various cytokines in the bone resorption process and indicate the importance of the availability of the 80-kDa IL-1 receptor for expression of tumor necrosis factor's bone-resorbing effects.

IL-1 also has powerful effects on osteoclastic bone resorption *in vivo*. Sabatini et al.[14] showed that when IL-1 was infused into normal mice subcutaneously for 3 d, profound hypercalcemia and increased osteoclastic bone resorption were found. Boyce et al.[15] used a different model in which IL-1 was given by repeated injections. They showed that after several injections of IL-1, intense osteoclastic bone resorption was stimulated. This continued for a period of about 7 to 10 d, and was followed by a period of intense new bone formation for 3 weeks. The injections were given in the subcutaneous tissue overlying the calvaria of normal intact mice. The calvarium was completely resorbed in a patchy distribution over a period of 7 d, and then completely replaced after 1 month by the intense new bone formation.

Boyce et al.[16] found that following a single injection of IL-1, mild hypocalcemia occurred in this model even when the mice were hypercalcemic due to repeated injections of IL-1. A transient phase of hypocalcemia lasting for several hours always occurred in response to a bolus injection of IL-1. This hypocalcemic response was blocked by treatment of the mice with indomethacin, suggesting that it was prostaglandin mediated.

It is likely that IL-1 exerts at least some of its effects on bone resorption through prostaglandin-mediated mechanisms. When indomethacin was given to mice treated by local injections of IL-1, there was markedly less inflammatory response at the same site as the injections.[15] However, osteoclastic bone resorption occurred at distant sites. In most physiological and pathological situations, it is likely that IL-1 acts in conjunction with other factors to mediate its effects on bone and calcium homeostasis. Dewhirst et al.[17] show that parathyroid hormone (PTH) and IL-1 produce synergistic effects on osteoclastic bone resorption in fetal rat long-bone cultures. Later, Sato et al.[18] showed that IL-1α and PTH-related protein (rP) produce synergistic responses by an increase in plasma calcium in rats. This may be an important observation because these tumor-related peptides may be produced together in some cancers associated with hypercalcemia. In this situation, IL-1 could be produced either directly by the tumor cells or indirectly by host immune cells in response to the presence of the tumor.

IL-1 may have an important role in a number of disease states. Sato et al.[18] showed that it is associated with some squamous cell carcinomas with hypercalcemia. The form of IL-1 found in this situation is IL-1α. Some groups have shown that it may be produced by cultured marrow cells derived from patients with myeloma. The form of IL-1 produced in this situation is IL-1β. In this situation, it may be working in conjunction with other cytokines such as lymphotoxin and IL-6. IL-1 is also likely to be important in local bone destruction which occurs in association with chronic inflammatory diseases. Diseases such as rheumatoid arthritis or periodontal disease are associated with accumulations of chronic inflammatory cells producing IL-1 together with other cytokines. These cytokines are present in the fluids produced in conjunction with the inflammatory deposits, and they may mediate local osteoclastic bone resorption locally in this manner.

Recently, Pacifici et al.[19,20] have shown increased constitutive production of IL-1 by peripheral blood mmonocytes of patients with a subset of osteoporosis. They have linked IL-1 to the bone loss associated with this condition.

In support of their data, they have shown that IL-1 production by peripheral blood monocytes may be decreased by treatment of postmenopausal women with estrogens.

As already indicated, IL-1 may also be important in several other situations. It could be important in the local bone destruction which occurs in conditions such as eosinophilic granuloma, where chronic inflammatory cells which are likely sources of IL-1 are present adjacent to bone-resorbing surfaces. As already indicated, Boyce et al.[16] have suggested that hypocalcemia occurs after bolus injections of IL-1. This mechanism could be responsible for the hypocalcemia that is sometimes associated with sepsis. Bolus entry of IL-1 into the circulation in this particular circumstance could be responsible for transient hypocalcemia.

III. INTERLEUKIN-1 RECEPTOR ANTAGONIST

Recently, a protein has been purified and molecularly cloned which has considerable homology with the IL-1 family, but inhibits some IL-1 activities.[10,21,22] This molecule is referred to as the IL-1 receptor antagonist or IL-1 RA. There is 25% homology at the amino acid level, but even more homology at the gene level with the IL-1 molecules. There is 26% homology with IL-1β and 19% homology with IL-1α, but an organization of its gene similar to that of the IL-1 family suggests that it is probably part of this fairly. This molecule has a molecular weight of 17 kDa and is comprised of 152 amino acids. It is produced by monocytes, but regulated separately from IL-1. This molecule turns out to bind to the 80-kDa high-affinity IL-1 receptor and has a mode of action similar to IL-1 receptor antibodies which block the 80-kDa receptor. It competes with IL-1 for cell surface binding on target cells, and it has an affinity for the 80-kDa receptor similar to that of IL-1. We have studied the effects of this factor on osteoclastic bone resorption using fetal rat long-bone cultures. It seems to have effects identical to those we have found with IL-1 receptor antibodies. It blocks osteoclastic bone resorption mediated by IL-1, tumor necrosis factor, or lymphotoxin.[8,23] It has no effects on osteoclastic bone resorption mediated by parathyroid hormone or 1,25-dihydroxyvitamin D. We have also tested the effects of this factor on calcium homeostasis in normal mice injected repeatedly with IL-1 or tumor necrosis factor. In both circumstances, blood-ionized calcium is reduced. However, hypercalcemia mediated by systemic factors such as PTH-rP is unaffected by treatment with IL-1 RA.

Since IL-1 is likely to be an important mediator of osteoclastic bone resorption in a number of pathological states, the IL-1 RA may have considerable therapeutic potential in those circumstances where IL-1 is the major mediator involved. Moreover, since IL-1 RA also blocks bone resorption stimulated by other cytokines, it may even turn out to be a useful therapeutic agent in all varieties of cytokine-mediated bone resorption.

IV. TUMOR NECROSIS FACTOR AND LYMPHOTOXIN

Tumor necrosis factor (TNF) and lymphotoxin are powerful bone-resorbing cytokines. These factors are likely to be involved, as is IL-1, in both normal bone remodeling and bone destruction associated with a number of disease states. Both factors are multifunctional cytokines. TNF is produced by activated monocytes and macrophages, and lymphotoxin (which is also called TNF-β) by activated T-lymphocytes. Lymphotoxin is also produced by established myeloma cells in culture and by lymphoid cell lines with B-cell surface characteristics.

TNF and lymphotoxin were first shown to resorb bones in fetal rat bone organ cultures.[24] They stimulated osteoclastic bone resorption in both fetal rat long bones and neonatal mouse calvarial cultures. Their effects are inhibited by indomethacin. They also stimulate osteoclastic bone resorption *in vivo* and, when produced in sufficient amounts, cause hypercalcemia.[25,26]

These factors probably increase osteoclast progenitor proliferation and differentiation. Pfeilschifter et al.[6] studied their effects on cultures of human bone marrow mononuclear cells. Both TNF and lymphotoxin increased the formation of cells with osteoclast characteristics in these culture systems. They acted on progenitor cells to stimulate proliferation and on later progenitors to promote differentiation and fusion of committed progenitors to form mature multinucleated cells. They also act on the mature osteoclast to stimulate bone-resorbing activity. However this effect may be indirect and mediated through stromal cells or cells with the osteoblast phenotype.[27]

TNF and lymphotoxin are likely involved in bone destruction, which is associated with a number of pathological conditions. In myeloma, long-term cultures of myeloma cells produce and express lymphotoxin.[28] Lymphotoxin may not be the only cytokine which is involved in the bone destruction associated with myeloma, since patients with myeloma also have increase circulating concentrations of IL-6, which is produced as a growth factor in myeloma by stromal cells, and cultures of marrow cells derived from patients with myeloma which contain mixtures of myeloma cells and monocytes have been shown to produce IL-1β.

TNF may be involved in the bone destruction and hypercalcemia which occurs in some solid tumors. Yoneda et al.[29,30] showed that hypercalcemia in a human squamous cell carcinoma model can be suppressed by passive immunization with neutralizing antibodies to TNF. In this model system, TNF is produced by host immune cells and not by the malignant cells. Similar phenomena have been shown by Sabatini et al. in the human A375 melanoma[31] and in the Leydig cell tumor.[32] In these latter two models, the tumor cells themselves also produce bone-resorbing factors. In the case of the A375 tumor, transforming growth factor-α (TFG-α) is expressed, and in the case of the Leydig cell tumor, both TGF-α and the parathyroid hormone-related protein are produced. The mechanism by which the human tumor cells increase TNF production is possibly by the production of soluble mediators which stimulate host macrophages to release TNF. One such soluble mediator is colony-stimulating factor of the granulocyte-macrophage series (GM-CSF), which has been shown by Sabatini et al.[31] to be the responsible mediator in the A375 melanoma. In the squamous cell carcinoma studied by Yoneda et al.,[29,30] one of the mediators may be GM-CSF, but in addition to GM-CSF, there is a separate mediator of approximately 40 kDa molecular weight which these workers have called the tumor necrosis factor-inducing factor (TNF-IF).

Recently, Pacifici et al.[19,20] have shown evidence that not only IL-1, but also TNF and GM-CSF can be produced in increased amounts constitutively by peripheral blood monocytes in patients with osteoporosis. This may be part of the mechanism by which estrogens regulate bone resorption in patients with this condition.

It is obvious that complex interactions occur between TNF and other cytokines such as IL-1 and IL-6. The effects of TNF on bone resorption in organ cultures are partially inhibited by antibodies to the IL-1 receptor or by the IL-1 RAs.[8,9] Moreover, the effects of TNF to cause increased bone resorption are partially inhibited by antibodies to IL-6.[9] Since TNF stimulates IL-6 production by bone cells, there are complex interactions between IL-1 and IL-6, and in some circumstances IL-6 leads to increased IL-1 production, there are multiple mechanisms by which these cytokines may interact and potentiate the effects of each other on osteoclastic bone resorption.

V. INTERLEUKIN-6

IL-6 is a recently described multifunctional cytokine which apparently has unique effects on bone cells. It not only affects bone cells, but is also produced by bone cells.[33] Its production by bone cells can be modulated by parathyroid hormone, IL-1, and TNF.[33] However, the effects of the cytokines on IL-6 production by bone cells is much greater than the effects of parathyroid hormone and 1,25-dihydroxyvitamin D. The effects of IL-6 on bone resorption and organ cultures have been controversial. Most workers have found no effects. However, Lowik et al.[34] showed that IL-6 could stimulate osteoclastic bone resorption in a mouse metatarsal system. Moreover, Kurihara et al.[35] showed that it increases the formation of cells with the osteoclast phenotype. We have studied the effects of IL-6 on bone resorption and calcium homeostasis *in vivo*.[9] We have used Chinese hamster ovary (CHO) cells which have been transfected with the human IL-6 gene. These cells express IL-6 in a stable, consistent manner. When these cells are inoculated in nude mice, the mice become profoundly hypercalcemic and have marked loss of trabecular bone. This loss of trabecular bone is associated with increased osteoclastic bone resorption. Whether these effects of IL-6 are direct or indirect is not clear. However, we know that mice which are transfected with CHO cells and express IL-6 in small amounts when given tiny amounts of IL-1 show a marked increase in blood calcium. This confirms other observations that IL-6 potentiates the effects of IL-1 on osteoclastic bone resorption *in vitro*. IL-1 stimulates bone resorption at concentrations of 10^{-13} and 10^{-14} *M* in the presence of IL-6 at 50 ng/ml. TNF is effective at 10^{-12} *M* in the presence of IL-6.

It is likely that IL-6 is an important bone-resorbing factor in patients with myeloma. Patients with severe myeloma and plasma cell leukemia have IL-6 concentrations in the circulation in the same general magnitude as those seen in the CHO models we have studied.[36,37] Since myeloma cells in culture have also been shown to express lymphotoxin, and mixtures of myeloma cells in the marrow cells produce IL-1β, the combined effects of these cytokines on osteoclasts may be responsible for the bone resorption and hypercalcemia which is seen in this condition.

VI. LEUKEMIA INHIBITORY FACTOR

Abe et al.[38] showed that a factor called differentiation-inducing factor (DIF), which was purified from Ehrlich's ascites cancer cells and influenced mmonocyte differentiation, stimulated osteoclastic bone resorption in organ cultures. Several years later, a closely related and probably identical factor was purified by Gearing et al.[39] This factor was called leukemia inhibitory factor (LIF) because of its capacity to inhibit leukemic cell differentiation. LIF or DIF is a stimulator of osteoclastic bone resorption, but is probably less potent than the other bone-resorbing cytokine. This factor is also produced by bone cells *in vitro* in response to factors such as PTH.[40] Although it is not known whether it causes hypercalcemia *in vivo*, there is one report suggesting that in an animal model in which it is overexpressed, the animals get bony abnormalities and possibly an increase in serum calcium.[41]

VII. γ-INTERFERON

γ-Interferon is an inhibitor of osteoclastic bone resorption *in vitro*,[42] although the significance of these observations to *in vivo* situations is unknown. It seems to be particularly effective in inhibiting bone resorption which is stimulated by IL-1 or TNF, and is probably more effective in inhibiting bone resorption stimulated by the cytokines than it is in inhibiting bone resorption stimulated by parathyroid hormone or 1,25-dihydroxyvitamin D. The target cell for the mode of action of γ-interferon has been examined in cell culture systems. It has been shown to inhibit both proliferation of marrow mononuclear cells which contain osteoclast precursors, and differentiation of these cells to fuse and form multinucleated cells.[43,44] It does not have morphological effects on the mature osteoclast in the way that calcitonin does. It is unlikely to be a powerful hypocalcemic agent, since we have found no effect on calcium homeostasis when it is injected *in vivo*. However, its local effects may be important as a modulator of cytokine-mediated bone resorption in the bone cell microenvironment.

VIII. TRANSFORMING GROWTH FACTOR-β

The most drammatic effects of TGF-β on bone cells are on bone formation. It is a powerful stimulator of bone formation, causing proliferation of osteoblast progenitors as well as differentiated function in more mature cells. It increases the number of matrix vesicles in cultured cells with the osteoblast phenotype. It may act synergistically in this regard with 1,25-dihydroxyvitamin D, at least in some systems.

TGF-β has complex effects on osteoclasts. In organ cultures of fetal rat long bones, it inhibits bone resorption, but its effects occur only at high concentrations.[45] It also inhibits the formation of osteoclasts from bone marrow progenitors *in vitro* in long-term marrow cultures.[46] However, in mouse calvarial systems, it stimulates prostaglandin generation, which in turn leads to increased osteoclast formation and bone resorption.[45,47] In isolated osteoclasts, it has very complex effects. Its overall effect probably depends on other factors which are present in the culture media. In some systems, it inhibits osteoclast activity when tartrate-resistant acid phosphatase content or staining with nitro-blue tetrazolium, which indicates oxygen radical formation,[48] are measured. However, under other circumstances, and particularly when cultured in the presence of serum, osteoclastic bone resorption may be enhanced. When injected *in vivo* into the subcutaneous tissue overlying the calvaria of rats or infused into the marrow cavity of long-bone shafts, osteoclast formation may be generated in adjacent marrow spaces.[49] This effect appears to be at least in part due to prostaglandin synthesis, because it can be inhibited by indomethacin. Thus, the effects of TGF-β on bone resorption are complex and may be dependent on other factors in the osteoclast microenvironment to which the osteoclasts have been exposed.

IV. COLONY-STIMULATING FACTORS

The CSFs for the formed elements of the blood, i.e., the granulocyte macrophage series (CSF-GM), the monocyte series (CSF-M or CSF-1), and the granulocyte series (CSF-G), probably do not have any direct effects on bone resorption. However, because the osteoclast is derived from a hematopoietic stem cell which it shares with granulocytes and monocytes, these factors may affect the osteoclast stem cell pool. Increased osteoclastic bone resorption is frequently found in solid tumors which are associated with leukocytosis and increased production of the CSFs.[29,30] There are a number of studies reporting the effects of CSFs on *in vitro* systems and the results have been discordant.[50,51] For example, CSF-M has been shown to either increase or decrease the formation of cells with osteoclastic characteristics. The explanation for these discrepancies almost certainly lies in the population of cells to which the CSF was added. For example, CSF-M is a powerful regulatory growth factor for cells in the monocyte lineage and will direct committed precursors down this lineage. Its effects on osteoclast formation are probably less prominent. However, the recent observation of Yoshida et al.[52] showed that osteopetrosis which occurs in the *op/op* mouse, a murine variant of osteopetrosis, is due to impaired production of this CSF. They demonstrated a single base-pair insertion in the coding region of the gene for CSF-M in these mice. Earlier. Wiktor-Jedrzejczak et al.[53] had shown that there was impaired production of CSF activity by stromal cells in these osteopetrotic mice. Thus, the CSFs almost certainly have effects on osteoclast formation, most of which are probably indirect. It seems likely that eventually a CSF for osteoclasts will be identified. This factor presumably will have characteristics similar to those of the other CSFs for their target cells, and be either lineage specific or relatively lineage restricted to the osteoclast lineage.

X. PROSTAGLANDINS

Prostaglandins are nonpeptide local factors which also have powerful effects on osteoclasts. Prostaglandins of the E series (PGEs) were first identified in organ culture systems as osteoclast stimulators.[54,55] They stimulate osteoclastic bone resorption in both organ cultures of fetal rat long bones and neonatal mouse calvaria.[54,55] Prostaglandins have not been shown to increase bone resorption *in vivo*. However, they may be mediators involved in the bone destruction which is associated with a number of other factors such as IL-1, TNF, and TGFs. At least part of the action of these factors on osteoclastic bone resorption is mediated by secondary generation of prostaglandins within bone.[15,24,25] When they are added to isolated cell culture systems, they may have differing effects. In isolated osteoclasts, PGEs have been shown to cause contraction of the cytoplasmic membrane which has been linked to inhibition of bone resorption.[56] It is likely that this effect is transient and that the overall effect of prostaglandins on bone resorption may be stimulatory.

There is also the possibility that prostaglandins may have differing effects in different species. In isolated rat osteoclasts, they cause morphological responses similar to those of calcitonin indicated above. However, when added to human long-term bone marrow cultures, they inhibit the formation of cells with osteoclast characteristics.[57] This is the reverse of what is seen in murine bone marrow cultures.[58] There, the formation of cells with osteoclast characteristics as indicated by tartrate-resistant acid phosphatase content, bone resorption, multinuclearity, and the presence of calcitonin receptors is inhibited by PGEs.

PGEs are not the only eicanosoids which modulate osteoclast activity. Recently, other eicanosoids have also been tested for their effects on bone resorption *in vitro*. The leukotriene compounds generated by the 5-lipoxygenase pathway have also been shown to stimulate bone resorption in neonatal mouse calvaria in one study.[59] All of the compounds which were tested were effective, and all were more potent than PGE_2. These compounds included LTB4, LTC4, LTD4, 5-HETE, and 12-HETE. These compounds could be important in bone loss associated with inflamed gingiva and rheumatoid synovia.[60-62]

ACKNOWLEDGMENTS

I am grateful to Nancy Garrett for her help in the preparation of this review. Part of the work reported here was supported by grants AR28149, CA-40035, DE-08569, AM-07464, and AR-39529 from the National Institutes of Health.

REFERENCES

1. **Gowen, M., Meikle, M. C., and Reynolds, J. J.,** Stimulation of bone resorption *in vitro* by a non-prostanoid factor released by human monocytes in culture, *Biochem. Biophys. Acta,* 762, 471, 1983.
2. **Heath, J. K., Saklatvala, J., Meikle, M. C., Atkinson, S. J., and Reynolds, J. J.,** Pig interleukin-1 (catabolin) is a potent stimulator of bone resorption *in vitro, Calcif. Tissue Int.,* 37, 95, 1985.
3. **Gowen, M. and Mundy, G. R.,** Actions of recombinant interleukin-1, interleukin-2 and interferon gamma on bone resorption *in vitro, J. Immunol.,* 136, 2478, 1986.
4. **Dewhirst, F. E., Stashenko, P. P., Mole, J. E., and Tsurumachi, T.,** Purification and partial sequence of human osteoclast-activating factor: identity with interleukin-1 beta, *J. Immunol.,* 135, 2562, 1985.
5. **Lorenzo, J. A., Sousa, S. L., Alander, C., Raisz, L. G., and Dinarello, C. A.,** Comparison of the bone resorbing activity in the supernatants from phytohemagglutinin stimulated human peripheral blood mononuclear cells with that of cytokines through the use of an antiserum to interleukin-1, *Endocrinology,* 121, 1164, 1987.
6. **Pfeilschifter, J., Mundy, G. R., and Roodman, G. D.,** Interleukin-1 and tumor necrosis factor stimulate the formation of human osteoclast-like cells *in vitro, J. Bone Miner. Res.,* 4, 113, 1989.
7. **Thomson, B. M., Saklatvala, J., and Chambers, T. J.,** Osteoblasts mediate interleukin-1 stimulation of bone resorption by rat osteoclasts, *J. Exp. Med.,* 164, 104, 1986.
8. **Garrett, I. R., Black, K. S., and Mundy, G. R.,** Interactions between interleukin-6 and interleukin-1 in osteoclastic bone resorption in neonatal mouse calvariae, *Calcif. Tissue Int.,* 46, A37, 1990.
9. **Black, K., Mundy, G. R., and Garrett, I. R.,** Interleukin-6 causes hypercalcemia *in vivo,* and enhances the bone resorbing potency of interleukin-1 and tumor necrosis factor by two orders of magnitude *in vitro, J. Bone Miner. Res.,* 5 (Suppl. 2), 787, 1990.

10. **Carter, D. B., Deibel, M. R., Jr., Dunn, C. J., Tomich, C. S. C., Laborde, A. L., Slightom, J. L., Berger, A. E., Bienkowski, M. J., Sun, F. F., McEwan, R. N., Harris, P. K. W., Yem, A. W., Waszak, G. A., Ghosay, J. G., Sieu, L. C., Hardee, M. M., Zurcher-Neely, H. A., Reardon, I. M., Heinrikson, R. L., Truesdell, S. E., Shelly, J. A., Eessalu, T. E., Taylor, B. M., and Tracey, D. E.,** Purification, cloning, expression and biological characterization of an interleukin-1 receptor antagonist protein, *Nature (London),* 344, 633, 1990.
11. **Benjamin, W. R., Tare, N. S., Hayes, T. J., Becker, J. M., and Anderson, T. D.,** Regulation of hemopoiesis in myelosuppressed mice by human recombinant IL-1 alpha, *J. Immunol.* 142, 792, 1989.
12. **Hannum, C. H., Wilcox, C. J., Arend, W. P., Joslin, F. G., Dripps, D. J., Heimdal, P. L., Armes, L. G., Sommer, A., Eisenberg, S. P., and Thompson, R. C.,** Interleukin-1 receptor antagonist activity of a human interleukin-1 inhibitor, *Nature (London),* 343, 336, 1990.
13. **Eisenberg, H., Pallotta, J., Sacks, B., and Brickman, A. S.,** Parathyroid localization, 3-dimensional modeling and percutaneous ablation techniques, *Endocrinol. Metab. C,* 18, 659, 1989.
14. **Sabatini, M., Boyce, B., Aufdemorte, T., Bonewald, L., and Mundy, G. R.,** Infusions of recombinant human interleukin-1 alpha and beta cause hypercalcemia in normal mice, *Proc. Natl. Acad. Sci. U.S.A.,* 85, 5235, 1988.
15. **Boyce, B. R., Aufdemorte, T. B., Garrett, I. R., Yates, A. J. P., and Mundy, G. R.,** Effects of interleukin-1 on bone turnover in normal mice, *Endocrinology,* 125, 1142, 1989.
16. **Boyce, B. R., Yates, A. J. P., and Mundy, G. R.,** Bolus injections of recombinant human interleukin-1 cause transient hypocalcemia in normal mice, *Endocrinology,* 125, 2780, 1989.
17. **Dewhirst, F. E., Ago, J. M., Peros, W. J., and Stashenko, P.,** Synergism between parathyroid hormone and interleukin-1 in stimulating bone resorption in organ culture, *J. Bone Miner. Res.,* 2, 127, 1987.
18. **Sato, K., Fujii, Y., Kasono, K., Ozawa, M., Imamura, H., Kanaji, Y., Kurosawa, H., Tsushima, T., and Shizume, K.,** Parathyroid hormone-related protein and interleukin-1α synergistically stimulate bone resorption *in vitro* and increase the serum calcium concentration in mice *in vivo, Endocrinology,* 124, 2172, 1989.
19. **Pacifici, R., Rifas, L., Teitelbaum, S., Slatopolsky, E., McCracken, R., Bergfeld, M., Lee, W., Avioli, L. V., and Peck, W. A.,** Spontaneous release of interleukin-1 from human blood monocytes reflects bone formation in idiopathic osteoporosis, *Proc. Natl. Acad. Sci. U.S.A.,* 84, 4616, 1987.
20. **Pacifici, R., Rifas, L., McCracken, R., Vered, I., McMurty, C., Avioli, L., and Peck, W. A.,** Ovarian steroid treatment blocks a postmenopausal increase in blood monocyte interleukin-1 release, *Proc. Natl. Acad. Sci. U.S.A.,* 86, 2398, 1989.
21. **Arend, W. P., Joslin, F. G., Thompson, R. C., and Hannum, C. H.,** An IL-1 inhibitor from human monocytes. Production and characterization of biologic properties, *J. Immunol.,* 15, 1851, 1989.
22. **Eisenberg, S. P., Evans, R. J., Arend, W. P., Verdeber, E., Brewer, M. T., Hannum, C. H., and Thompson, R. C.,** Primary structure and functional expression from complementary DNA of a human interleukin-1 receptor antagonist, *Nature (London),* 343, 341, 1990.
23. **Guise, T., Yoneda, T., Alsina, M., and Mundy, G. R.,** Synergism between parathyroid hormone associated related protein and TGFα, *Clin. Res.,* 38, 953A, 1990.
24. **Bertolini, D. R., Nedwin, G. E., Bringman, T. S., and Mundy, G. R.,** Stimulation of bone resorption and inhibition of bone formation *in vitro* by human tumour necrosis factor, *Nature (London),* 319, 516, 1986.

25. **Tashjian, A. H., Voelkel, E. F., Lazzaro, M., Goad, D., Bosma, T., and Levine, L.,** Tumor necrosis factor α (cachectin) stimulates bone resorption in mouse calvaria via a prostaglandin-mediated mechanism, *Endocrinology,* 120, 2029, 1987.
26. **Johnson, R. A., Boyce, B. F., Mundy, G. R., and Roodman, G. D.,** Tumors producing human TNF induce hypercalcemia and osteoclastic bone resorption in nude mice, *Endocrinology,* 124, 1424, 1989.
27. **Thomson, B. M., Mundy, G. R., and Chambers, T. J.,** Tumor necrosis factors α and β induce steoblastic cells to stimulate osteoclastic bone resorption, *J. Immunol.,* 138, 775, 1987.
28. **Garrett, I. R., Durie, B. G. M., Nedwin, G. E., Gillespie, A., Bringman, T., Sabatini, M., Bertolini, D. R., and Mundy, G. R.,** Production of the bone resorbing cytokine lymphotoxin by cultured human myeloma cells, *N. Engl. J. Med.,* 317, 526, 1987.
29. **Yoneda, Y., Aufdemorte, T. B., Nishimura, R., Nishikawa, N., Sakuda, M., Alsina, M. M., Chavez, J. B., and Mundy, G. R.,** Occurrence of hypercalcemia and leukocytosis with cachexia in a human squamous cell carcinoma of the maxilla in athymic nude mice. A novel experimental model of three concomitant paraneoplastic syndromes, *J. Clin. Oncol.,* in press.
30. **Yoneda, T., Mundy, G. R., and Roodman, G. D.,** Induction of differentiation of the human promyelocytic HL-60 cells into cells with the osteoclastic phenotype, *J. Bone Miner. Res.,* 4, 559, 1989.
31. **Sabatini, M., Chavez, J., Mundy, G., and Bonewald, L. F.,** Stimulation of tumor necrosis factor release from monocytic cells by the A375 human melanoma via granulocyte-macrophage colony stimulating factor, *Cancer Res.,* 50, 2673, 1990.
32. **Sabatini, M., Yates, A. J., Garrett, R., Chavez, J., Dunn, J., Bonewald, L., and Mundy, G. R.,** Increased production of tumor necrosis factor by normal immune cells in a model of the humoral hypercalcemia of malignancy, *Lab. Invest.,* 63, 676, 1990.
33. **Feyen, J. H. M., Elford, P., Dipadova, F. E., and Trechsel, U.,** Interleukin-6 is produced by bone and modulated by parathyroid hormone, *J. Bone Miner. Res.,* 4, 633, 1989.
34. **Lowik, C. W., van der Pluijm, G., Bloys, H., Hoekman, K., Bijvoet, O. L. M., Aarden, L. A., and Papapoulos, S. E.,** Parathyroid hormone (PTH) and PTH-like protein (PLP) stimulate interleukin-6 production by osteogenic cells: a possible role of interleukin-6 in osteoclastogenesis, *Biochem. Biophys. Res. Commun.,* 162, 1546, 1989.
35. **Kurihara, N., Chenu, C., Miller, M., Civin, C., and Roodman, G. D.,** Identification of committed mononuclear precursors for osteoclast-like cells formed in long term human marrow cultures, *Endocrinology,* 126, 2733, 1990.
36. **Bataille, R., Jourdan, M., Zhang, X. G., and Klein, B.,** Serum levels of interleukin-6, a potent myeloma cell growth factor, as a reflection of disease severity in plasma cell dyscrasias, *J. Clin. Invest.,* 84, 2008, 1989.
37. **Klein, B., Zhang, X. G., Jourdan, M., and Bataille, R.,** Cytokines involved in human multiple myeloma, *Monoclonal Gammapathies II,* 12, 55, 1989.
38. **Abe, E., Tanaka, H., Ishimi, Y., Chisato, M., Hayashi, T., Nagasawa, H., Tomida, M., Yamaguchi, Y., Hozumi, M., and Suda, T.,** Differentiation-inducing factor purified from conditioned medium of mitogen-treated spleen cell cultures stimulates bone resorption, *Proc. Natl. Acad. Sci. U.S.A.,* 83, 5958, 1986.
39. **Gearing, D. P., Gough, N. M., King, J. A., Hilton, D. J., Nicola, N. A., Simpson, R. J., Nice, E. C., Kelsa, A., and Metcalf, D.,** Molecular cloning and expression of cDNA encoding a murine myeloid leukaemia inhibitory factor (LIF), *EMBO J.,* 6, 3995, 1987.
40. **Reid, I. R., Lowe, C., Cornish, J., Gray, D. H., and Skinner, S. J. M.,** Adenylate cyclase blockers dissociate PTH-stimulated bone resorption from cAMP production, *Am. J. Physiol.,* 258, E708, 1990.

41. **Metcalf, D. and Gearing, D. P.,** Fatal syndrome in mice engrafted with cells producing high levels of the leukemia inhibitory factor, *Proc. Natl. Acad. Sci. U.S.A.,* 86, 5948, 1989.
42. **Gowen, M. and Mundy, G. R.,** Actions of recombinant interleukin-1, interleukin-2 and interferon gamma on bone resorption *in vitro, J. Immunol.,* 136, 2478, 1986.
43. **Takahashi, N., MacDonald, B. R., Hon, J., Winkler, M. E., Derynck, R., Mundy, G. R., and Roodman, G. D.,** Recombinant human transforming growth factor alpha stimulates the formation of osteoclast-like cells in long term human marrow cultures, *J. Clin. Invest.,* 78, 94, 1986.
44. **Takahashi, N., Mundy, G. R., and Roodman, G. D.,** Recombinant human interferon gamma inhibits formation of human osteoclast like cells, *J. Immunol.,* 137, 3544, 1986.
45. **Pfeilschifter, J. P., Seyedin, S., and Mundy, G. R.,** Transforming growth factor beta inhibits bone resorption in fetal rat long bones, *J. Clin. Invest.,* 82, 680, 1988.
46. **Chenu, C., Pfeilschifter, J., Mundy, G. R., and Roodman, G. D.,** Transforming growth factor beta inhibits formation of osteoclast-like cells in long-term human marrow cultures, *Proc. Natl. Acad. Sci. U.S.A.,* 85, 5683, 1988.
47. **Tashjian, A. H., Voelkel, E. F., Lazzaro, M., Singer, F. R., Roberts, A. B., Derynck, R., Winkler, M. E., and Levine, L.,** Alpha and beta transforming growth factors stimulate prostaglandin production and bone resorption in cultured mouse calvaria, *Proc. Natl. Acad. Sci. U.S.A.,* 82, 4535, 1985.
48. **Oreffo, R. O. C., Mundy, G. R., Seyedin, S., and Bonewald, L.,** Activation of the bone derived latent TGF beta complex by isolated osteoclasts, *Biochem. Biophys. Res. Commun.,* 158, 817, 1989.
49. **Marcelli, C., Yates, A. J. P., and Mundy, G. R.,** *In vivo* effects of human recombinant transforming growth factor beta on bone turnover in normal mice, *J. Bone Miner. Res.,* 5, 1087, 1990.
50. **MacDonald, B. R., Mundy, G. R., Clark, S., Wang, E. A., Kuehl, T. J., Stanley, E. R., and Roodman, G. D.,** Effects of human recombinant CSF-GM and highly purified CSF-1 on the formation of multinucleated cells with osteoclast characteristics in long term bone marrow cultures, *J. Bone Miner. Res.,* 1, 227, 1986.
51. **Hagennars, C. E., van der Kraan, A. A., Kawilarang-de Haas, E. W., Visser, J. W., and Nijweide, P. J.,** Osteoclast formation from cloned pluripotent hemopoietic stem cells, *Bone Miner.,* 6, 179, 1989.
52. **Yoshida, H., Hayashi, S. I., Kunisada, T., Ogawa, M., Nishikawa, S., Okamura, H., Sudo, T., Shultz, L. D., and Nishikawa, S. I.,** The murine mutation osteopetrosis is in the coding region of the macrophage colony stimulating factor gene, *Nature (London),* 345, 442, 1990.
53. **Wiktor-Jedrzejzcak, W., Ahmed, A., Szczylik, C., and Skelly, R. R.,** Hematological characterization of congenital osteopetrosis in op/op mouse, *J. Exp. Med.,* 156, 1516, 1982.
54. **Klein, D. C. and Raisz, L. G.,** Prostaglandins: stimulation of bone resorption in tissue culture, *Endocrinology,* 86, 1436, 1970.
55. **Tashjian, A. H., Voelkel, E. F., Levine, L., and Goldhaber, P.,** Evidence that the bone resorption-stimulating factor produced by mouse fibrosarcoma cells is prostaglandin E2: a new model for the hypercalcemia of cancer, *J. Exp. Med.,* 136, 1329, 1972.
56. **Chambers, T. J. and Ali, N. N.,** Inhibition of osteoclastic motility by prostaglandins I2, E1, E2 and 6-oxoE1, *J. Pathol.,* 139, 383, 1983.
57. **Chenu, C., Kukita, T., Mundy, G. R., and Roodman, G. D.,** Prostaglandin E2 inhibits formation of osteoclast-like cells in long-term human marrow cultures but is not a mediator of the inhibitory effects of transofrming growth factor β, *J. Bone Miner. Res.,* 5, 677, 1990.

58. **Akatsu, T., Takahashi, N., Debari, K., Morita, I., Murota, S., Nagata, N., Takatani, O., and Suda, T.,** Prostaglandins promote osteoclast like cell formation by a mechanism involving cyclic adenosine 3′,5′-monophosphate in mouse bone marrow cell cultures, *J. Bone Miner. Res.*, 4, 29, 1989.
59. **Meghji, S., Sandy, J. R., Scutt, A. M., Harvey, W., and Harris, M.,** Stimulation of bone resorption by lipoxygenase metabolites of arachidonic acid, *Prostaglandins,* 36, 139, 1988.
60. **El Attar, T. M. A. and Lin, H. S.,** Relative conversion of arachidonic acid through lipoxygenase and cyclo-oxygenase pathways by homogenates of diseased periodontal tissues, *J. Oral Pathol.*, 12, 7, 1983.
61. **El Attar, T. M. A., Lin, H. S., Killoy, W. J., Vanderhoek, J. Y., and Goodson, J. M.,** Hydroxy fatty acids and prostaglandin formation in diseased human periodontal pocket tissue, *J. Periodont. Res.*, 21, 169, 1986.
62. **Davidson, E. M., Rae, S. A., and Smith, M. J.,** Leukotriene B_4, a mediator of inflammation present in synovial fluid in rheumatoid arthritis, *Ann. Rheum. Dis.*, 43, 677, 1983.

Chapter 9

THE INFLUENCE OF MATRIX COMPONENTS ON OSTEOCLASTS

Julie Glowacki

TABLE OF CONTENTS

I. INTRODUCTION

Various experimental models have been developed for investigation of osteoclast biology. Because bone resorption is complex and because skeletal tissues cannot always be manipulated for analysis in ways that soft tissues can, different models have been designed to answer specific questions about the mechanisms and regulation of osteoclast differentiation and activity. There have been numerous studies on the regulation of osteoclast differentiation by soluble factors and fewer on the role played by matrix constituents. Different approaches may be needed to fully explore relationships between cells and insoluble factors.

It is realized that constituents in extracellular matrix are involved in the growth, differentiation, and morphogenesis of various tissues. In addition to extracellular structural molecules, some growth factors, such as basic fibroblast growth factor[1] and transforming growth factor-β (TGF-β)[2] function in the solid state in local regulation of cell proliferation and differentiation. Accordingly, many biological activities have been attributed to the numerous organic constituents of bone matrix. Evidence has been accumulating that bone matrix influences osteoclast physiology.

My colleagues and I have used the rat to develop models to define and modulate tissue responses to bone-derived and synthetic materials.[3] Healing[4] and nonhealing[5,6] intraosseous defects, onlay sites,[7] and heterotopic sites[8] provide different information about tissue and cellular responses to materials. In the course of our work to understand mechanisms of healing in osseous reconstructive surgery and to develop implants to predictably promote repair, we implanted pulverized materials in order to amplify surface interactions with recipient tissue beds, but too large for phagocytosis.

Conventional bone grafting procedures are often limited by the amount of fresh autogenous bone available for transplantation.[9] Alternative bone-derived and synthetic materials invoke different mechanisms of healing and incorporation.[3] Banked bone and large free segments of cortical bone do not contain viable osteoblasts or preosteoblasts capable of osteogenesis in the new site. Rather, they promote repair by the mechanism of osteoconduction, in which the nonvital bone provides a scaffolding for the ingrowth of bone. Burchardt clearly described the distinct repair processes with transplanted cancellous or cortical bone.[10] With fresh autogenous cancellous bone, revascularization is rapid and repair is initiated by viable osteoblasts from the transplant. The value of a cortical bone transplant does not depend upon the survival of its cells, but on the balance of new bony ingrowth and resorption of the dead bony scaffold. Thus, repair with cortical segments is initiated by osteoclasts.[10] There are other clinical situations in which osteoclastic removal of dead bone plays a beneficial role, e.g., removal of bony fragments and devitalized bone in fracture healing, incorporation of iliac marrow grafts that contain particulated fragments of trabecular bone, and repair of subclinical infarcts. In some sense, dead bone may act as a foreign body in triggering its removal. These processes raise important questions about the relationship between osteoclasts and inflammatory giant cells, their lineages, and interactions between inflammatory mediators and bone-active modulators. Available information suggests that common mechansims apply to bone repair, tissue response to osseous reconstructive interventions, and homeostatic and pathological skeletal remodeling, including the metabolic bone diseases. It is possible that the resorption of devitalized bone is regulated differently from vital bone which contains osteocytes, osteoblasts, and other cells capable of specific interactions. Nevertheless, it is clear that bone matrix components themselves interact with and influence osteoclasts.

II. ATTACHMENT OF OSTEOCLASTS TO BONE

A. ULTRASTRUCTURE

The light and electron microscopic features of rat and human osteoclasts have been reviewed recently by Holtrop,[11] Mills,[12] and in Chapter 6 of this volume. Osteoclasts are the multinucleated resorbing cells of bone found in contact with bone surfaces. They can be located in Howship's or resorption lacunae or can be wrapped around bone spicules. The nucleoli are prominent. The cytoplasm is characterized by multiple perinuclear Golgi complexes, unusually numerous mitochondria, scattered small, dense lysosomal granules, many vacuoles, giant centrospheres, and short and sparse rough endoplasmic cisternae. The face of the cell not in contact with bone may be relatively smooth or show numerous microvilli.

Osteoclasts make close contact with the bone surfaces being resorbed. The clear zone of attachment is devoid of cytoplasmic organelles and, at higher magnifications, displays dense parallel bands representing bundles of actin filaments perpendicular to the bone surface. The clear zone is not a specific osteoclast structure. Macrophages and giant cells in contact with foreign bodies can organize clear zones.

The most characteristic ultrastructural feature of an osteoclast is the membrane specialization adjacent to fraying bone called the ruffled border. It may assume different appearances, but always is an area of enormously extended foldings of the plasma membrane. Holtrop et al. reported that in random thin sections (60 nm) of cultured fetal rat bones, only 11% of the osteoclast profiles showed ruffled borders.[13] Ruffled borders can easily be missed because the average area of ruffled borders occupies only 3.4% of the area of an osteoclast's cytoplasm.[14] Fortuitous sectioning through a ruffled border parallel to the bone surface reveals that the ruffled area is encircled by a ring of clear zone.

The mechanism of bone resorption at the site of the ruffled border is discussed in Chapter 14 of this volume. Because morphological changes accompany changes in the resorptive activity of osteoclasts *in vivo* and *in vitro*, nuclear number, percent of cell profiles with ruffled borders, ruffled border area, and clear zone areas can be indices of osteoclastic functional activity.[11]

In addition, dramatic changes in osteoclast cell surface features take place during the egg cycle in quail.[15] During egg shell calcification, ruffled borders and vacuoles are well developed. After calcification, the osteoclasts appear more round; small clear zones abutting bone surfaces remain, but no ruffled borders are seen. The loss of ruffled borders is associated with the appearance of extensive membrane interdigitations on the side of the osteoclast away from the bone. From these observations, Miller speculated that the microvilli represent reservoirs of membrane between waves of active resorption. These structures have not been described for other osteoclasts; however, the egg cycle may be an accelerated situation where they can easily be seen. That they resemble microvilli of macrophages and macrophage polykaryons may pertain to the common machinery required for cellular functions rather than to the relationship between these cell types. Microvilli can be found on bone-resorbing cells *in vitro* and in other model systems, as shown below.

B. CELL SUBSTRATUM INTERACTIONS

The development of procedures for isolating osteoclasts has enabled investigators to understand more about the mechanisms of osteoclastic bone resorption. *In vitro* studies have also provided important information about osteoclast differentiation. (See Chapters 5, 8, and 21 of this volume.) Clearly, contact between bone and osteoclasts or their progenitor/precursors is important for these physiological processes. Osdoby et al.'s studies with chick cells provided evidence for specific cellular/matrix recognition and attachment events.[16,17] Their *in vitro* adhesion studies indicated that osteoclasts and monocytes selectively adhered to bone compared to hyaline cartilage and relative to fibroblasts or mesenchymmal cells. The ability of an osteoclast to adhere to, spread over, and move along the bone matrix will depend on many factors. Zambonin-Zallone et al. described multiple actin-containing "dots" with a belt of vimentin-type filaments in isolated chicken osteoclasts and monocytes.[18] Subsequently, the dots were found to correspond to a core of fimbrin and α-actinin surrounded by a rosette structure with vinculin and talin.[19] These attachment structures or podosomes can be modulated by conditions that alter osteoclast activity.[20]

Osteoclasts isolated from rabbits show a peripheral distribution of these fine microfilaments, characteristic of cells with highly motile cell surfaces.[21] A similar organization of adhesion sites has been described for large cells derived from human giant cell tumors.[22]

Using antibodies raised against osteoclasts, Horton's group found that they reacted with different epitopes on the α-subunit of integrin, a heterodimeric receptor for the extracellular matrix protein vitronectin.[24] Vitronectin and other integrin receptors are important in membrane activity and fusion processes. Some specificity in the osteoclast may exist in the distribution of extracellular matrix receptors. Zambonin-Zallone et al. showed colocalization of β_3-integrin with vinculin and talin in the podosomes of giant cells from human tumors, but not in macrophages.[25] (See Chapter 12 of this volume.)

Certain bone matrix proteins — osteopontin (also known as 44K bone phosphoprotein, bone sialoprotein 1 and 2ar), bone sialoprotein II, and thrombospondin — contain cell-binding sites with the typical RGD sequence. The attachment properties of osteopontin extend to bone cells,[26] gingival fibroblasts,[27] and other cell types.[27] The vitronectin receptor, having specificity for osteopontin, was localized at regions where osteoclasts were attached to bone.[28] Whether this molecule attracts or binds osteoclast progenitor or precursor cells is not yet known.

III. BONE MATRIX AND OSTEOCLAST DIFFERENTIATION

A. *IN VITRO* INVESTIGATIONS

Osteoclasts of fetal mouse calvaria are first observed at the beginning of diploe formation at 17 d *in utero*. Ko and Bernard therefore used osteoclast-free 13 to 15 d-old fetal calvarial explants for cocultivation *in vitro* with mononuclear bone marrow cells.[29] After 10 d, multinucleated, resorbing osteoclasts developed in contact with bone surfaces; none were present in calvaria cultured without marrow cells. Similar results were obtained by Burger et al. with fetal mouse metatarsal rudiments of calcified cartilage.[30] When devitalized rudiments were used, few osteoclasts developed, but the cartilage matrix was far less calcified. These authors reported that the location and extent of osteoclast differentiation on devitalized rudiments was related to the extent of matrix calcification, which varied from bone to bone.[31] There is no doubt that osteoblasts promote osteoclast differentiation in these systems, but the mineralized matrix is also required. Using similar methods, Lowik et al. examined the migration of progenitors toward, and their contact with, mineralized matrix. They concluded that bisphosphonates interfere with differentiation by interacting with a matrix factor that is required for attachment of osteoclast progenitors.[32]

Several investigators have reported that osteoclasts can be generated from cultured bone marrow cells from a great many species. (See Chapter 21 of this volume.) These models have been used to identify the osteoclast progenitor and its relationship to monocytes/macrophages. Use of these cultures has increased our understanding of the role of hormones, cytokines, and growth factors in osteoclastogenesis. Nevertheless, there are studies that support the view that bone matrix also provides regulatory stimuli. Using rabbit marrow, Fuller and Chambers found three-fold more cells reacting with their osteoclast-specific antibody after incubation on bone slices compared to plastic.[33] Graves and Jilka found that bone particles stimulated the development of cells positive for tartrate-resistant acid phosphatase and calcitonin binding from cell cultures prepared from calvaria of normal and microphthalmic mice.[34]

Other studies can be viewed in this light. Heterogeneous cultures of marrow will contain cells of stromal lineage, including osteoblasts. Osteoblast-enriched cultures or cell lines have the ability to support osteoclast formation from spleen cells by contact-mediated processes.[35-37] The relative contributions of cell surface and extracellular matrix molecules are not known. Recently, Long et al. reported subpopulations of adherent cells that produce the bone-related proteins osteonectin and bone γ-carboxyglutamic acid-containing protein, BGP, within 1 week after incubation of bone marrow from humans.[38]

B. CHICKEN CHORIOALLANTOIC MEMBRANE

Krukowski and Kahn used the chick chorioallantoic membrane (CAM) as an implantation site *in ovo* to examine the role of bone matrix on osteoclast differentiation.[39] They reported that osteoclast-like cells developed around fine (<25 μm) bone powders that had been deposited onto CAMs. Unlike the multinucleated cells formed in response to hydroxyapatite or eggshell powders, the bone-associated cells formed prominent clear zones and ruffled borders. In more recent studies, purified bone matrix components were incorporated into slow-delivery polymers and placed onto CAMs.[40] Osteopontin-containing implants and hydroxyapatite particles were associated with multinucleated cells. Those cells and the mononuclear cells recruited by osteocalcin or osteonectin were positive for 29C antigen expression and were therefore considered putative osteoclast precursors. They did not express the 121F osteoclast-specific antigen as did those multinucleated cells formed in response to mineralized bone matrix. These investigators proceeded to adsorb the bone matrix components to hydroxyapatite.[41] The expression of the 121F osteoclast antigen correlated with the extent of ruffled membranes on giant cells formed in CAM implants. Both features were most prominent in multinucleated cells associated with bone particles and with hydroxyapatite that had been preadsorbed with whole bone extract or purified osteocalcin. These results are consistent with the view that osteoclast cytodifferentiation may be regulated by the mineral matrix and the organic components of bone, including osteocalcin.

C. SUBCUTANEOUS IMPLANTS IN RATS

1. Regulation

Early studies comparing the fate of mineral-containing and demineralized bone, either as calvarial discs or powders from long bones, showed strikingly different responses in nonhealing cranial defects in rats. Whereas demineralized bone implants promoted endochondral bone repair in this membranous bone, the mineral-containing implants did not promote healing and were associated with multinucleated cells apparently mediating the resorption of the matrix.[5,42] This intraosseous model was used to demonstrate that the bone particles (BPs) were surrounded by multinucleated cells within 7 d after implantation and extensively resorbed within 3 weeks.[42] BP resorption was inhibited by systemic administration of dichloromethane diphosphonate (Cl_2MDP, 4 mg/kg/d).[8] Further, it was shown that fewer multinucleated cells were generated in parathyroidectomized rats.[8] These results suggested that the multinucleated resorbing cells may be related to osteoclasts. Attempts to characterize these cells ultrastructurally[43] were not definitive at 7 d after implantation of BPs in cranial defects because of the abundance of mononuclear cells at that time and because of the possibility that the osteoclastic cells that were seen around the particles had migrated from the margins of the defect. For this reason, subsequent studies used BPs implanted into heterotopic (nonbony) sites.[8,44]

The subcutaneous implantation of BPs in rats initiates a rapid and reproducible resorptive sequence that is complete within 28 d (Figure 1). Because mast cells populate the interstitial tissue between BPs, we speculated that mast cell products may regulate BP resorption.[44] Indeed, exogenous heparin stimulated resorption of implanted BP to 141% of control values and protamine, a heparin antagonist,[45] reduced resorption to 55% of control.

In addition, the subcutaneous model was used to demonstrate reduced resorption (73% of control, $p < 0.001$) in rats treated for 12 d with synthetic human calcitonin.[46] Modulation by calcitonin suggested that the cells that resorb BPs may be related to osteoclasts.

Other systemic factors influence resorption of bone particles. Rats maintained on diets deficient in manganese and/or copper showed reduced resorption of normal BPs.[47] Animals maintained on test diets for 6 months resorbed BPs at 60% of control values. After 12 months on test diets, when the animals showed great depletions in serum Mn and Cu, BP resorption was only 30% of control value. These findings, and the accompanying reduction in demineralized bone powder (DBP)-induced osteogenesis, help to explain the slow healing of a bone fracture in a prominent athlete with low serum concentrations of Mn, Cu, and Zn.[48]

This model has been used to examine the resorbability of different BPs. For example, BPs prepared from Warfarin-treated rats were devoid of osteocalcin and were resistant to resorption when implanted in normal rats.[49] These data led us to propose that osteocalcin may be a matrix signal that plays a role in osteoclast differentiation or activity. We also found that BPs presoaked in 10^{-8} M Cl_2MDP for 15 min prior to implantation were poorly resorbed compared to controls.[8] The adsorption of the diphosphonate to the mineral in BPs interfered with osteoclast differentiation. Recently, Donnelly et al. used this model to examine the mechanism whereby gallium nitrate exerts its antiresorptive action.[50] They reported that gallium-containing BPs were not resorbed as well as normal BPs and concluded that incorporation of gallium into bone matrix confers resistance to resorption.

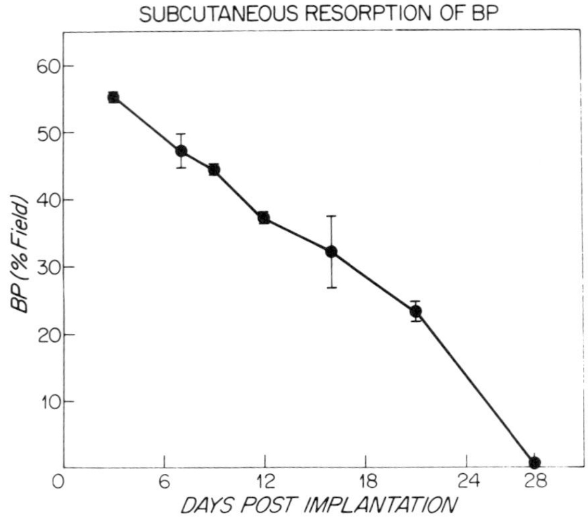

FIGURE 1. Resorption of bone particles (BPs) implanted subcutaneously in rats. BP (% field) represents the projected area occupied by BP as a percent of the histological field. (From Glowacki, J. and Lian, J. B., in *The Chemistry and Biology of Mineralized Tissues*, Butler, W. T., Ed., (Pub. Butler), Birmingham, AL, 1985, p. 164. With permission.)

2. Characterization of Multinucleated Cells

Tissue elicited 12 d after subcutaneous implantation of BPs was characterized cytochemically and ultrastructurally. At 12 d, resorption is active, with nearly half of the implant having been resorbed (Figure 1). Multinucleated cells are found associated with the particles, frequently in resorption lacunae on the surfaces of the BPs. Tartrate-resistant acid phosphatase (TRAP) activity can be localized in the multinucleated cells, expecially at the cell/bone interface.[51] Mast cells are seen vicinal to, but not in contact with the bone particles or multinucleated cells. Electron microscopic evaluation showed that many multinucleated cells in lacunae display features characteristic of *in osso* osteoclasts, including clear zones of attachment to the bone substrate and ruffled borders overlying bone with frayed collagen fibers (Figure 2). Other types of interactions are seen in random sections through the cell/BP interface.[51] Evidence of cell-associated disaggregation of collagen without classical ruffled borders may shed some light on mechanisms of bone resorption. Everts et al. have described phagocytosis of bone mineral and collagen fragments by osteoclasts in two patients with pycnodysostosis.[52] They proposed that in this disorder the extracellular mechanism of bone dissolution is inhibited, and thus the intracellular mechanism, normally too rapid to be detected, is the predominant finding.

These data show that devitalized bone matrix can be resorbed, at least in part, by mechanisms involving ruffled borders. Zambonin-Zallone et al. have also documented the resorption of dead bone at the site of ruffled borders in *in vitro* studies of isolated chicken osteoclasts.[53]

According to the Rodan and Martin hypothesis,[54] osteoblasts, because they possess PTH receptors, may mediate the regulation of osteoclastic bone resorption. No osteoblasts are present in the tissue elicited by BP implants. Thus, osteoblasts are not absolutely essential for osteoclast differentiation and activity in this model. Nevertheless, it is likely that osteoclasts could be stimulated by PTH treatment and by neighboring osteoblasts. This is supported by two findings. First, recruitment of osteoclastic cells by BP implants is impaired in parathyroidectomized rats.[8] Second, TRAP activity and BP resorption were stimulated in composite implants of BP and DBP (which stimulated islands of endochondral ossification among the BPs).[55] Further quantitative morphometric studies are needed to demonstrate whether modulation of ruffled borders in BP-elicited osteocasts by PTH is similar to that shown in *in osso* osteoclasts.[56]

The multinucleated cells elicited by subcutaneous implants of BPs can be distinguished from foreign body giant cells (FBGCs). Typical FBGCs were found associated with particles of polyethylene (PE) or polymethylmethacrylate (PMMA) 12 d after implantation.[57] The multinucleated cells around PE or PMMA particles did not stain for TRAP activity. Even in specimens of mixtures of BP/PE or BP/PMMA, cells on BPs were clearly different from the FBGCs in TRAP intensity.[57] Electron microscopic evaluation revealed that there were clear areas of attachment of the FBGCs adjacent to PE or PMMA particles. As expected, the cells in resorption lacunae on the surfaces of BPs displayed clear zones and ruffled borders. No ruffled borders were found on FBGCs.

Outgrowth cell cultures were prepared from tissue fragments from BP or PMMA implants.[58] TRAP activity was maintained in the multinucleated cells from BP-elicited tissue and was not striking in cells from PMMA implants.[59] This culture system was used to assess binding of ^{125}I-salmon calcitonin by autoradiography.[59] In replicate cultures, between 15 and 52% (mean 38%) of the multinucleated cells from BP-elicited tissues were positive for specific calcitonin binding. In cultures prepared from PMMA implants, 1 of 55 multinucleated cells (1.8%) demonstrated specific calcitonin binding. The variability in calcitonin binding among the multinucleated cells elicited by BPs, as well as the presence of CT receptor-negative multinucleated cells in these cultures, may be related to the loss of receptors from cells as they migrated away from the mineralized matrix. There may be subpopulations of osteoclasts that do not express functional receptors for this hormone. Another possibility is that in the 3-d culture period before labeling, CT receptor-negative polykaryons may have formed *in vitro* by fusion of mononuclear cells.

Thus, the cells that form in response to particulate plastics were typical FBGCs and distinct from the osteoclastic cells elicited by BPs. Another difference was detected between these two types of multinucleated cells. Conditioned media were collected from the second 24-h culture period of outgrowth cultures established from elicited tissues. Medium from BP out-

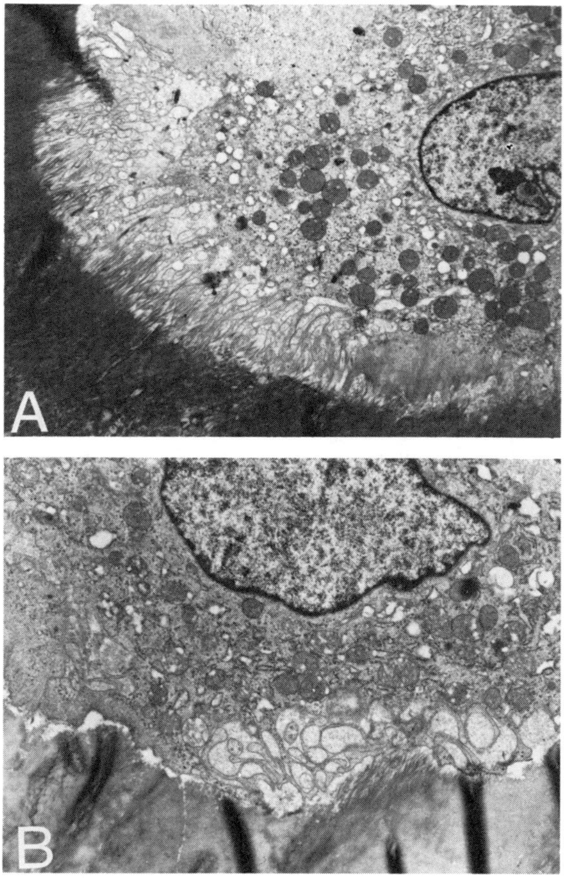

FIGURE 2. A. Membrane specializations, ruffled border, and clear zone, at interface of bone particle and elicited multinucleated cell 12 d after implantation. (Magnification × 3500.) B. Ultrastructure of membrane of another BP-elicited cell. (Magnification × 4900.) C. For comparison, electron micrograph of a rat osteoclast in tibial metaphysis. (Magnification × 1841.) D. Ultrastructure of membrane of a rat tibial osteoclast. (Magnification × 3640.)

growth cultures contained 269 pg/ml of prostaglandin E_2; medium from PMMA outgrowth cultures contained 2680 pg/ml. This tenfold difference attests that the nature of the cellular infiltrates was more "inflammatory" around PMMA than around BPs.

Relying on ultrastructural analysis of membrane specializations, Popoff and Marks described the multinucleated cells associated with bone implants as giant cells with extensive interdigitations between neighboring cells, but lacking the ruffled borders characteristic of metaphyseal osteoclasts.[60] There were three significant differences between their experiments and ours. Their

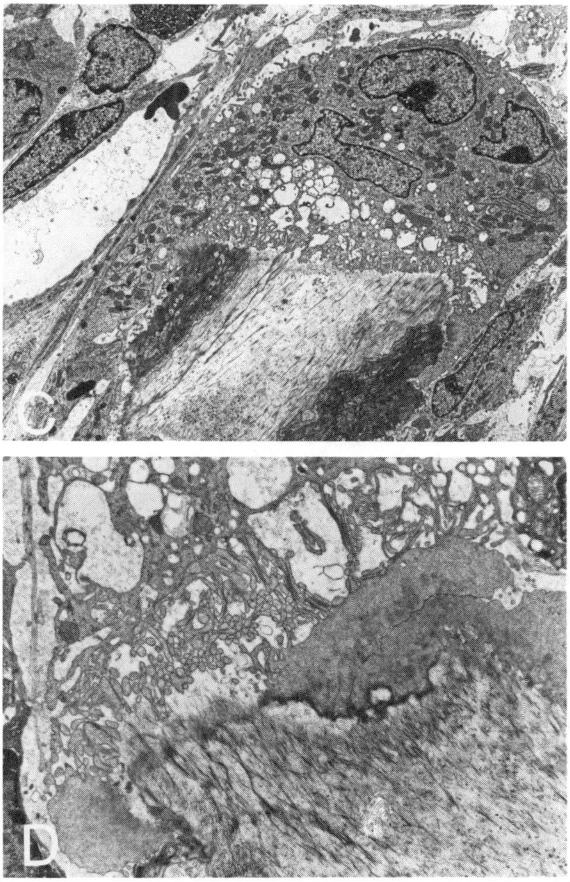

FIGURE 2C and D.

donor bone was obtained from suckling rats and mice, they used calvarial bone, and they implanted large (2.5 mm in diameter) molded pellets of bone powders.[60] We prepared bone particles from cortical diaphyses from 4- to 6-month-old rats and mice and reported that the extent of resorption was related to the surface area of the bone particles, smaller particles providing more area for interaction with resorbing cells. These differences in substrate preparation may account for the different results. That those investigators subsequently found no difference in giant cell reactions in osteopetrotic and normal animals[61,62] was confirmed by our studies showing intact giant cell responses to hydroxyapatite or PMMA by osteopetrotic mutants.[63] With our techniques, BP resorption was impaired in osteopetrotic mutants.[63]

Walters and Schneider reported that ^{45}Ca-prelabeled calvarial chips were degraded equally well when implanted into osteopetrotic *ia* or normal rats and that the multinucleated cells adjacent to the bone implants did not display ruffled borders in spite of evidence of ^{45}Ca release.[64] They concluded that subcutaneous implantation of bone chips resulted in the formation of phagocytic and macrophage polykaryons, and that the skeletal defects in *ia* mutants were not reproduced in the implants. Walters and Schneider then compared these cells with those induced by implants of half-tibia from normal or osteopetrotic *ia* rats and by small pieces of silk suture.[65] They described the multinucleated cells as similar in all cases, resembling macrophage polykaryons. The cells attached to the suture material appeared to be larger, with alignment of the nuclei at the periphery of the cells, as in Langhans cells. Walters and Schneider reported no difference in TRAP activity in the various elicited multinucleated cells and concluded that they did not resemble active osteoclasts.[66]

Their conclusions are in contrast to ours. We reported that multinucleated cells elicited by devitalized, mineral-containing bone particles can be distinguished from FBGCs. In studies of the modulation of BP resorption by bone-active agents, we found that generation of osteoclast-like cells was dependent upon the method of preparation of the bone particles and removal of potentially inflammatory cellular debris. This is accomplished by extensive washes in water and extractions in anhydrous ethanol followed by ethyl ether.[9] Further, Walters and Schneider implanted larger chips or half-tibiae and we used particles of a narrow range, 75 to 250 mm. Cellular reactions are related to particle size, smaller particles presenting greater surface areas for interaction with the tissue bed. The third point concerns the time of evaluation. It is not yet clear in what sequence osteoclast features are expressed in BP-elicited cells. The scarcity of ruffled borders on day 7 and their appearance by day 12 suggest that they may be a phenotypic feature of these cells that develops as the cells resorb bone.

Bagi and Miller, following our protocols, reported that osteoclast-like cells are involved in the resorption of devitalized mineral-containing bone particles in subcutaneous implants in rats.[67] Also following our protocol, Donnelly et al. made the qualitative statement that the cells resorbing bone particles appeared to have ruffled borders and clear zones.[50]

Other technical factors may influence the reproducibility of osteoclastic resorption of subcutaneously implanted BPs. We deposit BPs at the cephalic end of a narrow (5 mm) subcutaneous pocket or tunnel approximately 12 to 15 mm long, constructed by blunt dissection through a 5-mm skin incision over the diaphragm. With this technique, 50-mg aliquots of dry BPs are suitably compact so that they are readily encapsulated and invaded. We have found inconsistent results when the BPs are dispersed over the thoracic musculature. If skin sutures or wound clips invade the packet of BPs, they can provoke a confounding foreign body reaction. For these reasons, the BPs are slid into the blind end of the narrow tunnel at a distance from the incision so that the wound clip is 5 to 8 mm caudad.

3. The Role of Bone Matrix

We were struck by the remarkably different cellular responses to bone particles that were mineral containing and those that were decalcified by acid extraction. Mineral-containing bone particles induced osteoclasts, and after acid extraction, they induced endochondral osteogenesis.[42] Acid extraction is a decalcifying process, but it also removes some of the organic constituents of bone matrix, notably osteocalcin and a large amount of TFG-β. Thus, we reasoned that the osteoclastic response could be signaled by the acid-soluble constituents of bone. We first tested the tissue response to deorganified bone ash prepared by heating bone particles at 600°C for 24 h. This treatment did not appear to alter particle size and shape significantly. Ashed bone particles elicited multinucleated giant cells 12 d after implantation, but they lacked TRAP activity and membrane specializations. These observations showing that bone mineral alone did not initiate osteoclast differentiation raised the possibility that one or more of the organic components of the bone matrix signaled the differentiation of osteoclastic features in multinucleated cells.

The subcutaneous implantation model of bone resorption was used to define the effects of bone matrix components on the resorption process. Bone that was virtually devoid of the vitamin K-dependent protein osteocalcin was obtained from rats treated with Warfarin for 6 weeks. Vitamin K is necessary for the carboxylation of glutamate in osteocalcin to γ-carboxyglutamate and is inhibited by Warfarin treatment. Without the carboxylated residues, osteocalcin does not accumulate in bone matrix. Osteocalcin-deficient bone particles were resistant to resorption when implanted in normal rats.[49] The relative resorption was 60% of that of control bone, as assessed histomorphometrically. In addition, the number of multinucleated cells around the bone particles was reduced by 54%. Mononuclear cells were attracted to control bone particles soon after implantation, but cellularity was depressed around osteocalcin-deficient bone particles (35% of control cellularity).[68] The cellularity, TRAP activity, and number of multinucleated cells within osteocalcin-deficient implants never reached the maximum levels of control specimens.

The use of mixtures of normal and osteocalcin-deficient bone particles (with one or the other type prelabeled *in vivo* with tetracycline) permitted separate analysis of the substrate specificity of osteoclast recruitment.[69] Those studies revealed that in the mixed implants, osteoclasts were associated specifically with normal bone particles, but not with the adjacent osteocalcin-deficient bone particles. These data rule out the possibility that osteocalcin, released from normal bone particles in the course of their degradation, influences the resorption of adjacent osteocalcin-deficient bone particles.

The substrate specificity of the osteoclast induction suggested that osteocalcin functions as a matrix signal in this system and may have a role in bone resorption. To test this hypothesis directly, we evaluated tissue responses to microcrystalline bone-like apatite that we prepared with 0.1% osteocalcin, with bovine serum albumin, or without added protein.[70,71] Apatite particles, alone or with albumin, generated a mild foreign-body reaction. Multinucleated cells were in direct contact with the particles, but were negative for TRAP and lacked ruffled borders. In contrast, particles with osteocalcin were partially resorbed and were associated with more multinucleated cells. These were positive for TRAP and displayed ultrastructural features of osteoclasts, including ruffled borders, vacuoles, and lysosomes. These data support the hypothesis that osteocalcin may function as a matrix signal in the recruitment and differentiation of bone-resorbing cells.

Other evidence suggests that osteocalcin and other bone matrix constituents may act as diffusible chemoattractants. Malone et al.[72,73] and others[74] showed that peptides of type I collagen, α_2HS glycoprotein, osteocalcin, and the amino acid γ-carboxy glutamic acid evoke chemotaxis in monocytes.

Recent *in vitro* studies lend further support that osteocalcin acts in concert with other factors to regulate various stages of osteoclastogenesis. Kandemir et al. reported that osteocalcin increased the number of osteoclasts formed from marrow cells cultured with fetal metatarsal rudiments.[75] It is possible that the osteocalcin was adsorbed to the calcified cartilage in the rudiments. Using cultured murine hematopoietic stems cells, Liggett et al. reported that osteocalcin enhances osteoclastic differentiation in the presence of GM-CSF and M-CSF.[76]

D. SUBCUTANEOUS IMPLANTS IN MICE

Because of its synchrony and relative homogeneity in rats, the subcutaneous model of resorption of mineral-containing, devitalized BPs is useful for evaluating the recruitment, differentiation, and activity of bone-resorbing osteoclastic cells. Available strains of mice with specific, well-characterized mutations would be useful for analysis of the role of cell types such as B-lymphocytes, T-lymphocytes, and mast cells upon BP resorption.

We have begun to apply the rat model to normal mice and mutants in order to study aspects of the cytodifferentiation of osteoclasts. In rats, subcutaneous implantation of BP elicits a tissue reaction that supports the local differentiation of multinucleated cells with many features similar to *in osso* osteoclasts and which mediate the resorption of the BPs. Mononuclear and stromal cells invade the implant within 4 d postimplantation and TRAP-positive multinucleated cells surround the bone particles by day 7, after which time resorption is measurable by histomorphometric[49] or biochemical methods.[50,68] In normal strains of mice (C57BL/6J, CD-1, WBB6F1/J, and WCB6F1/J), stromal and osteoclastic responses were delayed by 3 d compared to rats.[77] In rats, BPs are completely resorbed within 1 month. Mice appear to be less exuberant in both osteoclastogenesis and BP resorption. Only 18.2% of the implanted BP was resorbed in normal mice by day 14 (in contrast to 35 to 45% by day 12 in normal rats).[73] Clearly, then, a more rigorous experimental design is required with validation by power analysis to avoid Type II or β-errors in incorrectly accepting the null hypothesis. Because of low resorption

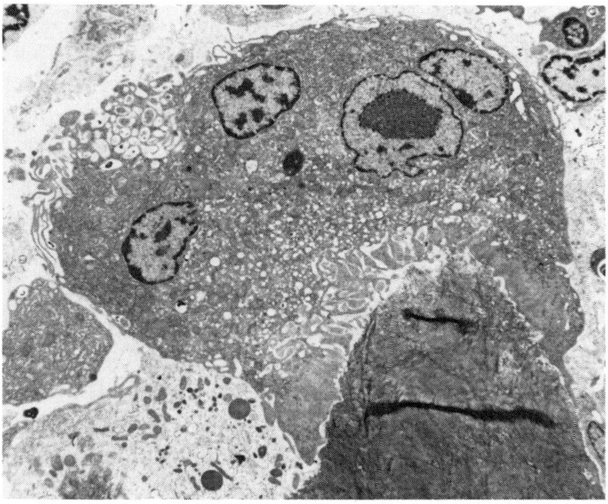

FIGURE 3. Electron micrograph of multinucleated cell elicited by bone particle in kelp bass adapted to diluted seawater. (Magnification × 3776.) (From Glowacki, J., Cox, K. A., O'Sullivan, J., Wilkie, D., and Deftos, L. F., *Proc. Natl. Acad. Sci. U.S.A.*, 83, 4104, 1986. With permission.)

of BP in normal mice, it may be more difficult to detect differences in mutants.[61,63] Nevertheless, we detected significant impairment in osteoclastic differentiation and BP resorption in microphthalmic osteopetrotic mice. The fact that this abnormality was reproduced by transplantation of *mi* spleen cells into normal irradiated mice[77] shows that the subcutaneous model can be applied in mice and may be useful for delineating the roles of different cellular components in skeletal physiology.

E. OTHER SPECIES

We have used subcutaneous implants to help explain how bony fish can maintain mineral homeostasis in the face of a chronic challenge.[78] Marine bass, *Paralabrax clathratus*, have an acellular bony skeleton and did not resorb implanted BPs. In dramatic contrast, BPs were resorbed by elicited osteoclasts in bass after adaptation to hyposalinity tank conditions. These bass were able to maintain normal serum chemistries despite a 23% reduction in environmental minerals and electrolytes. Electron microscopic analysis revealed large multinucleated cells with clear zones and ruffled borders apposing areas of the BP substrate that had the appearance of frayed collagen fibrils (Figure 3). Cartilaginous leopard sharks, *Triakis semifasciata,* did not develop multinucleated cells and did not resorb implanted BP under normal or hyposalinity conditions. These data show that bass have stem cells that can be stimulated to differentiate to osteoclasts. Cartilaginous sharks, however, appear to lack the cellular or humoral factors necessary for this recruitment.

Recently, Marks and Chambers reported that in rabbits, giant cells reacting to subcutaneous implants of bone did not express phenotypic markers of rabbit osteoclasts.[79] Whether their data are the result of differences in methodology or reflect species differences in unclear. Cortical bone particles were compressed into pellets rather than implanted as poweders. Differences in the surface area presented to host tissues may offer a physical barrier to rapid cellular reactions. Further, their materials were exposed to ultraviolet irradiation, which may alter the bone surfaces. Finally, it appears that each implant contained a xenogeneic bovine bone slice to enable analysis of pit excavation by scanning electron microscopy. Sampath and Reddi found that immunogenic, inflammatory, or inhibitory factors must be completely separated from human, monkey, and bovine demineralized bone matrix in order to show osteoinductive activity in rats.[80] It is likely that mineral-containing bovine bone slices, which are less processed than the demineralized bone matrix, would be inflammatory in rabbits.

IV. CONCLUSIONS

Many factors influence osteoclast differentiation and activity. Although the precise relationship between the differentiation pathways for osteoclasts and macrophage polykaryons is not fully elucidated, it is clear that striking differences and striking similarities exist in these two cell types. The osteoclast is the specialized multinucleated cell in bone whose bone-resorptive activity is highly regulated. In addition to the classical bone-active hormones, local factors from other bone cells, from the marrow microenvironment, or in the extracellular bone matrix are receiving more attention as regulators of osteoclast function. New experimental models are needed to define the mechanisms of their actions. Some may act as diffusible factors acting over short distances and others may act as insoluble agents requiring cell contact.

ACKNOWLEDGMENTS

Support is acknowledged from National Institutes of Health grants AR31330, CA45548, DE08798, and BRSG S07 RR05950. Excellent assistance was provided by Karen Oeschger.

REFERENCES

1. **Vlodarsky, I., Fuks, Z., Ishai-Michaeli, R., Bashkin, P., Levi, E., Korner, G., Bar-Shavit, R., and Klagsbrun, M.,** Extracellular matrix-resident basic fibroblast growth factor: implication for the control of angiogenesis, *J. Cell. Biochem.,* 45, 167, 1991.
2. **Paralkar, V. M., Vukicevic, S., and Reddi, A. H.,** Transforming growth factor b type 1 binds to collagen IV of basement membrane matrix: implications for development, *Dev. Biol.,* 143, 303, 1991.
3. **Glowacki, J.,** Cellular responses to bone-derived materials, in *Bone and Cartilage Allografts,* Friedlaender, G. E. and Goldberg, V. M., Eds., American Academy of Orthopedic Surgery, Park Ridge, IL, 1991, 55.
4. **Zelicof, S., Sledge, C. B., and Glowacki, J.,** The effects of particulate materials on spontaneous bone healing in a rat tibial defect model, *Orthop., Trans.,* 15, 585, 1991.
5. **Mulliken, J. B. and Glowacki, J.,** Induced osteogenesis for repair and construction in the craniofacial region, *Plast. Reconstr. Surg.,* 65, 553, 1980.
6. **Kaban, L. B. and Glowacki, J.,** Induced osteogenesis in the repair of experimental mandibular defects in rats, *J. Dent. Res.,* 60, 1356, 1981.
7. **Kaban, L. B. and Glowacki, J.,** Augmentation of the rat mandibular ridge with demineralized bone implants, *J. Dent. Res.,* 63, 998, 1984.
8. **Glowacki, J.,** Studies on the regulation of bone synthesis and bone resorption, in *Factors and Mechanisms Influencing Bone Growth,* Dixon, R. and Sarnat, B. G., Eds., Alan R. Liss, New York, 1982, 83.
9. **Glowacki, J. and Mulliken, J. B.,** Demineralized bone implants, *Clin. Plast. Surg.,* 12, 223, 1985.
10. **Burchardt, H.,** The biology of bone graft repair, *Clin. Orthop. Relat. Res.,* 174, 28, 1983.
11. **Holtrop, M. E.,** Light and electron-microscopic structure of osteoclasts, in *Bone,* Vol. 2, *The Osteoclast,* Hall, B. K., Ed., CRC Press, Boca Raton, FL, 1991, 1.
12. **Mills, B. G.,** Bone resorbing cells and human clinical conditions, in *Bone,* Vol. 2, *The Osteoclast,* Hall, B. K., Ed., CRC Press, Boca Raton, FL, 1991, 175.
13. **Holtrop, M. E., Raisz, L. G., and Simmons, H. A.,** The effects of parathyroid hormone, colchicine, and calcitonin on the ultrastructure and the activity of osteoclasts in organ culture, *J. Cell Biol.,* 60, 346, 1974.
14. **King, G. J., Holtrop, M. E., and Raisz, L. G.,** The relation of ultrastructural changes in osteoclasts to resorption in bone cultures stimulated with parathyroid hormone, *Metab. Bone Dis. Relat. Res.,* 1, 67, 1978.
15. **Miller, S. C.,** Osteoclast cell-surface changes during the egg-laying cycle in Japanese quail, *J. Cell Biol.,* 75, 104, 1977.
16. **Osdoby, P., Martini, M., and Caplan, A.,** Isolated osteoclasts and their presumed progenitor cells, the monocytes, in culture, *J. Exp. Zool.,* 224, 331, 1982.
17. **Osdoby, P., Martini, M., and Caplan, A. I.,** The development of long bones of the limb: cell and matrix interactions of osteoclasts and monocytes, in *Limb Development and Regeneration,* Alan R. Liss, New York, 1983, 229.
18. **Zambonin-Zallone, A., Teti, A., Primavera, M. V., Naldini, L., and Marchisio, P. C.,** Osteoclasts and monocytes have similar cytoskeletal structures and adhesion property *in vitro, J. Anat.,* 137, 57, 1983.
19. **Marchisio, P. C., Cirillo, D., Naldini, L., Primavera, M. V., Teti, A., and Zambonin-Zallone, A.,** Cell-substratum interaction of cultured avian osteoclasts is mediated by specific adhesion structures, *J. Cell Biol.,* 99, 1696, 1984.
20. **Teti, A., Blair, H. C., Schlesinger, P. H., Grano, M., Zambonin-Zallone, A., Kahn, A. J., Teitelbaum, S. L., and Hruska, K. A.,** Extracellular protons acidify osteoclasts, reduce cytosolic calcium and promote expression of cell-matrix attachment structures, *J. Clin. Invest.,* 84, 773, 1989.

21. **Turksen, K., Kanehisa, J., Opas, M., Heersche, J. N. M., and Aubin, J. E.,** Adhesion patterns and cytoskeleton of rabbit osteoclasts on bone slices and glass, *J. Bone Miner. Res.*, 3, 389, 1988.
22. **Zambonin-Zallone, A., Teti, A., Grano, M., Rubinacci, A., Abbadini, M., Gaboli, M., and Marchisio, P. C.,** Immunocytochemical distribution of extracellular matrix receptors in human osteoclasts: A b_3 integrin is colocalized with vinculin and talin in the podosome osteoclastoma giant cells, *Exp. Cell Res.*, 182, 645, 1989.
23. **Horton, M. A., Lewis, D., McNulty, K., Pringle, J. A. S., and Chambers, T. J.,** Monoclonal antibodies to osteoclastomas (giant cell bone tumours): definition of osteoclast-specific cellular antigens, *Cancer Res.*, 45, 5663, 1985.
24. **Davies, J. D., Warwick, J., Tutty, N., Philip, R., Helfrieck, M., and Horton, M. A.,** The osteoclast functional antigen is implicated in the regulation of bone resorption and biochemically related to the vitronectin receptor, *J. Cell Biol.*, 109, 1817, 1989.
25. **Zambonin-Zallone, A., Teti, A., Gaboli, M., and Marchisio, P. C.,** Beta 3 subunit of vitronectin receptor is present in osteoclast adhesion structures and not in other monocyte-macrophage derived cells, *Connect. Tissue Res.*, 20, 143, 1989.
26. **Oldberg, A., Franzen, A., and Heinegard, D.,** Cloning and sequence analysis of rat bone sialoprotein (osteopontin) cDNA reveals an Arg-Gly-Asp cell binding sequence, *Proc. Natl. Acad. Sci. U.S.A.*, 83, 8819, 1986.
27. **Somerman, M. J., Prince, C. W., Butler, W. T., Foster, R. A., Moehring, J. M., and Sauk, J. J.,** Cell attachment activity of the 44 kildalton bone phosphoprotein is not restricted to bone cells, *Matrix*, 9, 49, 1989.
28. **Reinholt, F. P., Hultenby, K., Oldberg, A., and Heinegard, D.,** Osteopontin — a possible anchor of osteoclasts to bone, *Proc. Natl. Acad. Sci. U.S.A.*, 87, 4473, 1990.
29. **Ko, J. S. and Bernard, G. W.,** Osteoclast formation *in vitro* from bone marrow mononuclear cells in osteoclast-free bone, *Am. J. Anat.*, 161, 415, 1981.
30. **Burger, E. H., Van der Meer, J. W. M., van de Gevel, J. S., Gribnau, J. C., Thesingh, C. W., and van Furth, R.,** *In vitro* formation of osteoclasts from long-term cultures of bone marrow mononuclear phagocytes, *J. Exp. Med.*, 156, 1604, 1982.
31. **Van de Wijngaert, F. P. Schipper, C. A., Tas, M. C., and Burger, E. H.,** Role of mineralizing cartilage in osteoclast and osteoblast recruitment, *Bone*, 9, 81, 1988.
32. **Lowik, C. W., van der Plujim, G., van der We Pals, L. J., van Treslong-De Groot, H. B., and Bijvoet, O. L.,** Migration and phenotypic transformation of osteoclast precursors into mature osteoclasts: the effect of a bisphosphonate, *J. Bone Miner. Res.*, 3, 185, 1988.
33. **Fuller, K. and Chambers, T. J.,** Bone matrix stimulates osteoclastic differentiation in cultures of rabbit bone marrow cells, *J. Bone Miner. Res.*, 4, 179, 1989.
34. **Graves, L., III and Jilka, R. L.,** Comparison of bone and parathyroid hormone as stimulators of osteoclast development and activity in calvarial cell cultures from normal and osteopetrotic (mi/mi) mice, *J. Cell. Physiol.*, 145, 102, 1990.
35. **Takahashi, N., Akatsu, T., Udagawa, N., Sasaki, T., Yamaguchi, A., Moseley, J. M., Martin, T. J., and Suda, T.,** Osteoblastic cells are involved in osteoclast formation, *Endocrinology*, 123, 2600, 1988.
36. **Udagawa, N., Takahashi, N., Akatsu, T., Sasaki, T., Yamaguchi, A., Kodama, H., Martin, T. J., and Suda, T.,** The bone marrow-derived stromal cell lines MC3T3-G2/PA6 and ST2 support osteoclast-like cell differentiation in cocultures with mouse spleen cells, *Endocrinology*, 125, 1253, 1989.
37. **Yamashita, T., Asano, K., Takahashi, N., Akatsu, T., Udagawa, N., Sasaki, T., Margin, T. J., and Suda, T.,** Cloning of an osteoblastic cell line involved in the formation of osteoclast-like cells, *J. Cell. Physiol.*, 145, 587, 1990.
38. **Long, M. W., Williams, J. L., and Mann, K. G.,** Expression of human bone-related proteins in the hematopoietic microenvironment, *J. Clin. Invest.*, 86, 1387, 1990.

39. **Krukowski, M. and Kahn, A. J.,** Inductive specificity of mineralized bone matrix in ectopic osteoclast differentiation, *Calcif. Tissue Int.*, 34, 474, 1982.
40. **Osdoby, P., Oursler, M. J., Salino-Hugg, T., and Krukowski, M.,** Osteoblast development: the cell surface and the bone environment, in, *Cell and Molecular Biology of Vertebrate Hard Tissues*, Ciba Foundation Symp. No. 136, John Wiley & Sons, Chichester, 1988, 108.
41. **Webber, D., Osdoby, P., Hauschka, P., and Krukowski, M.,** Correlation of an osteoclast antigen and ruffled border on giant cells formed in response to resorbable substrates, *J. Bone Miner. Res.*, 5, 401, 1990.
42. **Glowacki, J., Altobelli, D., and Mulliken, J. B.,** The fate of mineralized and demineralized osseous implants in cranial defects, *Calcif. Tissue Int.*, 33, 71, 1981.
43. **Holtrop, M. J., Cox, K. A., and Glowacki, J.,** Cells of the mononuclear phagocytic system resorb implanted bone matrix: a histologic and ultrastructural study, *Calcif. Tissue Int.*, 34, 488, 1982.
44. **Glowacki, J.,** The effects of heparin and protamine on resorption of bone particles, *Life Sci.*, 33, 1019, 1983.
45. **Sakamoto, S., Sakamoto, M., Goldhaber, P., and Glimcher, M. J.,** Studies on the interaction between heparin and mouse bone collagenase, *Biochim. Biophys. Acta*, 385, 41, 1975.
46. **Glowacki, J.,** Morphometric analysis in aging bone, in *Osteoporosis: Social and Clinical Aspects*, Gennari, C. and Segre, G., Eds., Excerpta Medica, Amsterdam, 1984, 103.
47. **Strause, L., Saltman, P., and Glowacki, J.,** The effect of deficiencies of manganese and copper on osteoinduction and on resorption of bone particles in rats, *Calcif. Tissue Int.*, 41, 145, 1987.
48. **Saltman, P.,** Trace minerals in health and disease, in *Frontiers in Longevity Research*, Charles C Thomas, Springfield, IL, 1984, 162.
49. **Lian, J. B., Tassinari, M., and Glowacki, J.,** Resorption of implanted bone prepared from normal and warfarin-treated rats, *J. Clin. Invest.*, 73, 1223, 1984.
50. **Donnelly, R., Bockman, R. S., Doty, S. B., and Boskey, A. L.,** Bone particles from gallium-treated rats are resistant to resorption *in vivo*, *Bone Miner.*, 12, 167, 1991.
51. **Glowacki, J. and Cox, K. A.,** Osteoclast features of cells that resorb bone implants in rats, *Calcif. Tissue Int.*, 39, 97, 1986.
52. **Everts, V., Aronson, D. C., and Beertsen, W.,** Phagocytosis of bone collagen by osteoclasts in 2 cases of pycnodysostosis, *Calcif. Tissue Int.*, 37, 25, 1985.
53. **Zambonin-Zallone, A., Teti, A., and Primavera, M. V.,** Resorption of vital or devitalized bone by isolated osteoclasts *in vitro*: the role of lining cells, *Cell Tissue Res.*, 235, 561, 1984.
54. **Rodan, G. A. and Martin, T. J.,** Role of osteoblasts in hormonal control of bone resorption: a hypothesis, *Calcif. Tissue Int.*, 33, 349, 1981.
55. **Glowacki, J., Libert, E., and Pettis, G.,** The influence of inflammation on osteogenesis induced by demineralized bone particles, *Calcif. Tissue Int.*, 46, A5, 1990.
56. **Holtrop, M. E., King, G. J., Cox, K. A., and Reit, B.,** Time-related changes in the ultrastructure of osteoclasts after injection of parathyroid hormone in young rats, *Calcif. Tissue Int.*, 27, 129, 1979.
57. **Glowacki, J., Jasty, M., and Goldring, S.,** Comparison of multinucleated cells elicited in rats by particulate bone, polyethylene, or polymethylmethacrylate, *J. Bone Miner. Res.*, 1, 327, 1986.
58. **Glowacki, J. and Goldring, S. R.,** Maintenance of osteoclastic features in cells isolated from subcutaneous implants of devitalized bone particles, *J. Cell Biol.*, 103, 104a, 1986.
59. **Goldring, S. R., Roelke, M., and Glowacki, J.,** Multinucleated cells elicited in response to implants of devitalized bone particles possess receptors for calcitonin, *J. Bone Miner. Res.*, 3, 117, 1988.

60. **Popoff, S. N. and Marks, S. C., Jr.,** Ultrastructure of the giant cell infiltrates of subcutaneously implanted bone particles in rats and mice, *Am. J. Anat.*, 177, 491, 1986.
61. **Anderson, J. M., Schmitz, P. E., and Marks, S. C., Jr.,** Bone particles from toothless (tl/tl) osteopetrotic rats not cured by bone marrow transplantation are resorbed in normal littermates, *Bone Miner.*, 1, 217, 1986.
62. **Schmitz, P. E., Marks, S. C., Jr., Wright, P., Seifert, M. F., and McGuire, J. L.,** Bone particles from osteopetrotic mice not cured by bone marrow transplants are resorbed in normal littermates, *Bone*, 7, 473, 1986.
63. **Glowacki, J., Cox, K. A., and Wilcon, S.,** Impaired osteoclast differentiation in subcutaneous implants of bone particles in osteopetrotic mutants, *Bone Miner.*, 5, 271, 1989.
64. **Walters, L. M. and Schneider, G. B.,** A functional and morphological study of cells adjacent to ectopic bone implants in rats, *Am. J. Anat.*, 173, 287, 1985.
65. **Walters, L. M. and Schneider, G. B.,** Cellular response to ectopically implanted silk sutures and osteopetrotic bone, *Cell Tissue Res.*, 248, 79, 1987.
66. **Walters, L. M. and Schneider, G. B.,** Tartrate resistant acid phosphatase activity in tibial osteoclasts and cells elicited by ectopic bone and suture implants in normal and osteopetrotic rats, *Bone Miner.*, 4, 49, 1988.
67. **Bagi, C. M. and Miller, S. C.,** Osteoclast features of cells that resorb demineralized and mineral-containing bone implants in rats, *Scanning Microsc.*, 3, 963, 1989.
68. **Glowacki, J. and Lian, J. B.,** Impaired recruitment and differentiation of osteoclast progenitors by osteocalcin-depleted bone implants, *Cell Diff.*, 21, 247, 1987.
69. **DeFranco, D. J., Glowacki, J., Cox, K. A., and Lian, J. B.,** Normal bone particles are preferentially resorbed in the presence of osteocalcin-deficient bone particles *in vivo, Calcif. Tissue Int.*, 49, 43, 1991.
70. **Glowacki, J., Rey, C., Cox, K. A., and Lian, J.,** Effects of bone matrix components on osteoclast differentiation, *Connect. Tissue Res.*, 20, 121, 1989.
71. **Glowacki, J., Rey, C., Glimcher, M. J., Cox, K. A., and Lian, J.,** A role for osteocalcin in osteoclast differentiation, *J. Cell. Biochem.*, 45, 292, 1991.
72. **Malone, J. D., Teitelbaum, S. L., Griffin, G. L., Senior, R. M., and Kahn, A. J.,** Recruitment of osteoclast precursors by purified bone matrix constituents, *J. Cell Biol.*, 92, 227, 1982.
73. **Malone, J. D. and Richards, M.,** a_2HS Glycoprotein is chemotactic for mononuclear phagocytes, *J. Cell. Physiol.*, 132, 118, 1987.
74. **Mundy, G. R. and Poser, J. W.,** Chemotactic activity of g-carboxy-glutamic acid-containing protein in bone, *Calcif. Tissue Int.*, 35, 164, 1983.
75. **Kandemir, N., Malone, J. D., and Kahn, A. J.,** Osteocalcin stimulates osteoclast development *in vitro, J. Bone Miner. Res.*, 5, S227, 1990.
76. **Liggett, W., Lian, J. B., Anklesaria, P., Greenberger, J. S., and Glowacki, J.,** Effects of osteocalcin on proliferation and differentiation of putative osteoclast progenitors from murine long-term bone marrow cultures, *J. Bone Miner. Res.*, 5, S78, 1990.
77. **Bleiberg, I., Greenberger, J. S., and Glowacki, J.,** Osteoclastogenesis in subcutaneous implants of bone matrix in normal, mutant, and chimeric mice, *J. Bone Miner. Res.*, 6, S93, 1991.
78. **Glowacki, J., Cox, K. A., O'Sullivan, J., Wilkie, D., and Deftos, L. F.,** Osteoclasts can be induced in fish having an acellular bony skeleton, *Proc. Natl. Acad. Sci. U.S.A.*, 83, 4104, 1986.
79. **Marks, S. C., Jr. and Chambers, T. J.,** The giant cells recruited by subcutaneous implants of mineralized bone particles and slices in rabbits are not osteoclasts, *J. Bone Miner. Res.*, 6, 395, 1991.
80. **Sampath, T. K. and Reddi, A. H.,** Homology of bone-inductive proteins from human, monkey, bovine and rat extracellular matrix, *Proc. Natl. Acad. Sci. U.S.A.*, 80, 6591, 1983.

Chapter 10

SIGNAL TRANSDUCTION MECHANISMS IN OSTEOCLASTS

Henry J. Donahue

TABLE OF CONTENTS

I. INTRODUCTION

It is now recognized that the osteoclast is the primary bone resorbing cell. Thus, the regulation of osteoclast activity by local, systemic, and physical factors is a critical step in the regulation of bone remodeling. As the list of factors which regulate osteoclast activity grows, it is becoming apparent that many factors which activate osteoclasts, do so through an interaction with osteoblasts, which communicate with osteoclasts through as yet undefined coupling factors.[1] On the other hand, factors which inhibit osteoclast activity, such as calcitonin (CT), extracellular calcium (EC-Ca^{2+}), and prostaglandins, do so through a direct effect on osteoclasts. In general, these factors do not cross the osteoclast cell membrane and thus rely on signal transduction, through the generation of second messengers, to affect osteoclast activity.

Considerable progress has been made regarding the mechansim of extracellular signal transduction in bone cells, most of this information being gained from studies of osteoblasts. Factors which control osteoblast activity, such as parathyroid hormone, (PTH), do so by stimulating the accumulation of two important second messengers, cAMP and cytosolic calcium ion (Ca^{2+}). PTH stimulates cAMP accumulation by increasing adenylate cyclase activity through activation of a stimulatory guanine nucleotide-binding protein (G-protein) coupled to the PTH receptor.[2,3] PTH mobilizes cytosolic Ca^{2+} both by activating the synthesis of inositol trisphosphate, which stimulates release of Ca^{2+} from cytosolic stores,[4] and by activating the influx of EC-Ca^{2+}, possibly by activating voltage-gated calcium channels.[5] These two actions are most likely also mediated by receptor coupled G-proteins.

Because of the difficulty in isolating and culturing osteoclasts, relatively less is known about signal transduction in these cells. However, the development of new isolation procedures as well as techniques to monitor cAMP accumulation and cytosolic calcium ion concentration, ($[Ca^{2+}]_i$), in single cells have led to important new insights into signal transduction in osteoclasts. The picture which is emerging is similar to that in osteoblasts, i.e., the cAMP and cytosloic Ca^{2+} pathways are important second-messengers in osteoclasts. The purpose of this chapter is to present an overview of signal transduction, with special reference to osteoclasts, and describe how these pathways are utilized by factors which regulate osteoclast activity.

II. OVERVIEW OF CELLULAR SIGNAL TRANSDUCTION

The initiating event in extracellular signal transduction is the interaction of an extracellular factor with the cell membrane. This often involves the binding of a ligand with its receptor. Once the ligand has bound to its receptor, several events can take place, one of which is the activation of adenylate cyclase and formation of cAMP. Since the discovery of cAMP by Sutherland and colleagues,[6,7] its role as an ubiquitous second messenger in mammalian cells has been realized. Additionally, the biochemical events which lead to the activation of adenylate cyclase, including the role of G-proteins, and the cell processes affected by cAMP, including activation of protein kinase A, have been elucidated.

Receptors coupled to adenylate cyclase are transmembrane proteins with four extracellular domains, seven membrane-spanning domains, and four intracellular domains. The ligand-activated receptors undergo transformational changes which allow intracellular domains of the receptor to interact with one of several G-proteins. These proteins are trimeric structures with an α-, β-, and γ-subunit. The cytosolic domain of the ligand-activated receptor interacts with the α-subunit of the G-protein, allowing it to exchange GTP for GDP. The α-subunit then disassociates from the other subunits and either activates or inhibits the catalytic subunit of adenylate cyclase, depending on the G-protein to which the receptor is coupled. If the receptor is coupled to an inhibitory G-protein (Gi), receptor binding will decrease adenylate cyclase activity. On the other hand, if the receptor is coupled to a stimulatory G-protein (Gs), receptor binding will increase adenylate cyclase activity. Cholera toxin activates Gs proteins,[8] while pertussis toxin inhibits Gi proteins;[9] thus, both are capable of activating adenylate cyclase and are useful tools for studying the role of G-proteins in signal transduction.

In addition to cAMP, cytosolic Ca^{2+} is an important second messenger in many cells. Douglas and Rubin[10] first advanced the concept of a central role for cytosolic Ca^{2+} in stimulus-secretion coupling, and Mitchel made the link between inositol phosphate metabolism and calcium signaling.[11] As is the case with cAMP, cytosolic Ca^{2+} functions as a second messenger in most cells. Just as ligand receptor-mediated regulation of adenylate cyclase activity is mediated by G-proteins, so is ligand receptor-activated cytosolic Ca^{2+} mobilization. In some cells, the ligand-activated receptor may activate a G-protein which, rather than interacting with adenylate cyclase, stimulates phospholipase C (PLC) activity. Once PLC is activated, it catalyzes the hydrolysis of phosphosinositol phosphate (PIP) to inosital trisphosphate (IP_3), and diacylglycerol (DAG). IP_3 can then bind to specific receptors on intracellular organelles, such as the endoplasmic reticulum or calciosome,[12] and stimulate the release of Ca^{2+} into the cytosol. DAG also serves as a second messenger by activating protein kinase C. The ligand-activated receptor may also increase $[Ca^{2+}]_i$ by interacting with G-proteins which activate calcium channels. These may be the same Gs-proteins which activate adenylate cyclase or other G-proteins, such as Go, which activates calcium channels in some neuronal cells.[13,14]

Other important signal transduction mechanisms include tyrosine kinase activity intrinsic to certain receptors such as those for epidermal growth factor and insulin-like growth factors. Additionally, certain oncogenes, such as *ras*, behave as α-units of G-proteins and thus are important in signal transduction. However, these pathways have not yet been investigated in osteoclasts and will not be reviewed in this chapter. Rather, this chapter will focus on the mechanism by which known regulators of osteoclast activity, including CT, prostaglandin, and EC-Ca^{2+}, affect cAMP accumulation and cytosolic Ca^{2+} mobilization in these cells.

III. SIGNAL TRANSDUCTION PATHWAYS

A. CALCITONIN

CT is a well-known inhibitor of bone resorption *in vivo*[15] and has been shown to decrease osteoclast mobility and cause podosome retraction *in vitro*.[16] Early studies of the mechanism by which CT interacts with bone suggested cAMP as a second messenger.[17,18] Chase and Aurbach[17] showed that both PTH and porcine CT increased cAMP accumulation in isolated calvaria, but it was unclear with which bone cells CT interacted. Subsequently, it was shown that PTH and CT affected cAMP accumulation in different cell types and that cells metabolically similar to osteoclasts respond to CT while those similar to osteoblasts respond to PTH.[19] Consistent with these findings were autoradiography studies which showed that salmon CT binds to rat osteoclasts *in vivo*, but not osteoblasts or osteocytes.[20] With the development of techniques capable of isolating small numbers of individual osteoclasts,[21-25] Nicholson et al. were able to show that rat osteoclasts possess large numbers of CT receptors and that salmon CT stimulates cAMP accumulation in these cells.[26] Thus, CT was believed to interact with rat osteoclasts through receptor-activated adenylate cyclase. This concept is supported by the finding of several groups that the membrane-permeable analog of cAMP, dibutryl cAMP, and forskolin, an activator of adenylate cyclase, produce changes in osteoclasts consistent with the inhibition of resorption.[25,27,28] Furthermore, recent evidence showing that cholera toxin inhibits osteoclast motility suggests that osteoclasts possess a G_s protein-linked adenylate cyclase,[29] as do osteoblasts.

While it is clear that cAMP serves as a second messenger for CT in rat osteoclasts, the role of cAMP in signal transduction in avian osteoclasts is less clear. Several groups have failed to detect an effect of CT on cAMP accumulation in avian osteoclasts.[30-33] Thus, it was believed that either avian osteoclasts do not have CT receptors or that CT interacts with osteoclasts through a signal transduction mechanism not involving cAMP. However, it has subsequently been shown that avian osteoclasts, when freshly isolated from severely hypocalcemic chickens, will respond to CT with an increase in cAMP formation.[34] Thus, the issue of whether avian osteoclasts possess an adenylate cyclase responsive to CT remains controversial, but at this point the data suggest that both rat and chicken osteoclasts possess CT receptors and utilize cAMP as a second messenger.

Recent advances in osteoclast isolation techniques, as well as microspectrofluometric techniques for measuring $[Ca^{2+}]_i$, have led to the discovery that cytosolic Ca^{2+} may be an important second messenger in osteoclasts, as has already been suggested in osteoblasts.[35-38] Malgaroli et al.[39] examined cytosolic Ca^{2+} in rat and avian osteoclasts loaded with the fluorescent calcium indicator fura-2-AM. These investigators showed that $10^{-6}M$ human CT elicited a slow, very modest increase in $[Ca^{2+}]_i$ in rat osteoclasts, but was without effect on avian osteoclasts, renewing the debate as to whether avian osteoclasts respond to CT. The CT-elicited response in rat osteoclasts disappeared in the absence of EC-Ca^{2+}, suggesting that CT was activating membrane Ca^{2+} channels. An effect very similar to that of CT was elicited by dibutryl cAMP, suggesting that the Ca^{2+} mobilizing effects of the hormone were mediated by a cAMP-dependent process. Zaidi et al.[40] also showed that exposure to CT caused an increase in $[Ca^{2+}]_i$ in rat osteoclasts. However, the response was much more dramatic than that of Malgaroli et al.[39] Zaidi et al.[40] showed that $10^{-9}M$ eel CT produced an approximately 600-nM increase in $[Ca^{2+}]_i$, while Malgaroli et al.[39] found a 19-nM increase in response to $10^{-6}M$ human CT. Additionally, Zaidi et al.[40] reported a biphasic response consisting of an initial peak and subsequent plateau. Furthermore, the absence of EC-Ca^{2+} did not affect the initial peak, but completely blocked the plateau response. These findings suggest that CT stimulates both the influx of EC-Ca^{2+} and release of Ca^{2+} from cytosolic stores. The differences in the magnitude of the response to CT that these investigators recorded may be due to the species of CT used. Malgaroli et al.[39] used human CT, while Zaidi et al.[40] used eel CT which, like salmon CT, is a more potent inhibitor of osteoclast activity. We have also shown that salmon CT increases $[Ca^{2+}]_i$ in isolated rat osteoclasts.[41] The effect occurred only in a subpopulation of osteoclasts, and was also much more dramatic than that reported by Malgaroli et al.,[40] the increase in $[Ca^{2+}]_i$ over basal being 147 nM.

Our finding that only a subpopulation of osteoclasts responds to CT is very interesting in light of our observation that all osteoclasts we examined responded to EC-Ca^{2+} with an increase in $[Ca^{2+}]_i$ (see below). In fact, since osteoblasts and fibroblasts do not respond to EC-Ca^{2+} with an increase in $[Ca^{2+}]_i$, one can use this response to identify osteoclasts in a mixed cell population. In doing so, we found that some cells which did respond to Ec-Ca^{2+}, and thus presumably were osteoclasts, did not respond to CT. While the reason for this is unclear, it is possible that osteoclasts which do not respond to CT are precursors or osteoclasts in a different stage of the cell cycle than those which respond to CT. Interestingly, there has been a report of cell cycle-dependent coupling of CT receptor to different G-proteins in kidney tubule cells, suggesting that CT may induce different biological responses in target cells, depending on the cell cycle.[42] If this is the case in osteoclasts, it might explain why we only observed a subpopulation of cells which responded to CT, whereas all osteoclasts respond to EC-Ca^{2+}.

Recent studies indicate that in those cells which respond to CT with an increase in $[Ca^{2+}]_i$, the response is G-protein mediated. Zaidi et al.[40] have described a biphasic effect of CT on rat osteoclasts which includes abolition of motility and cellular retraction. The former is mimicked by cholera toxin and dibutryl cAMP, and the latter by pertussis toxin, ionomycin, and increases in EC-Ca^{2+}. Aluminum fluoride, which activates G-proteins in general, mimicked both phases of the response to CT. Thus, these investigators suggest that CT increases cAMP through an interaction with Gs proteins and mobilizes cytosolic Ca^{2+} through an interaction with a G-protein which is not Gs. Possible ways in which CT could activate Ca^{2+} channels include inhibition of inhibitory G-proteins or activation of a Go-protein directly linked to Ca^{2+} channels, such as is the case in some excitable cells.[13,14] Thus, CT may activate two separate second-messenger pathways (cAMP and cytosolic Ca^{2+}) by binding with two receptors or one receptor with two distinct binding sites. A similar model has been proposed to explain PTH activation of adenylate cyclase and mobilization of cytosolic Ca^{2+} in osteoblasts.[43-45] Recently, both the CT and PTH receptor have been cloned and shown to be strikingly homologous with one another, but not with other G-protein linked receptors.[46,47] While no evidence of separate receptors was found, the receptors were shown to possess short cytosolic loops not similar to corresponding regions in other adenylate-cyclase coupled receptors. This difference might partly explain the ability of the CT receptor and the PTH receptor to activate both adenylate cyclase and cytosolic Ca^{2+} mobilization.

While it is apparent that the osteoclast response to CT begins with receptor binding, activation of cAMP accumulation, and cytosolic Ca^{2+} mobilization, and ends with contraction of osteoclastic cellular processes, a decrease in cell motility, and inhibition of resorption, the intervening sequence of events is unclear. An attractive hypothesis is that CT first stimulates an increase in cAMP formation. Subsequent protein kinase A activation leads to phosphorylation of integrin proteins within the membrane favoring detachment of the osteoclast membrane from extracellular matrix proteins. At the same time, increases in $[Ca^{2+}]_i$ facilitate contraction of cytoskeletal proteins, much as Ca^{2+} activates actin and myosin sliding in muscle cells. This results in retraction of the cell and a cessation of resorptive activity.

B. EXTRACELLULAR CALCIUM

The observation that increasing EC-Ca^{2+} will decrease osteoclast podosome formation, an indication of decreased resorptive capacity, suggests that EC-Ca^{2+} is an important regulator of osteoclast activity.[48] An increase in EC-Ca^{2+} increases $[Ca^{2+}]_i$ in both chicken and rat osteoclasts,[39,40,49-51] but not in osteoblasts or fibroblasts.[41] Additionally, agents which increase $[Ca^{2+}]_i$, such as ionomycin, decrease osteoclastic bone resorption and release of acid phosphatase.[49] Thus, EC-Ca^{2+} appears to inhibit bone resorption through a mechanism utilizing cytosolic Ca^{2+} as a second messenger. Also implicating cytosolic Ca^{2+} as a second messenger important in controlling osteoclast activity is the finding that an increase in proton concentration stimulates osteoclastic resorption while decreasing $[Ca^{2+}]_i$ in individual rat osteoclasts.[52] Thus, $[Ca^{2+}]_i$ appears to be inversely related to osteoclast resorptive activity. We have also observed that following exposure to EC-Ca^{2+}, there are sometimes oscillations in $[Ca^{2+}]_i$.[42] This suggests that frequency as well as amplitude modulation of the Ca^{2+} signal may be important in controlling osteoclast behavior.

The mechanism by which EC-Ca^{2+} regulates $[Ca^{2+}]_i$ in osteoclasts is not clear. However, studies on other cell types in which EC-Ca^{2+} regulates $[Ca^{2+}]_i$ provide clues. For example, in a given population of parathyroid cells[53,54] and placental trophoblasts,[55] some, but not all of the cells respond to EC-Ca^{2+} with an increase in $[Ca^{2+}]_i$. Additionally, antibodies have been generated to parathyroid cells which bind to some parathyroid cells and some placental trophoblasts. What is more interesting is that those cells to which the antibody binds, whether parathyroid cells or trophoblasts, are the same cells that respond to EC-Ca^{2+} with an increase in $[Ca^{2+}]_i$. Furthermore, binding of the antibody blocks the ability of EC-Ca^{2+} to increase $[Ca^{2+}]_i$ in these cells. Thus, the antibody appears to bind to a calcium sensor on parathyroid cells and trophoblasts. Recent evidence suggests that this calcium sensor is a 500-kDa membrane-bound protein.[53] Considering the similarity between the ability of EC-Ca^{2+} to regulate $[Ca^{2+}]_i$ in parathyroid cells, placental trophoblasts, and osteoclasts, it is possible, indeed likely, that such a calcium sensor exists in osteoclasts, although it is yet to be identified.

Divalent ions other than calcium, including cadmium and barium, also increase $[Ca^{2+}]_i$ in osteoclasts, suggesting that the cation receptor on osteoclasts is not selective between closely related cations. Furthermore, it has been reported that $CdCl_2$ is able to elicit a cytosolic Ca^{2+} response in the absence of EC-Ca^{2+}, although the response is less than in the presence of EC-Ca^{2+}, suggesting that binding of cations to their receptor activates mobilization of intracellular Ca^{2+} from cytosolic stores as well as influx of EC-Ca^{2+}. Contrary to these findings are those of Zaidi et al.,[49] which show that nickel, which blocks transmembrane flux of Ca^{2+}, completely blocks EC-Ca^{2+}-stimulated cytosolic Ca^{2+} mobilization. Thus, it is not yet clear whether EC-Ca^{2+} stimulates both intracellular release of Ca^{2+} and influx of extracellular Ca^{2+}, but it seems certain that entry of EC-Ca^{2+} is part of the mechanism.

The way in which EC-Ca^{2+} enters the osteoclast is not clear. Verapamil and nifedipine, which block voltage-gated calcium channels (VGCCs), fail to affect Ca^{2+}-induced changes in $[Ca^{2+}]_i$ in osteoclasts growing on glass coverslips.[39,50,51] Membrane depolarization by exposure to high KC1, which presumambly would open VGCCs, fails to increase $[Ca^{2+}]_i$[50] in osteoclasts on coverslips, and actually decreases $[Ca^{2+}]_i$ in osteoclasts growing in the presence of bone chips.[52] These findings suggest that Ca^{2+} enters the osteoclast by a mechanism other than VGCCs. This idea is supported by the failure of electrophysiological studies to detect Ca^{2+} currents in osteoclasts.[56,57] On the other hand, Miyauchi et al.[58] reported that KC1 produces dose-dependent increases of $[Ca^{2+}]_i$ in osteoclasts on glass coverslips, and this effect was blocked by dihydroperidine and potentiated by Bay K 8644, a VGCC agonist. However, these VGCCs were inactivated by the attachment of osteoclasts to bone, leading these investigators to suggest that a rapid phenotypic change in the Ca^{2+} signaling mechanism is associated with adhesion of osteoclasts to their resorptive surface. Clearly, additional studies are required to explain the mechanism by which EC-Ca^{2+} enters osteoclasts and whether VGCCs are involved.

The role of cytosolic Ca^{2+} in osteoclastic bone resorption is now emerging. As osteoclasts attach to bone, they form a sealed-off space between the bone surface and the osteoclast membrane known as the clear zone. The ruffled border, characteristic of the osteoclast cell membrane, then forms within the clear zone. This sealed-off microenvironment is where bone resorption takes place. Proton-pumping ATPases located in the ruffled border region pump protons into the clear zone, thus lowering the pH in this microenvironment. This not only favors dissolution of bone by lysommal enzymes secreted into the clear zone, but, by lowering $[Ca^{2+}]_i$ and inhibiting podosome retraction, also assures continued adhesion of the osteoclast to the bone surface. As the bone mineral dissolves, the calcium concentration in the clear zone microenvironment increases to 8 to 20 mmol/1,[59] thus increasing osteoclastic $[Ca^{2+}]_i$, which in turn stimulates retraction of the osteoclast, possibly through interaction with cytoskeletal proteins. As the osteoclast detaches, the clear zone becomes continuous with the extracellular fluid, thus diluting the high Ca^{2+} ion environment, allowing the osteoclast to form a new clear zone and continue the resorptive process. Thus, a negative feedback loop is created whereby EC-Ca^{2+} created by the resorption process inhibits further resorption.

C. PROSTAGLANDINS

Prostaglandins have complex effects on bone resorption, generally stimulating resorption *in vivo*[60,61] while inhibiting osteoclast motility *in vitro*.[62,63] The latter effect is believed to be via a direct interaction with osteoclasts. Thus, prostaglandins have biological activity *in vitro* similar to that of CT. Not surprisingly, prostaglandin also stimulates cAMP accumulation in both chicken and rat osteoclasts.[33] suggesting that, as was the case for CT, cAMP serves as a second messenger for prostaglandins. Further support for the concept that cAMP is a second messenger for the effects of both CT and prostaglandins on osteoclasts comes from studies showing that phosphodiesterase inhibitors such as isobutylmethylxanthine, which prevent the breakdown of cAMP, augment CT- and prostaglandin-induced cytosplasmic spreading.[62]

While both prostaglandin and CT increase cAMP accumulation in osteoclasts, they appear to interact with adenylate cyclase in a different manner. Nicholson et al.[64] showed that while CT induced homologous desensitization and persistent activation of adenylate cyclase in rat osteoclasts, prostaglandin treatment did not cause persistent activation of adenylate cyclase and induced only partial homologous densensitization. Thus, the two activators of adenylate cyclase differed significantly in their action upon the enzyme complex. Furthermore these investigators showed that pertussis toxin does not affect the cAMP response to prostaglandin, suggesting that these factors do not act via an inhibitory G-protein. While prostaglandins have been shown to mobilize cytosolic Ca^{2+} ion in fibroblasts[65] and osteoblasts,[66] there is no evidence that this pathway is activated by prostaglandins in osteoclasts. In fact, Malgaroli et al.[39] reported that 10^{-6} *M* PGE_2 did not increase $[Ca^{2+}]_i$ in fura-2-loaded osteoclasts.

D. EXTRACELLULAR BONE MATRIX PROTEINS

It has recently been suggested that the presence of bone fragments inactivates VGCCs in osteoclasts.[52] Furthermore, osteopontin, an extracellular bone matrix protein, decreases $[Ca^{2+}]_i$ in osteoclasts.[67] These findings suggest that extracellular bone matrix proteins may alter signal transduction pathways in osteoclasts. One way in which bone matrix proteins could do this is through an interaction with the cytoskeletal architecture. Signal transduction through cytoskeletal proteins is possible due to the dynamic nature of the intracellular matrix proteins, a characteristic which has been termed tensegrity.[68] In this model, extracellular matrix proteins are linked to transmembrane receptors of the integrin family, which bind proteins containing arginine-glycine-asparagine sequences, such as fibronectin and osteopontin. The integrins are linked to talin, vinculin, and actin. These cytoskeletal proteins are in turn linked to nuclear matrix proteins, including lamins, which can control DNA replication. Thus, through their interaction with the cytoskeleton, extracellular matrix proteins can ultimately control gene expression.

The regulation of signal transduction in osteoclasts by extracellular matrix proteins is suggested by several observations. First, the ability of osteopontin to decrease $[Ca^{2+}]_i$ in osteoclasts is blocked by anti-integrin antibodies,[67] suggesting that osteopontin regulates signal transduction through integrins, which in turn are linked to the cytoskeleton. Second, osteoclast podosomes contain gelsolin, a calcium-dependent, actin-regulating protein.[69] When gelsolin binds calcium, it fragments preexisting microfilaments[70,71] and nucleates new ones,[72] thus altering the cytoskeletal network. This could affect podosome contraction[52,58] and perhaps transduce a signal to the nuclear matrix and DNA. Third, EC-Ca^{2+} can stimulate cytosolic Ca^{2+} oscillation in osteoclasts[5] and induce cell retraction. Interestingly, Getzenberg et al.[68] suggest a relationship between cytoskeletal protein oscillation and oscillations in $[Ca^{2+}]_i$. An intriguing possibility is that the oscillations in cytosolic calcium alter cytoskeletal

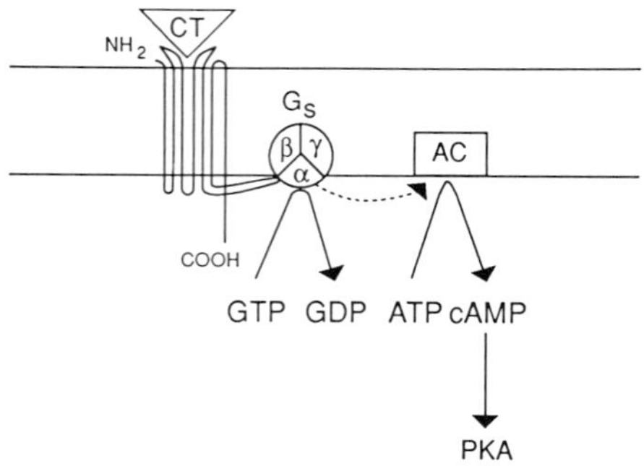

FIGURE 1. CT activation of adenylate cyclase in osteoclasts. CT-activated receptors interact with the α-subunit of a stimulatory G-protein (Gs). The α-subunit then exchanges guanosine 5′-diphosphate (GDP) for guanosine 5′-triphosphate (GTP) and activates the catylytic subunit of adenylate cyclase (AC). AC then catalyzes the conversion of adenosine triphosphate (ATP) to cyclic 3′,5′-adenosine monophosphate (cAMP). The cAMP then activates protein kinase A (PKA).

proteins and thereby induce cell retraction. Thus, EC-Ca^{2+} may decrease osteoclastic resorption by increasing $[Ca^{2+}]_i$ oscillations, which activates gelsolin and alters microfilament structure. This would induce podosome contraction, thus disrupting the clear zone, which would prevent further resorption. On the other hand, the binding of osteopontin to integrin leads to activation of Ca-MgATPase, decreasing $[Ca^{2+}]_i$[67] and stabilizing the cytoskeletal network, which favors continued attachment of the osteoclast to bone matrix and thus continued resorption. Therefore, there exists in osteoclasts a putative link between the cytosolic Ca^{2+} signal transduction pathway and the intracellular matrix, which in turn may be affected by bone matrix proteins.

IV. SUMMARY

The signal transduction mechanisms operating in osteoclasts are now beginning to be understood. Both cytosolic Ca^{2+} and cAMP are important second messengers in osteoclasts. CT affects osteoclasts by both activating adenylate cyclase and mobilizing cytosolic Ca^{2+}, while prostaglandin appears to work through activation of adenylate cyclase. EC-Ca^{2+} affects osteoclast by increasing $[Ca^{2+}]_i$, but its effect on adenylate cyclase has yet to be investigated. Extracellular matrix proteins alter osteoclast behavior through binding to integrin, which in turn mediates an effect on cell Ca^{2+} and perhaps also interacts with the cytoskeleton.

Figures 1 through 4 summarize the signal transduction pathways believed to be operating in osteoclasts. While considerable progress has been made in the last several years, many issues are unresolved. For instance, there is little direct information regarding inositol phosphate metabolism or protein kinase activity in osteoclasts. The issue of whether avian osteoclasts have physiologically significant calcitonin receptors is unresolved, as are the signal trans-

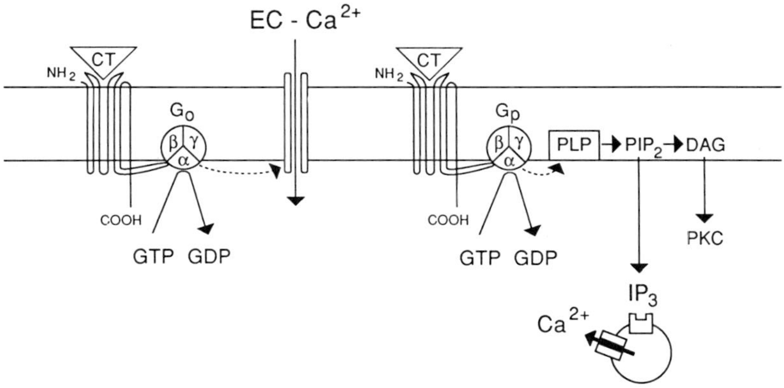

FIGURE 2. Possible mechanisms by which CT stimulates cytosolic Ca^{2+} mobilization in osteoclasts. CT binds to its receptor, which then activates one or more G-proteins. The α-subunits of Gp activate phospholipase C, which converts phosphoinositolphosphate (PIP_2) to diacylglycerol (DAG) and inositoltriphosphate (IP_3). IP_3 stimulates the release of Ca^{2+} from cytoplasmic stores, while DAG activates protein kinase C (PKC). The CT-receptor complex may also activate another G-protein, possibly Go, which interacts with membrane Ca^{2+} channels, increasing the influx of EC-Ca^{2+}.

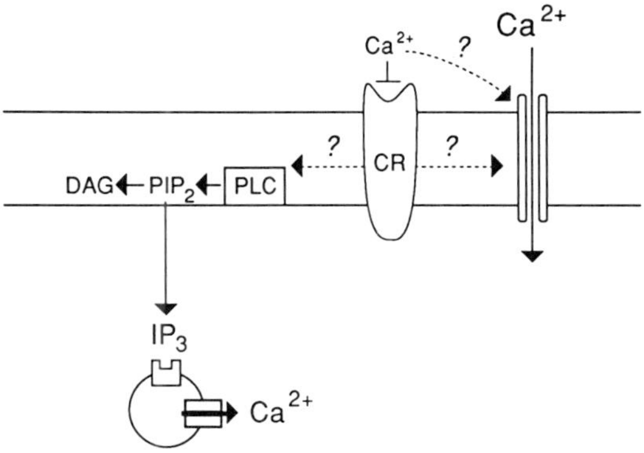

FIGURE 3. Possible mechanisms by which EC-Ca^{2+} might mobilize cytosolic Ca^{2+} in osteoclasts. EC-Ca^{2+} may increase $[Ca^{2+}]_i$ by directly interacting with membrane Ca^{2+} channels. Alternatively, EC-Ca^{2+} may interact with a cation receptor (CR), which in turn activates a membrane Ca^{2+} channel and/or activates phospholipase C, both of which would mobilize cytosolic Ca^{2+}.

duction pathways they may activate. Additionally, the signal transduction pathways involved in osteoblast-osteoclast communication have yet to be examined. Finally, the events occurring between either cAMP or cytosolic Ca^{2+} activation and alterations in osteoclast activity are unknown. As isolation techniques for osteoclasts continue to improve and osteoclast-like cell lines are developed, the answer to these questions and several others will begin to emerge.

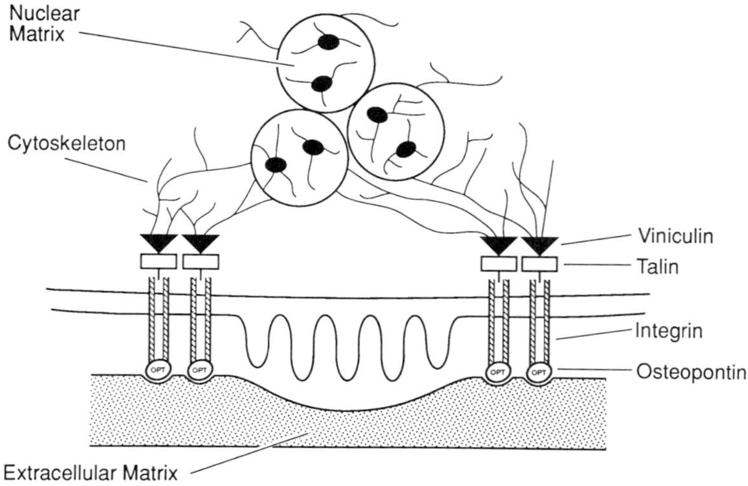

FIGURE 4. Interaction of extracellular matrix proteins with osteoclasts. Osteopontin and other arginine-glycine-asparagine-containing proteins within the extracellular matrix interact with transmembrane integrins. Integrins are linked to the cytoskeletal network via talin and viniculin. The cytoskeleton can interact with nuclear matrix proteins, which in turn can modulate gene expression. In this manner, a signal from the extracellular matrix can be transduced to the genome.

REFERENCES

1. **Rodan, G. A. and Martin, T. J.,** Role of osteoblasts in hormonal control of bone resorption — a hypothesis, *Calcif. Tissue Int.*, 33, 349, 1981.
2. **Rodan, S. B. and Rodan, G. A.,** Dexamethasone effects on β-adrenergic receptors and adenylate cyclase regulatory proteins G_s and G_i in ROS 17/2.8 cells, *Endocrinology,* 118, 2510, 1986.
3. **Pines, M., Santora, A., Gierschik, P., Menczel, J., and Spiegel, A.,** The inhibitory guanine nucleotide regulatory protein modulates agonist-stimulated cAMP production in rat osteosarcoma cells, *Bone Miner.*, 1, 15, 1986.
4. **Dunlay, R. and Hruska, K.,** PTH receptor coupling to phospholipase C is an alternate pathway of signal transduction in bone and kidney, *Am. J. Physiol.*, 258, F223.
5. **Donahue, H. J. and Brink, P. R.,** Regulation of calcium channels in ROS 17/2.8 cells: PTH activates t-type calcium currents, in 1st St. Louis Symp. Bone Metabolism, April 30-May 2, 1991.
6. **Sutherland, E. W., Rall, T. W., and Menon, T.,** Adenyl cyclase. I. Distribution, preparation, and properties, *J. Biol. Chem.*, 237(4), 1220, 1962.
7. **Rall, T. W., Sutherland, E. W., and Berthet, J.,** The relationship of epinephrine and glucagon to liver phosphorylase, *J. Biol. Chem.*, 224, 463, 1956.
8. **Cassel, D. and Selinger, Z.,** Mechanism of adenylate cyclase activation by cholera toxin: inhibition of GTP hydrolysis at the regulatory site, *Proc. Natl. Acad. Sci. U.S.A.*, 74, 3307, 1977.
9. **Katada, T. and Ui, M.,** Direct modification of the membrane adenylate cyclase system by islet-activating protein due to ADP-ribosylation of a membrane protein, *Proc. Natl. Acad. Sci. U.S.A.*, 79, 3129, 1982.

10. **Douglas, W. W. and Rubin, R. P.,** The role of calcium in the secretory response of the adrenal medulla to acetylcholine, *J. Physiol.,* 159, 40, 1961.
11. **Mitchell, R. H.,** Inositol phospholipids and cell surface receptor function, *Biochem. Biophys. Acta,* 415, 81, 1975.
12. **Krause, K.-H., Pittet, D., Volpe, P., Pozzan, T., Meldolesi, J., and Lew, D. P.,** Calciosome, a sarcoplasmic reticulum-like organelle involved in intracellular Ca^{2+}-handling by non-muscle cells: studies in human neutrophils and HL-60 cells, *Cell Calcium,* 10, 351, 1989.
13. **Gilman, A. G.,** G proteins: transducers of receptor-generated signals, *Annu. Rev. Biochem.,* 56, 615, 1987.
14. **Hescheler, J., Rosenthal, W., Trautwein, W., and Schultz, G.,** The GTP-binding protein, Go, regulates neuronal calcium channels, *Nature (London),* 325, 445, 1987.
15. **Marks, S. S. and Popoff, S. N.,** Bone cell biology: the regulation of development, structure, and function in the skeleton, *Am. J. Anat.,* 183, 1, 1988.
16. **Chambers, T. J., Athanasou, N. A., and Fuller, K.,** Effect of parathyroid hormone and calcitonin on the cytoplasmic spreading of isolated osteoclasts, *J. Endocrinol.,* 102, 281, 1984.
17. **Chase, L. R. and Aurbach, G. D.,** The effect of parathyroid hormone on the concentration of adenosine 3′,5′-monophosphate in skeletal tissue *in vitro, J. Biol. Chem.,* 245(7), 1520, 1970.
18. **Johannes, N., Heersche, M., Marcus, R., and Aurbach, G. D.,** Calcitonin and the formation of 3′,5′-AMP in bone and kidney, *Endocrinology,* 94(1), 241, 1974.
19. **Luben, R. A., Wong, G. L., and cohn, D. V.,** Biochemical characterization with parathormone and calcitonin of isolated bone cells: provisional identification of osteoclasts and osteoblasts, *Endocrinology,* 99(2), 526, 1976.
20. **Warshawsky, H., Goltzman, D., Rouleau, M. F., and Bergeron, J. J. M.,** Direct *in vivo* demonstration by radioautography of specific binding sites for calcitonin in skeletal and renal tissues of the rat, *J. Cell Biol.,* 85, 682, 1980.
21. **Zambonin-Zallone, A., Teti, A., and Primavera, M. V.,** Isolated osteoclasts in primary culture: first observations on structure and survival in culture media, *Anat. Embryol.,* 165, 405, 1982.
22. **Zambonin-Zallone, A., Teti, A., and Primavera, M. V.,** Resorption of vital or devitalized bone by isolated osteoclasts *in vitro*: the role of lining cells, *Cell Tissue Res.,* 235, 561, 1984.
23. **Chambers, T. J.,** Osteoblasts release osteoclasts from calcitonin-induced quiescence, *J Cell Sci.,* 57, 147, 1982.
24. **Chambers, T. J., Athanasou, N. A., and Fuller, K.,** Effect of parathyroid hormone and calcitonin on the cytoplasmic spreading of isolated osteoclasts, *J. Endocrinol.,* 102, 281, 1984.
25. **Murrills, R. J. and Dempster, D. W.,** The effects of stimulators of intracellular cyclic AMP on rat and chick osteoclasts *in vitro*: validation of a simplified light microscope assay of bone resorption, *Bone,* 11, 333, 1990.
26. **Nicholson, G. C., Moseley, J. M., Sexton, P. M., Mendelsohn, F. A. O., and Martin, T. J.,** Abundant calcitonin receptors in isolated rat osteoclasts, *J. Clin. Invest.,* 78, 355, 1986.
27. **Arnett, T. R. and Dempster, D. W.,** A comparative study of disaggregated chick and rat osteoclasts *in vitro*: effects of calcitonin and prostaglandins, *Endocrinology,* 120(1), 602, 1987.
28. **Lerner, U. H., Ransjo, M., Klaushofer, K., Horandner, H., Hoffmann, O., Czerwenka, E., Koller K., and Peterlik, M.,** Comparison between the effects of forskolin and calcitonin on bone resorption and osteoclast morphology *in vitro, Bone,* 10, 377, 1989.

29. **Ransjo, M., Lerner, U. H., and Heersche, J. N. M.,** Calcitonin-like effects of forskolin and choleratoxin on surface area and motility of isolated rabbit osteoclasts, *J. Bone Miner. Res.*, 3(6), 611, 1988.
30. **Ito, M. B., Schraer, H., and Gay, C. V.,** The effects of calcitonin, parathyroid hormone and prostaglandin E_2 on cyclic AMP levels of isolated osteoclasts, *Comp. Biochem. Physiol. A,* 81(3), 653, 1985.
31. **Felix, R., Rizzoli, R., Fleisch, H., Zambonin, A., Teti, A., and Primavera, M. V.,** Isolation, culture and hormonal response of chicken medullary bone osteoclasts, *Calcif. Tissue Int.*, 35, 641a, 1983.
32. **Miyaura, C., Nagata, M., and Suda, T.,** Failure to demonstrate the stimulative effect of calcitonin on cyclic AMP accumulation in avian bone *in vitro, Endocrinol. Jpn.*, 28, 403, 1981.
33. **Nicholson, G. C., Livesey, S. A., Moseley, J. M., and Martin, T. J.,** Actions of calcitonin, parathyroid hormone, and prostaglandin E_2 on cyclic AMP formation in chicken and rat osteoclasts, *J. Cell Biochem.*, 31, 229, 1986.
34. **Rifkin, B. R., Auszmann, J. M., Kleckner, A. P., Vernillo, A. T., and Fine, A. S.,** Calcitonin stimulates cAMP accumulation in chicken osteoclasts, *Life Sci.*, 42, 799, 1987.
35. **Donahue, H. J., Fryer, M. J., Eriksen, E. F., and Heath, H.,** III, Differential effects of parathyroid hormone and its analogues on cytosolic calcium ion and cAMP levels in cultured rat osteoblast-like cells, *J. Biol. Chem.*, 263(27), 13522, 1988.
36. **Yamaguchi, D. T., Hahn, T. J., Lida-Klein, A., Kleeman, C. R., and Muallem, S.,** Parathyroid hormone-activated calcium channels in an osteoblast-like clonal osteosarcoma cell line, *J. Biol. Chem.*, 262, 1, 1987.
37. **Herrmann-Erlee, M. P. M., Van Der Meer, J. M., Lowik, C. W. G. M., Van Leeuwen, J. P. T. M., and Boonekamp, P. M.,** Different roles for calcium and cyclic AMP in the action of PTH: studies in bone explants and isolated bone cells, *Bone,* 9, 93, 1988.
38. **Reid, I. R., Civitelli, R., Halstead, L. R., Avioli, L. V., and Hruska, K. A.,** Parathyroid hormone acutely elevates intracellular calcium in osteoblast-like cells, *Am. J. Physiol.*, 253, E45, 1987.
39. **Malgaroli, A., Meldolesi, J., Zambonin-Zallone, A., and Teti, A.,** Control of cytosolic free calcium in rat and chicken osteoclasts: the role of extracellular calcium and calcitonin, *J. Biol. Chem.*, 264(24), 14342, 1989.
40. **Zaidi, M., Datta, H. K., Moonga, B. S., and MacIntyre, I.,** Evidence that the action of calcitonin on rat osteoclasts is mediated by two G proteins acting via separate post-receptor pathways, *J. Endocrinol.*, 126, 473, 1990.
41. **Donahue, H. J., Iijima, K., Goligorsky, M., Rubin, C. T., and Rifkin, B.,** Altered regulation of cytoplasmic calcium ion concentration in tetracycline treated osteoclasts, *Trans. Orthop. Res. Soc.*, 16(2), 404, 1991.
42. **Chakraborty, M., Chatterjee, D., Kellokumpu, S., Rasmussen, H., and Baron, R.,** Cell cycle-dependent coupling of the calcitonin receptor to different G proteins, *Science,* 251, 1078, 1991.
43. **Donahue, H. J., Fryer, M. J., and Heath, H.,** III, Structure-function relationships for full-length recombinant parathyroid hormone-related peptide and its amino-terminal fragments: effects on cytosolic calcium ion mobilization and adenylate cyclase activation in rat osteoblast-like cells, *Endocrinology,* 126(3), 1471, 1989.
44. **Lowik, C. W. G. M., van Leeuwen, J. P. T. M., van der Meer, J. M., van Zeeland, Scheven, B. A. A., and Herrmann-Erlee, M. P. M.,** A two-receptor model for the action of parathyroid hormone on osteoblasts: a role for intracellular free calcium and cAMP, *Cell Calcium,* 6, 311, 1985.
45. **Civitelli, R., Hruska, K. A., Shen, V., and Avioli, L. V.,** Cyclic AMP-dependent and calcium-dependent signals in parathyroid hormone function, *Exp. Gerontol.*, 25, 223, 1990.

46. **Lin, H. Y., Harris, T. L., Flannery, M. S., Aruffo, A., Kaji, E. H., Gorn, A., Kolakowski, L. F., Lodish, H. F., and Goldring, S. R.,** Expression and cloning of an adenylate cyclase-coupled calcitonin receptor, *Science,* 254, 1022, 1991.
47. **Juppner, H., Abou-Sama, A. B., Freeman, M., Kong, X. F., Schipani, E., Richards, J., Kolakowski, L. F., Hock, J., Potts, J. T., Kronenberg, H. M., and Segre, G. V.,** A G protein-linked receptor for parathyroid hormone and parathyroid hormone-related peptide, *Science,* 254, 1024, 1991.
48. **Teti, A. and Zambonin-Zallone, A.,** A working hypothyesis: calcium concentration controls directly osteoclast activity, in *Calcium Regulation and Bone Metabolism: Basic and Clinical Aspects,* Vol. 9, Cohn, D. V., Martin, T. J., and Meunier, P. J., Eds., Elsevier, Amsterdam, 1987, 358.
49. **Zaidi, M., Datta, H. K., Patchell, A., Moonga, B. S., and MacIntyre, I.,** Calcium-activated intracellular calcium elevation: a novel mechansim of osteoclast regulation, *Biochem. Biophys. Res. Commun.,* 163, 1461, 1989a.
50. **Datta, H. K., MacIntyre, I., and Zaidi, M.,** Intracellular calcium in the control of osteoclast function. I. Voltage-insensitivity and lack of effects of nifedipine, BAYK8644 and diliazem, *Biochem. Biophys. Res. Commun.,* 167(1), 183, 1990.
51. **Zaidi, M., MacIntyre, I., and Datta, H.,** Intracellular calcium in the control of osteoclast function. II. Paradoxical elevation of cytosolic free calcium by verapamil, *Biochem. Biophys. Res. Commun.,* 167(2), 807, 1990.
52. **Teti, A., Blair, H. C., Schlesinger, P., Grano, M., Zambonin-Zallone, A., Kahn, A. J., Teitelbaum, S. L., and Hruska, K. A.,** Extracellular protons acidify osteoclasts, reduce cytosolic calcium, and promote expression of cell-matrix attachment structures, *J. Clin. Invest.,* 84, 773, 1989.
53. **Nemeth, E. F.,** Regulation of cytosolic calcium by extracellular divalent cations in C-cells and paraphyroid cells, *Cell Calcium,* 11, 323, 1990.
54. **Juhlin, C., Johansson, H., Holmdahl, R., Gylfe, E., Larsson, R., Rastad, J., Akerstrom, G., and Klareskog, L.,** Monoclonal anti-parathyroid antibodies interfering with a Ca^{2+}-sensor of human parathyroid cells, *Biochem. Biophys. Res. Commun.,* 143(2), 570, 1987.
55. **Juhlin, C., Lundgren, S., Johansson, H., Lorentzen, J., Rask, L., Larsson, E., Rastad, J., Akerstrom, G., and Klareskog, L.,** 500-kilodalton calcium sensor regulating cytoplasmic Ca^{2+} in cytotrophoblast cells of human placenta, *J. Biol. Chem.,* 265(14), 8275, 1990.
56. **Ravesloot, J. L., Ypey, D. L., Vrijheid-Lammers, T., and Nijweide, P. J.,** Voltage-activated K^+ conductances in freshly isolated embryonic chicken osteoclasts, *Proc. Natl. Acad. Sci. U.S.A.,* 86, 6821, 1989.
57. **Sims, S. M. and Dixon, S. J.,** Inwardly rectifying K^+ current in osteoclasts, *Am. J. Physiol.,* C1277, 1989.
58. **Miyauchi, A., Hruska, K. A., Greenfield, E. M., Duncan, R., Alvarez, J., Barattolo, R., Colucci, S., Zambonin-Zallone, A., and Teitelbaum, S. L.,** Osteoclast cytosolic calcium, regulated by voltage-gated calcium channels and extracellular calcium, controls podosome assembly and bone resorption, *J. Cell Biol.,* 111, 2543, 1990.
59. **Silver, I. A., Murrills, R. J., and Etherington, D. J.,** Microelectrode studies on the acid microenvironment beneath adherent macrophages and osteoclasts, *Exp. Cell Res.,* 175, 266, 1988.
60. **Klein, D. C. and Raisz, L. G.,** Prostaglandins: stimulation of bone resorption in tissue culture, *Endocrinology,* 86, 1436, 1970.
61. **Schelling, S. H., Wolfe, H. J., and Tashjian, A. H., Jr.,** Role of the osteoclast in prostaglandin E_2-stimulated bone resorption, *Lab. Invest.,* 42, 290, 1980.
62. **Chambers, T. J. and Ali, N. N.,** Inhibition of osteoclastic motility by prostaglandins I_2, E_1, E_2 and 6-oxo-E_1, *J. Pathol.,* 139, 383, 1983.

63. **Chambers, T. J. and Dunn, C. J.,** The effect of parathyroid hormone, 1,25-dihydroxycholecalciferol and prostaglandins on the cytoplasmic activity of isolated osteoclasts, *J. Pathol.,* 137, 193, 1982.
64. **Nicholson, G. C., Moseley, J. M., Yates, A. J. P., and Martin, T. J.,** Control of cyclic adenosine 3′,5′-monophosphate production in osteoclasts: calcitonin-induced persistent activation and homologous desensitization of adenylate cyclase, *Endocrinology,* 120(5), 1902, 1987.
65. **Yamashita, T. and Takai, Y.,** Inhibition of prostaglandin E_1-induced elevation of cytoplasmic free calcium ion by protein kinase C-activating phorbol esters and diacylglycerol in Swiss 3T3 fibroblasts, *J. Biol. Chem.,* 262(12), 5536, 1987.
66. **Yamaguchi, D. T., Green, J., Merritt, B. S., Kleeman, C. R., and Muallem, S.,** Modulation of osteoblast function by prostaglandins, *Am. J. Physiol.,* 257, F755, 1989.
67. **Miyauchi, A., Alvarez, J., Greenfield, E. M., Teti, A., Grano, M., Colucci, S., Zambonin-Zallone, A., Ross, F. P., Teitelbaum, S. L., Cheresh, D., Hruska, K. A.,** Recognition of osteopontin and related peptides by an $\alpha v \beta_3$ integrin stimulates immediate cell signals in osteoclasts. *J. Biol. Chem.,* 266(30), 20369, 1991.
68. **Getzenberg, R. H., Pienta, K. J., and Coffey, D. S.,** The tissue matrix: cell dynamics and hormone action, *Endocr. Rev.,* 11(3), 399, 1990.
69. **Marchisio, P. C., Cirillo, D., Naldini, L., Primavera, M. V., Teti, A., and Zambonin-Zallone, A.,** Cell-substratum interaction of cultured avian osteoclasts is mediated by specific adhesion structures, *J. Cell Biol.,* 99, 1696, 1984.
70. **Bryan, J. and Coluccio, L. M.,** Kinetic analysis of F-actin depolymerization in the presence of platelet gelsolin and gelsolin-actin complexes, *J. Cell Biol.,* 101, 1236, 1985.
71. **Chaponnier, C., Yanmey, P. A., and Yin, H. L.,** The actin filament-severing domain of plasma gelsolin, *J. Cell Biol.,* 103, 1473, 1986.
72. **Selve, N. and Wegner, A.,** pH-dependent rate of formation of gelsolin-actin complex from gelsolin and monomeric actin, *Eur. J. Biochem.,* 168, 111, 1987.

Chapter 11

ELECTROPHYSIOLOGY OF OSTEOCLASTS

Stephen M. Sims, Melanie E. M. Kelly, Stuart A. Arkett, and S. Jeffrey Dixon

TABLE OF CONTENTS

I. INTRODUCTION

Knowledge of the movement and distribution of ions across plasma membranes is essential for understanding the function and regulation of all cells. The resorption of mineralized substrates by osteoclasts is thought to involve a number of different ion transporters, including a proton (H^+) ATPase,[1-4] a chloride-bicarbonate exchanger,[5,6] and a chloride channel.[7,8] Electrophysiological measurements of membrane potential and ionic currents provide a direct approach to studying the movement of ions across the osteoclast membrane. Yet, compared to many other cell types such as neurons and muscle cells, relatively little is known about the electrophysiological properties of osteoclasts. This paucity of information may be attributed to difficulties in obtaining osteoclasts for *in vitro* studies and in applying electrophysiological recording techniques to the cells. The development of protocols for the isolation (see Chapter 1 of this volume) and short-term culture of osteoclasts[9,10] has overcome the first of these problems. Similarly, tight-seal patch-clamp recording techniques[11] have eliminated some of the difficulties encountered in early studies employing intracellular microelectrodes. In this chapter, we review the electrophysiological properties of osteoclasts and their significance in osteoclast function and regulation. Some of the questions that remain to be answered concerning the roles of ion channels in osteoclasts are also considered.

II. TECHNIQUES FOR STUDYING OSTEOCLAST ELECTROPHYSIOLOGY

A. INTRACELLULAR RECORDING WITH MICROELECTRODES

The first electrophysiological recordings from osteoclasts utilized intracellular microelectrodes to measure the membrane potential.[12,13] This technique, which involves impaling cells with small-diameter electrodes (<1-μm tip), is of limited utility because of damage to the cell membrane at the site of impalement, resulting in leakage of current around the electrode. In small cells, this damage can result in membrane potentials that are artificially low. To avoid this problem, more recent investigations have employed tight-seal patch-clamp recording techniques.

B. PATCH-CLAMP/VOLTAGE-CLAMP TECHNIQUES

Voltage-clamp techniques enable one to measure current flowing through ion channels or electrogenic carriers while using feedback circuitry to maintain membrane potential at a designated voltage. Several different configurations (Figure 1) of the patch-clamp technique[11] are available for recording ion currents using the voltage-clamp mode, as well as recording membrane potential using the current-clamp mode.[8,14,15] Patch-clamp recording is initiated by pressing a low-resistance (<10 MΩ, 1-2-μm tip diameter), heat-polished

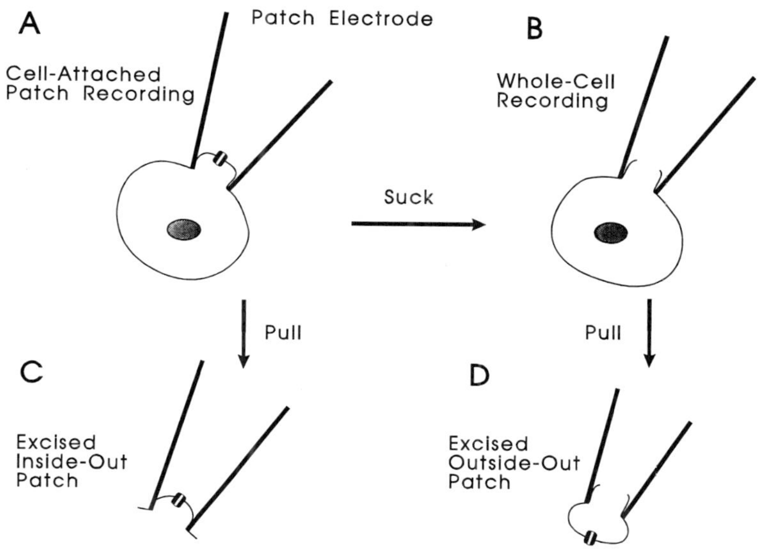

FIGURE 1. Patch-clamp recording configurations. A glass microelectrode is pressed against the membrane of a cell to establish a gigaohm seal between the glass and the membrane of the cell. (A) currents flowing through individual ion channels in the intact cell can be recorded in this cell-attached patch configuration. (B) The patch of membrane can be ruptured by suction, establishing the whole-cell recording configuration. Membrane potential can be recorded under current clamp, or membrane currents flowing across the entire cell membrane can be studied under voltage clamp. (C) Patches of membrane may be excised to provide access to the cytosolic surface of the membrane (inside-out patch). (D) Alternatively, a small patch of outside-out membrane can be excised.

glass microelectrode (''patch electrode'') against the membrane of the cell being studied (Figure 2). Application of gentle suction to the back of the pipette creates a very high resistance (>1 GΩ, i.e., $>10^9\Omega$) seal between the electrode glass and the membrane. In contrast to conventional intracellular microelectrode techniques, there is usually very little leakage of current around the electrode. In the cell-attached configuration (Figure 1A), one can record the currents flowing through individual ion channels in the patch of membrane beneath the recording electrode. Whole-cell configuration is achieved by applying a brief pulse of stronger suction to the electrode, rupturing the patch of membrane beneath the electrode. The cytoplasm and electrode are now contiguous (Figure 1B) and one can record the net ionic current across the entire plasma membrane through all channels and electrogenic carriers. Two approaches can be used to isolate and study the currents through different classes of ion channels. One is to separate currents based upon their voltage activation range (i.e., the range of membrane potentials over which the channels are open). The second is to use ion substitution and channel blockers to selectively eliminate certain currents, thereby isolating the current of interest.

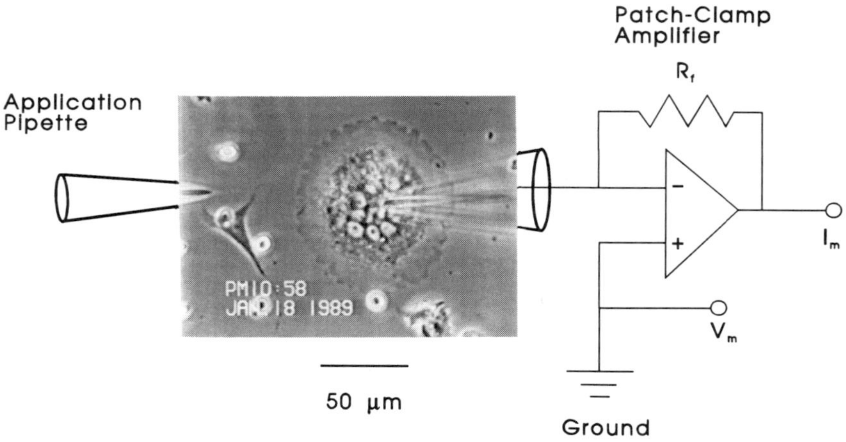

FIGURE 2. Patch-clamp recording from rat osteoclast. Phase contrast video micrograph shows a freshly isolated rat osteoclast on glass. The cell is flattened and roughly circular, with more than ten nuclei. The recording electrode, seen extending to the right of the cell, is attached to a patch-clamp amplifier that holds the cell membrane potential (V_m) at a desired level. Membrane current (I_m) is measured from the voltage drop across the feedback resistor (R_f). Test substances can be applied to cells by pneumatic pressure ejection from an application pipette seen at the left of the field.

An advantage of the whole-cell configuration is that it enables one to dialyze the cytoplasm with known concentrations of ions or drugs. In some cases, dialysis may be undesirable since soluble cytoplasmic proteins and other molecules may be lost, leading to loss of channel activity. This problem can be avoided using the ''perforated patch'' technique,[16] in which a gigaohm seal is first established on the cell surface. The pore-forming antibiotic nystatin, which is in the patch electrode, then permeabilizes the membrane to monovalent ions. In this way, multivalent ions such as Ca^{2+} and large molecules remain within the cytoplasm, and membrane currents from the whole cell can be recorded without disruption of second-messenger systems.

One can also isolate very small patches of membrane in either the inside-out (Figure 1C) or outside-out (Figure 1D) configuration. In either of these configurations, the activity of one to several ion channels can be recorded, depending upon the size of the membrane patch and the density of the channels in the membrane. Using the cell-attached (Figure 1A) or excised patch (Figure 1C, D) configurations, it is possible to determine single-channel conductances and study the kinetic behavior of single ion channels.

III. MEMBRANE POTENTIAL AND PASSIVE PROPERTIES

A. MEMBRANE POTENTIAL

Mears[12] was the first to utilize an electrophysiological approach to study osteoclasts and their responses to hormones. Using intracellular microelectrode techniques, Mears recorded membrane potential from rabbit osteoclasts both *in vivo* and *in vitro*. He noted two resting potentials (-10 and -23 mV), while Ferrier and co-workers,[13] also using intracellular electrodes to record from rabbit osteoclasts, reported only a single resting potential (approximately -17 mV). Calcitonin, which is known to interact directly with osteoclasts to inhibit bone resorption,[17,18] was reported to cause hyperpolarization of osteoclasts.[12,13] Whereas Mears reported that parathyroid hormone caused rapid depolarization, Ferrier and co-workers found that the response to parathyroid hormone and dibutyryl cyclic AMP was variable.[13]

Using the tight-seal patch-clamp recording technique in whole-cell configuration (Figure 1B) and in the current-clamp mode, it is also possible to monitor the membrane potential of isolated osteoclasts. In both rat and rabbit osteoclasts, membrane potentials as negative as -75 to -70 mV have been reported for some cells, while other cells are more depolarized, with membrane potentials of ≈ -20 to 0 mV.[8,15,19] A similar bimodal distribution of membrane potentials was observed in chicken osteoclasts, where 18% of the cells studied initially exhibited a resting potential of -66 mV, with the remaining cells close to -11 mV.[14] The membrane potentials of chicken osteoclasts diminished to 0 mV within the first minute of recording, a change attributed by the authors to the development of nonselective leakage current.

Evidence indicates that a K^+ conductance is responsible for setting the membrane potential of those osteoclasts having membrane potentials close to -70 mV. Ba^{2+}, which blocks the inwardly rectifying K^+ conductance (see below), depolarizes osteoclasts.[15] Furthermore, in studies conducted using the voltage-clamp technique, the zero-current potential (the potential at which there is no net current across the membrane, i.e., the resting membrane potential) shifted positive with increased concentration of external K^+ ($[K^+]_o$).[8] This shift is predicted by the Nernst relation for K^+-selective channels. These findings are consistent with the interpretation that, in the absence of other conductances being activated, outward current through K^+ channels is responsible for setting the osteoclast membrane potential close to -70 mV.

As reviewed above, populations of osteoclasts display bimodal distributions of membrane potentials. In some cases, two stable membrane potentials have been demonstrated in the same cell (rat,[15] rabbit[19]), and this property can develop over time during recording. A series of traces illustrating this phenomenon is shown in Figure 3A. Beginning 50 s after establishing whole-cell configuration, the cell exhibited a membrane potential of -70 mV. A 1-s pulse of positive current injected through the recording electrode elicited large but transient depolarization, with the cell repolarizing to -70 mV several seconds later. After recording from this cell for 8 min, an identical pulse of current caused a similar large depolarization, but this time the cell remained at -10 mV for 10 s, at which time a negative pulse of current caused the cell to switch back to -70 mV. After a period of recording, the cell could remain at either of two stable membrane potentials. With continued

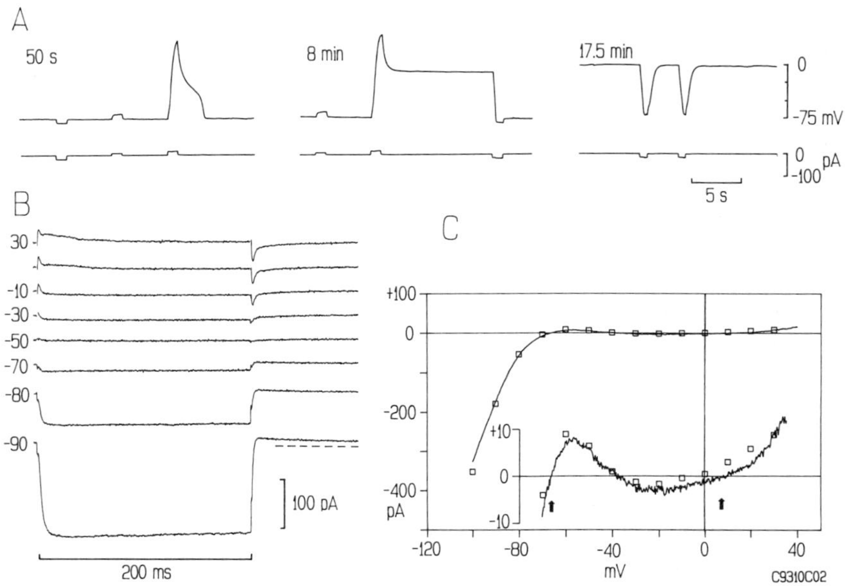

FIGURE 3. Rat osteoclasts can exhibit two stable membrane potentials. (A) Membrane potential was measured at various times after the whole-cell recording configuration (Figure 1B) was established. Time is indicated to the left of each trace (membrane potential above and current monitor below). Initially, at 50 s, membrane potential was −70 mV, and a depolarizing current step elicited large, but transient depolarization. After recording for 8 min, the depolarization persisted, and the cell remained steadily at −10 mV for 10 s until negative current caused repolarization (middle panel). By 17.5 min, the cell had depolarized to − 10 mV. (B, C) The current-voltage relationship measured under voltage clamp during the time the cell exhibited two stable membrane potentials. Current traces, measured at 7 min, are shown in B, with the command potential indicated to the left of the traces. The I-V for the steady-state current, measured at 200 ms, is shown in C (open boxes). The continuous line is the current-voltage relationship measured with a ramp command, where the membrane potential was changed from −100 to + 40 mV over 1 s (measured at 9 min). The I-V is N-shaped and intersects the zero-current level at three different potentials, giving rise to two stable membrane potentials (indicated by arrows). The zero-current level in B, as indicated by the dashed line to the right of the bottom current trace, was ≈10 pA negative to the holding current for all traces. There is a small discrepancy between the steady-state I-V and that determined from the ramp in C, amounting to a negative shift of the ramp current by 2 pA. Traces in A were filtered (−3 dB) at 30 Hz and digitally sampled at 85 Hz. The cell was on glass substrate.

recording, the cell assumed the depolarized level of −10 mV (17.5 min of recording). The I-V relationship, determined under voltage clamp during the period when the cell exhibited two stable membrane potentials (Figure 3C), was N-shaped and intersected the zero-current level at three different potentials. This form of I-V can account for the two stable membrane potentials, since resting membrane potentials occur where the I-V intersects the voltage axis and has a positive slope conductance (Figure 3C, arrows). The type of

I-V shown in Figure 3C can be explained by the summation of inwardly rectifying K^+ current (described below), with another current having a reversal potential close to 0 mV. The gradual development of this additional conductance, which can account for the transition from a cell with a single resting potential of −70 mV to a cell with two stable membrane potentials, was spontaneous during this recording. Evidence suggests that development of an anion conductance may be responsible for the inward current at negative potentials that leads to the second membrane potential (see below). Two stable membrane potentials have been observed in other cells with inwardly rectifying currents, including cardiac Purkinje fibers[20] and macrophages.[21]

B. PASSIVE MEMBRANE PROPERTIES

The biophysical characteristics of rat osteoclasts have been studied by characterizing their passive membrane properties.[8] The time course of charging of the osteoclast membrane in response to small current steps was shown to be a single exponential process, indicating that the membrane charges as a simple parallel resistor-capacitor network.[22] This provides direct evidence that the osteoclast membrane is isopotential, and thus the voltage-clamp technique can be applied to the cells. From measurements of cell capacitance and cell area, specific membrane capacitance was estimated to be 1.2 $\mu F/cm^2$. Specific membrane resistance was determined to be 17 $k\Omega/cm^2$.[8] These passive membrane characteristics are in the range found for other cells.[23] Membrane properties were the same for osteoclasts plated on permeable collagen substrate as for cells plated on glass,[8] suggesting that cells do not establish high-resistance seals to glass, which might give rise to extracellular voltage gradients. This finding was unexpected in light of a previous report that osteoclasts adhering to glass substrate can establish an extracellular pH gradient.[24]

IV. IONIC CURRENTS IN OSTEOCLASTS

A. K^+ CURRENTS

1. Inwardly Rectifying K^+ Currents

A summmary of the ion channels identified in osteoclasts to date is presented in Table 1. A common feature of chicken and rat osteoclasts is the presence of inwardly rectifying K^+ current.[14,15] This current has been identified in whole-cell configuration of voltage clamp (Figure 1B) based upon voltage activation properties and blockade by various ions. Hyperpolarizing commands to voltages negative to the K^+ equilibrium potential (E_K) elicit large inward currents, indicating a large conductance at negative potentials (Figures 3 through 5). As shown in the recordings from rat osteoclasts illustrated in Figures 3 and 4, the conductance decreases rapidly at voltages positive to E_K, where only small outward currents are evident. This nonlinear current-voltage relationship, where there is more inward current than outward current,

TABLE 1
Summary of Ion Channels in Osteoclasts

Channel	Species	Unitary conductance	Blockers	Ref.
Inward rectifier	Rat, rabbit	25 pS	Ba^{2+}, Cs^{+}	8, 15, 19
K^{+} current	Chicken	—	Cs^{+}	14
Transient outward K^{+} current	Chicken	—	4-Aminopyridine	14
Outward rectifier K^{+} current	Chicken	—	—	14
ATP-activated[a]	Chicken	—	—	48
Cl^{-}	Rat, rabbit	—	SITS, DIDS	8, 19
	Rat	340 pS	—	35
	Chicken[b]	—	DIDS	7
Ca^{2+}	Chicken[c]	—	Dihydropyridines	31
	Rat[c]	—	Ni^{2+}	42

[a] Selectivity of this conductance was not reported.
[b] Channel protein isolated biochemically and activity assessed using $^{36}Cl^{-}$ uptake.
[c] Inferred from fluorescence measurements.

is characteristic of an inwardly (or anomalously) rectifying conductance. Evidence that the inwardly rectifying conductance is selective for K^{+} comes from experiments where $[K^{+}]_o$ is altered. When $[K^{+}]_o$ is raised by replacing NaCl with KCl, the whole-cell I-V relation shifts in the depolarizing direction (Figure 4), as predicted for K^{+} current by the Nernst relation. Furthermore, the inwardly rectifying K^{+} current of osteoclasts is blocked by low concentrations of extracellular Ba^{2+} (Figure 4)[15] and Cs^{+}.[14] In these respects, the inward rectifier of rat and chicken osteoclasts resembles the classical inward rectifier K^{+} currents of skeletal muscle,[25] cardiac muscle,[26] and oocytes.[27] A similar channel is variably expressed in macrophages and macrophage-like cells (reviewed in References 28 and 29).

Recent studies have confirmed the presence of inwardly rectifying K^{+} current in rabbit osteoclasts,[19] where, as in rat osteoclasts, the current is responsible for setting the resting membrane potential of cells close to the equilibrium potential for K^{+}. Thus, inwardly rectifying K^{+} current has been identified in osteoclasts from all species studied to date, suggesting that this conductance may be important in osteoclast function or regulation.

Single-channel currents underlying the whole-cell currents described above have been recorded from rat osteoclasts using the cell-attached patch configuration (Figure 1A). With 130 m*M* KCl in the recording pipette, inward currents are recorded at potentials more negative than E_K. No outward current is detected at depolarized potentials, indicating the rectifying nature of the channel.[8] The single-channel conductance measured from the current-voltage relationship of single channel currents was 25 pS, which is in the range of values reported for inward rectifier K^{+} channels in other cell types (e.g., cardiac muscle,[26] macrophages,[28] and endothelial cells[30]).

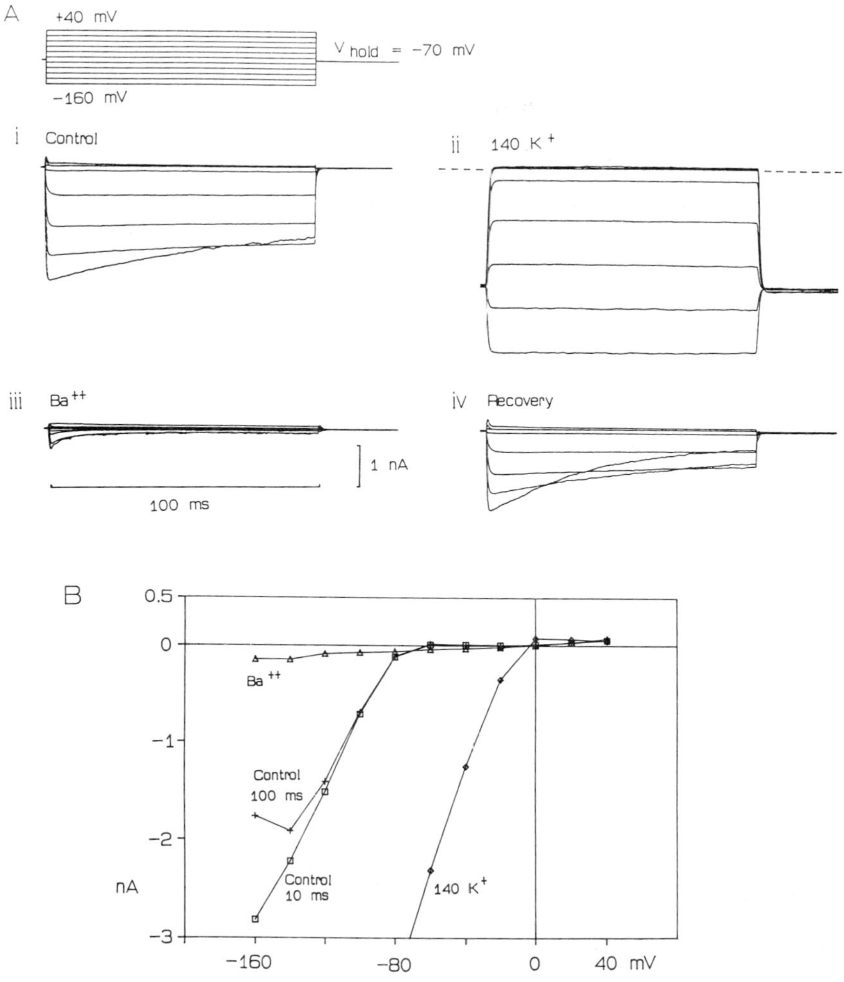

FIGURE 4. Inwardly rectifying K^+ current in rat osteoclast on glass substrate. The cell was held under voltage clamp at −70 mV (V_{hold}) and stepped to various potentials, as indicated above control response. All traces are from the same cell. (Ai) Hyperpolarizing commands elicited an inward current across the membrane (downward deflections) that peaked and then declined with time. Depolarizing commands caused only a small outward current (small upward deflections). (Aii) 140 m*M* K^+ was applied to the cell by application pipette, causing an inward current at the holding level and increasing the magnitude of the inward currents. Traces are shown for −100 mV and more positive. (Aiii) Barium (1 m*M* in an application pipette) suppressed the inward rectifier K^+ current. (Aiv) Recovery from Ba^{2+}. The zero-current level is indicated by the dashed lines alongside the current traces in Aii. (B) Current-voltage relationship for responses shown above. 140 m*M* K^+ shifted the current-voltage relationship to the right along the voltage axis. Ba^{2+} suppressed the inward K^+ current. Current amplitudes are shown at 10 and 100 ms for the control cell, but only the 100-ms currents are shown for the other conditions. (From Sims, S. M. and Dixon, S. J., *Am. J. Physiol.*, 256, C1277, 1989. With permission.)

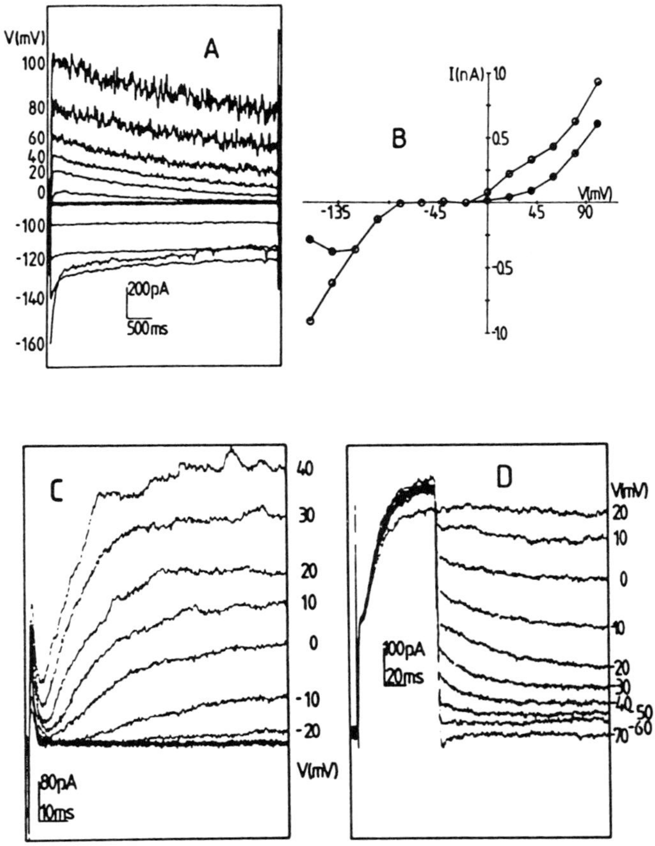

FIGURE 5. Three types of voltage-dependent currents in chicken osteoclasts on glass substrate. (A) Whole-cell currents measured at voltages indicated to the left of the current traces. Hyperpolarizing voltage commands caused inward current (downward deflections) that peaked, then declined over time. Depolarizing voltage commands elicited outward current (upward deflections). A transient outward current was apparent at 0 mV and positive. The conductance underlying this current is termed G_{kto}. The increased current noise on traces recorded at potentials positive to + 40 mV indicates activation of another conductance, termed G_{ko}. Currents between − 80 and − 20 mV are not displayed. (B) Current-voltage relationship for the cell illustrated in A, with open symbols indicating the peak current levels and solid symbols the current measured at 4.5 s. (C) G_{kto} activated with delay. The cell was held at − 70 mV and commanded to the potentials indicated to the right of the current traces. (D) Evidence that the outwardly rectifying current is due to K^+. The cell was held at − 70 mV and commanded to + 20 mV to activate outward current. Membrane potential was then commanded to negative potentials to observe decay of the current. Reversal of the current relaxations close to − 60 mV suggests that the outward current is due to activation of K^+ channels. All currents illustrated except those in D were corrected for leakage. (Adapted from Ravesloot, J. H., Ypey, D. L., Vrijheid-Lammers, T., and Nijweide, P. J. *Proc. Natl. Acad. Sci. U.S.A.*,, 86, 6821, 1989. With permission.)

2. Outwardly Rectifying K^+ Currents

In addition to the inwardly rectifying K^+ current mentioned above, two types of outwardly rectifying K^+ currents have been described in avian osteoclasts by Ravesloot and co-workers.[14] A transient outward current was activated by depolarization to potentials more positive than −20 mV in 72% of the chick osteoclasts studied (Figure 5). This transient outward current decayed with an exponential time course (time constants of 1.5 to 2 s) and was reduced by the K^+ channel blocker 4-aminopyridine. An additional outward K^+ current was activated in all chicken osteoclasts at potentials more positive than +40 mV. This noninactivating current (Figure 5A) was not blocked by external 4-aminopyridine or cesium.[14]

Outwardly rectifying currents have also been identified in mammalian osteoclasts,[8,19] but these have been shown to arise from an anion-selective conductance (see Section IV.B). While there is not yet any evidence for outwardly rectifying K^+ currents in mammalian osteoclasts, further studies are required. It is possible that cells studied under different experimental conditions (such as different substrates) may express different currents (see, e.g., Reference 31).

B. ANION CURRENTS

Anion-selective channels are present in a variety of cell types and subserve a number of functions, including regulation of cell volume, electrolyte transport,[32] and stabilization of membrane potential.[33] Recent patch-clamp studies have demonstrated the presence of a voltage-regulated chloride conductance in osteoclasts.[8,19] While inwardly rectifying K^+ current predominates at negative potentials in mammalian osteoclasts (see above), whole-cell recordings have revealed that a proportion of cells (30 to 60%) also show outward rectification at more positive potentials.[8,19] In these cells, depolarization elicits a steady outward current. With symmetrical Cl^- ($[Cl^-]_i = [Cl^-]_o$), this current shows pronounced rectification around 0 mV. Unlike the outwardly rectifying Cl^- current described in airway epithelial cells,[34] however, the outward current in osteoclasts shows little evidence of time-dependent inactivation at positive potentials. In some cells, where outward current was not apparent during the initial stages of recording, this current was observed to develop over time during recording.[19]

Evidence that this outwardly rectifying current in rat and rabbit osteoclasts is attributable to a Cl^- conductance was provided by the following findings.[8,19] (1) The outward current persisted in the presence of internal Cs^+ and TEA^+ and external Ba^{2+}, conditions under which all K^+ currents are blocked. (2) This outward current was reversibly blocked by the anion transport inhibitors 4,4′-diisothiocyanostilbene-2,2′-disulfonic acid (DIDS) and 4-acetamido-4′-isothiocyanostilbene-2,2′-disulfonic acid (SITS). (3) Under symmetrical Cl^- conditions, the DIDS- and SITS-sensitive current reversed direction at 0 mV, the predicted reversal potential for Cl^-. (4) When the transmembrane Cl^- gradient was altered, the reversal potential shifted in the manner expected for a Cl^- current.[8,19] Preliminary results[64] suggest that a low-conductance ($\approx$20 pS) channel underlies this macroscopic outward Cl^- current.

Evidence for the presence of Cl^--selective channels has also been provided by Schoppa and co-workers,[35] who reported a large-conductance (340 pS) Cl^--selective channel in inside-out patches (Figure 1C) from rat osteoclasts. This channel was voltage dependent, activating between -20 and $+20$ mV. The purification of a stilbene-sensitive chloride channel protein from chicken osteoclast membranes has been reported recently by Blair and Schlesinger.[7]

The chloride currents of osteoclasts resemble those identified in other cell types. For example, stilbene disulfonates block Cl^- currents in osteoblasts[36] and astrocytes.[37] A large-conductance Cl^- channel in mouse macrophages has been shown to be dependent on both voltage and $[Ca^{2+}]_i$.[38] In osteoblasts,[36] cardiac cells,[39] and airway epithelium,[34] hormones and neurotransmitters that increase cAMP have been shown to activate Cl^- conductances. These effects may be due to activation of cAMP-dependent protein kinase and phosphorylation of regulatory sites, possibly the channels themselves. Little is known about the mechanisms by which Cl^- currents in osteoclasts are regulated. The spontaneous appearance and disappearance of this conductance during whole-cell recording (see above) suggests that these channels are regulated. It would be of interest to examine the effects of calcitonin on osteoclast chloride channels, since calcitonin is thought to act primarily through elevation of cAMP.[40]

C. OTHER CURRENTS

1. Ca^{2+} Currents

Calcium plays an important role in regulating the activity of virtually all cells. Cytosolic free Ca^{2+} ($Ca^{2+}{}_i$) of osteoclasts has been the focus of several recent studies.[31,41-43] In osteoclasts, the contribution of plasma membrane Ca^{2+} channels to the regulation of $[Ca^{2+}]_i$ has been investigated using either patch-clamp recording or fluorescent Ca^{2+} indicator dyes. Using electrophysiological techniques, no evidence has been found for voltage-activated Ca^{2+} or Na^+ currents in rat,[8,15] chicken,[14] or rabbit osteoclasts.[19] A number of different strategies were used to test for Ca^{2+} currents in rat osteoclasts.[8] No evidence was found for Ca^{2+} currents when using the perforated patch technique to maintain the integrity of the cytosol and help preserve labile currents. Furthermore, the dihydropyridine BAY K 8644, which enhances L-type Ca^{2+} channel activity in other cell types,[44] had no effect on membrane currents of rat osteoclasts.[8] Similarly, there are no reports of voltage-activated Ca^{2+} channels in macrophages, a related cell type whose electrophysiological properties have been thoroughly characterized.[28,29]

Fluorescent Ca^{2+} indicator dyes have also been used to test for the presence of voltage-activated Ca^{2+} channels in osteoclasts, with conflicting results. Using an increase of $[K^+]_o$ to depolarize cells, no elevation of $[Ca^{2+}]_i$ was reported in some studies,[45,46] supporting the electrophysiological data described above. More recent studies, however, have reported elevation of $[Ca^{2+}]_i$ upon depolarization, which was interpreted as evidence for voltage-activated Ca^{2+} channels in osteoclasts.[31] It has been suggested that the substrate to which osteoclasts adhere influences the expression of voltage-activated Ca^{2+} channels, with channels evident in cells studied on glass but not on bone.[31]

Regardless of whether voltage-activated Ca^{2+} channels exist, Ca^{2+} entry pathways must be present in the plasma membrane of osteoclasts. In unstimulated cells, elevation of external $[Ca^{2+}]$ has been shown to increase $[Ca^{2+}]_i$. Sensitivity of this response to the Ca^{2+} channel blockers Ni^{2+} (Reference 42) and La^{3+} (Reference 31) suggests that increased $[Ca^{2+}]_i$ in response to elevation of $[Ca^{2+}]_o$ arises, at least in part, through channel-mediated influx of Ca^{2+}. These observations led to the proposal that a low-affinity receptor for external Ca^{2+} exists on the osteoclast membrane and that receptor occupancy is linked to activation of a Ca^{2+} channel and release of Ca^{2+} from internal stores.[31,47] However, the possibility also exists that elevation of $[Ca^{2+}]_o$ simply increases the influx of Ca^{2+} through existing leak pathways without activation of a channel or a Ca^{2+} release mechanism. Consistent with this possibility, we have observed that removal of extracellular Ca^{2+} reduces the basal $[Ca^{2+}]_i$ of fura-2-loaded rat osteoclasts. ($[Ca^{2+}]_i$ in the presence of 1 m*M* $Ca^{2+}{}_o$ was 81 ± 9 n*M* and was reduced to 47 ± 7 n*M* in Ca^{2+}-free medium with 0.5 m*M* EGTA).[65]

2. ATP-ACTIVATED CURRENT

A preliminary report has described a current in chicken osteoclasts that is activated by extracellular ATP.[48] This current reversed direction close to 0 mV, but its ionic selectivity was not reported. ATP is known to activate a nonselective cation conductance in mouse macrophages.[49]

V. SIGNIFICANCE OF CURRENTS IN OSTEOCLAST FUNCTION AND REGULATION

A. MEMBRANE POTENTIAL

Osteoclastic bone resorption involves chemotaxis, motility, secretion of enzymes, and transport of H^+. The ability of the osteoclast to switch between two stable membrane potentials (see Section III.A, Figure 3) could influence each of these processes. For example, depolarization of osteoclasts by 60 mV would facilitate electrogenic pumping of H^+ out of the cell and into the resorption lacuna, thereby stimulating bone resorption. A 60-mV change of membrane potential is equivalent to a tenfold change in concentration gradient

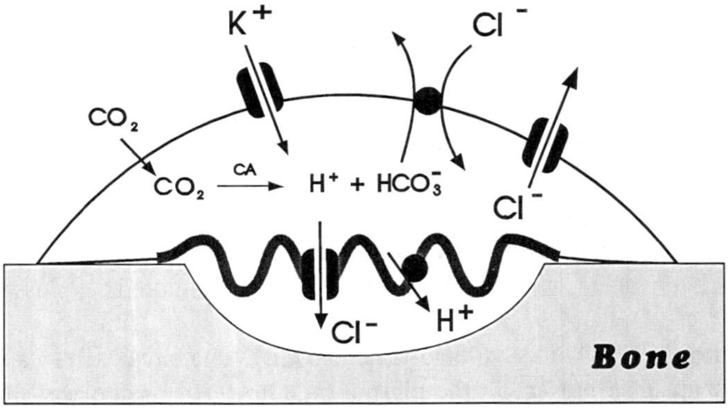

FIGURE 6. Model for the possible role of K^+ and Cl^- channels in osteoclast proton transport. Protons are generated in the cytosol through the action of carbonic anhydrase, and transported across the ruffled membrane (thickened region of membrane) by electrogenic H^+ pumps. (See Chapter 13 of this volume for further details). Electrogenic efflux of protons can only be maintained with the appropriate movement of counterions. Inwardly rectifying K^+ channels and Cl^- channels provide conductive pathways able to dissipate the potential arising from electrogenic pump activity. The inward rectifier K^+ conductance allows inward movement of K^+ at hyperpolarized potentials, preventing excessive hyperpolarization while minimizing efflux of K^+ at depolarized potentials. The Cl^- channels would function in a similar manner, providing an anion conductance that compensates for H^+ movement. In addition, Cl^- efflux would maintain the Cl^- concentration gradient that drives Cl^--HCO_3^- exchange.

for protons or other univalent ions, or a 100-fold change in concentration gradient for divalent ions like Ca^{2+}. Thus, changes in membrane potential from one level to the other could be involved in signal transduction by altering, e.g., $[Ca^{2+}]_i$. If voltage-activated Ca^{2+} channels exist in osteoclasts (see Section IV.C.1), then depolarization would open these Ca^{2+} channels, elevating $[Ca^{2+}]_i$. On the other hand, if Ca^{2+} entry pathways are voltage insensitive, then depolarization would decrease the electrochemical gradient driving Ca^{2+} influx, reducing $[Ca^{2+}]_i$. The membrane potential directly regulates the activity of voltage-gated ion channels such as the inwardly rectifying K^+ channels and outwardly rectifying K^+ and Cl^- channels found in osteoclasts (described in Section IV).

Recent evidence indicates that acidification of the resorption lacuna is mediated by the vacuolar class of ATP-driven proton pump (H^+-ATPase),[2-4,50] which is electrogenic.[51] Transport of positively charged ions (H^+) across the osteoclast ruffled membrane can therefore only be maintained with the appropriate movement of counterions. Both the inwardly rectifying K^+ conductance and the Cl^- conductance of osteoclasts provide conductive pathways able to compensate for electrogenic H^+ transport (Figure 6). The inwardly rectifying K^+ conductance allows inward movement of K^+ at hyperpolarized potentials, making it ideally suited to counteract the change in membrane potential arising from activity of the H^+ pump, preventing excessive hyperpolarization while minimizing efflux of K^+ at depolarized potentials. The Cl^- conductance may play a similar role, providing a parallel Cl^- conductive pathway that can compensate for H^+ movement across the ruffled membrane. Indeed, there is evidence that the activity of vacuolar and osteoclastic H^+-ATPase is strictly dependent upon Cl^-.[51,52] This dependence on Cl^- has led Forgac[51] to suggest that control of vacuolar acidification may not be achieved at the level of the transporter itself, but rather by modification of other accessory proteins, such as the Cl^- channel. Recent studies provide additional support for the involvement of Cl^- transport in osteoclast activity. The Cl^- transport blockers DIDS and SITS reduced bone resorption,[6,53] an effect attributed to inhibition of the chloride-bicarbonate exchanger. However, the possibility also exists that DIDS and SITS inhibited electrogenic H^+ movement by blocking Cl^- channels. Cl^- currents in other cells are regulated by cytosolic second messengers, including cyclic AMP and intracellular Ca^{2+},[34,36,39] but it is not known if these factors are involved in the gradual activation of the current seen in some cases during recording from osteoclasts.[19] Physiological regulation of the osteoclast Cl^- conductance may modulate bone resorption, and the 30 to 60% of mammalian osteoclasts that exhibit outwardly rectifying current[8,19] may represent osteoclasts that are actively engaged in H^+ transport.

B. ION MOVEMENT

In addition to providing pathways allowing redistribution of charge across the plasma membrane of osteoclasts, ion channels mmay also provide pathways for the movement of specific ionic species, each having essential roles in osteoclast function.

1. K^+ Channels

It has been proposed that, in depolarized osteoclasts, outwardly rectifying K^+ channels mediate K^+ movement from the cytoplasm into the resorption lacuna.[14] A gastric-type H^+,K^+-ATPase located in the ruffled membrane would then mediate exchange of external K^+ for internal H^+, leading to acidification of the resorption lacuna. Early reports suggested the existence of the gastric-type H^+ pump in osteoclasts.[1,54] However, more recent evidence is consistent with the presence of the vacuolar-type H^+-ATPase,[2-4,50] an electrogenic H^+ pump that does not translocate K^+ and, therefore, is not dependent on K^+ efflux.

2. Cl^- Channels

Plasma membrane Cl^- channels may play an essential role in maintaining the cytoplasmic pH of osteoclasts. It has been suggested that H^+ are generated in osteoclasts through the action of carbonic anhydrase that catalyzes the hydration of CO_2, forming H^+ and HCO_3^- (reviewed in Reference 55; Figure 6). H^+ are then pumped across the ruffled membrane, while cytosolic pH is maintained by anion exchangers that mediate the efflux of HCO_3^- across the dorsal surface of the cell in exchange for Cl^-.[5] Cl^- channels would allow for Cl^- efflux, thus maintaining the inward Cl^- concentration gradient that drives Cl^--HCO_3^- exchange, which in turn is required for cytosolic pH homeostasis (Figure 6).

3. Ca^{2+} Channels

In osteoclasts, as in other cell types, cytosolic Ca^{2+} is thought to be an important second messenger regulating a number of functions, including motility and resorption.[31,43,56] Although the existence in osteoclasts of voltage-activated Ca^{2+} channels is unresolved, Ca^{2+} entry pathways must exist in the plasma membrane (see Section IV.C.1). It has been suggested that elevation of $[Ca^{2+}]_i$ in response to physiological increases of $[Ca^{2+}]_o$ inhibits osteoclast activity, serving as a feedback mechanism to limit the extent of bone resorption.[31,47]

No voltage-dependent Ca^{2+} currents have yet been identified in osteoclasts using electrophysiological techniques.[8,14,15] However, evidence is accumulating from fluorescence studies for the existence of other types of Ca^{2+} channels activated by ligands. Calcitonin has been reported to elevate $[Ca^{2+}]_i$ by a mechanism completely[41] or partially[57] dependent on external Ca^{2+}. In the former study, Malgaroli and co-workers[41] observed similar elevations of $[Ca^{2+}]_i$ in response to the permanent cAMP analog 8Br-cAMP, suggesting that calcitonin-induced influx of Ca^{2+} is secondary to activation of adenylate cyclase and elevation of cAMP. Recent fluorescence studies from our laboratory have established that platelet-activating factor, a potent inflammatory mediator, induces biphasic elevation of osteoclast $[Ca^{2+}]_i$.[43] The initial transient phase of the platelet-activating factor-induced $[Ca^{2+}]_i$ elevation was not dependent on external Ca^{2+}, and therefore likely arose by release of Ca^{2+} from intracellular stores. In contrast, the subsequent sustained elevation of $[Ca^{2+}]_i$ above baseline was entirely dependent on $[Ca^{2+}]_o$, consistent with activation of a Ca^{2+}-permeable channel by platelet-activating factor. Although the electrophysiological properties and functional significance of such Ca^{2+} entry pathways have yet to be investigated, they will likely be found to play critical roles in osteoclast regulation.

VI. PROSPECTS

A. IDENTIFICATION OF ADDITIONAL MEMBRANE CURRENTS

As reviewed above in Section V.B, results of fluorescence studies have led to the suggestion that ligand-activated Ca^{2+} channels exist in osteoclasts. Patch-clamp electrophysiological studies are required to identify currents activated by ligands such as calcitonin, platelet-activating factor, and ATP or by elevation of external Ca^{2+}. Simultaneous fluorescence measurement of $[Ca^{2+}]_i$ and electrophysiological recording would be valuable in determining the contribution of plasma membrane ion channels to receptor-mediated elevations in $[Ca^{2+}]_i$.

Patch-clamp techniques have been used to detect currents arising from the activity of electrogenic pumps such as the Na^+,K^+-ATPase in cardiac myocytes.[58,59] As discussed above, recent evidence suggests that acidification of the resorption lacuna is mediated by the vacuolar class of H^+-ATPase located in the ruffled membrane of the osteoclast. This H^+ pump is electrogenic, and transport of H^+ from the cytosol into the resorption lacuna (Figure 6) will therefore give rise to net outward current (i.e., net flow of positive charge out of the cell). If of sufficient magnitude, pump currents can be detected by applying a selective inhibitor of the pump while monitoring whole-cell currents (Figure 1B). In the case of the Na^+,K^+-ATPase, ouabain and its more readily reversible analog strophanthidin have been used.[58,59] Bafilomycin A_1 is claimed to be a relatively selective blocker of the vacuolar class of H^+-ATPase[60] and may therefore be useful in detecting osteoclast H^+ pump current. Pump current would be apparent as a net inward current arising from inhibition of *outward* pump current. The ability to monitor osteoclast H^+ transport would enable direct investigation of its physiological and pharmacological regulation. Furthermore, the presence or absence of H^+ pump current would indicate whether the cell under study was, at that time, actively resorbing or quiescent.

B. REGULATION OF IONIC CONDUCTANCES

Little is known regarding the physiological regulation of membrane ionic currents in osteoclasts. Many of the currents described to date are voltage dependent (Section IV), and therefore their activity is strongly influenced by membrane potential. As described above (Section III.A), the membrane potentials of osteoclasts are bimodally distributed, and in some cases, the same cell can shift between two stable levels of membrane potential (Figure 3). In cells with more hyperpolarized membrane potentials (-75 to -65 mV), inwardly rectifying K^+ current will be active, while at depolarized resting potentials (-20 to 0 mV), this conductance will be inactive. As discussed, above, this voltage dependence may be functionally important, since electrogenic H^+ transport will tend to hyperpolarize the cell, activating compensatory inward K^+ current. In contrast, at more depolarized potentials, this conductance is inactive, preventing excessive efflux of K^+. On the other hand, *outwardly* rectifying currents (e.g., the K^+ currents in avian osteoclasts and Cl^- current in mammalian osteoclasts; Section IV) will be activated upon depolarization of the membrane potential.

In addition to voltage-dependent gating, other mechanisms exist to regulate ionic conductances. These include regulation of the numbers of channels in the membrane or of channel activity by various pathways, such as guanine nucleotide-binding proteins and cytosolic second messengers.[61] Regarding osteoclasts, there is evidence that at least the Cl^- conductance of mammalian cells is regulated, since it is present in only 30 to 60% of cells.[8,19] In some cells, this current is initially absent but develops over time during recording, indicating some mechanism for short-term channel regulation. The pathways regulating this Cl^- conductance are presently under investigation in our laboratory.

It is conceivable that, as in other cells, ionic currents in osteoclasts will be found to be regulated by a variety of agents, including humoral mediators (e.g., calcitonin, platelet-activating factor, and ATP), second messengers (e.g., cAMP, $Ca^{2+}{}_i$), and cell-cell and cell-matrix interactions. Furthermore, it is likely that osteoclasts, like osteoblastic cells,[62,63] will be found to possess mechanosensitive ion channels that may mediate the responsiveness of osseous tissues to mechanical stimuli.

Just as little is known about the physiological regulation of osteoclast currents, our understanding of their pharmacology is still meager (Table 1). Investigating the sensitivity of these channels to pharmacological blockers and activators is important for a number of reasons. First, along with defining other properties such as ion selectivity and unitary conductance, pharmacological sensitivity is useful in identifying the channels that are present. For example, the inwardly rectifying K^+ channel (Section IV.A.1) appears in all respects to be similar to that described previously in skeletal and cardiac muscle (see references above). The Cl^- channel, on the other hand, has not yet been characterized sufficiently to assess whether it is similar to the anion channels described in other cell types. Ultimately, osteoclast channels will be identified using biochemical and molecular approaches combined with patch-clamp recording. An important starting point has been the recent purification and functional reconstitution of a stilbene-sensitive Cl^- channel from avian osteoclasts.[7]

The second reason for investigating the pharmacology of osteoclast ion channels is to identify selective blockers and activators that may then be used as tools for determining the precise roles of these conductances in osteoclast function (see Section VI.C). Last, in other systems, membrane ion channels are targets for drug therapy. It is conceivable that agents will be identified that interact selectively with osteoclast channels to inhibit cell function. Appropriate analogs of these agents might be useful clinically to inhibit bone resorption in a variety of metabolic and inflammatory bone diseases.

C. ROLES OF IONIC CONDUCTANCES IN OSTEOCLASTS

Assays are now available for assessing *in vitro* a number of osteoclast functions, including differentiation, chemotaxis, and resorption of mineralized substrate. As our knowledge of membrane currents in these cells expands, a number of experimental approaches will become available to test directly the involvement of ion channels in osteoclast in osteoclast function. For example, ion substitution or blockers could be used to selectively eliminate a particular conductance and the effect of elimmination assessed using an *in vitro* functional assay. Similarly, the functional role of membrane potential could be addressed using ion substitution or ionophores to chemically "clamp" the membrane potential at different levels. Such studies will undoubtedly provide novel insights into the basic biology of osteoclasts and perhaps suggest new strategies for therapeutic intervention in resorptive disorders.

ACKNOWLEDGMENTS

We thank Dr. Z. G. Zheng for help in the preparation of the manuscript. The work from our laboratory reviewed in this chapter was funded by The Arthritis Society and Medical Research Council of Canada. S. M. Sims is supported by a Scholarship, M. E. M. Kelly by a Fellowship, and S. J. Dixon by a Development Grant from the Medical Research Council of Canada.

NOTE ADDED IN PROOF

Since preparation of this chapter, there have been two additional reports on the electrophysiology of osteoclasts:

- **Kelly, M. E. M., Dixon, S. J., and Sims, S. M.,** Inwardly rectifying potassium current in rabbit osteoclasts: a whole-cell and single-channel study, *J. Membrane Biol.*, 126, 171, 1992.
- **Arkett, S. A., Dixon, S. J., and Sims, S. M.,** Substrate influences rat osteoclast morphology and expression of potassium conductances, *J. Physiol.*, in press.

REFERENCES

1. **Baron, R., Neff, L., Louvard, D., and Courtoy, P. J.,** Cell-mediated extracellular acidification and bone resorption: evidence for a low pH in resorbing lacunae and localization of a 100-kD lysosomal membrane protein at the osteoclast ruffled border, *J. Cell Biol.*, 101, 2210, 1985.
2. **Blair, H. C., Teitelbaum, S. L., Ghiselli, R., and Gluck, S.,** Osteoclastic bone resorption by a polarized vacuolar proton pump, *Science*, 245, 855, 1989.
3. **Bekker, P. J. and Gay, C. V.,** Biochemical characterization of an electrogenic vacuolar proton pump in purified chicken osteoclast plasma membrane vesicles, *J. Bone Miner. Res.*, 5, 569, 1990.
4. **Sundquist, K., Lakkakorpi, P., Wallmark, B., and Väänänen, K.,** Inhibition of osteoclast proton transport by bafilomycin A_1 abolishes bone resorption, *Biochem. Biophys. Res., Commun.*, 168, 309, 1990.
5. **Teti, A., Blair, H. C., Teitelbaum, S. L., Kahn, A. J., Koziol, C., Konsek, J., Zambonin-Zallone, A., and Schlesinger, P. H.,** Cytoplasmic pH regulation and chloride/bicarbonate exchange in avian osteoclasts, *J. Clin. Invest.*, 83, 227, 1989.

6. **Hall, T. J. and Chambers, T. J.,** Optimal bone resorption by isolated rat osteoclasts requires chloride bicarbonate exchange, *Calcif. Tissue Int.*, 45, 378, 1989.
7. **Blair, H. C. and Schlesinger, P. H.,** Purification of a stilbene sensitive chloride channel and reconstitution of chloride conductivity into phospholipid vesicles, *Biochem. Biophys. Res. Commun.*, 171, 920, 1990.
8. **Sims, S. M., Kelly, M. E. M., and Dixon, S. J.,** K^+ and Cl^- currents in freshly isolated rat osteoclasts, *Pflugers Arch.*, 1419, 358, 1991.
9. **Chambers, T. J. and Magnus, C. J.,** Calcitonin alters behaviour of isolated osteoclasts, *J. Pathol.*, 136, 27, 1982.
10. **Boyde, A., Ali, N. N., and Jones, S. J.,** Resorption of dentine by isolated osteoclasts *in vitro, Brit. Dent. J.*, 156, 216, 1984.
11. **Hamill, O. P., Marty, A., Neher, E., Sakmann, B., and Sigworth, F. J.,** Improved patch-clamp techniques for high-resolution current recordings from cells and cell-free membrane patches, *Pflügers Arch.*, 391, 85, 1981.
12. **Mears, D. C.,** Effects parathyroid hormone and thyrocalcitonin on the membrane potential of osteoclasts, *Endocrinology,* 88, 1021, 1971.
13. **Ferrier, J., Ward, A., Kanehisa, J., and Heersche, J. N. M.,** Electrophysiological responses of osteoclasts to hormones, *J. Cell. Physiol.*, 128, 23, 1986.
14. **Ravesloot, J. H., Ypey, D. L., Vrijheid-Lammers, T., and Nijweide, P. J.,** Voltage-activated K^+ conductances in freshly isolated embryonic chicken osteoclasts, *Proc. Natl. Acad. Sci. U.S.A.*, 86, 6821, 1989.
15. **Sims, S. M. and Dixon, S. J.,** Inwardly rectifying K^+ current in osteoclasts, *Am. J. Physiol.*, 256, C1277, 1989.
16. **Horn, R. and Marty, A.,** Muscarinic activation of ionic currents measured by a new whole-cell recording method, *J. Gen. Physiol.*, 92, 145, 1988.
17. **Warshawsky, H., Goltzman, D., Rouleau, M. F., and Bergeron, J. J. M.,** Direct *in vivo* demonstration by radioautography of specific binding sites for calcitonin in skeletal and renal tissue of the rat, *J. Cell Biol.*, 85, 682, 1980.
18. **Chambers, T. J., Revell, P. A., Fuller, K., and Athanasou, N. A.,** Resorption of bone by isolated rabbit osteoclasts, *J. Cell Sci.*, 66, 383, 1984.
19. **Kelly, M. E. M., Arkett, S. A., Dixon, S. J., and Sims, S. M.,** Potassium and anion currents in rabbit osteoclasts, *Biophys. J.*, 59 (Abstr.), 461A, 1991.
20. **Gadsby, D. C. and Cranefield, P. F.,** Two levels of resting potential in cardiac Purkinje fibers, *J. Gen. Physiol.*, 70, 725, 1977.
21. **Gallin, E. K.,** Voltage clamp studies in macrophages from mouse spleen cultures, *Science,* 214, 458, 1981.
22. **Jack, J. J. B., Noble, D., and Tsien, R. W.,** *Electric Current Flow in Excitable Cells,* Clarendon Press, Oxford, 1975.
23. **Hille, B.,** *Ionic Channels of Excitable Membranes,* Sinauer, Sunderland, MA, 1984.
24. **Silver, I. A., Murrills, R. J., and Etherington, D. J.,** Microelectrode studies on the acid microenvironment beneath adherent macrophages and osteoclasts, *Exp. Cell Res.*, 175, 266, 1988.
25. **Standen, N. B. and Stanfield, P. R.,** A potential- and time-dependent blockade of inward rectification in frog skeletal muscle fibres by barium and strontium ions, *J. Physiol.*, 280, 169, 1978.
26. **Sakmann, B. and Trube, G.,** Conductance properties of single inwardly rectifying potassium channels in ventricular cells from guinea-pig heart, *J. Physiol.*, 347, 641, 1984.
27. **Hagiwara, S. and Jaffe, L. A.,** Electrical properties of egg cell membranes, *Annu. Rev. Biophys. Bioeng.*, 8, 385, 1979.
28. **Gallin, E. K. and McKinney, L. C.,** Potassium conductances in macrophages, in *Cell Physiology of Blood,* Gunn, R. B. and Parker, J. C., Eds., Rockefeller University Press, New York, 1988, 315.

29. **Gallin, E. K. and McKinney, L. C.,** Ion transport in phagocytes, in *The Neutrophil: Cellular Biochemistry and Physiology,* Hallett, M. B., Ed., CRC Press, Boca Raton, FL, 1989, 243.
30. **Takeda, K., Schini, V., and Stoeckel, H.,** Voltage-activated potassium, but not calcium currents in cultured bovine aortic endothelial cells, *Pflügers Arch.,* 410, 385, 1987.
31. **Miyauchi, A., Hruska, K. A., Greenfield, E., Duncan R., Alvarez, J., Barattolo, R., Colucci, S., Zambonin-Zallone, A., Teitelbaum, S. L., and Teti, A.,** Osteoclast cytosolic calcium, regulated by voltage-gated calcium channels and extracellular calcium, controls podosome assembly and bone resorption, *J. Cell Biol.,* 111, 2543, 1990.
32. **Greger, R., Kynzelmann K., and Gerlach, L.,** Mechanisms of chloride transport in secretory epithelia, *Ann. N. Y. Acad. Sci.,* 574, 403, 1989.
33. **Bretag, A. H.,** Muscle chloride channels, *Physiol. Rev.,* 67, 618,. 1987.
34. **McCann, J. D. and Welsh, M. J.,** Regulation of Cl^- and K^+ channels in airway epithelium, *Annu. Rev. Physiol.,* 52, 197, 1990.
35. **Schoppa, N. A., Su, Y., Baron, R., and Boulpaep, E. L.,** Identification of single ion channels in neonatal rat osteoclasts, *J. Bone Miner. Res.,* 5 (Suppl. 2) (Abstr.), 204, 1990.
36. **Chesnoy-Marchais, D. and Fritsch, J.,** Chloride current activated by cyclic AMP and parathyroid hormone in rat osteoblasts, *Pflügers Arch.,* 415, 104, 1989.
37. **Gray, P. T. A. and Ritchie, J. M.,** A voltage-gated chloride conductance in rat cultured astrocytes, *Proc. R. Soc. London Ser. B,* 228, 267, 1986.
38. **Schwarze, W. and Kolb, H. A.,** Voltage dependent kinetics of an anionic channel of large unit conductance in macrophages and myotube membranes, *Pflügers Arch.,* 402, 281, 1984.
39. **Harvey, R. D., Clark, C. D., and Hume, J. R.,** Chloride current in mammalian cardiac myocytes, *J. Gen. Physiol.,* 95, 1077, 1990.
40. **Nicholson, G. C., Livesey, S. A., Moseley, J. M., and Martin, T. J.,** Actions of calcitonin, parathyroid hormone, and prostaglandin E_2 on cyclic AMP formation in chicken and rat osteoclasts, *J. Cell Biochem.,* 31, 229, 1986.
41. **Malgaroli, A., Meldolesi, J., Zambonin-Zallone, A., and Teti, A.,** Control of cytosolic free calcium in rat and chicken osteoclasts, *J. Biol. Chem.,* 264, 14342, 1989.
42. **Zaidi, M., Datta, H. K., Patchell, A., Moonga, B., and MacIntyre, I.,** 'Calcium-activated' intracellular calcium elevation: a novel mechanism of osteoclast regulation, *Biochem. Biophys. Res. Commun.,* 163, 1461, 1989.
43. **Wood, D. A., Hapak L. K., Sims, S. M., and Dixon, S. J.,** Direct effects of platelet-activating factor on isolated rat osteoclasts: rapid elevation of intracellular free calcium and transient retraction of pseudopods, *J. Biol. Chem.,*. 266, 15369, 1991.
44. **Bean, B. P.,** Classes of calcium channels in vertebrate cells, *Annu. Rev. Physiol.,* 51, 367, 1989.
45. **Teti, A., Blair, H. C., Schlesinger, P., Grano, M., Zambonin-Zallone, A., Kahn, A. J., Teitelbaum, S. L., and Hruska, K. A.,** Extracellular protons acidify osteoclasts, reduce cytosolic calcium and promote expression of cell-matrix attachment structures, *J. Clin. Invest.,* 84, 773, 1989.
46. **Datta, H. K., MacIntyre, I., and Zaidi, M.,** Intracellular calcium in the control of osteoclast function. I. Voltage-insensitivity and lack of effects of nifedipine, BAY K 8644 and diltiazem, *Biochem. Biophys. Res. Commun.,* 167, 183, 1990.
47. **Zaidi, M.,** "Calcium receptors" on eukaryotic cells with special reference to the osteoclast, *Biosci. Rep.,* 10, 493, 1990.
48. **Ravesloot, J. H., Ypey, D. L., Nijweide, P. J., Buisman, H. P., Weidema, F., and Vrijheid-Lammers, T.,** Three voltage activated K^+ conductances and a by extracellular ATP activated conductance in freshly isolated embryonic chick osteoclasts, *J. Bone Miner. Res.,* 4 (Suppl. 1) (Abstr.), 200, 1989.

49. **Buisman, H. P., Steinberg, T. H., Fischbarg, J., Silverstein, S. C., Vogelzang, S. A., Ince, C., Ypey, D. L., and Leijh, P. C. J.,** Extracellular ATP induces a large nonselective conductance in macrophage plasma membranes, *Proc. Natl. Acad. Sci. U.S.A.*, 85, 7988, 1988.
50. **Väänänen, H. K., Karhukorpi, E.-K., Sundquist, K., Wallmark, B., Roininen, I., Hentunen, T., Tuukkanen, J., and Lakkakorpi, P.,** Evidence for the presence of a proton pump of the vacuolar H^+-ATPase type in the ruffled borders of osteoclasts, *J. Cell Biol.*, 111, 1305, 1990.
51. **Forgac, M.,** Structure and function of vacuolar class of ATP-driven proton pumps, *Physiol. Rev.*, 69, 765, 1989.
52. **Blair, H. C., Teitelbaum, S. L., Tan, H.-L., Koziol, C. M., and Schlesinger, P. H.,** Passive chloride permeability charge coupled to H^+-ATPase of avian osteoclast ruffled membrane, *Am. J. Physiol.*, 260, C1315, 1991.
53. **Klein-Nulend, J. and Raisz, L. G.,** Effects of two inhibitors of anion transport on bone resorption in organ culture, *Endocrinology,* 125, 1019, 1989.
54. **Anderson, R. E., Woodbury, D. M., and Jee, W. S. S.,** Humoral and ionic regulation of osteoclast acidity, *Calcif. Tissue Int.*, 39, 252, 1986.
55. **Arnett, T. R. and Dempster, D. W.,** Perspectives: protons and osteoclasts, *J. Bone Miner. Res.*, 5, 1099, 1990.
56. **Datta, H. K., MacIntyre, I., and Zaidi, M.,** The effect of extracellular calcium elevation on morphology and function of isolated rat osteoclasts, *Biosci. Rep.*, 9, 747, 1989.
57. **Zaidi, M., Datta, K., Moonga, B. S., and MacIntyre, I.,** Evidence that the action of calcitonin on rat osteoclasts is mediated by two G proteins acting via separate post-receptor pathways, *J. Endocrinol.*, 126, 473, 1990.
58. **Nakao, M. and Gadsby, D. C.,** Voltage dependence of Na translocation by the Na^+/K^+ pump, *Nature (London),* 323, 628, 1986.
59. **De Weer, P., Gadsby, D. C., and Rakowski, R. F.,** Voltage dependence of the Na-K pump, *Annu. Rev. Physiol.*, 50, 225, 1988.
60. **Bowman, E. J., Siegers, A., and Altendorf, K.,** Bafilomycins: a class of inhibitor of membrane ATPases from microorganisms, animal cells, and plant cells, *Proc. Natl. Acad. Sci. U.S.A.*, 85, 7972, 1988.
61. **Kaczmarek, L. K. and Levitan, I. B.,** *Neuromodulation: The Biochemical Control of Neuronal Excitability,* Oxford University Press, New York, 1987.
62. **Duncan, R. and Misler, S.,** Voltage-activated and stretch-activated Ba^{2+} conducting channels in an osteoblast-like tumor cell line (UMR 106), *FEBS Lett.*, 251, 17, 1989.
63. **Davidson, R. M., Tatakis, D. W., and Auerbach, A. L.,** Multiple forms of mechanosensitive ion channels in osteoblast-like cells, *Pflügers Arch.*, 416, 646, 1990.
64. **Kelly, M. E. M., Dixon, S. J. and Sims, S. M.,** unpublished observations.
65. **Wood, D. A., Sims, S. M., and Dixon, S. J.,** unpublished observations.

Chapter 12

OSTEOCLAST CYTOSKELETON AND ATTACHMENT PROTEINS

Anna Teti and Alberta Zambonin-Zallone

TABLE OF CONTENTS

I. INTRODUCTION

The ability of osteoclasts, as well as other cells, to adopt different shapes, polarize, coordinate specific movements, and adhere to the extracellular matrix depends on the cytoskeleton, a complex network of protein filaments that extends throughout the cytoplasm. In addition, osteoclast behavior is influenced by extracellular matrix molecules, which communicate with the cytoskeleton across the plasma membrane and modify fundamental cellular events. For these reasons, in this chapter we focus on osteoclast cytoskeletal organization and adhesion properties, pointing out their importance in the cell function. We also update some novel findings concerning the recognition of bone matrix and the influence that bone proteins have on osteoclast activity.

II. OSTEOCLAST MORPHOLOGY

Osteoclast morphology depends upon the functional status of the cell.[1] Considerable evidence indicates that to resorb bone, osteoclasts have to express functional polarization. Resting osteoclasts, noninvolved in bone resorption, show variable morphology, minor, if any, adhesion to the bone surface, and a diffuse distribution of their organelles. The plasma membrane arrays several folds along the entire perimeter of the cell, probably representing a membrane reservoir to be utilized when the cell activates.[2]

Osteoclasts engaged in bone resorption rapidly modify their morphology and become polarized cells. The plasmalemma organizes into the basolateral and apical membranes, causing the cell to functionally resemble a tight epithelium.[1] The basolateral membrane faces the vascular side and is in close proximity to the blood capillaries. The apical membrane faces the bone surface and is organized into two distinct domains. The inner domain consists of a number of finger-like membrane protrusions which, similar to microvilli, accomplish the task of increasing the membrane surface, and retain the capability of acidifying the extracellular space[3,4] and enabling receptor-mediated secretion and reuptake of lysosomal enzymes.[5,6] The peripheral domain represents the osteoclast-to-substrate adhesion apparatus, and is named "the sealing membrane". The cytoplasm above the sealing membrane consists of an organelle-free area containing polyribosomes and cytoskeletal proteins, and is called the "clear zone" or "sealing zone". The sealing membrane interacts with the bone surface[7] through specific adhesions, the podosomes,[8] which seal the external space beneath the cell where ruffled border folds spread and bone matrix dissolution takes place. This extracellular space is called the "resorbing compartment" or "resorption lacuna".

The shift of the osteoclast from the resting (nonpolarized) to the active (polarized) status has been proposed to be a key event in the regulation of cell activity.[9] Critical for this dynamism is the alteration of the organization of the cytoskeleton, which underlies subsequent morphological reorganization.

III. THE CYTOSKELETON

Movements of cells and subcellular structures and the generation of cell shape involve the organization of a complex set of protein fibers found in the cytoplasm, the cytoskeleton. Eukaryotic cells contain three major classes of cytoskeletal fibers: 7-nm-diameter actin microfilaments, 24-nm-diameter microtubules, and 10-nm-diameter intermediate filaments.[10-13]

Actin microfilaments and microtubules are formed by polymerization of protein subunits called G-actin and tubulin, respectively. The polymerization and depolymerization of these fibers are finely regulated and lead to the organization of permanent features of the cell (i.e., actin filaments in the contractile apparatus of muscle and in microvilli, and microtubules in cilia) or temporary structures (e.g., mitotic spindle) formed in certain functional stages. Most of the eukaryotic cells also contain one or more types of intermediate filaments, each of which is built from specific protein subunits.

The features of the cytoskeletal proteins in osteoclasts have been extensively studied *in vitro* in both avian and mammalian species. All three classes of cytoskeletal structures — microtubules, intermediate filaments, and microfilaments — have been characterized, and their features are discussed in the following sections.

A. MICROTUBULES

Microtubules are formed by tubulin, αβ-heterodimers (molecular weight, 100,000) that, although encoded by multiple genes, display equivalent functions. Their polymerization is a guanosine triphosphate (GTP)-dependent process and starts from specific microtubule organizing centers (MTOCs), amorphous regions surrounding the centrioles. Mature microtubules have a cylinder-like shape with a 4- to 5-nm-thick wall formed by 13 αβ, "head to tail"-arranged tubulin protofilaments. Microtubules form diverse structures in specialized cells. They may act as tracks along which organelles and vesicles move or, in proliferating cells, form the mitotic spindle for the segregation of chromosomes.

In cultured osteoclasts, microtubules form a solid framework consisting of bundles of radially oriented fibers which originate from single or multiple MTOCs[14,15] (Figure 1A). They radiate close, but not directly into, and often run parallel to the plasma membrane, but appear to be excluded from membrane ruffles, motile pseudopodia, and retraction fibers.[15] Microtubules may extend into cytoplasmic bridges which connect osteoclasts to wide cellular protrusions containing a nucleus surrounded by an organelle-containing cytosol.[14] These protrusions could correspond to fusing, or to detaching mononuclear cells which are seen to develop in close relationship with the osteoclast.

Concerning their role in osteoclast activity, similar to other cell types, we can hypothesize that microtubules may be involved in the movement of the cell and subcellular structures, and in morphological changes occurring

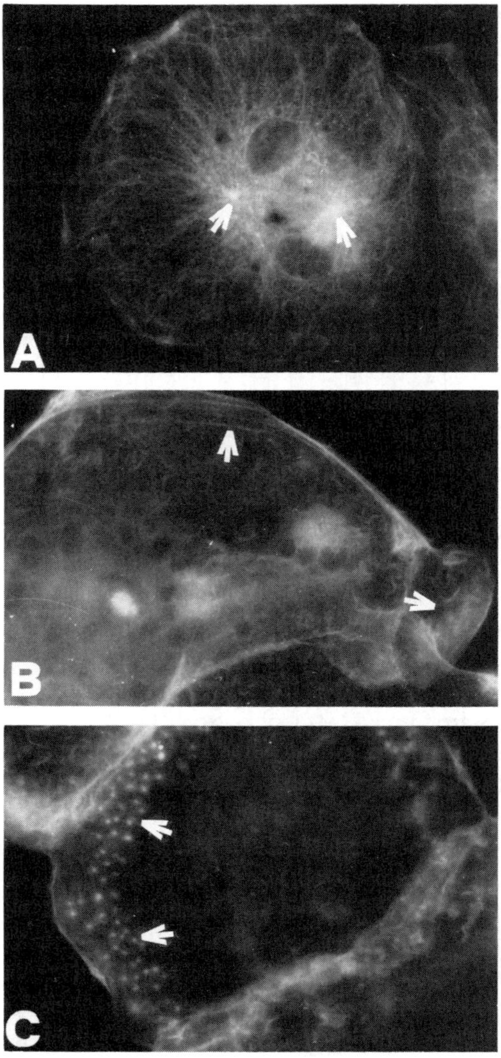

FIGURE 1. Chicken osteoclasts in culture have been evaluated for cytoskeletal proteins by fluorescence techniques. (A) Immunofluorescence staining of α- and β-tubulin. Microtubules appear to originate from microtubule-organizing centers (arrows) and radiate to the cell periphery, where they fold and run parallel to the cell edge, or return toward the cell center. (B, C) Decoration of microfilaments with rhodamine-conjugated phalloidin. In B, the osteoclast presents a motile aspect, with membrane rufflings and diffuse microfilament network (arrows). In C, the cell appears disk-shaped and microfilaments are mainly arranged in a punctate pattern at the cell edge (arrows), forming the axis of the podosomes (see text). (Original magnification × 1000.)

in different cellular phases. Experimental demonstrations in this regard are, however, still limited. For instance, retinol-treated osteoclasts show reversible microtubule depolymerization.[16] The significance of this effect is still unclear. However, since retinol has been demonstrated to increase bone resorption both *in vivo*[17] and *in vitro*,[16] we can speculate that these tubulin modifications could be related to activation of the cell function. However, despite morphological changes consisting of cell retraction and arrest of motility which occur in calcitonin-treated osteoclasts, microtubule appearance and arrangement in this circumstance are not altered.[15] Therefore, further experimental work is needed to elucidate the role of microtubules in osteoclast function.

B. INTERMEDIATE FILAMENTS

Intermediate filaments are key determinants of cellular structures and comprise at least five subclasses. Desmin connects adjacent Z-disks in muscle cells. Keratin interacts with spot desmosomes in epithelial cells. Neurofilaments and glial filaments form a framework in neurons and glial cells, respectively. Vimentin is present in cells of mesenchymal origin, such as fibroblasts and adipocytes. All the intermediate filaments have a similar structure. A highly conserved, 40-nm-long central rod-like domain is formed by coiling α-helical sections of two polypeptides around each other. The N- and C-terminal domains are globular. Dimers form antiparallel tetramers, which in turn form long protofilaments; eight of these form a 10-nm-diameter intermediate filament.

Intermediate filaments have been studied in avian osteoclasts *in vitro*. As expected in accordance with their mesenchymal origin from the leukocyte lineage, osteoclasts display an array consisting of vimentin filaments. Vimentin is a 57,000 molecular weight protein. In several cell types, vimentin fibers often terminate at the nuclear membrane and at desmosomes or adhesions on the plasma membrane, or are closely associated with microtubules. In osteoclasts, vimentin forms radially oriented filaments which terminate in a sort of reinforcing circumferential vimentin network in the paramarginal area.[18] This area, in well-developed, clear zone-presenting osteoclasts *in vitro*, contains the specialized structures that mediate osteoclast adhesion to the substrate (see below). Therefore, vimentin filaments may play a role in the interaction that occurs between the cell and the substrate to be resorbed. This possible role remains, however, to be elucidated.

C. MICROFILAMENTS

Microfilaments, or F-actin, are polymers of a globular protein subunit (molecular weight, 42 000) called globular actin or G-actin. Microfilaments are dumbbell shaped. Each actin subunit has a defined polarity, and the subunits polymerize head-to-tail. As a consequence, F-actin also has a defined polarity, and all subunits point in the same direction.

Microfilaments represent an important component of the osteoclast cytoskeleton, and their arrangement is modified according to the functional status of the cell. *In vivo*, microfilaments are particularly concentrated in the clear zone,[7] where, by decoration with heavy meromyosin, they have been found to form a fine network. Microfilaments are also organized in the axis of short protrusions of the plasma membrane, the podosomes,[8] which are specialized in the attachment of the sealing membrane to the bone surface (see below). From each axis, microfilamentous material extends laterally, linking together adjacent podosomes.

In vitro, microfilaments have been studied by immunofluorescence or by decoration of F-actin with fluorescent phalloidin. In mammalian as well as avian osteoclasts, a thin network of microfilamentous material has been described during the early stage of cell adhesion, particularly concentrated in membrane rufflings[19] (Figure 1B). In several osteoclasts, a peculiar dot-like organization subsequently evolves. Membrane ruffling diminishes and microfilaments organize into discrete bundles, 0.3 to 0.5 μm in diameter, perpendicular to the substrate.[18,20] These discrete bundles are initially diffusely distributed, then are predominantly arranged in several rings in the paramarginal area, in a zone resembling the sealing zone (Figure 1C). Broad stress-fibers are never observed. However, as in *in vivo* osteoclasts, short and thin stress fiber-like filaments may be seen connecting adjacent podosomes at the cell edge. Several reports[8,21-23] described a peripheral distribution of microfilaments in a clear zone-resembling area also in resorbing osteoclasts. Taken together, these observations demonstrate the importance of such actin organization in cell function.

Warshafsky et al.[15] demonstrated that filament organization is regulated in rat osteoclasts by the inhibiting hormone calcitonin. Calcitonin induces cell shrinking and arrests motility in disaggregated osteoclasts.[24] In this circumstance, F-actin extends in retraction fibers, which, however, exclude myosin. In osteoclasts as well as other cell types, myosin is associated with the microfilaments. Calcitonin also alters myosin distribution, inducing changes from a dotty or reticular pattern, which does not extend to the the cell periphery, to a more diffuse organization.[15] These results suggest that the cytoskeleton pattern of cultured osteoclasts is a target for regulation, and accounts for morphological modifications that occur when the cell changes shape and activity.

IV. ADHESION

Eukaryotic cells are capable of interacting with the surrounding environment.[25] Two classes of interactions may be distinguished: cell-cell and cell-matrix.

Adhesion of cells of a particular type is a primary feature of the architecture of many tissues. A number of cell-cell adhesion molecules are now known; a single cell type may use more than one of such molecules to mediate homotypic adhesion. In certain tissues, cell-cell adhesion leads to the organization of specific cell junctions that permit mechanical or chemical relationships. In bone tissue, cell-cell adhesion has been studied only from a morphological point of view and found to permit osteoblast-osteoblast, osteoblast-osteocyte, and osteocyte-osteocyte interaction via typical cell junctions.[26]

In osteoclasts, homotypic or heterotypic cell-cell interactions have not been extensively investigated. Osteoclasts do not form epithelioid sheets, as do osteoblasts; thus, mechanical interaction with other cells does not seem to be required. However, in specific circumstances, the capability of osteoclasts to form cell junctions has been described.[27] Medullary bone osteoclasts, at the end of a resorbing phase accentuated by a hypocalcemic diet, detach from the bone surface and migrate in the extracellular boundary in the hemopoietic bone marrow spaces. Such osteoclasts are surrounded by a number of mononuclear cells with which they establish specific cell junctions. Since the origin of these cells has not been elucidated, it is not clear whether such cell-cell interactions are homotypic or heterotypic, nor is its functional significance understood.

The second class of interaction occurs between cell and extracellular matrix both *in vivo* and *in vitro*. Many cells are capable of secreting extracellular proteins that serve as the adhesion substrate. For example, fibroblasts secrete most of the proteins present in the connective extracellular matrix and simultaneously display specific relationships with the same proteins.

A typical fibroblast adhesion is termed focal adhesion,[25] and consists of a wide area in which both cellular and extracellular proteins form a specific structure. Intracellularly, bundles of microfilaments, forming typical stress fibers, terminate close to the plasma membrane. Subunits of attachment proteins, such as α-actinin, vinculin, and talin, bind microfilaments and connect them to receptors for the extracellular matrix. These receptors are *trans*-membrane proteins capable of binding specific extracellular macromolecules and display, at the same time, recognition and adhesion properties.

As mentioned above, osteoclasts behave as polarized epithelia, since they are capable of segregating separated extracellular spaces. Unlike epithelial cells, however, osteoclasts fulfill this segregation through cell-matrix rather than cell-cell interactions. Such adhesion is unusual and is critical for specific cell function in bone resorption. The cell accomplishes the contact *in vitro* in two different ways. Freshly isolated osteoclasts have elevated membrane ruffling. Due to the similarity with what has been described for other cell types, including monocyte-macrophages, we interpret this as a motile aspect. Cells, in fact, present broad lamellipodia, probably extending in the direction of the movement, and retraction fibers are organized on the opposite side. The exact organization of the adhesions in such ''motile'' osteoclasts has not been investigated. However, due to the presence of a network of microfilaments at the level of lamellipodia,[19] it is reasonable to assume that this may be the site where adhesion to the substrate occurs. The molecular anatomy of such adhesion remains, however, to be elucidated.

Most cultured osteoclasts develop a second type of cell-matrix adhesion. This is represented by podosomes, modified focal adhesions displayed by cells that migrate and invade the extracellular matrix.[25,28] From a morphological point of view, podosomes correspond to short, foot-like protrusions of the ventral membrane, initially diffusely distributed and subsequently concentrated in the cell periphery.[18] By interference reflection microscopy, it has been observed that the membrane protrusion contacts the substrate.[18,20] In the maximal membrane-substrate vicinity, an extracellular space, less than 10 nm thick and containing an unknown material, is observed. The plasma membrane then withdraws, reaching a maximal distance from the substrate of more than 100 nm. An enormous number of protrusions are also present in the clear zone of an osteoclast. This could account for a sort of seal formed by a complicated labyrinth of cell-to-substrate contacts similar in function to the epithelial tight junctions. This peculiar "cell-matrix tight junction" probably is impermeable to most molecules and segregates two different compartments, the resorbing lacuna and the extracellular space.[9]

The adhesion property of podosomes has been confirmed by immunofluorescence study of their molecular anatomy.[9,16,20] The axis of the membrane protrusions is formed by F-actin microfilaments which (as mentioned above), due to their orientation perpendicular to the substrate in intact cultured cells viewed from the top, appear as dot-like structures. F-actin is associated with the actin-binding proteins fimbrin and α-actinin, which organize microfilaments in bundles, and with the regulatory protein gelsolin, which regulates filament length. The microfilamentous core is surrounded by a ring-like structure formed by α-actinin, vinculin, and talin, adhesion proteins involved in the attachment of the microfilaments to the plasma membrane.[25] These proteins are present in focal adhesions as well, but their organization in podosomes is quite unusual. In fact, in focal adhesions, they are superimposed to the terminal region of the microfilaments contacting the substrate. In podosomes, these proteins surround the terminal region of microfilaments, forming a ring-like structure.

Podosomes are dynamic structures which assemble and disassemble in cultured osteoclasts within minutes.[23] Since they represent the active part of the sealing zone that induces the tight cell-to-substratum interaction, we assume that the seal provided by these structures is dynamic as well. This suggests that, according to the functional demand, regulative phenomena may affect the adhesion property of the cell and the organization of the sealing membrane (see below). Since morphological changes occur when osteoclasts shift from the resting (nonpolarized) to the active (polarized) conformation, we assume that the regulation of the cytoskeletal and adhesion structures may represent a key event in the control of bone resorption.[9]

This hypothesis is supported by a consistent body of experimental evidence. As mentioned above, calcitonin, the inhibiting hormone which blocks bone resorption, alters osteoclast morphology and the cytoskeleton.[15] Cell retraction, as well as podosome disassembly, has also been observed in osteoclasts incubated in high-Ca^{2+}-containing medium.[29-31] These alterations are also related to changes in bone resorption, since inhibition of the resorbing activity has been demonstrated as well. By contrast, induction of cell polarization and increase of podosome number[19] have been noticed in osteoclasts stimulated to increase their resorptive function by a drop in cell pH[32] or by treatment with retinol.[16]

V. INTRACELLULAR SIGNALS REGULATING CYTOSKELETAL ORGANIZATION

According to recent reports, the cytoskeletal arrangement in osteoclasts is regulated by two different signal transduction mechanisms. These are consistent with the phosphoinositide-diacylglycerol pathway, activated by elevated $[Ca^{2+}]_o$,[33] and the adenylate cyclase pathway, activated by calcitonin.[34]

Both pathways are receptor mediated and G-protein dependent. The former is a dual-branched pathway which induces the breakdown of phosphoinositol-4,5-bisphosphate in inositol-1,4,5-trisphosphate (IP_3) and diacylglycerol (DAG). IP_3, through a specific receptor-mediated mechanism, induces the release of Ca^{2+} from intracellular deposits and causes an increase of the cytosolic free calcium concentration, $[Ca^{2+}]_1$. This, in turn, acts as a second messenger activating a number of Ca^{2+}-sensitive mechanisms, including Ca^{2+}/calmodulin-dependent Ca^{2+}-ATPase, which rapidly restores the basal $[Ca^{2+}]_i$. DAG activates protein kinase C (PKC), which in turn phosphorylates serine/threonine-containing proteins and affects cellular events. The adenylate cyclase pathway depends upon receptor-mediated activation of the membrane adenyl cyclase, which in turn converts ATP into cyclic AMP (cAMP). cAMP activates cAMP-sensitive protein kinase (PKA), which also phosphorylates serine/threonine-containing proteins, inducing changes in their function.

All these second messengers — $[Ca^{2+}]_i$, PKC, and cAMP — affect cytoskeletal and adhesion proteins. Elevated $[Ca^{2+}]_o$, in fact, promotes podosome disassembly and the organization of microfilaments into a thin network, particularly concentrated in membrane ruffles.[31] PKC induces overassembly of microfilaments in large bundles which are present in the podosomes, or spread throughout the cytoplasm as stress fibers.[45] Finally, cAMP induces cell retraction and modifications of actin and myosin distribution.[15,24] Interestingly, all these modifications are related to the inhibition of bone resorption.[31,35,45]

VI. RECOGNITION OF BONE MATRIX

Recognition of extracellular proteins occurs via cell membrane receptors belonging to the family of integrins.[36] Integrins are complicated arrays of $\alpha\beta$-heterodimers in which the α- and β-subunits may arrange according to several combinations, and account for an enormous number of specific receptors for extracellular matrix proteins. Osteoclasts present at least two different integrins, a β_1-containing integrin diffusely distributed in the plasma membrane[37] and the $\alpha_v\beta_3$ or vitronectin receptor[38] located at the level of the podosomes.[37,39]

The presence of the β_1-containing integrin belonging to the collagen receptor family suggests that osteoclasts may recognize collagen. Collagen, the most abundant constituent of bone matrix, is indeed specifically recognized at least by chicken osteoclasts *in vitro*.[45]

The $\alpha_v\beta_3$ receptor recognizes the extracellular protein vitronectin. This protein is highly concentrated in serum, but is not present in bone. However, in other cell types such as platelets[40] and osteosarcoma cells,[41] the $\alpha_v\beta_3$ integrin has been found to be a promiscuous receptor capable of recognizing more than one extracellular protein. In osteosarcoma cells, for example, this receptor recognizes bone sialoprotein II (BSP II),[41] a protein widely represented in bone matrix.

Osteoclasts seem to employ this receptor to recognize at least two bone matrix proteins, osteopontin and BSP II.[42] Interestingly, osteoclasts are also capable of synthesizing BSP II,[43] indicating that they participate in the production of their own adhesive substrate. Both proteins, after binding with their receptor, are capable of activating osteoclast function, altering the $[Ca^{2+}]_i$.[42]

This body of evidence indicates that the critical stimulus for osteoclasts to resorb bone is present in the bone matrix itself. Experimental data, in fact, indicate that osteoclasts are programed to resorb bone in the absence of exogenous factors, except for contact with an available bone surface. Unstimulated osteoclasts, cultured *in vitro* in serum-free media,[31] spontaneously and efficiently digest bone matrix. This suggests that the critical condition for bone resorption to take place is the presentation of a cell-free bone surface, as has been demonstrated *in vitro*.[44] Exogenous factors could, therefore, influence bone resorption by modulating this spontaneous resorbing function, including the ability to change the polarized conformation. Furthermore, these factors could also affect osteoclast differentiation from committed precursors. Although fascinating, this hypothesis still needs experimental support.

VI. CONCLUSIONS

Osteoclast function is dependent upon the cytoskeleton and adhesion, which play important roles in affecting both cell morphology and activity, and represent critical targets for cell regulation. Further work is, however, necessary to fully elucidate these roles.

REFERENCES

1. **Baron, R.,** Molecular mechanisms of bone resorption by the osteoclast, *Anat. Rec.,* 224, 317, 1989.
2. **Miller, S. C.,** Osteoclast cell-surface changes during egg-laying cycle in Japanese quail, *J. Cell Biol.,* 75, 104, 1977.
3. **Baron, R., Neff, L., Louvard, D., and Courtoy, P. J.,** Cell-mediated extracellular acidification and bone resorption: evidence for a low pH in resorbing lacunae and localization of a 100-kDa lysosomal membrane protein on the osteoclast ruffled border, *J. Cell Biol.,* 101, 2210, 1985.
4. **Blair, H. C., Teitelbaum, S. L., Ghiselli, R., and Gluck, S.,** Osteoclastic bone resorption by a polarized vacuolar pump, *Science,* 245, 855, 1989.
5. **Baron, R., Neff, L., Brown, W., Courtoy, P. J., Louvard, D., and Farquhar, M. G.,** Polarized secretion of lysosomal enzymes: co-distribution of cation-independent mannose-6-phosphate receptors and lysosomal enzymes along the osteoclast exocytic pathway, *J. Cell Biol.,* 106, 1863, 1988.
6. **Blair, H. C., Teitelbaum, S. L., Schimke, P. A., Konsek, J. D., Koziol, C. M., and Schlesinger, P. H.,** Receptor mediated uptake of a mannose-6-phosphate bearing glycoprotein by isolated chicken osteoclasts, *J. Cell. Physiol.,* 137, 476, 1988.
7. **Holtrop, M. E. and King, G. J.,** The ultrastructure of osteoclast and its functional implications. *Clin. Orthop. Relat. Res.,* 123, 177, 1977.
8. **Zambonin-Zallone, A., Teti, A., Carano, A., and Marchisio, P. C.,** The distribution of podosomes in osteoclasts cultured on bone laminae: effect of retinol, *J. Bone Miner. Res.,* 3, 517, 1988.
9. **Teti, A., Marchisio, P. C., and Zambonin-Zallone, A.,** The clear zone in the osteoclast function. The role of podosomes in the regulation of the bone resorbing activity, *Am. J. Physiol.,* in press.
10. **Bershadsky, A. D. and Vasiliev, J. M.,** *Cytoskeleton,* Plenum Press, New York, 1988.
11. **Lackie, J. M.,** *Cell Movement and Cell Behavior,* Allen and Unwin, London, 1985.
12. **Schliwa, M.,** *The Cytoskeleton,* Springer-Verlag, New York, 1986.
13. **Sheterline, P.,** *Mechanisms of Cell Motility,* Academic Press, London, 1983.
14. **Zambonin-Zallone, A., Teti, A., Primavera, M. V., Naldini, L., and Marchisio, P. C.,** Osteoclasts and monocytes have similar cytoskeletal structures and adhesion property *in vitro. J. Anat.,* 137, 57, 1983.
15. **Warshafsky, B., Aubin, J. E., and Heersche, J. N. M,.** Cytoskeleton rearrangements during calcitonin-induced changes in osteoclast motility, *Bone,* 6, 179, 1985.
16. **Oreffo, R. O. C., Teti, A., Francis, M. J. O., Triffitt, J. T., Carano, A., and Zambonin-Zallone, A.,** Effect of vitamin A on bone resorption: evidence of direct stimulation of isolated chicken osteoclasts by retinol and retinoic acid, *J. Bone Miner. Res.,* 3, 203, 1988.
17. **Barnicott, N. A. and Datta, S. P.,** Vitamin A and bone, in *Biochemistry and Physiology of Bone,* Bourne, G. H., Ed., Academic Press, New York, 1972, 197.
18. **Marchisio, P. C., Naldini, L., Cirillo, D., Primavera, M. V., Teti, A., and Zambonin-Zallone, A.,** Cell-substratum interaction of cultured avian osteoclasts is mediated by specific adhesion structures, *J. Cell Biol.,* 99, 1696, 1984.
19. **Teti, A., Blair, H. C., Schlesinger, P. H., Grano, M., Zambonin-Zallone, A., Kahn, A. J., Teitelbaum, S. L., and Hruska, K. A.,** Extracellular protons acidify osteoclasts, reduce cytosolic calcium and promote expression of cell-matrix attachment structures, *J. Clin. Invest.,* 84, 773, 1989.
20. **Marchisio, P. C., Cirillo, D., Teti, A., Zambonin-Zallone, A., and Tarone, G.,** Rous sarcoma virus transformed fibroblasts and cells of monocytic origin display a peculiar dot-like organization of cytoskeletal proteins involved in microfilament-membrane interactions, *Exp. Cell Res.,* 169, 202, 1987.

21. **Turkesen, K., Kanehisa, J., Opas, M., Heersche, J. N. M., and Aubin, J. E.,** Adhesion patterns and cytoskeleton of rabbit osteoclasts on bone slices and glass, *J. Bone Miner. Res.*, 3, 389, 1988.
22. **Lakkakorpi, P., Tuukkanen, J., Hentunen, T., Jarvelin, K., and Vaananen, H. K.,** Organization of osteoclast microfilaments during the attachment to bone surface *in vitro, J. Bone Miner. Res.*, 4, 817, 1989.
23. **Kanehisa, J., Yamanaka, T., Doi, S., and Heersche, J. N. M.,** A band of F-actin containing podosomes is involved in bone resorption by osteoclasts, *J. Bone Miner. Res.*, 11, 287, 1990.
24. **Chambers, T. J., Chambers, J. C., Symond, J., and Darby, J. A.,** The effect of human calcitonin on the cytoplasmic spreading of rat osteoclasts, *J. Clin. Endocrinol. Metab.*, 63, 1080, 1986.
25. **Burridge, K., Fath, K., Kelly, T., Nuckolls, G., and Turner, C.,** Focal adhesions: transmembrane junctions between the extracellular matrix and the cytoskeleton, *Annu. Rev. Cell Biol.*, 4, 487, 1988.
26. **Kahn, A., Fallon, M. D., and Teitelbaum, S. L.,** Structure-function relationships in bone: an examination of events at the cellular level, in *Bone and Mineral Research, Annual 2,* Peck, W. A., Ed., Elsevier, Amsterdam, 1984, 125.
27. **Zambonin Zallone, A. and Teti, A.,** The osteoclasts of hen medullary bone under hypocalcaemic conditions, *Anat. Embryol.*, 162, 379, 1981.
28. **Chen, W.-T.,** Proteolytic activity of specialized surface protrusions formed at rosette contact sites of transformed cells, *J. Exp. Zool.*, 251, 167, 1989.
29. **Malgaroli, A., Meldolesi, J., Zambonin Zallone, A., and Teti, A.,** Control of cytosolic free calcium in rat and chicken osteoclasts. The role of extracellular calcium and calcitonin, *J. Biol. Chem.*, 264, 14342, 1989.
30. **Zaidi, M., Datta, H. K., Patchell, A., Moonga, B., and MacIntyre, I.,** Calcium-activated intracellular calcium elevation: a novel mechanism of osteoclast regulation, *Biochem. Biophys. Res. Comun.*, 163, 1461, 1989.
31. **Miyauchi, A., Hruska, K. A. , Greenfield, E. D., Duncan, R., Alvarez, J., Barattolo, R., Colucci, S., Zambonin-Zallone, A., Teitelbaum, S. L., and Teti, A.,** Osteoclast cytosolic calcium, regulated by voltage gated calcium channels and extracellular calcium, controls podosome assembly and bone resorption, *J. Cell Biol.*, 111, 2543, 1990.
32. **Arnett, T. R. and Dempster, D. W.,** Effect of pH on bone resorption by rat osteoclasts *in vitro, Endocrinology,* 119, 1986.
33. **Miyauchi, A., Greenfield, E. M., Alvarez, J., Bar-Shavit, Z., Huskey, M., Teitelbaum, S., and Hruska, K.,** The osteoclast calcium/divalent cation receptor stimulates subcellular Ca^{2+} gradients and phospholipase C activation, *J. Bone Miner. Res.*, 5 (Suppl.) (Abstr.), 299, 1990.
34. **Nicholson, G. C., Moseley, J. M., Sexton, J. M., Mendelson, F. A. O., and Martin, T. J.,** Abundant calcitonin receptors in isolated rat osteoclasts, *J. Clin. Invest.*, 78, 355, 1986.
35. **Chambers, T. J., McSheey, P. M. J., Thomson, B. M., and Fuller, K.,** The effect of calcium-regulation hormones and prostaglandins on bone resorption by osteoclasts disaggregated from neonatal rabbit bones, *Endocrinology,* 116, 234, 1985.
36. **Ruoslahti, E. and Pierschbacher, M. D.,** New perspectives in cell adhesion: RGD and integrins, *Nature (London),* 238, 491, 1987.
37. **Zambonin-Zallone, A., Teti, A., Grano, M., Rubinacci, A., Abbadini, M., Gaboli, M., and Marchisio, P. C.,** Immunocytochemical distribution of extracellular matrix receptors in human osteoclasts: a β_3 integrin is colocalized with vinculin and talin in the podosomes of osteoclastoma giant cells, *Exp. Cell Res.*, 182, 645, 1989.
38. **Davies, J., Warwick, J., Totty, N., Philip, R., Helfrich, M., and Horton, M.,** The osteoclast funtional antigen, implicated in the regulation of bone resorption activity, is biochemically related to the vitronectin receptor, *J. Cell Biol.*, 109, 1817, 1989.

39. **Sato, M., Sardana, M. K., Grasser, W. A., Garsky, V. M., Murray, J. M., and Gould, R. J.,** Echistatin is a potent inhibitor of bone resorption, *J. Cell Biol.*, 111, 1713, 1990.
40. **Nurden, A. T., George, J. N., and Phillips, D. R.,** in *Biochemistry of Platelets*, Phillips, D. R. and Scuman, M. S., Eds., Academic Press, New York, 1988, 111.
39. **Oldberg, A., Franzen, A., and Heinegard, D.,** Identification of a bone sialoprotein receptor in osteosarcoma cells, *J. Biol. Chem.*, 263, 19433, 1988.
42. **Teti, A., Ross, P., Miyauchi, A., Teitelbaum, S., Hruska, K., Alvarez, J., Grano, M., Colucci, S., Gehron-Robey, P., and Zambonin-Zallone, A.,** RGD containing sequences of osteopontin (RGD-OP) and BSP II (RGD-BSP) produce immediate cell signals and affect osteoclast (OC) adhesion, *J. Bone Miner. Res.*, 5 (Abstr.), 7, 1990.
43. **Bianco, P., Fisher, L. W., Young, M. F., Termine, T. J., and Gehron-Robey, P.,** Expression of bone sialoprotein in human developing tissues, *Calcif. Tissue Int.*, in press.
44. **Zambonin-Zallone, A., Teti, A., and Primavera, M. V.,** Resorption of vital and devitalized bone by isolated osteoclasts *in vitro*, *Cell Tissue Res.*, 235, 561, 1984.
45. **Teti, A. and Zambonin-Zallone, A.,** unpublished observations.

Chapter 13

THE MECHANISM OF OSTEOCLAST ACIDIFICATION

Harry C. Blair and Paul H. Schlesinger

TABLE OF CONTENTS

Man is but earth;
'Tis true, but earth is the centre.
John Donne

I. INTRODUCTION

The correct function of bone is vital for life, both as the supporting structure for the body and as the reservoir for the body's calcium. However, many of the bone's cellular processes, including its modeling and remodeling, have remained enigmatic, while complex cellular processes of other organs have been clarified. On the other hand, bone matrix, without its cellular constituents, is relatively simple as biological products go, being composed almost exclusively of type I collagen and a hard, amorphous phosphate and hydroxide salt of calcium, hydroxyapatite. By considering the chemistry of this inorganic matrix, the cellular process that places bone matrix in solution may be shown to require net proton transport. This chapter shows how this reasoning developed and has been confirmed by direct demonstration of osteoclastic acid transport, and reviews recent work extending this line of reasoning to identification of specific biochemical steps that, in sum, produce the vectorial net acid transport required. These newly identified biochemical processes will, it is hoped, in turn provide quantifiable probes for study of the remaining central questions of bone physiology, which include such basic issues as what differentiates an osteoclast from other cells capable of acid-dependent degradative processes, and how is bone resorption controlled? The answers to these questions may, in turn, allow better and systematic treatment of crippling diseases. Indeed, before concluding, we discuss proposals for use of what has been learned about osteoclastic acid transport for rational drug design and testing *in vitro*. The acid transport data presented here, although still incomplete, are testament to decades of dedicated work by dozens of researchers besides ourselves who have struggled to understand a delicate cell that cannot, in its mature form, be propagated in tissue culture and which, in its natural place, is surrounded by what amounts to solid rock.

II. HISTORY OF THE ACID BONE RESORPTION THEORY

The acid secretion hypothesis of bone mineral dissolution developed incrementally, and hence its origin cannot be placed with certainty, but some mention should be made of the pioneering work that led to its development. This, incidentally, has also provided us with the sometimes quaint terminology still used to describe the process of bone degradation. Appreciation that bone is absorbed as part of normal development is generally attributed to the great British surgeon John Hunter (1728-1793).[1,2] One of Hunter's contemporaries, William Cruickshank (1745-1800), an anatomist with particular interest in absorptive processes, considered the process of bone absorption at length. He studied anatomy by experimental methods, his principal technique being mercury injection to trace vascular and lymphatic connections. Cruickshank's opinion, on completing his studies, was that absorption of bone differed from lymphatic absorption, and he became perhaps the first proponent of acid degradation by stating in his great work, *Anatomy of Absorbing Vessels* (1790):[3]

> I have observed, in internal exfoliations of cylindric bones, where a very large portion of the old bone has died throughout, that the cancelli were destroyed, and that the bone must have been thinned as much upon the inside as it was upon the outside; nearly the whole length of the tibia was in this way reduced to a tube whose sides were not thicker than a wafer, and resembled portions of cylindric bones which had been exposed for some time to the gastric juice in the stomach of a leopard.

John Howship (1781-1841; *Howship's lacuna*) systematized bone pathology and studied its physiology with astonishing sophistication, describing, in 1816, metabolic disease in growing bone as "dependent upon the more or less perfect removal of the phosphate of lime from the ossific texture".[2] He recognized, without the benefit of effective staining techniques, the importance of the membranous (cellular) component in bone removal, described roughened regions of bone in regions being absorbed, and, regarding the mechanism for bone removal, quoted Cruikshank[3] that "previous to the absorption of a solid, the parts immediately to be absorbed may be broke[n] down, mixed with, or even converted into fluids".

Rudolf Virchow (1821-1902) first clearly stated the cellular nature of bone and established, through observations of growth and pathological resorption, that cells lay down a "basis substance, or osteoid" to which mineral is later added, and the "very much enlarged osseous corpuscles" appear with excavations of bone. He first demonstrated use of specific strong acids, including hydrochloric acid, to artificially demineralize sections, simultaneously advancing the study of bone degradation and making useful histological sections of bone possible, as reported in his *Cellular Pathology,* published in 1858.[4] Virchow's achievement was remarkable considering the limited biochemical knowledge available at the time, although in a number of respects his detailed conclusions from histology have fared less well, including that of the hypothetical derivation of the osteoclast from the osteocyte. A direct statement of the hypothesis that the osteoclast is the bone-degrading cell (and the modern name of the cell itself, οστοχλαστηζ, the bone-breaker, formerly known as Robin's myeloplax or Virchow's Riesenzellen) was the work of Albert Kölliker, a Würzburg anatomist. Kölliker used bones of animals fed madder (containing alizarin dyes, coloring the bone matrix red) to demonstrate that the osteoclast removed bone matrix, publishing his results in 1873.[1] However, it would be another three quarters of a century before the elements of acid demineralization and cellular physiology were finally merged.

Further early findings in bone and degradative cells that are important background to current theory were Metchnikoff's work on macrophage function and acidification of the phagosome in the final quarter of the nineteenth century,[5] and the discovery of the necessity of vitamin D for bone remodeling in the first quarter of the twentieth century,[6] which overlapped with the discovery of parathyroid hormone (PTH) and its modulation of cell-mediated bone resorption and osteoclast number.[7] These facts, though not obviously related (and certainly not thought to be related when discovered), together set the stage for understanding the osteoclast as a regulated, macrophage-like, acid-producing cell derived from a circulating precursor.

Contemporaneous with the identification, in increments, of cell-mediated acidic degradative processes and the role of vitamin D and PTH in controlling bone degradation during approximately 1880 to 1930, accurate analysis of electrolytes was developed and applied to bone metabolism. One consequence of this was discovery of increased calcium excretion in the tragic natural model of diabetic ketoacidosis, then untreatable,[8] and in experimental acid treatment of animals,[9] which established that negative calcium and phosphate balance is seen with extreme acid loading of the body. These findings were further refined by study of mineral balance in humans fed acid equivalents as NH_4Cl, which caused calcium loss[10] and, when taken to the extreme in animals, caused osteoporosis, an effect identified by Jaffe and Bodansky as being PTH mediated.[11] However, before the mid-twentieth century, the cellular mechanisms presiding over bone resorption were not directly measurable. Therefore, the hypothesis that systemic acid loading and osteoclast PTH-mediated bone resorption both resulted from increased proton transport for bone resorption at the cellular level, in that bone is the body's largest store of both base equivalents and calcium, awaited experimental verification.

III. BIOCHEMISTRY OF BONE MATRIX DETERMINES REQUIREMENTS FOR MATRIX DISSOLUTION

By the 1950s, progress in biochemistry and physical chemistry provided a sophisticated statement of bone matrix content, organic and inorganic. By this time, it was clear that bone mineral is essentially hydroxyapatite, $(Ca_3(PO_4)_2)_3Ca(OH)_2$, and that the major organic extracellular component is collagen. The formula for hydroxyapatite can be written in different, but equivalent, ways, e.g., $Ca_{10}(PO_4)_6(OH)_2$, and is, as applied to bone, approximate.[12] However, as regards apatite crystals, the structure has been definitively determined by X-ray diffraction.[13] It bears mention, however, that some of the minor inorganic components of bone matrix were also established in the 1950s, including carbonate, fluoride, and citrate anions, which may replace anions of hydroxyapatite. Similarly, magnesium, radium, and strontium (discovered as an unpleasant consequence of the production of ^{90}Sr in atomic fission) are divalent cations that appear as replacements for calcium in the crystal lattice.

These, and other more recently described substitutions, replacements in the hydroxyapatite matrix or at the crystalline surface are all lesser, stoichiometrically unimportant components of bone. As minor components, they are of no consequence to the overall buffering capacity of the skeleton. Hence, their importance lies largely in their potential effects on surrounding cells, perhaps by carcinogenesis or by physical effects on the bone's function. In the case of mineral dissolution, any such bone-bound substances would be released locally. If the bone-bound substance were an inhibitor of an osteoclastic process, then the rate of bone resorption would be decreased. The first recorded attempt to exploit this mechanism was the use of strontium to treat osteoporosis in the early 1950s, an approach abandoned due to reported association with nephrolithiasis.[14,15] On the other hand, as discussed in detail below after description of the specific biochemical processes of osteoclast acid secretion, newer bone-binding inhibitory compounds are promising candidates for pharmacological control of bone absorption.

Understanding of the physical nature of bone mineral allowed its solubility properties to be established. Replacement of Ca^{2+} by $2H^+$ in crystal dissolution was demonstrated, and the physical possibilities for crystal dissolution were elegantly stated by Neuman as "changes in ion composition altering solubility, addition of chelators, or reduced environmental pH".[12] The central fact, however (now, in retrospect, seeming obvious), is that by whatever means bone matrix is to be dissolved, the physical formula of the inorganic salts require that ~1.5 equivalents of base be neutralized for each mole of Ca^{2+} placed in solution, if the final solution is to have an approximately neutral pH, since OH^- and PO_4^{3-} cannot exist in solution at pH 7 in significant quantities. Thus, at some point in bone resorption, a massive net acid equivalent production, equal to 1.5 times the calcium solubilized, *must* take place, whether as an organic acid like citrate or as a direct proton production (see below).

IV. THE ACIDIC BONE DEGRADATION SITE: MICROSCOPIC, HISTOCHEMICAL, AND PHYSICAL DATA

In 1947, Barnicot showed neutral red uptake by osteoclasts, which was attributed to the lysosomal content of the cells, although in retrospect his photographs appear to clearly demonstrate uptake of this basic (acidophilic) dye at the entire cell-matrix contact.[16] In 1951, Cretin more correctly interpreted this result when he reported that use of supravital staining showed the osteoclastic attachment site to be acidic and the osteoblastic attachment alkaline.[17] Cretin published a series of careful studies of bone histochemistry, pathology, and experimental observations at the cellular level from 1923 to 1961; his findings are now rarely credited, although it has in recent years become clear that his observations were, in some respects, ahead of the field. Skepticism concerning the possibility of extracellular acidification caused other possibilities for control of bone mineral flux to be extensively explored for over 30 years before a return to the simple acid secretion model.

Advances in histochemistry and microscopy added corollary data concerning osteoclastic acid secretion during the 1950s and 1960s. Direct experimental analysis of many subcellular transport processes, including H^+ transport, was not yet possible, but important work proceeded with tissue explant and whole animal models. The electron microscope had by 1961 revealed that the osteoclast's oddly staining bone-apposed "brush" border was in fact a region of highly convoluted cell membrane renamed, appropriately, the ruffled membrane,[18] which has lately been characterized as the cell's unique acid-secreting organ, as is discussed below in more detail. The new antibiotic tetracycline was found by Frost (1962) to bind to calcium phosphate at the site of bone mineralization.[19] Tetracycline, which in addition to binding calcium is a weak base (i.e., acidophilic), is mentioned in the context of proton-mediated bone resorption (rather than its usual role, measurement of bone deposition rate) because recent work suggests that it may be useful in controlling bone resorption. Rifkin et al. have shown that tetracyclines inhibit ^{45}Ca release from bone by osteoclasts (i.e., acid secretion, *vide infra*) in a dose-dependent manner unrelated to PTH-mediated recruitment of osteoclasts.[20] Tartrate-resistant acid phosphatase activity as an osteoclastic marker was developed.[21] This enzyme, with its low pH optimum, was localized, by the new technique of electron microscopy of histochemical reaction products, to the osteoclast's ruffled border and vacuolar system,[22] providing compelling, though indirect, evidence for low pH at the bone resorption site.

Beginning in the mid-1950s, considerable experimental work was performed regarding secretion of weak acids such as citrate in bone. These were hypothesized to be capable of mineral solubilization *in vivo* by calcium chelation at bone resorption sites.[23] This hypothesis has been supplanted in recent years by the transmembrane proton transport model, for which the evidence is outlined below. On the other hand, the recent work focused on H^+ transport should not cause this earlier work to be forgotten: the trivalent organic acid citrate dramatically increases the solubility of hydroxyapatite at concentrations as low as 2 m*M*; conversely, fluoride substitution substantially increases the resistance of apatite to acid dissolution.[24] These facts have, to date, not been of significant practical use in medical control of bone mass, but remain potentially important. Similar crystal absorption and solubility effects of pyrophosphates led to the discovery of the skeletal effects of the bisphosphonates in the early 1960s.[25,26] However, as discussed in detail elsewhere in this volume, although pyrophosphates affect hydroxyapatite deposition at 2 to 3 μM, the solubility effects of the bisphosphonates occur at millimolar concentrations, while, *in vivo*, the metabolic inhibition of bone resorption, at micromolar bisphosphonate solution concentrations, is largely due to osteoclastic acid-mediated release of bone-bound bisphosphonates.[27]

Thus, by about 1970, there was a clear reason to expect, from the chemistry of bone matrix and observations of bone cells, that net acid transport was central to bone catabolism. However, traditional methods for the study of bone were not suitable for application of modern biochemical methods to dissect the process further. This suggested the need for a suitable experimental system for definitive study of osteoclast degradative mechanisms, the limiting factor being the availability of isolated bone-degrading cells. The degradative role of the osteoclast and its morphological and histochemical features, as well as its origin from a circulating precursor,[28] suggested that the macrophage would be useful as a model for osteoclast activity. Indeed, peritoneal macrophages,[29] and the HL60 line of myelomonocytic precursor cells which in the presence of 1,25-dihydroxyvitamin D_3 differentiate into bone-degrading, osteoclast-like multinucleated cells,[30] provided considerable information concerning the bone degradation process. While these findings have been largely superseded by study of isolated osteoclasts, recent description of cell surface H^+ secretion processes by macrophages resembling those of osteoclasts,[31,32] as well as recent studies of osteoclast ontogeny suggesting a relationship between the macrophage and osteoclast,[33-35] indicate that macrophage-derived, bone-degrading cells are a good model for osteoclastic activity. Results from these studies relevant to current theory of osteoclastic acid secretion include increased efficiency of bone degradation with cell size in multinucleation[36] and dissociation of bone degradation from enzyme secretion into cell supernatants.[37] Other workers used bone explants or whole animals to study bone metabolism, but little detailed information bearing on osteoclast physiology resulted, since direct biochemical measurements on explants result from summed effects on several types of cells. Exceptions included Cretin's early study using indicator dyes, discussed above, and studies suggesting a relationship between systemic pH and rate of bone resorption,[38] which have added some detail to the earlier *in vivo* studies.[8-11] However, with the introduction of techniques by which bona fide osteoclasts could be isolated, by several workers (including Zambonin-Zallone, Teti, and Primavera; Athanasou, Chambers et al.; and Caplan, Osdoby, and Martini),[39-41] the direct study of osteoclastic physiology and biochemistry at the subcellular level was greatly facilitated. On the other hand, a technique by which murine cells indistinguishable from osteoclasts can be generated *in vivo* from marrow monocyte precursors has recently been developed[42,43] which, if confirmed, may provide a source of mammalian osteoclasts in large numbers, with distinct advantages over the use of primary cultures of isolated avian or mammalian cells.

Michael Fallon, using a microprobe, achieved direct physical measurements of acid pH at the osteoclastic attachment site in 1984,[44] one of many accomplishments of a brilliant young physician-scientist whose recent, untimely death is a great loss to the field. The use of the weak base acridine orange to demonstrate osteoclastic attachment *in situ,* reported shortly thereafter independently by Green et al.[45] and Baron et al.,[46] has proved to be of considerable value in studying the mechanism of osteoclast acidification due to the concentration of the (basic) dye in acidic compartments and its fluorescence, which makes it easily quantifiable.[27] At the same time, using newly available isolated osteoclasts, it was established that osteoclasts degrade both the organic and inorganic components of bone by a mechanism involving acidification.[47] This result correlated well with the results of Vaes and coworkers, which had shown that inhibitors of lysosomal proteases eliminated bone collagen degradation.[48,49]

V. BIOCHEMICAL ELEMENTS CONSTITUTING VECTORIAL HCl TRANSPORT BY THE OSTEOCLAST, AND CELLULAR PROCESSES THAT ARE LINKED TO ACID TRANSPORT

A. THE MEMBRANE H^+ TRANSPORTER

Three major types of acid-secreting transmembrane transport processes were described as mediating cellular pH homeostasis or H^+ transport in most organs other than bone. These were thought to constitute the main possibilities for the osteoclast's H^+-secreting activity. These consisted of Na^+-H^+ exchange, a nonenergy-dependent process that functions in most cell types to control intracellular pH;[50,51] the gastric-type (E_1E_2) H^+,K^+-ATPase;[52] and the vacuolar (F_1F_0-like) electrogenic H^+-ATPase.[53]

Initial studies by immunoelectron microscopy showed that an antigen related to the gastric H^+,K^+-ATPase was expressed at the cell's ruffled membrane, suggesting to Reggio et al. that the process might be analogous to H^+ secretion by the stomach.[54] Ouabain binding and immunoelectron microscopic evidence suggesting presence of high concentrations of Na^+,K^+-ATPase on the osteoclast's cell membrane was presented as a further argument for a cation exchanger by Baron's group.[55] However, while recent electrophysiological work has shown an inward rectifying K^+ conductance in osteoclast cell membranes,[56] direct measurements of H^+ transport properties of osteoclasts have shown no measurable potassium-dependent proton production, as is discussed in more detail below.

Regarding the possibility of an H^+-Na^+ exchanger on the osteoclast, there is more evidence: Hall and Chambers have shown that amiloride inhibits pit formation by osteoclasts,[57] Teti et al. found that recovery from intracellular acidification of osteoclasts was reduced in the absence of Na^+.[58] On the other hand, amiloride is also a potent inhibitor of protein synthesis,[59] suggesting that the effect seen by Hall and Chambers[57] may not reflect solely the effect of the drug on acid production. The other data supporting the presence of a sodium-hydrogen exchanger on osteoclasts[58] are related to regulation of intracellular pH, a typical role for the Na^+-H^+ exchanger,[50,51] rather than to net vectorial proton pumping.

In contrast, direct measurements of proton translocation by osteoclast membrane vesicles, which was made possible by the Zambonin-Zallone et al. technique for isolation of gram quantities of greatly purified osteoclasts from calcium-depleted laying hens, has produced definitive results showing that the major proton-transporting process of the osteoclast is a vacuolar-type electrogenic (i.e., translocating protons without balancing anions) H^+-ATPase.[60] This pump, related to F_1F_0-ATPases, was known, from the elegant work of Al-Awqati and Gluck,[61,62] to be responsible for net vectorial H^+ transport in the kidney, where its activity is regulated by pCO_2.[63] In other systems, however, $[Ca^{2+}]$ is known to be a principal regulator of vacuolar-type proton pumps, suggesting that the pump is differently regulated in various organs.[64] Isolation of this complex (i.e., multisubunit) proton pump has been achieved by Bowman and Bowman et al. from *Neurospora*[65] and by Gluck et al. from bovine kidney.[66-68] Antibodies to the bovine vacuolar H^+-ATPase made it possible to directly confirm the osteoclast vesicular transport measurements using Western blotting and immune staining.[60] Furthermore, these results have been independently confirmed by Gay's group using a different osteoclast membrane preparation,[69] and by Väänänen et al., who used the antibodies of the Bowmans et al. for immunoelectron microscopy.[70]

These results raised more intellectual problems than they solved, including how is cellular acid-base balance reconciled with acid secretion, what anion balances the electrogenic pump, and why does the cell also express a rectifying inward K^+ conductance? Demonstration of vacuolar H^+-ATPase as the driving force behind net acid transport directly implied at least two additional transport processes. This is because basic physical theory requires that the cell must secrete as much base as acid to remain at its physiologic pH and, similarly, it must transport balanced quantities of anions and cations to retain electrical balance. Furthermore, each step of the overall acid secretion process that is characterized adds one additional possibility for specific characterization of bone diseases, and for pharmacologic mediation of bone mass. Thus, understanding each step is of potential practical importance.

It is also noteworthy that the flux of H^+ required for bone matrix demineralization is so large that it has no apparent biological parallel. In other processes, pH is managed in solutions having $[H^+]$ generally on the order of 10^{-7} to 10^{-8} *M* and buffering capacities of ~10 to 100 m*M*. Even in the lysosome, enzymes with pH optima of ~4 imply $[H^+]$ of 10^{-4} *M*, and that in a space only a few microns, at most, in diameter. However, if bone, largely solid hydroxyapatite, is considered for the moment as a hypothetical solid solution of OH^- and PO_3^{3-}, the daunting task of the osteoclast is apparent: to transport *molar* concentrations of H^+ into an interface that often measures over 10 μm across, with sufficient speed to dig a lacuna several microns deep in a few hours. Put another way, to move 1 g (25 m*M*) of skeletal calcium in a day (a large, but physiological rate), a minimum of ~0.04 mol of protons are moved at osteoclast-bone interfaces. By way of perspective, 0.04 mol of protons is ~5000 times the total intra- and extracellular H^+ of an average person. These calculations, while approximations, leave little doubt as to why osteoclasts are absolutely bursting with mitochondria: the energy dissipated in H^+ transport requires them. Similarly, the process of acid transport must be simple and efficient, or the drain on the body's resources would be a distinct evolutionary disadvantage; as discussed below, the remaining elements of bone degradation conform to this prediction. Furthermore, the stoichiometry of bone mineral removal explains the necessity of the osteoclast's characteristic organelle, the ruffled membrane: its expanded surface area is necessitated by the large numbers of copies of H^+ pump required for this most active of all acid transport processes.

B. BASE EQUIVALENT TRANSPORT BY OSTEOCLASTS

The only extensive study, to date, regarding handling of base equivalents by the osteoclast is that of Teti et al.,[58] who studied the intracellular pH of isolated osteoclasts to show that the cells possess a remarkably high capacity for Cl^--HCO_3^- exchange. This was shown to be necessary as well as sufficient to maintain the cell's internal pH in the face of massive acid loading. In keeping with these observations of Cl^--HCO_3^- exchange in avian osteoclasts, Baron et al. have exploited antibodies produced to react with partial sequences of the mammalian erythrocytic band 3 base exchanger to stain mammalian osteoclasts for immunoelectron microscopy. Their results strongly suggested that mammalian osteoclasts possess a similar Cl^--HCO_3^- exchanger that is localized to their basolateral surfaces.[71] This basolateral location of the base exchanger had been predicted from studies of cytoplasmic pH, assuming net vectorial secretion of acid, in the avian model.[58] The necessity of HCO_3^- exchange for osteoclastic bone resorption has recently been confirmed by Hall and Chambers in the rat.[72]

Demonstration of the high-capacity base exchanger of the osteoclast, which appears to be regulated,[58] indicates a function for the high content of carbonic anhydrase type II in the osteoclast.[73] To maintain neutrality, there must be one HCO_3^- exchanged for each H^+ transported. Hence, a ready source of both ions is required, implying the rapid interconversion of CO_2 to H_2CO_3, which is the function of this enzyme. Billecocq et al. have recently reported that expression of carbonic anhydrase II is regulated by vitamin D, suggesting that the elements of the H^+-transporting apparatus of the osteoclast may be coregulated, a hypothesis that will not be fully testable until specific probes for more elements of the acid-secreting apparatus are available.[74] With rapid respiration required to support the H^+-ATPase, it is not surprising that the ubiquitous biological buffer, carbonic acid, is used as the apparently single source of acid and base equivalents in the system; other buffers would require their own (energy-dependent) transport systems.

C. CHLORIDE TRANSPORT ELECTRICALLY COUPLED TO THE VACUOLAR H^+-ATPase OF THE OSTEOCLAST'S RUFFLED MEMBRANE

The vacuolar H^+-ATPase has an absolute requirement for either a cation antiport or an anion synport to maintain charge balance. The Na^+,K^+-ATPase, hypothesized by Baron et al. to be present on the osteoclast in large quantities, is not a candidate for this role, since its net charge transport is in the same cation-outward direction as the H^+-ATPase (although this second electrogenic pump may very well be involved in moment-to-moment regulation of H^+-ATPase activity, as discussed below).

The problem of a charge-balancing transport step for the H^+-ATPase was studied directly, using membrane visicles produced from avian osteoclasts by explosive decompression and differential centrifugation.[75] These vesicles contained vacuolar ATPase activity, but no detectable H^+ cation-exchange process. Anion transport consisted of a Cl^- channel, its activity shown to be coupled to the membrane charge, whether due to the vacuolar H^+-ATPase or an artificially created K^+ potential, even in the complete absence of ATP. Dissipation of the H^+ gradient with NH_4^+ left the Cl^- transport unchanged. In contrast to the properties of a chloride conductance in acid-secreting rabbit kidney, dependence of H^+ transport on $[Cl^-]$ was distinctly nonlinear.[76] Competitive binding of the diisothiocyanate anion transport inhibitor DIDS enabled isolation, by affinity chromatography, of the avian osteoclast's Cl^- channel.[77] This yielded a single 60-kDa protein which reconstituted DIDS-inhibitable Cl^- conductance in artificial phospholipid vesicles. A 64-kDa chloride channel has recently been isolated from kidney by Akabas et al.,[78] which by its physical properties and molecular weight appears to be a slightly different molecule.

The molecular weight of the osteoclastic chloride channel suggests that it belongs to the family of single-barreled anion transporters that includes the $GABA_A$ channel, containing four hydrophobic domains hypothesized to form a channel by interaction of common, hydrophilic anion-specific portions of the otherwise hydrophobic α-helical transmembrane domains.[79] This is in contrast to the ~100-kDa double-barreled Cl^- channel,[80] hypothesized to be involved in water transport, that has recently been cloned from *Torpedo* electroplax by cDNA expression in oocytes,[81] which contains additional hydrophobic domains.

There is considerable precedent for involvement of anion conductances in regulated acidifying processes.[82,83] The anion conductance prevents, by its activity, production of a large transmembrane potential by the electrogenic H^+ pump that would stop further H^+ transport. The Cl^- transporter may be subject to regulation by transmembrane potential,[84] pH, or other chemical regulators (such as a hormone,[85] [ATP], or [Ca^{2+}]).[76] If its conductance is so regulated, this would provide an obvious mechanism for short-term control of acid excretion.

D. MODEL FOR OSTEOCLAST ACID SECRETION

The model we propose is strictly conservative. The same anion, Cl^-, is electrically coupled to the osteoclast's H^+ pump and exchanged for HCO_3^- at the cell's basolateral surface. This yields a balanced transport equation for acid, base, and charge production without additional supporting transport systems or energy input (Figure 1). CO_2, hydrated to H_2CO_3, dissociates to produce acid equivalents (H^+), which are pumped across the ruffled membrane, and bicarbonate (HCO_3^-), which is exchanged for Cl^- at the basolateral membrane. Imported Cl^- leaves the cell by electroneutral cotransport, coupled to the H^+ pump electrostatically. This scheme permits the osteoclast to transport massive quantities of H^+ without disrupting intracellular pH or electrical balance, and requires only one ATP-dependent step, the vacuolar H^+-ATPase. Note that the anion exchange step on the osteoclasts basolateral surface is passive, but favored by the physiological inward chloride gradient (Cl_i^- ~10 m*M*). Chloride transport at the cell-bone interface is also passive, but, since it is selectively conductive, it can be driven against the Cl^- gradient by the electrogenic H^+-ATPase also present in the ruffled membrane. Thus, the transport of HCl across the osteoclast's bone-apposed cell membrane requires one ATP-consuming step, is electrically balanced, and does not alter cytoplasmic pH.

1. Other Processes Required for Osteoclastic Acid Transport

While not included in this simple model, which focuses solely on the principal ion transport steps of HCl secretion, there are several other requirements for effective bone resorption by the osteoclast. Chief among them is a tight annular attachment of the cell to its underlying matrix, about which

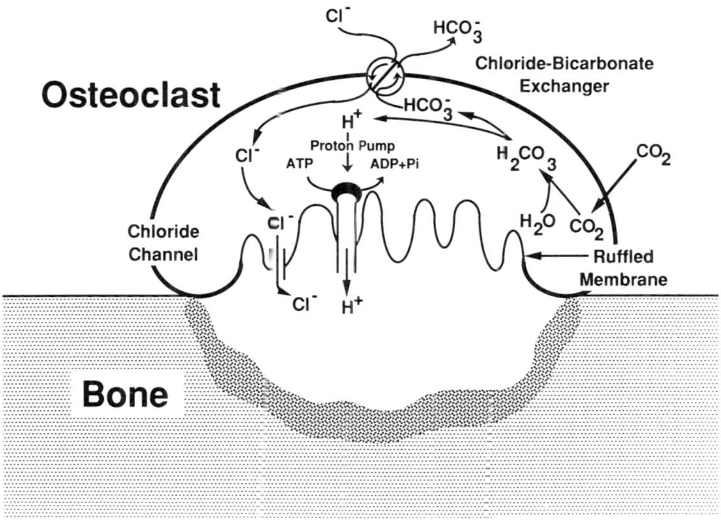

FIGURE 1. Model for the secretion of acid equivalents by osteoclasts. CO_2 is hydrated in a reaction occurring spontaneously, but accelerated dramatically by carbonic anhydrase, which is abundant in the osteoclast. Carbonic acid dissociates rapidly to form HCO_3^- and H^+. Proton equivalents are pumped across the ruffled membrane by the vacuolar-type H^+-ATPase. While ATP hydrolysis may affect pH, [ATP] is held roughly constant by the cell's metabolism, and thus its hydrolysis has no net effect on intracellular pH (i.e., [ADP] + [P_i] and [ATP] are in steady state). The net charge of the electrogenic H^+ pump drives cotransport (against its chemical gradient) of Cl^-. The cell's internal pH balance is preserved, and [Cl^-] maintained, by an anion exchanger at the osteoclast's basolateral cell membranes. In this exchange, both ion movements are energetically favored. Thus, one ATP-consuming process results in net HCl transport, with cellular homeostasis assured by the direct coupling of proton movement to anion and base transport.

we know little. This attachment process probably involves integrin binding, and may be modified by intracellular [Ca^{2+}].[86,87] This in turn may be related to detachment of the osteoclast from a mature resorption pit to allow escape of degraded collagen fragments, phosphates, and calcium itself. There is direct experimental data for retraction of osteoclasts resulting from high extracellular [Ca^{2+}].[88] Supporting this possibility, it has recently been observed, by direct microelectrode measurement, that extracellular calcium at the cell attachment site is ~20 m*M*.[89]

Another element required for effective HCl secretion is cellular fusion to create the multinucleated giant cell. This has the effect, regarding the HCl transport-dependent degradation process as being across a surface bounded by a circular border, of increasing the ratio of the area included in the attachment site to the length of the border structure confining the acid compartment. This is so because the area is proportional to the radius squared, while the circumference is proportional to the radius (the volume included would vary as the cube of the radius, but the energy-requiring acid transport is, as a practical matter, across a surface, albeit a rather complex surface). Hence, the effect of any constant leak rate at the low/neutral pH interface is minimized by creating a larger degradative site. This property, in which the osteoclast resembles the foreign body giant cell, perhaps for similar reasons of efficiency, has been shown by Fallon et al. to be directly related to the efficiency of bone degradation.[36]

There are certainly additional specializations of the osteoclast, such as that leading to the construction of the ruffled membrane, and to directed synthesis of the specific transporters involved in acid transport. At present, nothing specific is known about these elements. Furthermore, the factors controlling osteoclastic acid secretion are unknown; some possibilities for the control of acid secretion, and the relationship of collagenolysis to HCl secretion by osteoclasts, are discussed below.

E. OTHER MEMBRANE TRANSPORTERS OF THE OSTEOCLAST: POSSIBLE RELATIONSHIP TO H^+ TRANSPORT

Mears reported a bimodal distribution of osteoclast transmembrane potential in 1971,[90] a result that was not repeatable in the hands of Heersche et al.,[91] using osteoclasts produced by the method of Chambers et al. from rabbits,[40] although transient depolarizations were seen in response to hormones. In contrast, Sims and Dixon, using rat osteoclasts, have recently confirmed Mears's initial observation, finding two stable states of osteoclast membrane potential at ~ -15 and -70 mV, resulting from an inward rectifying potassium current.[92] These studies, together, establish the presence of at least one membrane ion transporter in addition to those described above, simply because the proton translocating processes do not involve or require K^+.

Two stable states of osteoclast membrane potential suggest that the cell may have two activity states as well, related to the change in membrane potential. In the absence of a definitive study relating secretory activity to membrane potential, any such statement must be regarded as speculative. However, it should be noted that the process of H^+ secretion by the osteoclast is driven by an electrogenic pump, and thus it must be sensitive to changes in the cell membrane potential. The osteoclast's membrane potential would undoubtedly be influenced by the Na^+,K^+-ATPase, as reported by Baron et al.,[55] although, as in other cells, rapid transitions and medium speed changes in membrane potential are more commonly the result of channel activity. Rectifying channels, similar to those noted by Sims and Dixon,[92] are frequently involved in stabilizing membrane potential in response to, or for regulation of, cellular electrical activity. In addition, the presence of a high-conductance Cl^- channel was recently noted by Schoppa et al., who performed patch-clamping of isolated rat osteoclasts.[93] This channel, if its existence is confirmed, would likely be found on the cell's basolateral surface (the only surface generally accessible to patch clamping).

Recent studies with the isolated ruffled-membrane chloride channel indicate that its conductance is voltage sensitive, with much less activity at ~0 mV than at ~−70 mV.[94] The existence of a voltage gate that would switch the osteoclast's transmembrane potential is hypothetical, and the effects of various hormones on this potential, V_M, are controversial.[90-92] It does, however, seem logical that the Cl^- channel, if its voltage dependence is confirmed, could tie these elements purposefully together: the H^+ pump, central to skeletal catabolism, is absolutely dependent on the activity of the Cl^- transporter to which it is electrically linked.[75] Hence, if the Cl^- channel is active at ~−70 mV, and inactive at ~−0 mV, H^+ transport would be up- or downregulated, almost immediately, by the membrane potential (hypotheses regarding elements possibly controlling osteoclastic acid secretion over a longer time scale are considered below). Membrane potential is discussed in greater detail in Chapter 11 of this volume.

It must be noted that these complicated conclusions require comparing results from a number of species (chicken, rat, and mouse), and that there are definitely species differences in bone metabolism among them. Thus, the specific elements involved in control of proton transport may well be unique to a given species, and a consensus mechanism for a single species is not yet available.

F. TRANSPORT OF PROTEASES TO THE OSTEOCLASTIC ATTACHMENT SITE DEPENDS ON HCl SECRETION

In our early work with avian osteoclasts, we considered whether collagenase, a neutral protease dependent on zinc chelation, mediated osteoclastic bone collagen resorption, as had been hypothesized.[95] When we had been unable to identify collagenase in isolated osteoclasts,[96] we turned our attention to the possibility of acidic proteases. For this purpose, we studied bone degradation by isolated osteoclasts to which 5 m*M* NH_4Cl was added. Ammonia is a weak base that diffuses into acidic compartments, neutralizing them.[97] We found that this treatment reversibly eliminated bone collagen degradation by osteoclasts, an effect that could be reproduced by inhibitors of cysteine or aspartate acid proteases to the same experimental system (Figure 2). These simple experiments confirmed the importance of acid secretion to degradation of the organic as well as the inorganic components of bone,[47] and showed that the results of Vaes[49] reflected direct involvement of acid proteases at the bone resorption site.

While definitive identification of the enzyme or enzymes secreted into the acidic bone degradation site is not yet in hand, the leading candidates are cysteine proteases of the papain family, including cathepsins B and L,[98,99] and the aspartate protease cathepsin D.[100] On the other hand, results from histochemical studies have not shown definite involvement of cysteine or aspartate proteases in extracellular matrix dissolution,[101,102] and the possibility of involvement of neutral metalloproteases in bone collagen degradation is still controversial.[96,103] (Also see Chapters 14 and 15 of this volume.)

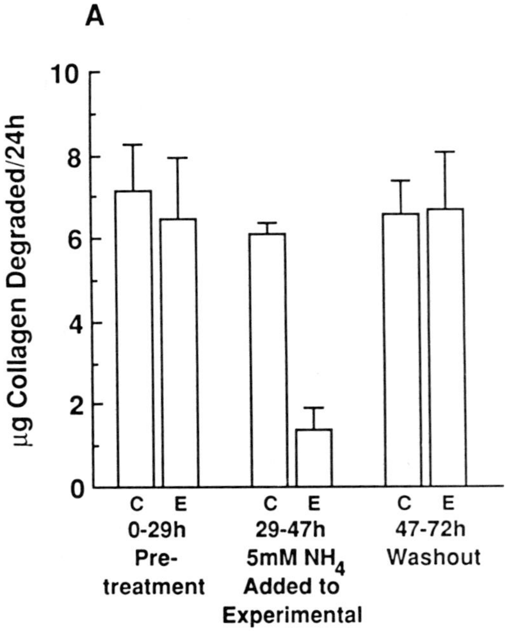

FIGURE 2. Inhibition of bone collagenolysis by addition of weak bases (A) or inhibitors of lysosomal proteases (B) to osteoclast cultures. Solubilization of ^{3}H-proline-labeled bone by 5000 osteoclasts in 2 cm^2 tissue culture wells was measured as described.[41] (A) The weak base NH_4Cl (5 m*M*) reversibly eliminates bone collagen degradation. Left, baseline bone resorption; middle, experimental but not control treated with the weak base; right, reversal of effect after washout ($n = 6$, mean ± SD). Weak bases, such as ammonia, accumulate in acidic compartments, neutralizing them. The results in this experiment suggests that bone collagen degradation is dependent on an acidic compartment. (B) Addition of several protease inhibitors (50 μ*M*) reduces collagen degradation by isolated osteoclasts. When compared with control, both pepstatin (an aspartate protease inhibitor) and two cysteine protease inhibitors, leupeptin and antipain, inhibit bone collagen degradation ($n = 8$, mean ± SD). After washout, there was complete recovery of collagen degradative activity (not illustrated), as with the ammonia inhibition experiment. Thus, inhibitor studies do not definitively show what class or classes of acid collagenase degrade bone collagen; this issue will be resolved when the protease or proteases are isolated.

Besides acid secretion itself, one respect in which an osteoclastic acid transport-related process appears to be consistent across species is that of extracellular delivery of lysosomal-type, mannose-6-phosphate-labeled degradative enzymes to the osteoclast's acidic extracellular bone attachment site. Mannose-6-phosphate modification of carbohydrates attached to glycoproteins generally results in their transport to the vacuolar degradative system. However, in the osteoclast, which has a unique acidic extracellular degradative site, Baron et al., studying mammalian cells by immunoelectron microscopy, and our group, using avian bone to localize the mannose-6-phosphate-inhibitable transport of isolated ^{125}I-labeled mannose-6-phosphate-containing pro-

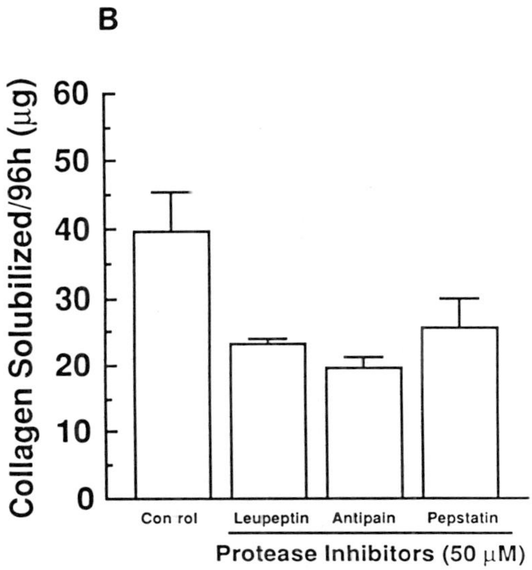

FIGURE 2B.

teins, found independently that in each of these two species mannose-6-phosphate-labeled proteins are targeted to the extracellular bone attachment site.[104,105]

The mechanism by which this transport process, which generally delivers degradative enzymes to lysosomes, is modified for delivery of acid hydrolyses to the extracellular bone attachment site is unknown, but presumably related to the acidification of that site. This is so because, in the presence of calcium (a certainty at the acidic bone degradation site),[89] binding by the mannose-6-phosphate receptor is known to be reversible only by low pH.[106] As discussed above, the osteoclast's bone attachment is not, however, an "extracellular lysosome", since it has orders of magnitude more H^+ transport capacity than a lysosome, as well as unique structural components that produce the ruffled membrane, a remarkable resistance to calcium, and its matrix attachment apparatus.

A consequence of the acid dependence of mannose-6-phosphate receptor-ligand release would be that, if the bone degradation site were changed to neutral pH, the mannose-6-phosphate receptor would not release the acidic protease at the cell surface, but could perhaps scavenge free enzyme. Indeed, we have found mannose-6-phosphase receptor-mediated uptake by osteoclasts.[105] This element fits well with the hypothesis that the osteoclast releases the elements of degraded bone matrix (Ca^{2+}, HPO_4^{-2}, peptides, and amino acids) by detaching from the bone surface and reattaching to form a new acidic, degradative pit.[107] With the added ability to retrieve matrix-degrading enzymes, this model results in a conservative and very efficient system.

G. OSTEOCLASTIC H^+ TRANSPORT AND BONE DISEASE

There are, as yet, scant data regarding the relationship between osteoclastic acid transport and bone disease. However, several basic facts deserve exposition, since, as the process of bone resorption is better understood, it is hoped that these results will allow effective treatment of abnormal bone modeling in childhood, or remodeling in adulthood. Abnormalities in acidic bone degradation are likely to be important to abnormal skeletal development in osteopetrosis, in the course of abnormal skeletal turnover in Paget's disease, and in osteoporosis. There are also a number of other arthritic, metabolic, or inflammatory conditions that involve components of bone degradation locally or throughout the body. These are not considered here in detail, but may possibly be benefited by drugs directly influencing acidic bone degradation.

Osteopetrosis, the partial or complete inability of osteoclasts to remove mineralized cartilage or bone for remodeling,[108] may be caused by defects in a number of osteoclastic processes, including attachment, acid secretion, enzyme secretion, or cellular control. It is clear that the disease is polymorphous, and that it represents a cellular defect curable by bone marrow transplant.[109] It remains to be seen which types of the disease result from inability of the osteoclast to secrete acid. Abnormalities in acid secretion by osteoclasts are definitely present in some forms of osteopetrosis, since, in some cases, there is complete lack of ruffled membranes.[108] One would expect that other forms of osteopetrosis would be due to abnormalities of the mechanism controlling acid secretion, or in isolating the acidic compartment from the general extracellular space.

Paget's disease (adult type) represents a regional excess in bone degradation.[110] As such, if the basic theory of acidic bone degradation is valid, this disease must represent excessive activity of osteoclastic acid secretion. Whether this excessive activity is intrinsic, or represents interference with the cell's normal control mechanism, is as yet unknown. However, as Paget's disease results in a distinct increase in the number of osteoclasts, it certainly involves defective control at a level above the single cell.

Osteoporosis is an insufficient quantity of mineralized bone matrix, of any etiology. This malady clearly takes several forms, from mild to severe. No case of primary osteoporosis has, as yet, been characterized at the molecular or biochemical levels.[111] The victims of osteoporosis are, in general, apparently able to remodel bone normally. Hence, it is likely that the disease, which obviously represents an excess of bone matrix removal over bone matrix deposition, represents a defect in effective coupling of osteoclastic activity to the precise bone formation rate, rather than in a fundamental inability to form or degrade bone.

With regard to most cases of osteoporosis, and to many other diseases related to abnormal bone removal through osteoclastic acid secretion, the important unanswered question is, how is bone formation normally coupled to bone degradation? Attempts to understand the issue through measurements of factors, such as estrogen, that have been associated with osteoporosis have proved generally unsatisfactory, although estrogen therapy definitely slows postmenopausal bone loss.[112] Attempts to influence osteoporosis by mass action, such as by increasing systemic calcium input, have proved of limited value, highlighting awareness that bone remodeling is not generally controlled by diet, but by specific cellular processes.[113] Hence, until specific probes can be used to measure the activity of individual biochemical steps of bone deposition and resorption in response to isolated bone matrix or cellular factors, the reason that osteoclastic acid secretion (net calcium loss) exceeds mineral deposition in osteoporosis will likely remain enigmatic.

VI. MATRIX AND HCl TRANSPORT CHEMISTRY DELIMIT THE DESIGN OF RESORPTION-MODIFYING DRUGS

Pharmacologic mediators of osteoclastic HCl secretion must be targeted to the osteoclast in some manner. Since acidic compartments are vital to many cell types, lysosomotropic or acidotropic agents would seem to be impractical, as they would have too many serious side effects.

There are, however, other possibilities. For instance, the natural control of the osteoclast may be exploited, agonists upregulating osteoclast acidification and antagonists downregulating activity. Thus, bone-seeking hormones are obvious candidates to control osteoclast activity. Paradoxically, however, the mature osteoclast appears to lack vitamin D receptors, and response to calcitonin appears to differ in magnitude between species. Since osteoblasts do respond to PTH, a second intercellular messenger induced by PTH has been hypothesized, which would be another possibility for manipulation of osteoclastic activity. Unfortunately, such specific factors, if they exist, are unknown; indeed, the recently confirmed relationship of the osteoclast to the macrophage[33-35,42] suggests that any systemically administered factor regulating bone degradation may also have effects on other immune system cells, a potentially prohibitive disadvantage. Recent work suggesting that tumor necrosis factor-α and -β, interleukin-1α and -1β, and interleukin-6 increase the rate of bone degradation, and that γ-interferon decreases the rate of bone degradation, support this possibility.[114-116] These and a wide variety of other factors affecting bone resorption but not known to be related to the acid secretion hypothesis, have been reviewed by Vaes.[117] (See also Chapter 14 of this volume.)

Contrariwise, another possible and theoretically fairly specific means of reducing acid secretion by osteoclasts would be to bind selected acid degradation-inhibiting compounds to bone matrix. This could, perhaps, be engineered through exploitation of the essentially unique biological occurrence of hydroxyapatite in bone. Since several biochemical steps of the HCl secretion apparatus are known, if these biochemical steps were to be shown to be susceptible to a hydroxyapatite-bound compound, good candidates for antiosteoporotic drugs could be evaluated without resorting to difficult to use (and expensive) *in vivo* models. These hypothetical compounds would reduce osteoclastic activity directly, bypassing the hypothetical supercellular control processes, and thus hopefully would not adversely affect macrophage precursors or have other unacceptable side effects.

To affect osteoclasts, but not other cells, via bone matrix binding, two properties are necessary. First, the hypothetical agent must bind to the bone matrix tightly enough that it is sequestered effectively. Thus, the body's other cells, including osteoblasts, would not be exposed to inhibitory drug concentrations. Second, the agent must be released, at effective concentrations, by acidification of the matrix at the active site of the osteoclast. To date, three types of compounds have been described which appear likely to fulfill these conditions: bisphosphonates, group III metals, and tetracyclines.

Tetracyclines have been known to bind to newly deposited bone mineral since the early 1960s;[19] indeed, this property of the antibiotic has revolutionized the study of bone matrix. One of the unfortunate side effects of the tetracyclines, however, is abnormalities of developing skeleton and teeth, principally related to retardation of growth, and to hypoplasia of dental enamel.[118] These adverse effects suggested an inhibition of the skeletal cells by bound tetracycline. Rifkin et al. have recently shown reduction of ^{45}Ca release from labeled bone by tetracyclines.[20] While the specific biochemical step targeted by the tetracyclines has not been identified, the fact that mineral removal is reduced suggests an effect on osteoclastic function. (Tetracyclines are discussed further in Chapters 15 and 17 of this volume.)

Bisphosphonates, compounds containing the enzymatically nonhydrolyzable PCP group, are analogs of pyrophosphates with a number of potentially useful properties, including a high affinity for binding hydroxyapatite (i.e., bone mineral). They are also known to inhibit a number of enzymes that bind ATP. We have recently demonstrated that three bisphosphonates inhibit osteoclastic bone degradation, and that, in the case of 1-hydroxy-1,1-ethanebisphosphonate (EHDP, Didronel), this occurs via a mechanism involving reduced acid secretion.[27] (See Chapter 13 of this volume for a discussion of bisphosphonates.)

Bone-binding metals, such as gallium, have recently been proposed as agents for reducing bone degradation,[119] and calcium release from bone in particular,[120] by a mechanism that affects isolated osteoclasts directly.[121] We have recently determined that Ga at micromolar levels inhibits acidification of osteoclast membrane vesicles, supporting this possibility.[122]

There may well be many other classes of bone resorption-modifying agents yet undiscovered. The ideal agent should either uniquely affect the osteoclast or be concentrated on hydroxyapatite to such an extent that only an osteoclast degrading the mineral is exposed to a high level of the agent. The agent must then also, directly or indirectly, reduce the osteoclast's ability to acidify and solubilize its substrate, without unacceptable side effects.

VII. CONTROL OF OSTEOCLASTIC H^+ TRANSPORT: HYPOTHESES

Osteoclastic HCl-dependent bone resorption appears to be regulated at more than one level: PTH results in almost immediate changes in serum [Ca^{2+}], but over a longer period PTH results in an absolute increase in osteoclast number. Hence, we hypothesize that HCl secretion is controlled in at least two ways: by cell differentiation (or senescence) and by a more rapid means that turns acid secretion on and off rapidly.

A. SHORT-TERM MODULATION OF HCl TRANSPORT

We have observed that visualization of the vacuolar-type H^+-ATPase in avian osteoclasts shows two classes of osteoclasts, one with the proton pump polarized to the bone attachment site, the other with the acid-transporting pump distributed, fairly uniformly, in the cytoplasm in a vacuolar pattern (Plate 1*). This indicates that, as is established for renal acid transport, the osteoclast's acid-transporting apparatus exists in inactive (intracellular) and active (polarized) configurations. However, in contrast to the kidney, high pCO_2 does not appear to increase or decrease osteoclastic acid secretion significantly.[63,123] Thus, if acid secretion is modulated by redistribution of H^+-ATPase in the mature osteoclast, the stimulus is not pCO_2.

It is also possible that control of acid secretion occurs *in situ* at the ruffled membrane. There is a strong precedent for modulation of vacuolar acidification by chloride permeability.[82,83,124,125] The chloride channel isolated from avian osteoclast ruffled membranes appears to be different from the bovine kidney H^+ transport-coupled Cl^- channel.[76,77] This variation in chloride channels, each associated with proton transport, reflects their varied biological function and is consistent with the many types of Cl^- channels reported.[80,85] Detailed analysis of chloride channel characteristics is not presently available, although some properties may be inferred from physical measurements of channel activity in membranes. The osteoclast's Cl^- channel has a nonlinear relationship to [H^+] transport,[75] which is hypothesized to be due to sensitivity of chloride permeability to V_M, the cell membrane potential. This would provide a regulation of acid secretion that is strongly influenced by V_M in a cell reported to express a rectifying K^+ conductance and two very active electrogenic pumps in its plasma membrane; this hypothesis awaits verification.

B. CONTROL OF HCl SECRETION VIA DIFFERENTIATION OF CELLS

The osteoclast's phylogeny may hold further clues to control of its differentiation. What, then, is the origin of this cell? It appears in evolution with

* Plate 1 appears at the end of this chapter.

the transition of vertebrate life from sea to fresh water, where maintenance of serum calcium cannot be accomplished by regulated transport across membranes in contact with calcium-rich sea water. Later, in the transition from water to land, where the weight of a solid skeleton (present in bony fishes) would be a distinct disadvantage, acellular bone is replaced by the characteristic cellular bone of the land-living vertebrate, i.e., the blood-derived,[28] acid-secreting osteoclast precedes the osteocyte in evolution. Adaptation of the macrophage for bone degradation seems obvious, in that this cell has long been known to be specialized for acidic degradation,[5] and migration from the circulation of macrophages to form osteoclasts is seen in bony fishes. Indeed, teleost fish, while producing acellular bone,[126] are capable of hormonally regulated osteoclastic bone degradation for serum calcium activity regulation in fresh water.[127] Curiously, in teleost fish a pituitary product stimulates osteoclasts, and a separate [Ca^{2+}]-sensitive ultimobranchial body also regulates skeletal calcium mobilization.[127] Recent findings on osteoclast ontogeny strongly support the hypothesis of a macrophage-family osteoclast precursor.[33-35]

Hence, osteoclast differentiation would be expected to respond to macrophage factors. It remains to be seen whether osteoclast-controlling factors fit this prediction, as few facts are available. However, vitamin D, identified as an antirachitic factor, clearly affects normal macrophage development.[30,128] Thus, it would seem most likely that factors directly increasing osteoclastic number in response to PTH will, when finally defined, be known immune modulators.[115,116] The factors causing more osteoclasts to be produced would, consequently, upregulate overall skeletal acid secretion capacity over a period of days or weeks.

C. DOWNREGULATION OF ACID TRANSPORT

The mechanism by which the osteoclast's acid secretion is terminated is another question about which discussion is, at present, speculative. The commonly discussed candidate for this function, calcitonin, has limited effects in many species, including man.[129] It is possible that, in species not responding well to calcitonin, lack of a positive signal — say, from the hypothetical second intercellular messenger of PTH — may result in cessation of extracellular HCl production.

The ultimate fate of the osteoclast is unknown. Clearly, though, osteoclasts do disappear from bone. Hence, either osteoclasts have a naturally limited HCl secretion cycle and then disintegrate (programed senescence), or an osteoclast apoptotic factor exists, or the continued presence of an osteoclast maintenance factor maintains cellular vitality. The osteoclast's options appear to be limited to further fusion, stable metabolic activity, or death. Reversal of cell fusion was widely hypothesized in earlier work, but appears unlikely as multinucleated cells are now generally regarded, in the absence of reliable contrary observations, to be end-stage cells. We have observed vacuolar degeneration and fragmentation of cultured osteoclasts, perhaps representing the end stage of the osteoclast's acid degradation cycle. Disintegration of cultured osteoclasts is more commonly observed in the absence of significant numbers of mesenchymal cells,[122] suggesting that a maintenance factor may be involved.

ACKNOWLEDGMENTS

The assistance of the Rare Books Department of the Washington University Medical Library in obtaining historical material is gratefully acknowledged. We thank David Lacey, M.D., Department of Pathology, Washington Univerity School of Medicine, for reviewing portions of the manuscript, and Mary Sorrells, M.F.A., for illustrations.

REFERENCES

1. **Kölliker, A.,** *Die normale Resorption des Knochengewebes und ihre bedeutung für die entstehung der typischen knockenformen,* Vogel, Leipzig, 1873, 1.
2. **Howship, J.,** Observations on the morbid structure of bone and an attempt at an arrangement of their diseases, *Medico-Chir. Trans.*, 8, 57, 1816.
3. **Cruickshank, W.,** *The Anatomy of Absorbing Vessels of the Human Body,* 2nd ed., Nicol, London, 1790, p. 49, p. 108.
4. **Virchow, R.,** *Cellular Pathology as based upon Physiological and Pathological Histology* (transl.), J & A Churchill, London, 1860, p. 77, p. 407.
5. **Metchnikoff, E.,** Untersuchungen uber die mesodermalen phagocyten einiger wirbeltiere, *Biol. Zentralbl.*, 3, 560, 1883.
6. **Steenbock, H. and Black, A.,** Fat soluble vitamins. XVII. The induction of growth-promoting and calcifying properties in a ration by exposure to ultraviolet light, *J. Biol. Chem.*, 61, 405, 1924.
7. **Albright, F.,** A page out of the history of hyperparathyroidism, *J. Clin. Endocrinol. Metab.*, 8, 637, 1948.
8. **Tenbaum, F.,** Ueber Kalkausscheidung durch den harn bei diabetes, *Z. Biol.*, 33, 351, 1896.
9. **Fitz, R., Alsberg, C. L., and Henderson, L. J.,** Concerning the excretion of phosphoric acid during experimental acidosis in rabbits, *Am. J. Physiol.*, 18, 113, 1907.
10. **Aub, J. C.,** Calcium and phosphorus metabolism, *Harvey Lect.*, 23, 151, 1929.
11. **Jaffe, H. L. and Bodansky, A.,** Experimental fibrous osteodystrophy (osteitis fibrosa) in hyperparathyroid dogs, *J. Exp. Med.*, 52, 669, 1930.
12. **Neuman, W. F. and Neuman, M. W.,** *The Chemical Dynamics of Bone Mineral*, University of Chicago Press, Chicago, 1958, chap. 3, 4.
13. **Mehmel, M.,** Beziehungen zwischen Kristallstruktur und chemischer Formel des Apatits, *Z. Physi. Chem. Abt. B*, 15, 223, 1932.
14. **Shorr, E. and Carter, A. C.,** The usefulness of strontium as an adjuvant to calcium, *Bull. Hosp. Joint Dis.*, 13, 59, 1952.
15. **Arendt, E. C. and Pattee, C. J.,** Strontium therapy in the paraplegic patient with osteoporosis and hypercalciuria, *JAMA*, 156, 1193, 1954.

16. **Barnicot, N. A.,** The supravital staining of osteoclasts with neutral-red: their distribution on the parietal bone of normal growing mice, and a comparison with the mutants grey-lethal and hydrocephalus-3, *Proc. R. Soc. London Ser.* B, 134, 467, 1947.
17. **Cretin, A.,** Contribution histochimique a l'étude de la construction et de la destruction osseuse, *La Presse Med.,* 59, 1240, 1951.
18. **Gonzales, F. and Karnovsky, K. M. J.,** Electron microscopy of osteoclasts in healing fractures of rat bone, *J. Biophys. Biochem. Cytol.,* 9, 299, 1961.
19. **Frost, H. M.,** Tetracycline labeling of bone and the zone of demarcation of osteoid seams, *Can. J. Biochem. Pharmacol.,* 40, 485, 1962.
20. **Rifkin, B., Gonos, B., Ramamurthy, N. S., McNamara, T. F., and Golub, L. M.,** Effect of a chemically-modified tetracycline (CMT) on bone resorption tissue culture: preliminary observations, *J. Cell Biol.,* 105, 216a, 1987.
21. **Warner, S. P.,** Hydrolytic enzymes in osteoclasts cultured *in vitro, J. R. Microsc. Soc.,* 83, 397, 1964.
22. **Göthlin, G. and Ericsson, J. L. E.,** Fine structural localization of acid phosphomonoesterase in the brush border region of osteoclasts, *Histochemie,* 28, 337, 1971.
23. **Neuman, W. F., Firschen, H., Chen, P. S., Mulryan, B. J., and DeStefano, V.,** On the mechanism of action of parahormone, *J. Am. Chem. Soc.,* 78, 3863, 1956.
24. **Armstrong, W. D. and Singer, L.,** *In vitro* uptake and exchange of bone citrate, in *Bone Structure and Metabolism,* Wolstenholme, G. E. W. and O'Connor, C. M., Eds., Little, Brown, Boston, 1956, 309.
25. **Fleisch, H., Russell, G. G., and Francis, M. D.,** Diphosphonates inhibit hydroxyapatite dissolution *in vitro* and bone resorption in tissue culture and *in vivo, Science,* 165, 1262, 1969.
26. **Jung, A., Bisaz, S., and Fleisch, H.,** The binding of pyrophosphate and two diphosphonates by hydroxyapatite crystals, *Calcif. Tissue Res.,* 11, 269, 1973.
27. **Carano, A., Teitelbaum, S. L., Konsek, J. D., Schlesinger, P. H., and Blair, H. C.,** Bisphosphonates directly inhibit the bone resorption activity of isolated avian osteoclasts *in vitro, J. Clin. Invest.,* 85, 456, 1990.
28. **Kahn, A. J. and Simmons, D. J.,** Investigation of cell lineage in bone using a chimera of chick and quail embryonic tissue, *Nature (London),* 258, 325, 1975.
29. **Kahn, A. J., Stewart, C. C., and Teitelbaum, S. L.,** Contact mediated bone resorption by human monocytes, *in vitro, Science,* 199, 988, 1978.
30. **Bar-Shavit, Z., Teitelbaum, S. L., Reitsma, P., Hall, A., Pegg, L. E., Trial, J., and Kahn, A. J.,** Induction of monocytic differentiation and bone resorption by 1,25-dihydroxyvitamin D_3, *Proc. Natl. Acad. Sci. U.S.A.,* 80, 5907, 1983.
31. **Swallow, C. J., Grinstein, S., and Rotstein, O. D.,** A vacuolar type H^+ ATPase regulates cytoplasmic pH in murine macrophages, *J. Biol. Chem.,* 265, 7645, 1990.
32. **Swallow, C. J., Grinstein, S., and Rotstein, O. D.,** Cytoplasmic pH regulation in macrophages by an ATP-dependent and N,N′-dicyclohexylcarbodiimide-sensitive mechanism, *J. Biol. Chem.,* 263, 19558, 1988.
33. **Kukita, T., McManus, L. M., Miller, M., Civin, C., and Roodman, G. D.,** Osteoclast-like cells formed in long-term human marrow cultures express a similar surface phenotype as authentic osteoclasts, *Lab. Invest.,* 60, 532, 1989.
34. **Hagenaars, C. E., Van der Kraan, A. A. M., Kawailrang-de Haas, E. W. M., Visser, J. W. M., and Nijweide, P. J.,** Osteoclast formation from cloned hemopoietic stem cells, *Bone Miner.,* 6, 179, 1989.
35. **Alvarez, J. I., Teitelbaum, S. L., Blair, H. C., Greenfield, E. M., Athanasou, N. A., and Ross, F. P.,** Generation of avian cells resembling osteoclasts from mononuclear phagocytes, *Endocrinology,* in press.
36. **Fallon, M. D., Teitelbaum, S. L., and Kahn, A. J.,** Multinucleation enhances macrophage-mediated bone resorption, *Lab. Invest.,* 49, 159, 1983.

37. **Blair, H., Kahn, A. J., Sadow, J., Malone, J. D., and Teitelbaum, S. L.,** Disassociation of lysosomal enzyme secretions and macrophage-mediated bone resorption, *Biochem. Biophys. Res. Commun.*, 100, 959, 1981.
38. **Bushinsky, D. A., Krieger, N. S., Geisser, D. I., Grossner, E. B., and Coe, F. L.,** Effects of pH on bone calcium and phosphorus *in vitro, Am. J. Physiol.*, 245, F204, 1983.
39. **Zambonin-Zallone, A., Teti, A., and Primevera, M. V.,** Isolated osteoclasts in primary culture: first observations on structural and survival in culture media, *Anat. Embryol.*, 165, 405, 1982.
40. **Chambers, T. J., Revell, P. A., Fuller, K., and Athanasou, N. A.,** Resorption of bone by isolated rabbit osteoclasts, *J. Cell Sci.*, 66, 383, 1984.
41. **Osdoby, P., Martini, M. C., and Caplan, A. I.,** Isolated osteoclasts and their presumed progenitor cell, the monocyte in culture, *J. Exp. Zool.*, 224, 331, 1982.
42. **Udagawa, N., Takahashi, M., Akatsu, T., Sasaki, T., Yamaguchi, A., Kodama, H., Martin, T. J., and Suda, T.,** The bone marrow-derived stromal cell lines MC3T3-G2/PA6 and ST2 support osteoclast-like cell differentiation in cocultures with mouse spleen cells, *Endocrinology,* 125, 1805, 1989.
43. **Kurihara, N., Suda, T., Miura, T. T., Nakauchi, H., Kodama, H., Hiura, Z. K., Hakeda, Y., and Kumegawa, M.,** Generation of osteoclasts from isolated hematopoietic progenitor cells, *Blood,* 74, 1295, 1989.
44. **Fallon, M. D.,** Bone resorbing fluid from osteoclasts is acidic: an *in vitro* micropuncture study, in *Endocrine Control of Bone and Calcium Metabolism,* Vol. 8A, Cohn, C. V., Fujita, T., Potts, J. T., Jr., and Talmadge, R. V., Eds., Elsevier, Amsterdam, 1984, 144.
45. **Green, P. M., Marshall, M. J., and Nisbet, N. W.,** A study of osteoclasts on calvaria of normal and osteopetrotic *(mi/mi)* mice by vital staining with acridine orange, *Br. J. Exp. Pathol.*, 67, 85, 1986.
46. **Baron, R., Neff, L., Louvard, D., and Courtoy, P. J.,** Cell-mediated extracellular acidification and bone resorption: evidence for a low pH in the resorbing lacunae and localization of a 100-kD lysosomal membrane protein at the osteoclast ruffled border, *J. Cell Biol.*, 101, 2210, 1985.
47. **Blair, H. C., Kahn, A. J., Crouch, E. C., Jeffrey, J. J., and Teitelbaum, S. L.,** Isolated osteoclasts resorb the organic and inorganic components of bone, *J. Cell Biol.*, 102, 1164, 1986.
48. **Delaisse, J. M., Boyde, A., Maconnachie, E., Ali, N. N., Sear, C. H. J., Eeckhout, Y., Vaes, G., and Jones, S. J.,** The effects of inhibitors of cysteine-proteases and collagenase on the resorptive activity of isolated osteoclasts, *Bone,* 8, 305, 1987.
49. **Vaes, G.,** Collagenase, lysosomes and osteoclastic bone resorption, in *Collagenase in Normal and Pathological Connective Tissue,* Woollen, D. S. and Evanson, J. M., Eds., John Wiley & Sons, Chichester, 1980, 185.
50. **Muallem, S., Blissard, D., Cragoe, E. J., and Sachs, G.,** Activation of the Na^+/H^+ and Cl^-/HCO_3^- exchange by stimulation of acid secretion in the parietal cell, *J. Biol. Chem.*, 263, 14703, 1988.
51. **Breyer, M. D. and Jacobson, H. R.,** Regulation of rabbit medullary collecting duct cell pH by basolateral Na^+/H^+ and Cl^-/base exchange, *J. Clin. Invest.*, 84, 996, 1989.
52. **Maeda, M., Oshiman, K., Tamura, S., and Futai, M.,** Human gastric (H^+ + K^+)-ATPase gene, *J. Biol. Chem.*, 265, 9027, 1990.
53. **Schneider, D. L.,** The proton pump ATPase of lysosomes and related organelles of the vacuolar apparatus, *Biochim. Biophys. Acta,* 895, 1, 1987.
54. **Reggio, H., Bainton, D., Harms, E., Coudrier, E., and Louvard, D.,** Antibodies against lysosomal membranes reveal a 100,000-mol-wt protein that cross reacts with purified H^+,K^+ ATPase from gastric mucosa, *J. Cell Biol.*, 99, 1511, 1984.

55. **Baron, R., Neff, L., Roy, C., Boisvert, A., and Caplan, M.,** Evidence for a high and specific concentration of (Na^+,K^+) ATPase in the plasma membrane of the osteoclast, *Cell,* 46, 311, 1986.
56. **Ravesloot, J. H., Ypley, D. L., Vrijheid-Lammers, T., and Nijweide, P. J.,** Voltage-activated K^+ conductances in freshly isolated embryonic chicken osteoclasts, *Proc. Natl. Acad. Sci. U.S.A.,* 86, 6921, 1989.
57. **Hall, T. J. and Chambers, T. J.,** Na^+/H^+ antiporter is the primary proton transport system used by osteoclasts during bone resorption, *J. Cell. Physiol.,* 142, 420, 1990.
58. **Teti, A., Blair, H. C., Teitelbaum, S. L., Kahn, A. J., Koziol, C., Konsek, J., Zambonin-Zallone, A., and Schlesinger, P. H.,** Cytoplasmic pH regulation and chloride/bicarbonate exchange in avian osteoclasts, *J. Clin. Invest.,* 83, 227, 1989.
59. **Lubin, M., Cahn, F., and Countermarsh, A.,** Amiloride, protein synthesis, and activation of quiescent cells, *J. Cell. Physiol.,* 113, 247, 1982.
60. **Blair, H. C., Teitelbaum, S. L., Ghiselli, R., and Gluck, S.,** Osteoclastic bone resorption by a polarized vacuolar proton pump, *Science,* 245, 855, 1989.
61. **Gluck, S. and Al-Awqati, Q.,** An electrogenic proton-translocation adenosine triphosphatase from bovine kidney medulla, *J. Clin. Invest.,* 73, 1704, 1984.
62. **Hirsch, S., Strauss, A., Masood, K., Lee, S., Sukhatme, V., and Gluck, S.,** Isolation and sequence of a cDNA clone encoding the 31-kDa subunit of bovine vacuolar H^+-ATPase, *Proc. Natl. Acad. Sci. U.S.A.,* 85, 3004, 1988.
63. **Cannon, C., van Adelsberg, J., Kelly, S., and Al-Awqati, Q.,** Carbon dioxide-induced exocytic insertion of H^+-pumps in turtle bladder luminal membrane: role of cell pH and calcium, *Nature (London),* 314, 443, 1985.
64. **van Adelsberg, J. and Al-Awqati, Q.,** Regulation of cell pH by Ca^{+2}-mediated exocytic insertion of H^+-ATPases, *J. Cell Biol.,* 102, 1638, 1986.
65. **Bowman, B. J., Allen, R. A., Wechser, M. A., and Bowman, E. J.,** Isolation of genes encoding the *Neurospora* vacuolar ATPase, *J. Biol. Chem.,* 263, 14002, 1988.
66. **Caldwell, J. and Gluck, S.,** Immunoaffinity purification and characterization of vacuolar H^+ ATPase from bovine kidney, *J. Biol. Chem.,* 262, 15780, 1987.
67. **Brown, D., Gluck, S., and Hartwig, J.,** Structure of the novel membrane-coating material in proton-secreting epithelial cells and identification as an H^+ ATPase, *J. Cell Biol.,* 105, 1637, 1987.
68. **Yurko, M. A. and Gluck, S.,** Production and characterization of a monoclonal antibody to vacuolar H^+ ATPase of renal epithelia, *J. Biol. Chem.,* 262, 15770, 1987.
69. **Bekker, P. J. and Gay, C. V.,** Biochemical characterization of an electrogenic vacuolar proton pump in purified chicken osteoclast plasma membrane vesicles, *J. Bone Miner. Res.,* 5, 569, 1990.
70. **Väänänen, H. K., Karhukorpi, E.-K., Sundquist, K., Wallmark, B., Roininen, I., Hentunen, T., Tuukkanen, J., and Lakkakorpi, P.,** Evidence for the presence of a proton pump of the vacuolar H^+-ATPase type in the ruffled borders of osteoclasts, *J. Cell Biol.,* 111, 1305, 1990.
71. **Kellokumpu, S., Tschundi, C., Neff, L., Jamsa-Kellokumpu, S., Kapito, R., Ullu, E., and Baron, R.,** Isolation of a cDNA clone encoding one golgi anion exchanger in ROS17/2.8 osteoblasts, *J. Bone Miner. Res.,* 4S, S260, 1989.
72. **Hall, T. J. and Chambers, T. J.,** Optimal bone resorption by isolated rat osteoclasts requires chloride/bicarbonate exchange, *Calcif. Tissue Int.,* 45, 378, 1989.
73. **Gay, C. V., Ito, M. B., and Schraer, H.,** Carbonic anhydrase activity in isolated osteoclasts, *Metab. Dis. Relat. Res.,* 5, 33, 1984.
74. **Billecocq, A., Emanuel, J. R., Levenson, R., and Baron, R.,** 1α,25-dihydroxyvitamin D_3 regulates the expression of carbonic anhydrase II in nonerythroid avian bone marrow cells, *Proc. Natl. Acad. Sci. U.S.A.,* 87, 6470, 1990.
75. **Blair, H., Teitelbaum, S. L., Tan, H.-L., Koziol, C. M., and Schlesinger, P. H.,** Passive chloride permeability charge-coupled to the H^+-ATPase of avian osteoclast ruffled membrane, *Am. J. Physiol.,* 260, C1315, 1991.

76. **Hilden, S. A., Johns, C. A., and Madias, N. E.,** Cl^--dependent ATP-driven H^+ transport in rabbit renal cortical endosomes, *Am. J. Physiol.,* 255, F885, 1988.
77. **Blair, H. C. and Schlesinger, P. H.,** Purification of a stilbene sensitive Cl^- channel with reconstitution of chloride conductivity into phospholipid vesicles, *Biochem. Biophys. Res. Commun.,* 171, 920, 1990.
78. **Akabas, M., Edelman, A., Redhead, C., Land, D., and Al-Awqati, Q.,** Characterization of a purified chloride channel, *Biophys. J.,* 57, 320a, 1990.
79. **Montal, M.,** Channel protein engineering: an approach to identification of molecular determinants of function in voltage-gated and ligand-regulated channel proteins, in *Ion Channels,* Narahashi, T., Ed., Plenum Press, New York, 1989, 1.
80. **Barnard, E. A., Darlison, M. G., and Seeburg, P.,** Molecular biology of the $GABA_A$ receptor: the receptor/channel superfamily, *Trends Neurosci.,* 10, 502, 1987.
81. **Jentsch, T. J., Steinmeyer, K., and Schwarz, G.,** Primary structure of *Torpedo marmorate* chloride channel isolated by expression cloning in *Xenopus* oocytes, *Nature (London),* 348, 510, 1990.
82. **Glickman, J., Croen, K., Kelly, S., and Al-Awqati, Q.,** Golgi membranes contain an electrogenic H^+ pump in parallel to a chloride conductance, *J. Cell Biol.,* 97, 1303, 1983.
83. **Van Dyke, R.,** Proton pump-generated electrochemical gradients in rat liver multivesicular bodies. Quantitation and effects of chloride, *J. Biol. Chem.,* 263, 2603, 1988.
84. **Block, M. and Moody, W. J.,** A voltage-dependent chloride current linked to the cell cycle in ascidian embryos, *Science,* 247, 1090, 1990.
85. **Schofield, P. R., Darlinson, M. G., Fujita, N., Burt, D. R., Stephenson, F. A., Rodriguez, H., Rhee, L. M., Ramachandran, J., Reale, V., Glencorse, T. A., Seeburg, P. H., and Barnard, E. A.,** Sequence and functional expression of the $GABA_A$ receptor shows a ligand-gated receptor super-family, *Nature (London),* 328, 221, 1987.
86. **Marachiso, P. C., Cirillo, D., Naldini, L., Primavera, M. V., Teti, A., and Zambonin-Zallone, A.,** Cell substratum interaction of cultured avian osteoclasts is mediated by specific adhesion structures, *J. Cell Biol.,* 99, 1696, 1984.
87. **Teti, A., Ross, P., Miyauchi, A., Teitelbaum, S., Hruska, K., Alvarez, J., Grano, M., Colucci, S., Gehron-Robey, P., and Zambonin-Zallone, A.,** RGD containing sequences of osteopontin and BSP II produce immediate cell signals and affect osteoclast adhesion, *J. Bone Miner. Res.,* 5 (Suppl. 2), 75, 1990.
88. **Zaidi, M., Datta, H. K., Patchell, A., Moonga, B. S., Moss, D. W., Makgoba, M. W., and MacIntyre, I.,** Calcium-activated calcium entry: a novel feedback mechanism of osteoclast control, *J. Bone Miner. Res.,* 4 (Suppl. 1), 200, 1989.
89. **Silver, A., Murrills, R. J., and Etherington, E. J.,** Microelectrode studies on the acid microenvironment beneath adherent macrophages and osteoclasts, *Exp. Cell Res.,* 175, 266, 1988.
90. **Mears, D. C.,** Effects of parathyroid hormone and thyrocalcitonin on the membrane potential of osteoclasts, *Endocrinology,* 88, 1021, 1971.
91. **Ferrier, J., Ward, A., Kanehisa, J., and Heersche, J. N. M.,** Electrophysiological responses of osteoclasts to hormones, *J. Cell. Physiol.,* 128, 23, 1986.
92. **Sims, S. M. and Dixon, S. J.,** Inwardly rectifying K^+ current in osteoclasts, *Am. J. Physiol.,* 256, C1277, 1989.
93. **Schoppa, N. E., Su, Y., Baron, R., and Boulpaep, E. L.,** Identification of single ion channels in neonatal rat osteoclasts, *J. Bone Miner. Res.,* 5 (Suppl. 2), 204, 1990.
94. **Blair, H. C. and Schlesinger, P. H.,** unpublished results.
95. **Heath, J. K., Atkinson, S. J., Meilke, M. C., and Reynolds, J. J.,** Mouse osteoblasts synthesize collagenase in response to bone resorbing agents, *Biochim. Biophys. Acta,* 802, 151, 1984.

96. **Blair, H. C., Dean, D. D., Howell, D. S., Teitelbaum, S. L., and Jeffrey, J. J.,** Hypertrophic chondrocytes produce immunoreactive collagenase *in vivo, Connect. Tissue Res.,* 23, 65, 1989.
97. **DeDuve, C., DeBarsy, T., Poole, B., Trouet, A., Tulkens, P., and van Hoof, F.,** Lysosomotropic agents, *Biochem. Pharmacol.,* 23, 2495, 1974.
98. **Maciewicz, R. A., Wotton, S. F., Etherington, D. J., and Duance, V. C.,** Susceptibility of the cartilage collagens types II, IX, and XI to degradation by the cysteine proteinases, cathepsins B and L, *FEBS Lett.,* 269, 189, 1990.
99. **Barrett, A. J. and Kirschke, H.,** Cathepsin B, cathepsin H, and cathepsin L, *Methods Enzymol.,* 80, 535, 1981.
100. **Lee, C. H., Bonewald, L. F., Wo, G., Chirgwin, J., Rochefort, H., Capony, F., and Muny, G. R.,** Activation of isolated osteoclasts by lysosomal procathepsin D and related molecules, *J. Bone Miner. Res.,* 5 (Suppl. 2), 204, 1990.
101. **Ribari, O. and Sziklai, I.,** Cathepsin D activity in otosclerotic bone and perilymph, *Acta Oto-Laryngol.,* 105, 549, 1988.
102. **Sannes, P. L., Schofield, B. H., and McDonald, D. F.,** Histochemical localization of cathepsin B, dipeptidyl peptidase I, and dipeptidyl peptidase II in rat bone, *J. Histochem. Cytochem.,* 34, 983, 1986.
103. **Baron, R., Eeckhout, Y., Neff, L., Francois-Gillet, C., Henriet, P., Delaisse, J. M., and Vaes, G.,** Affinity purified antibodies reveal the presence of (pro)collagenase in the subosteoclastic bone resorbing compartment, *J. Bone Miner. Res.,* 5 (Suppl. 2), 203, 1990.
104. **Baron, R., Neff, L., Brown, W., Courtoy, P. J., Louvard, D., and Farquahr, M. G.,** Polarized secretion of lysosomal enzymes: co-distribution of cation-independent mannose-6-phosphate receptors and lysosomal enzymes along the osteoclast exocytic pathway, *J. Cell Biol.,* 106, 1863, 1988.
105. **Blair, H. C., Teitelbaum, S. L., Schimke, P. A., Konsek, J. D., Koziol, C. M., and Schlesinger, P. H.,** Receptor-mediated uptake of a mannose-6-phosphate bearing glycoprotein by isolated chicken osteoclasts, *J. Cell. Physiol.,* 137, 476, 1988.
106. **Sly, W. S. and Fischer, H. D.,** The phosphomannosyl recognition system for intracellular and intercellular transport of lysosomal enzymes, *J. Cell. Biochem.,* 18, 67, 1982.
107. **Jones, S. J., Boyde, A., Ali, N. N., and Maconnachie, E.,** Variation in the size of resorption lacunae made *in vitro, Scanning Electron Microsc.,* 4, 1571, 1986.
108. **Whyte, M. P.,** Sclerosing bone dysplasias, in *Primer on the Metabolic Bone Diseases and Disorders of Mineral Metabolism,* Favus, M. J., Ed., American Society of Bone and Mineral Research, Kelseyville, CA, 1990, 215.
109. **Coccia, P. F., Krivit, W., Cervenka, J., et al.,** Successful bone marrow transplantation for infantile malignant osteopetrosis, *N. Engl. J Med.,* 302, 701, 1980.
110. **Siris, E. S.,** Paget's disease of bone, in *Primer on the Metabolic Bone Diseases and Disorders of Mineral Metabolism,* Favus, M. J., Ed., American Society of Bone and Mineral Research, Kelseyville, CA, 1990, 253.
111. **Whyte, M. P., Bergfeld, M. A., Murphy, W. A., Avioli, L. V., and Teitelbaum, S. L.,** Postmenopausal osteoporosis: a heterogeneous disorder as assessed by histomorphometric analysis of iliac crest bone from untreated patients, *Am. J. Med.,* 72, 193, 1982.
112. **Riis, B. J., Rødbro, P., and Christianson, C.,** The role of serum concentrations of sex steroids and bone turnover in the development and occurrence of postmenopausal osteoporosis, *Calcif. Tissue Int.,* 38, 318, 1986.
113. **Riggs, B. L., Wahner, H. W., Melton, L. J., Richelson, L. S., Judd, H. L., and O'Fallon, W. M.,** Dietary calcium intake and rates of bone loss in women, *J. Clin. Invest.,* 80, 979, 1987.
114. **Bertolini, D. R., Nedwin, G. E., Bringman, T. S., Smith, D. D., and Mundy, G. R.,** Stimulation of bone resorption and inhibition of bone formation *in vitro* by human tumor necrosis factor, *Nature (London),* 319, 516, 1986.

115. **Gowen, M. and Mundy, G.,** Actions of recombinant interleukin 1, interleukin 2, and interferon-γ on bone resorption *in vitro, J. Immunol.,* 136, 2478, 1986.
116. **Conca, W., Auron, P., Welgus, H., Krane, S., and Gehrke, L.,** Transcription of early genes (c-*jun, jun*-B, IL-6) induced by interleukin-1β dissociated from effects on expression of late genes (procollagenase, prostromelysin), *J. Bone Miner. Res.,* 5 (Suppl. 2), 205, 1990.
117. **Vaes, G.,** Cellular biology and biochemical mechanism of bone resorption, *Clin. Orthop. Relat. Res.,* 231, 239, 1888.
118. **Francke, E. L. and Neu, H. C.,** Chloramphenicol and tetracycline, *Med. Clin. North Am.,* 71, 1155, 1987.
119. **Warrell, R. P., Bockman, R. S., Coonley, C. J., Isaacs, M., and Staszewski, H.,** Gallium nitrate inhibits calcium resorption from bone and is effective treatment for cancer-related hypercalcemia, *J. Clin. Invest.,* 73, 1487, 1984.
120. **Matkovic, V., Apseloff, G., Shepard, D. R., and Gerber, N.,** Use of gallium to treat Paget's disease of bone: a pilot study, *Lancet,* 335, 72, 1990.
121. **Hall, T. J. and Chambers, T. J.,** Gallium inhibits bone resorption by a direct effect on osteoclasts, *Bone Miner.,* 8, 211, 1990.
122. **Blair, H. C. and Schlesinger, P. H.,** unpublished observations.
123. **Carano, A., Teitelbaum, S., Schlesinger, P., and Blair, H.,** Acidosis augments osteoclast-mediated bone resorption *in vitro, J. Bone Miner. Res.,* 4 (Suppl. 1), 248, 1989.
124. **Barasch, J., Gershon, M. D., Nunez, E. A., Tamir, H., and Al-Awqati, Q.,** Thyrotropin induces acidification of the secretory granules of parafollicular cells by increasing the chloride conductance of the granular membrane, *J. Cell Biol.,* 107, 2137, 1988.
125. **Landry, D. W., Reitman, M., Cragoe, E. J., and Al-Awqati, Q.,** Epithelial chloride channel, *J. Gen. Physiol.,* 90, 779, 1987.
126. **Lopez, E. and Martelly-Bagot, E.,** L'os cellulaire d'un poisson téléostéen, *Anguilla anguilla* L. III. Etude histologique et histophysique au cours de la maturation provoquée par injections d'extrait hypophysaire de Carpe, *Z. Zellforsch.,* 117, 176, 1971.
127. **Moss, M. L.,** The biology of the acellular bone of teleost fish, *Acta Anat.,* 60, 262, 1965.
128. **Clohisy, D. R., Bar-Shavit, Z., Chappel, J. C., and Teitelbaum, S. L.,** 1,25-dihydroxyvitamin D_3 modulates bone marrow macrophage precursor proliferation and differentiation. Up-regulation of the mannose receptor, *J. Biol. Chem.,* 262, 15922, 1987.
129. **McDermott, M. T. and Kidd, G. S.,** The role of calcitonin in the development and treatment of osteoporosis, *Endocr. Rev.,* 8, 377, 1987.

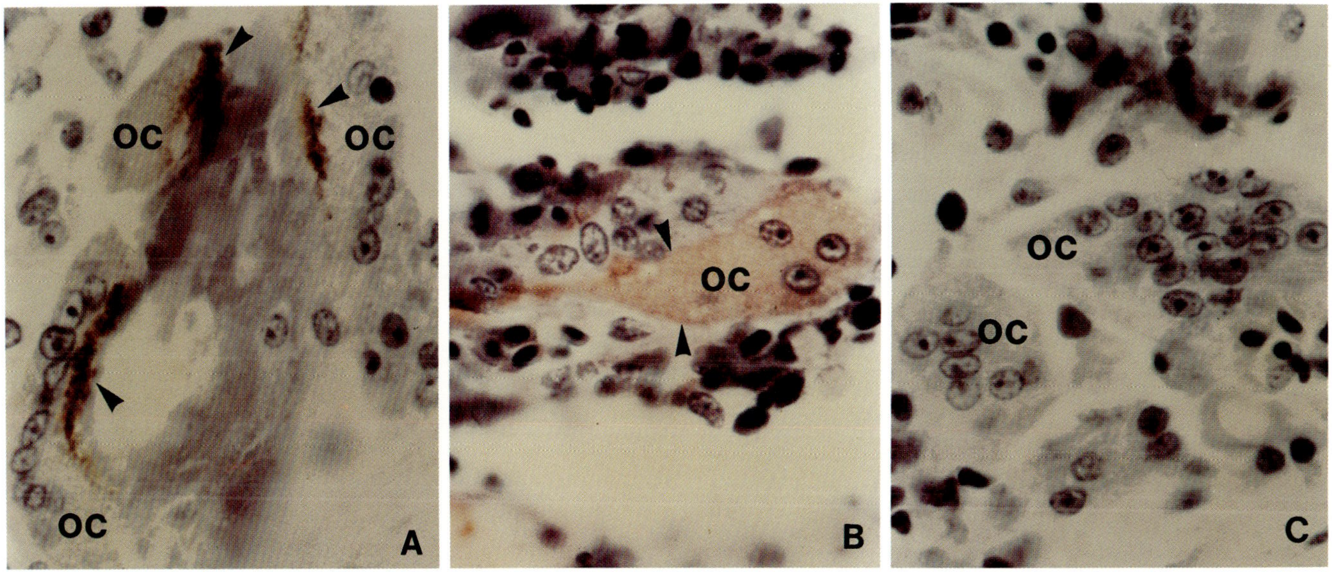

PLATE 1. Staining of avian medullary bone for vacuolar ATPase by the peroxidase anti-peroxidase technique. Gluck's antibody to the 31- kDa subunit of the bovine pump was used to stain medullary bone of calcium-deprived laying hens (*Gallus domesticus*) as reported.[51] OC, osteoclasts; arrows, H^+ pump staining. Osteoclast nuclei are ~5 μm in diameter. (A) Osteoclasts attached to mineralized bone are seen to express extremely high concentrations of the H^+ pump at the bone interface (in this circumstance, the proton pump has been shown to be on the ruffled membrane by electron microscopy).[51] (B) Osteoclasts not attached to mineralized bone have, in some cases, a uniform intracellular distribution of the proton pump. The 5-μm section transects the cell, indicating that the reaction product is not on the cell membrane. It is thought that the pump is in vacuolar membranes, but this cannot be established definitively at the light-microscopic level. (C) Preimmune serum control, establishing specificity of the reaction. These results indicate that the osteoclast is capable of redistributing its HCl-transporting apparatus, a possible means of control. The polarized pump is observed only in cells attached to mineralized bone. Uniform pump distribution through the cytoplasm (vacuolar system?) is seen in a minority of cross sections of osteoclasts not showing attachment to mineralized bone.

Chapter 14*

MECHANISM OF MINERAL SOLUBILIZATION AND MATRIX DEGRADATION IN OSTEOCLASTIC BONE RESORPTION

Jean-Marie Delaissé and Gilbert Vaes

TABLE OF CONTENTS

* Abbreviations used: CI-1, *N*-[3-*N*-(benzyloxycarbonyl)-amino-1-(R)-carboxypropyl]-L-leucyl-*O*-methyl-L-tyrosine *N*-methylamide; E-64, L-trans-epoxysuccinyl-L-leucylamido(4-guanidino)butane; MMP, matrix metalloproteinase; PTH, parathyroid hormone; Z-, benzyloxycarbonyl-.

I. INTRODUCTION

From their very first description,[1] osteoclasts were proposed to be the cells that are directly responsible for removal of the mineralized osseous substance during the process of bone resorption. This substance is known today to consist of an organic matrix, made mainly of insoluble type 1 collagen fibers,[2] embedded in a heavy deposit of solid calcium-phosphate salt, primarily calcium hydroxyapatite, $Ca_{10}(PO_4)_6(OH)_2$.[3] Despite occasional suggestions that the action of the osteoclast could be restricted to mineral solubilization only, it appears that activated osteoclasts resorb both the mineral and organic components of mineralized bone, even if other cell types might possibly cooperate in removal of the latter (reviewed in Reference 4). It appears also[4] that during the osteoclastic resorption of mineralized bone, removal of the mineral precedes degradation of the collagen fibers, and even conditions this degradation, as it is necessary to relieve the physical barrier which prevents the diffusion of collagenolytic enzymes into the bone matrix. Scanning electron micrographs of pits made by osteoclasts in cortical bone (Figure 1) nicely illustrate these two successive steps in osteoclast erosive action.

Instead of extensively reviewing the literature relevant to the cellular and biochemical mechanisms of osteoclastic bone resorption (see our previous reviews[4,6-8]), in this chapter we focus on recent views and document what we believe, perhaps more subjectively, to be the most probable, although still hypothetical, model to account for the action of the osteoclast in bone resorption (see Figure 12 at the end of this chapter).

II. MECHANISM OF BONE MINERAL SOLUBILIZATION

The solid state of bone mineral results from the low solubility of hydroxyapatite and other calcium-phosphate bone salts at physiological pH. The equilibrium established between the solid mineral phase and the concentrations of Ca^{2+} and $HPO_4^{-2}/H_2PO_4^-$ ions in the extracellular fluids can be displaced toward mineral solubilization (see Reference 9) either by addition of chelators, which bind calcium ions into soluble complexes, or by addition of acid, which increases considerably the solubility of calcium hydroxyapatite in the zone between pH 7 and 5 and favors the formation of soluble Ca^{2+} and HPO_4^{-2} ions following:

$$Ca_{10}(PO_4)_6(OH)_2 + 8\,H^+ \rightarrow 6\,HPO_4^{-2} + 10\,Ca^{2+} + 2\,H_2O \qquad (1)$$

As discussed elsewhere,[4] the secretion of citrate, which forms poorly dissociated complexes with calcium and whose production is increased during PTH-induced bone resorption,[10-15] was once thought to be operative in the

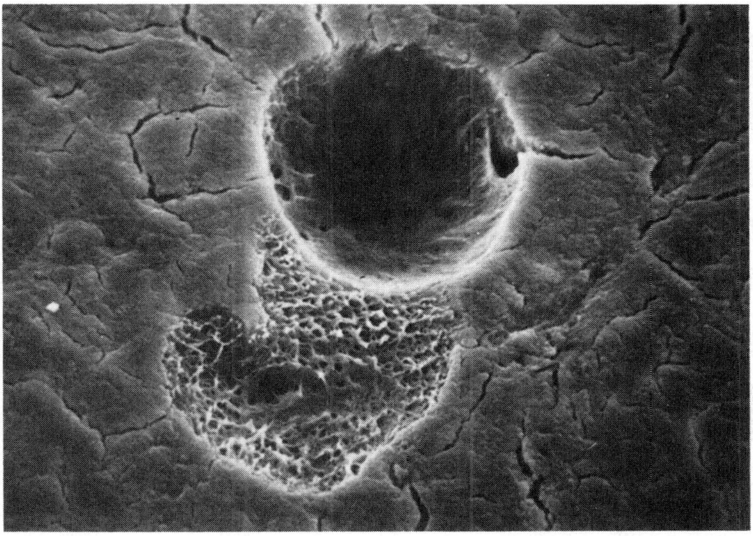

FIGURE 1. Scanning electron micrograph of pits made by human osteoclast in bovine cortical bone slice. Note exposed collagen fibrils in the lower, shallower pit contrasting with the smooth base of the upper, deeper pit. It may be presumed that, while moving on the bone surface, the osteoclast first superficially eroded the lower pit, removing mineral but not much collagen, before leaving the site and settling to erode in more depth the mineral and collagen of the upper pit. (From Murrills, R. J., Shane, E., Lindsay, R., and Dempster, D. W., *J. Bone Miner. Res.*, 4, 259, 1989. With permission.)

mobilization of bone mineral during its resorption, but the amounts of citrate so produced appear to be too low to account for the process. Most of today's evidence points, on the contrary, toward a major role of acid production and secretion by the osteoclasts.

A. EXCRETION OF ACID BY ACTIVATED OSTEOCLASTS

Early evidence had established that bone resorption in PTH-treated bone explants is paralleled by an increased release of acid.[13] It was only recently, however, that activated osteoclasts were shown to secrete acid in the subosteoclastic extracellular compartment where bone resorption occurs, a compartment which is segregated between the ruffled border of the osteoclasts and the bone extracellular matrix to which the osteoclasts are attached through an annular sealing portion of their membrane (see References 6 to 8 and Chapters 6 and 12 of this volume). Acridine orange, a weak base whose unprotonated form diffuses freely through intact biological membranes and is sequestered in the protonated form in closed acidic compartments, was shown to accumulate at the level of osteoclasts[16-18] and, apparently more specifically, in the subosteoclastic resorption compartment,[16] thus establishing its acidic nature (Figure 2). Acridine orange fluorescence, an index of local

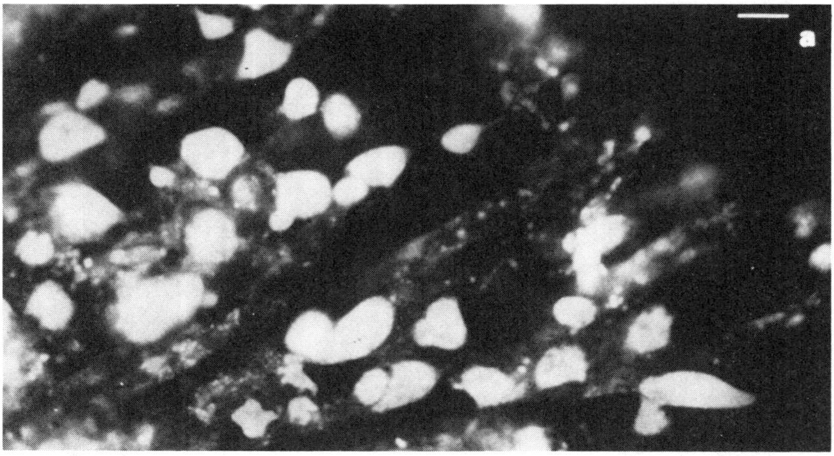

FIGURE 2. Accumulation of acid at the level of osteoclasts. Bones from calcium-deprived chicks were incubated with acridine orange, and their endosteal surface then observed in epifluorescence. The large (>50 μm in diameter), uniformly brilliant, orange fluorescent discs correspond to osteoclasts which have accumulated the dye in the underlying acid resorption compartments. These discs can no longer be seen when the osteoclasts are detached from the bone substrate, presumably due to pH neutralization subsequent to the opening of these compartments to the surrounding, well-buffered extracellular fluids. The bar corresponds to 0.1 mm. (From Baron, R., Neff, L., Louvard, D., and Courtoy, P. J., *J. Cell Biol.*, 101, 2210, 1985. With permission.)

acidity, was increased by PTH[18-20] and inhibited when the activation of the osteoclasts was inhibited by calcitonin.[19,20] Direct pH measurement at the level of the subosteoclastic resorption zone, by either micropuncture[21] or microelectrode studies,[22] provided pH values of 4.5 to 4.8, which shifted toward neutrality under the action of calcitonin,[21] providing further evidence that this local acidity is linked to the active resorptive function of the osteoclast.

B. INVOLVEMENT OF A PROTON PUMP IN THE OSTEOCLAST RUFFLED BORDER

The mechanism of the acidification of the subosteoclastic bone-resorbing compartment (see Chapter 13 of this volume) involves the action of a polarized proton pump which is located at the level of the ruffled border of the osteoclast. Proteins sharing common antigenic determinants with purified H^+,K^+-ATPase from gastric mucosa were initially localized at that level,[16] suggesting similarities with the gastric parietal cells active in H^+ secretion, an hypothesis further supported by the fact that omeprazole, an inhibitor of the gastric H^+,K^+-ATPase proton pump, inhibited bone resorption in cultured explants[23] as well as the osteoclast acidification manifested by acridine orange fluorescence.[19] Further studies, however, have established that the osteoclast proton pump possesses several characteristics in common with the vacuolar type of proton-transport ATPases (H^+-ATPases), and is thus more similar to the electrogenic proton pumps found in endosomes and lysosomes or in renal tubular cells, where it is responsible for the acidification of urine.[24-26] The inhibition of bone resorption in disaggregated osteoclast cultures by bafilomycin A1, a highly specific inhibitor of the vacuolar type H^+-ATPases, supports the proposal that this type of proton-transport ATPase is directly implicated in osteoclastic bone resorption processes.[27]

C. ORIGIN OF THE EXCRETED PROTONS AND ROLE OF CARBONIC ANHYDRASE

On the basis that bone resorption induced by PTH is paralleled by increased production of lactate,[11-13,15,28] it had been proposed that the acidification accompanying the conversion of glucose to lactate (or citrate; see above) could be the source of the H^+ ions required for the solubilization of bone mineral during its resorption. It was even observed[13] that the amount of H^+ ions released as lactic acid under the action of PTH in an organ culture system of resorbing bone was quantitatively sufficient to account for both the solubilization of the mineral and the increased acidification of the culture medium. The fact that calcitonin reduces the lactate production of PTH-stimulated explanted bones[29,30] (as well as the increased accumulation of citrate in the medium surrounding PTH-treated bone explants[31]) suggests that at least part of the PTH-induced lactate (or citrate) release in bone is due to increased production by activated osteoclasts. It is thus possible that the H^+ ions produced together with these organic anions within the osteoclasts may contribute to the feeding of the osteoclast proton pump on the cytosolic site of the ruffled border. However, the hydration of CO_2 into H_2CO_3, a slow process unless catalyzed by carbonic anhydrase, is likely to provide most of the H^+ ions to be excreted by the pump, as it does in kidney tubular cells or gastric parietal cells (see below). Besides providing the ATP necessary for the action of the ruffled border's proton pump, and of the other ion pumps required for the homeostasis of the cell interior (see Chapter 13 of this volume), the abundant mitochondria present in the osteoclasts (see Chapter 6 of this volume) suggest an important local production of CO_2 that could be used for this purpose. Interestingly, PTH-treated bones were shown to produce, under certain conditions, more CO_2 from glucose than untreated bones,[32] and bone resorption induced by PTH *in vitro* appeared to be dependent on the ambient concentration of CO_2.[33]

Siegmund, Dulce, and co-workers were the first to suggest that the activity of carbonic anhydrase was necessary for an osteoclastic secretion of hydrogen ions that was presumably required for bone demineralization. They observed that acetazolamide, an inhibitor of carbonic anhydrase, given *in vivo,* inhibits bone resorption,[34-36] and they demonstrated the presence of carbonic anhydrase in bone tissue.[37] At the same time, carbonic anhydrase was shown to be present in osteoclasts, but not in osteoblasts.[38] These early data and hypotheses have been amply confirmed and extended. The carbonic anhydrase found in osteoclasts[39-45] corresponds to one of the six isoenzymes known to exist in mammals, isoenzyme II.[42,43,45] Genetic evidence implicating this isoenzyme in bone resorption is provided by a congenital deficiency of carbonic anhydrase II, which has been identified in an autosomal recessive syndrome of osteopetrosis, a disease characterized by a failure of bone resorption.[46,47] Also, the inhibition of bone resorption by acetazolamide and other inhibitors of carbonic anhydrase was further observed under various *in vivo* conditions[48-52] or *in vitro,* in cultures of either whole-bone explants[33,53-58] or isolated osteoclasts.[59] The effectiveness of the inhibition of bone resorption *in vitro* by a series of carbonic anhydrase inhibitors was shown to be parallel to their potency as carbonic anhydrase inhibitors;[53] it was also inversely related to the ambient CO_2 concentration.[33,58] Moreover, acetazolamide was shown to inhibit the acidity (acridine orange fluorescence) of osteoclasts *in situ*[18] or in culture.[20,59]

II. MECHANISM OF ORGANIC MATRIX DEGRADATION

In spite of the fact that activated osteoclasts may produce oxygen radicals[60] (see also Chapter 15 of this volume), which might well contribute to some extracellular degradative processes (either directly, or indirectly, by activating the zymogens of collagenase and related MMPs[61,62] or by increasing the proteolytic susceptibility of proteins to be degraded[63]), proteinases appear to be the major agents which are required to degrade into soluble fragments the structural macromolecules, mainly type 1 collagen fibers, constitutive of the organic matrix of bone.

Specific collagenases are necessary to degrade at neutral pH the triple helical body of native collagen molecules (see reviews in Reference 64). These enzymes are produced by several cell types, including many connective tissue cells and macrophages. They are members of the MMP family (see Reference 65) and are often produced together with an efficient protein inhibitor, the tissue inhibitor of metalloproteinases (TIMP). They are released under a latent zymogen form, procollagenase, and activated into fully active collagenase by limited proteolysis, which can be achieved *in vitro* by several proteinases, including trypsin, plasmin, kallikrein, and the lysosomal cathepsin B.[66-69] The mechanism of this activation is, however, complex (see Reference 70); it appears to involve the preliminary activation of an endogenous proactivator,[67-69,71] now identified with the zymogen of another MMP, prostromelysin.[72,73] Little is known about the mode of activation of these zymogens *in vivo.* Plasmin, generated from plasminogen by the plasminogen activators, may well be the most significant activator under many physiological situations. Although collagenase still has an important activity around pH 6,[66] its optimum action is exerted around pH 7.5.

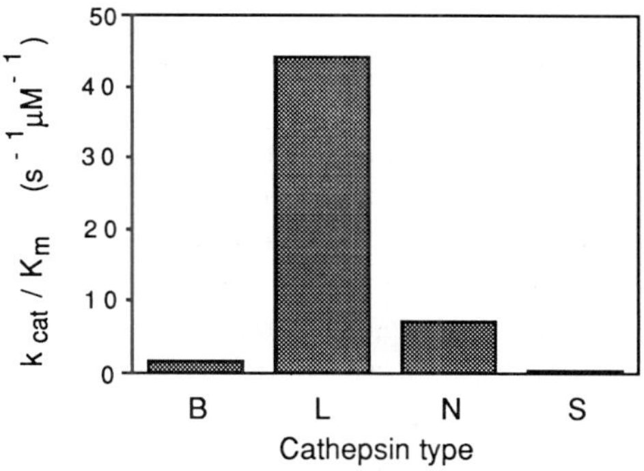

FIGURE 3. Comparative collagenolytic activities of cysteine proteinases. The efficacy of collagen degradation by the main collagenolytic cysteine proteinases (cathepsins B, L, N, and S) is expressed by the ratio k_{cat}/K_m. (Data from Reference 85.)

Various nonspecific lysosomal proteinases belonging to the cysteine proteinase family, and including cathepsins L,B, and N, can efficiently degrade collagen at acid pH.[74-78] These enzymes first cleave the nonhelical telopeptide extensions of the collagen molecules on the helical side of the intermolecular cross-links, which are responsible for collagen fiber polymerization, allowing depolymerization and solubilization of the collagen molecules.[74] These molecules then rapidly denature at acid pH and physiological temperature,[79] allowing the cysteine proteinases, and possibly other lysosomal proteinases, to further rapidly degrade the collagen α-chains to peptides of low molecular weight. These enzymes are responsible for the intracellular digestion of fragments of collagen fibers when they are phagocytosed by macrophages or other cells;[80-82] the digestion then occurs within phagolysosomes, at about pH 5.0.[83,84] Among these enzymes, cathepsin L is particularly active on collagen (Figure 3).

Both types of enzymes, the lysosomal cysteine proteinases and collagenase, appear to be implicated in the degradation of collagen related to bone resorption. It is presently unknown whether the degradation of some other structural macromolecules present in much lower amounts in the matrix, such as the bone noncollagenous proteins or proteoglycans, may represent a limiting factor in bone resorption. Their removal might, however, be accomplished either at acid pH by the lysosomal glycosidases and proteinases (including the cysteine proteinases) or at neutral pH by other, less specific MMPs such as stromelysin (proteoglycanase) or the gelatinases.

A. LYSOSOMAL ENZYMES AND THE NEED FOR CYSTEINE PROTEINASES

In our early studies,[6,13,86,87] the development of bone resorption in a model of PTH-stimulated bone explants was found to be closely correlated with the excretion of a series of lysosomal enzymes. Excretion and bone resorption were induced simultaneously after a few hours of culture,[13] both were inhibited by calcitonin,[87] and their extent was parallel under a wide range of experimental conditions. This led to the proposal[6,13] that the acid hydrolases of the lysosomes are active in the resorption of the organic matrix of bone. These hydrolases exhibit optimal activity at an acid pH corresponding to the pH that exists at the level of the subosteoclastic resorption zone, and it was postulated that they were secreted at that level by the osteoclasts, together with acid, to degrade the collagen and other organic components of the bone extracellular matrix. Various biochemical or ultrastructural studies (reviewed in References 4, 7, and 8) provided considerable support for that hypothesis, establishing in particular[16,88] that lysosomal enzymes, synthesized by the osteoclast, are transported via primary lysosomes toward the ruffled border membrane and are directionally secreted at that level into the bone-resorbing lacuna. Digestion of bone matrix could then occur in that lacuna, producing residues that may diffuse away through cell membranes. Endocytosis for further digestion inside the osteoclast phagolysosomes would then not be required for resorption of the organic matrix, although it could sometimes occur.

More recently, we identified cysteine proteinases as the major lysosomal proteinases whose activity is required for the degradation of collagen in the subosteoclastic bone-resorbing lacuna. Two cysteine proteinases, cathepsins L and B, have been identified in homogenates of bone tissue, together with a cathepsin L-like proteinase of larger molecular weight whose identity is still unclear; all these enzymes were able to depolymerize collagen at acid pH.[89] A good correlation was found in bone cultures between bone resorption and the accumulation of cathepsin B in the culture medium,[90] indicating that collagenolytic cysteine proteinases were present within the "cocktail" of lysosomal enzymes released during bone resorption. More significantly, however, we found[90-92] that the addition of low molecular weight, active site-directed cysteine proteinase inhibitors belonging to three distinct molecular families — leupeptin, antipain Z-Phe-Ala-CHN_2, Z-Phe-Tyr(O-*t*-Bu)-CHN_2, and E-64 — to the culture medium of bone explants considerably inhibits the bone resorption induced by PTH or other agents (Figure 4). This inhibition was fully reversible, indicating that it was not due to irreversible damage to the cells. As the inhibitors did not prevent the PTH-enhanced secretion of lysosomal enzymes,[90,91] it appears likely that they act by blocking the extracellular activity of excreted cysteine proteinases. Given *in vivo* to calcium-deprived rats,[90] inhibitors E-64 and leupeptin induced a concomitant fall in both the serum calcium level and the urinary excretion of hydroxyproline (Figure 5), most likely due to an inhibition of bone resorption. Furthermore, it was shown that the cysteine-proteinase inhibitors acted directly on the osteoclasts, as they reduced to less than 25% of the resorptive activity of isolated osteoclasts cultured on dentine slices by inhibiting not only the numbers, but also the depths and volumes of the resorption pits bored by the cells in this bone-like substrate[98] (Figure 6). Interestingly, these inhibitors had no

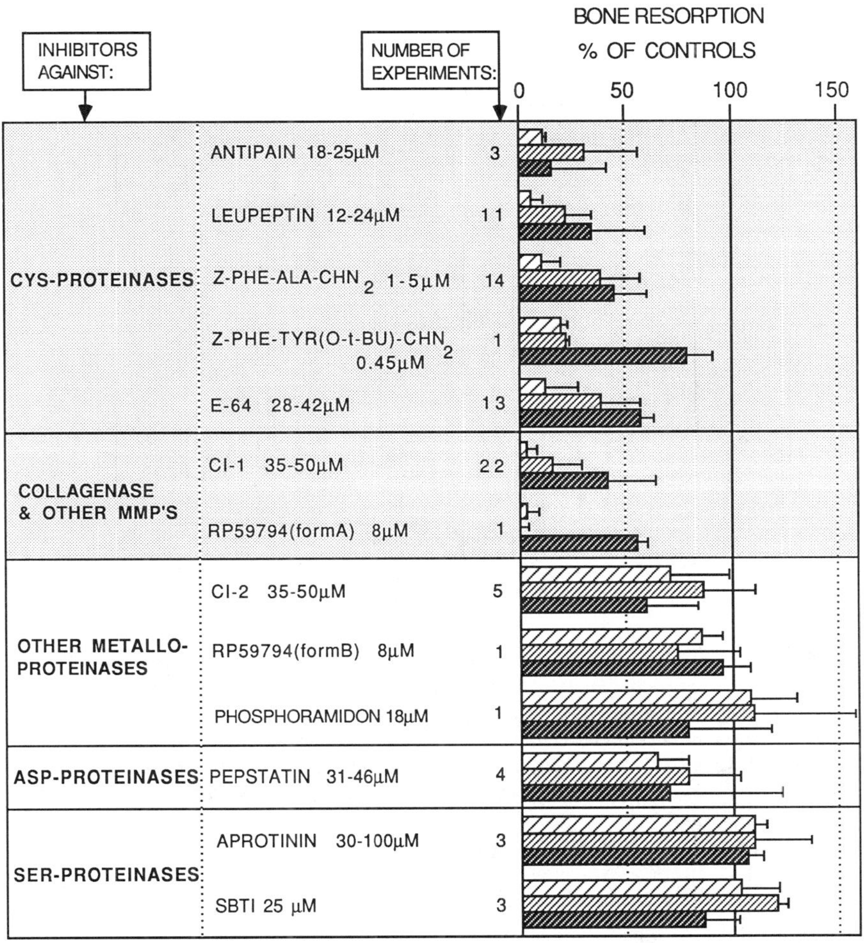

FIGURE 4. Effect of inhibitors of the various proteinase types on the resorption of explanted calvaria in culture. Bone resorption was induced by PTH in mouse embryonic (day 19) calvaria cultured for 2 to 4 d with or without the indicated concentrations of inhibitors. It was then evaluated by the extension of the resorption lacunae (▨, upper bars within each group), by the amount of hydroxyproline left in the explants (▨, central bars), and by the amount of calcium released in the medium (▩, lower bars). The effects of the inhibitors are shown as percent ratios of the controls cultured without inhibitors. The results are presented as means ± SD for the indicated number of experiments (each involving of the four cultures of two calvaria). CI-1[93] and RP59794 (form A)[95] are highly specific inhibitors of collagenase and other MMPs; CI-2 and RP59794 (form B) are their respective stereoisomers, about 200-fold less active on collagenase. For references on the other inhibitors, see References 96 and 97. RP59794 is *N*-[3-(hydroxyamino)carbonyl-2-isobutyl-1-oxopropyl]valine benzylamide. (Data from References 90 to 94 and unpublished observations.)

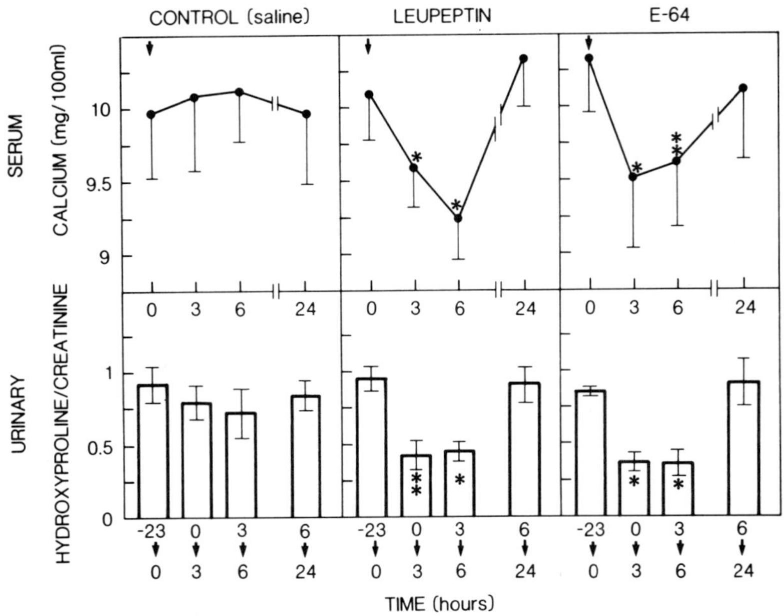

FIGURE 5. *In vivo* effects of cysteine proteinase inhibitors E-64 and leupeptin on serum calcium and urinary hydroxyproline. Rats (80 to 90 g) fed on a low-calcium diet were injected intraperitoneally at time zero with saline supplemented or not with 4 mg of leupeptin or 2 mg of E-64. The serum calcium concentrations and the urinary hydroxyproline/creatinine ratios were measured at the time (or within the time interval) indicated on the abscissae, before or after injection; they are expressed as means ± SD (n = 6 or 10). Comparison to their respective controls: $^{**}p < 0.0025$; $^{*}p < 0.005$ (the other differences are nonsignificant). (From Delaissé, J. M., Eeckhout, Y., and Vaes, G., *Biochem. Biophys. Res. Commun.*, 125, 441, 1984. With permission.)

effect on the osteoclast erosion of the purely mineral layer covering the inner surface of oyster shells.[98]

These studies thus indicate the need for cysteine proteinases in the osteoclastic resorption of bone. The proposal that these enzymes act in the subosteoclastic resorption zone to remove collagen after its initial demineralization is further supported by the demonstration that bone explants treated with cysteine-proteinase inhibitors exhibit under the ruffled border of osteoclasts relatively broad fringes of collagenous matrix, free of mineral crystallites, which are not seen in control bones[99] (Figure 7). This suggested that demineralization had proceeded up to a certain point under the osteoclast, whereas matrix degradation had been retarded. Also, in PTH-treated explants, the degradation of calcified collagen, but not that of uncalcified collagen, is inhibited not only by calcitonin (which suggests its mediation by osteoclasts), but also by leupeptin, an inhibitor of cysteine proteinases.[100]

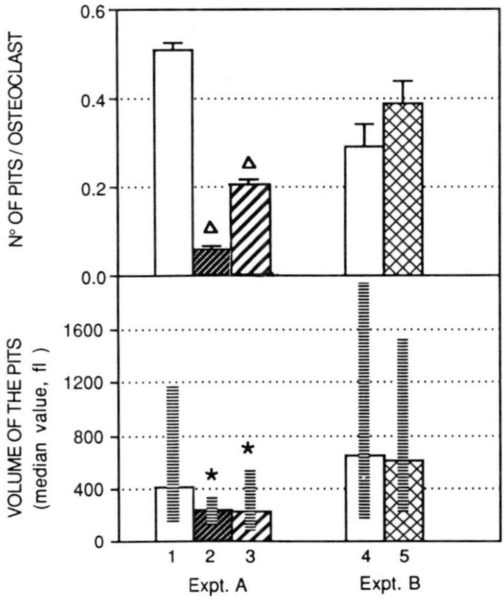

FIGURE 6. Effects of inhibitors of cysteine proteinases and of collagenase on the resorptive activity of isolated osteoclasts. Disaggregated chick osteoclasts were cultured for 1 or 2 d on dentine slices in the absence (☐, blocks 1 or 4) or presence of either the cysteine proteinase inhibitors Z-Phe-Ala-CHN_2, 12.5 μM (▨, blocks 2), and E-64, 40 or 60 μM (▨, blocks 3), or the collagenase inhibitor CI-1, 40 or 100 μM (⊠, blocks 5). The ratios (means ± SD) between the number of resorption pits and the number of osteoclasts are presented in the upper graph. The median volumes of the pits are shown in the lower graph, with vertical bars corresponding to the interquartile range, i.e., the range covering 25% of the measured values on each side of the median. The results presented are from two series of experiments, A and B. Comparison of their respective controls: $^{\Delta}p < 0.001$; $^{*}p < 0.05$ (the other differences are nonsignificant). (Data from Reference 98.)

B. THE NEED FOR COLLAGENASE

In sharp contrast to the behavior of lysosomal enzymes during bone resorption, we could establish, in our early studies, no consistent correlations between resorption in explanted bones and the amounts of either collagenase or procollagenase that accumulated in the culture fluid,[101] even when collagenase was assayed by an immunoreactive method that detected not only native collagenase and procollagenase, but also catalytically inactive forms such as collagenase inhibitor (presumably TIMP) complexes or partially denatured or degraded enzyme.[102] It was observed that procollagenase and collagenase were sometimes released by the bone explants in the absence of any process of collagen resorption[101] and that, on the contrary, bone resorption could be induced by PTH in the absence of immunodetectable (pro)collagenase in the culture fluids.[102]

However, more recently, Eeckhout et al.[103] developed a procedure that allowed the extraction and assay of procollagenase from as little as one fetal

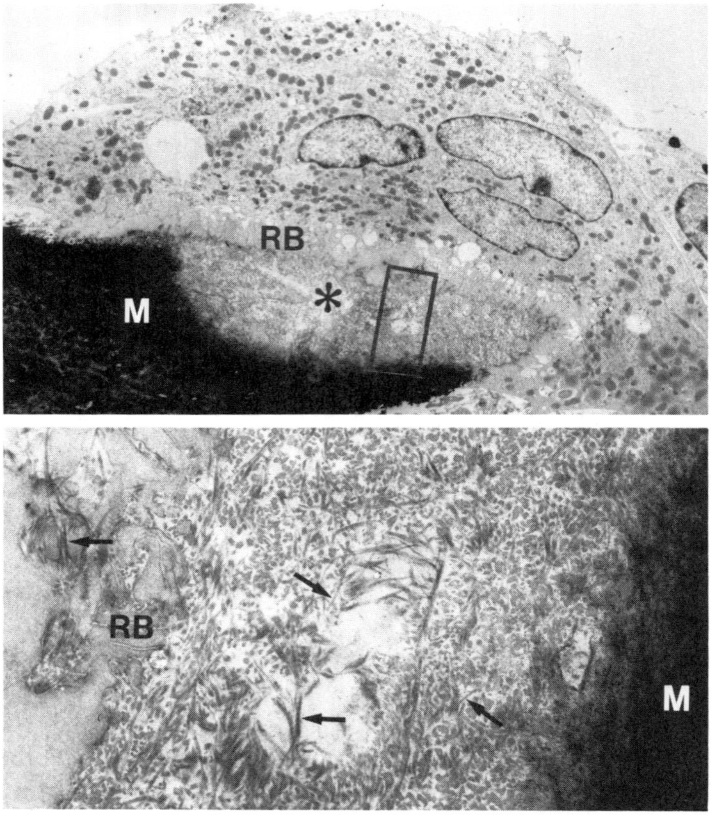

FIGURE 7. Effect of the cysteine proteinase inhibitor leupeptin on the ultrastructure of the osteoclastic resorption zone. Upper: osteoclast of a bone explant cultured for 24 h in the presence of 65 μM leupeptin. (Magnification × 1920.) Note the mineral-free area (asterisk) of bone matrix adjacent to the ruffled border (RB). M, mineralized bone surface. Lower: high magnification (Magnification × 19,200) of mineral-free collagenous matrix (see rectangle in upper part). Arrows indicate collagen fibrils. (From Everts, V., Beertsen, W., and Schroder, R., *Calcif. Tissue Int.*, 43, 172, 1988. With permission.)

(19-d-old) mouse calvarium (about 5 mg wet weight). In nonstimulated bones, procollagenase was found to be mainly associated with the mineralized matrix and required demineralization for its extraction.[94,103] Its activity appears to be sufficient to degrade within a few days most of the collagen matrix of the bone explants once it had been demineralized (Figure 8). Using this procedure, it was established that bone resorption, induced in culture by a variety of bone-resorbing agents such as PTH, 1,25$(OH)_2$ vitamin D_3, prostaglandin E_2, interleukin-1, "catabolin" (an interleukin-1), tumor necrosis factor-α, retinoic acid, and endotoxin,[94] or induced *in vivo* by PTH,[94,104] accompanied a further accumulation of procollagenase inside the bone tissue (Figure 9).

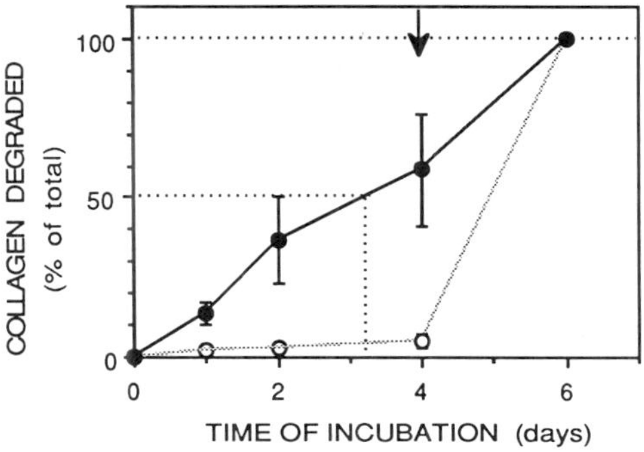

FIGURE 8. Efficiency of the endogenous collagenase present in embryonic calvaria to degrade the collagenous matrix to which it is associated. Calvaria from 3-d-old mice were demineralized[103] in a dialysis bag and further incubated in the medium of these bags for the times indicated at 37°C and pH 7.3 with (○) or without (●) the collagenase inhibitor CI-1 (85 μ*M*). Collagen degradation was monitored by the release of soluble hydroxyproline. A large excess of bacterial collagenase was added to the media on day 4 (arrow) to allow determination of the total amount of collagen (100%) present in the tissue. Each point is the mean ± SD for five calvaria. Note that 50% of the collagen matrix was degraded in about 3 d by the endogenous, CI-1-inhibitable collagenase. (100% corresponds to 69 ± 12 and 74 ± 9 μg hydroxyproline per calvarium for the series incubated, respectively, without or with CI-1).

Under PTH stimulation *in vitro,* neither this accumulation nor bone resorption was affected by indomethacin, suggesting that prostaglandins did not mediate these effects. However, calcitonin, acetazolamide, or bisphosphonates, while inhibiting bone resorption, did not prevent PTH-induced procollagenase synthesis (Figure 10). This suggested that the synthesis was unlikely to be due to activated osteoclasts, but that osteoblasts, or osteoblast-related cells, were more probably the source of the enzyme. Indeed, collagenase has been immunolocalized in osteoblasts *in situ,* but not in osteoclasts,[105] and osteoblast production of procollagenase is enhanced by PTH and various other bone-resorbing agents.[106-109]

Most significantly, studies with specific enzyme inhibitors indicated that collagenase is necessary for the resorption of bone tissue (at least for that of the actively growing, woven-type of bone considered in all our experimental models of living bone, as discussed further below). Indeed, in our studies, potent, specific synthetic inhibitors of collagenase and related MMPs were found to inhibit the PTH-induced resorption of explanted fetal mouse calvaria, preventing predominantly the degradation of bone collagen and partially inhibiting bone demineralization[93,110] (Figure 4; see also Figure 11). Similar inhibitors also inhibited bone resorption in newborn mice *in vivo.*[110] When bone explants were treated with these inhibitors, broad fringes of demineralized collagenous matrix could be observed under the ruffled border of the osteoclast, similar to those found (see Figure 7) when the explants were treated with inhibitors of cysteine proteinases.[111] This suggested that collagenase was

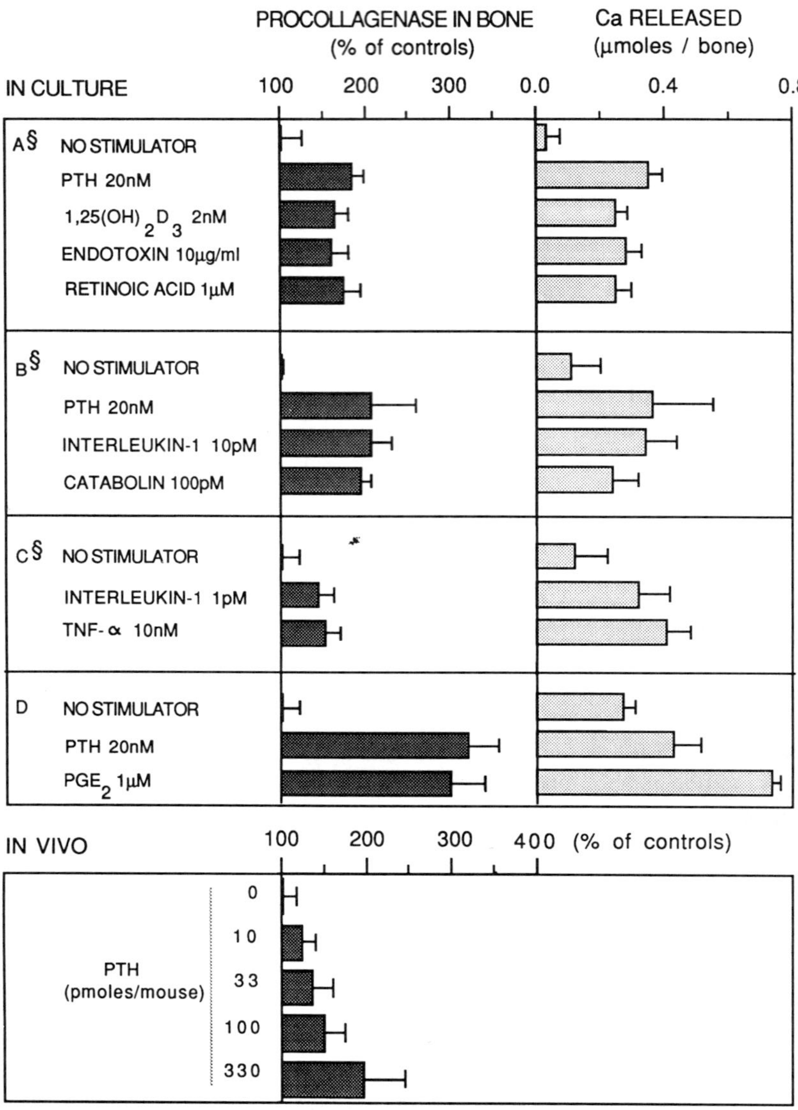

FIGURE 9. *In vitro* and *in vivo* effects of stimulators of bone resorption on the accumulation of procollagenase in mouse calvaria. (1) *In vitro* effects.[94] Calvaria from 19-d-old fetal mice were cultured for 2 d with bone-resorbing agents as indicated. The results of a series of experiments (A, B, C, and D) are presented in which bone resorption was monitored by the release of calcium in the media, and procollagenase assayed in the explants at the end of the cultures. (§, cultures done in the presence of 14 μM indomethacin). (2) *In vivo* effects.[94,104] Three-d-old mice were injected twice daily for 3 d with the indicated amounts of PTH. Procollagenase was then assayed in their calvaria. Data are shown as means ± SD (n = 4 or 5). When compared to their respective controls, all the effects of the bone-resorbing agents, except those of 10 or 33 pmol PTH *in vivo,* were significant at $p < 0.05$. (Data from References 94 and 104.)

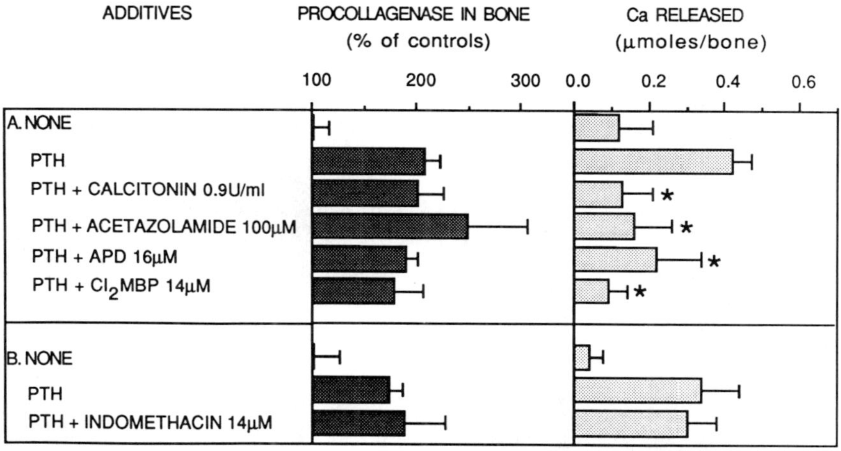

FIGURE 10. Effects of calcitonin and various pharmacological agents on PTH-induced bone resorption and accumulation of procollagenase in cultured calvaria. Calvaria from 19-d-old fetal mice were cultured for 2 d with or without PTH (2×10^{-8} *M*) in the presence or not in the presence of one of the following agents at the indicated concentrations: calcitonin, acetazolamide, 3-amino-1-hydroxypropylidene-1,1-bisphosphonate (APD), dichloromethylene-bisphosphonate (Cl_2MBP), or indomethacin. Bone resorption was monitored by the release of calcium in the media, and procollagenase assayed in the explants at the end of the cultures. The results of two experiments, A and B, are presented, each involving four cultures of two calvaria (means ± SD). Comparison to the respective PTH groups: *p <0.05 (the other differences are nonsignificant). (Data from Reference 94.)

active at that level, together with the cysteine proteinases, in the resorption of the collagenous matrix. In support of this, the recent demonstration that affinity-purified antibodies reveal the presence of (pro)collagenase in the subosteoclastic bone-resorbing compartment in the growth plates of newborn mouse, at the interface between osteoclasts and the bone matrix,[112] is particularly interesting. Although this study provides no clue as to the origin of the enzyme (the simplest hypothesis would be a secretion by the osteoclasts themselves, but the antigen was not seen in these cells, nor in other types of cells; see, however, the Note Added in Proof in Reference 134), it strongly suggests participation of collagenase in the osteoclastic resorption of bone collagen once this collagen has been demineralized by the osteoclast-secreted protons. However, there does not seem to exist an absolute requirement for collagenase activity for bone erosion at the level of the osteoclast. Indeed, the direct resorptive activity exerted by isolated osteoclasts on dentine slices (devitalized lamellar bone-like surfaces devoid of any lining layer of nonmineralized collagen) was not affected by the presence of the same collagenase inhibitor (CI-1) that was found to inhibit bone resorption when acting *in vitro,* in organ cultures of bone, or *in vivo*[98] (Figure 6).

Thus, it appears that collagenase participates in the bone resorptive processes, but that osteoclasts are nevertheless probably able to resorb bone (at least devitalized bone; see Reference 113) without the participation of col-

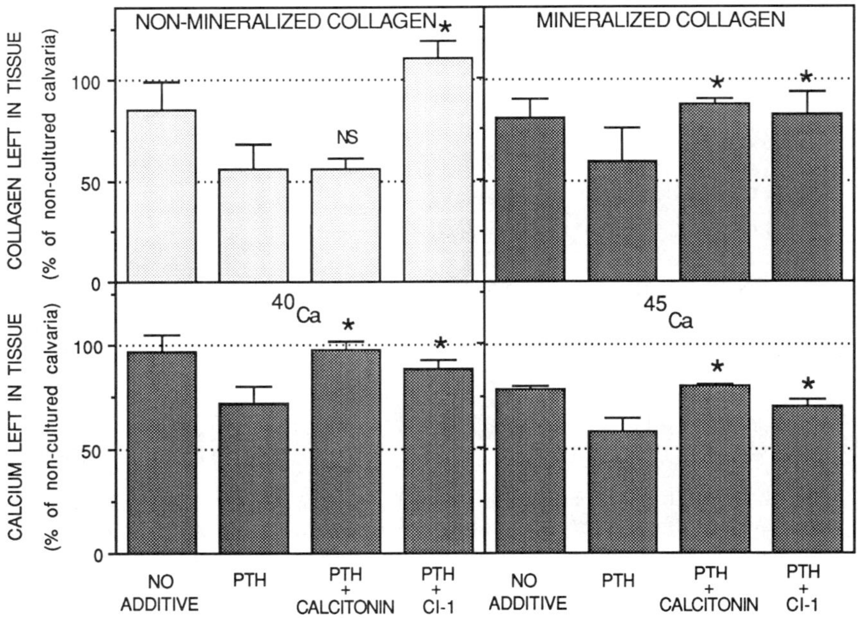

FIGURE 11. Effect of calcitonin and the collagenase inhibitor CI-1on PTH-induced resorption of nonmineralized and mineralized bone collagen. ^{45}Ca-labeled calvaria from 19-d-old fetal mice (see Reference 13) were cultured for 4 d either without additives or with PTH (2×10^{-8} *M*) in the presence or not in the presence of calcitonin (0.9 U/ml) or CI-1 (50 μ*M*). The total amounts (100%) of nonmineralized or mineralized collagen present in the tissue at the start of the cultures (i.e., the amounts of collagen available or not available for degradation by bacterial collagenase; see Reference 94), as well as those of ^{40}Ca and ^{45}Ca, were evaluated on a series of paired, noncultured calvaria. At the end of the cultures, the corresponding amounts left in the calvaria were measured and expressed as a percent of those present in the noncultured calvaria. Each point is the mean ± SD of four cultures involving two calvaria each. Comparison to the PTH group: $^{*}p < 0.05$, and N.S., nonsignificant. It is apparent that PTH induces a loss of nonmineralized collagen from the explants which is unaffected by calcitonin (and thus presumably not linked to the activation of osteoclasts), but completely inhibited by CI-1 and thus collagenase dependent. However, the loss of calcium and mineralized collagen induced by PTH is inhibited by both calcitonin and CI-1. This may possibly indicate that the inhibition of collagenase, by interfering with the removal of the nonmineralized collagen covering bone surfaces, prevents (as does calcitonin) the activation of osteoclasts and their action on mineralized bone. Alternatively, however, it may indicate that collagenase is also required for the removal of collagen in the subosteoclastic bone-resorbing compartment (see text).

lagenase (a fact also supported by Blair et al.[114]), in all likelihood due to the sole actions of acid and lysosomal cysteine proteinases.

C. COOPERATION BETWEEN CYSTEINE PROTEINASES AND COLLAGENASE IN BONE RESORPTION

Admitting that lysosomal cysteine proteinases and collagenase are both required for bone resorption, how could their actions be integrated within a plausible, although still largely hypothetical model (Figure 12)?

It has been proposed by Chambers et al.[115] that collagenase, produced by osteoblasts, could exert a permissive role for the activity of the osteoclasts. This collagenase would be responsible for the removal of nonmineralized collagen which, whenever present on bone surfaces, is considered to prevent osteoclast activation and adherence to mineralized bone, two processes which condition their resorptive action (reviewed in Reference 4). Such a mechanism might well be operative in resorption of the actively growing, rapidly turning over, woven-type of bone, where the preliminary removal of surface layers of unmineralized matrix (''osteoid'') by enzymes secreted by the osteoblasts could well be required to allow osteoclastic resorption of the subjacent, newly formed mineralized bone matrix. In our culture studies[94] (see also Reference 116), collagenase indeed appeared to be responsible for the removal of non-mineralized collagen during the PTH-induced resorption of explanted fetal mouse calvaria. This removal, which was prevented by inhibitors of collagenase, was not inhibited by calcitonin and is thus likely to be independent of the osteoclasts (Figure 11). Latent procollagenase could be activated at that level by a plasminogen activator/plasmin cascade,[67-69] as PTH and other bone-resorbing agents are able to induce the production of plasminogen activator by isolated osteoblast-like cells in culture[117] as well as its accumulation in explanted bone tissue.[118] However, as discussed elsewhere,[119] the universal presence of such a ''shield'' of unmineralized organic material on all resorbable bone surfaces, particularly in mature lamellar bone, is open to question, and, thus, so is the need for a preliminary action of collagenase at that level under all situations of bone resorption.

When considering the direct action of activated osteoclasts in the resorption of mineralized bone, it now appears likely that both collagenase and lysosomal cysteine proteinases cooperate in the removal of the organic matrix achieved under the osteoclast ruffled border (Figure 12). As mentioned above, lysosomal enzymes are probably secreted at that level, but the origin of collagenase is still unclear. Despite the fact that collagenase has not been detected so far in osteoclasts,[105,120] (see, however, the Note Added in Proof at Reference 134) it could nevertheless be produced and secreted by these cells under specific states of differentiation or activation, similar to the situation in monocytes/macrophages.[121-124] Conceivably, this activation state could depend not only on exogenous mediators such as cytokines or other mediators released by nearby cells, but also on messages provided to the osteoclast by the nature of the matrix on which it adheres or by some of its preliminary degradation products. Interestingly, stromelysin, a matrix metalloproteinase closely related to collagenase, has been recently immunolocalized in osteoclasts.[125] Alternatively, procollagenase could be a product of other neighboring cell types, possibly lining cells, osteocytes, or other osteoblast-related cells,[106,107,126] and reach the resorptive zone by diffusion. It could even be deposited by osteoblasts, together with collagen, into the bone matrix during its formation, and later activated under the osteoclasts to serve for its resorption.[94,103,127,128]

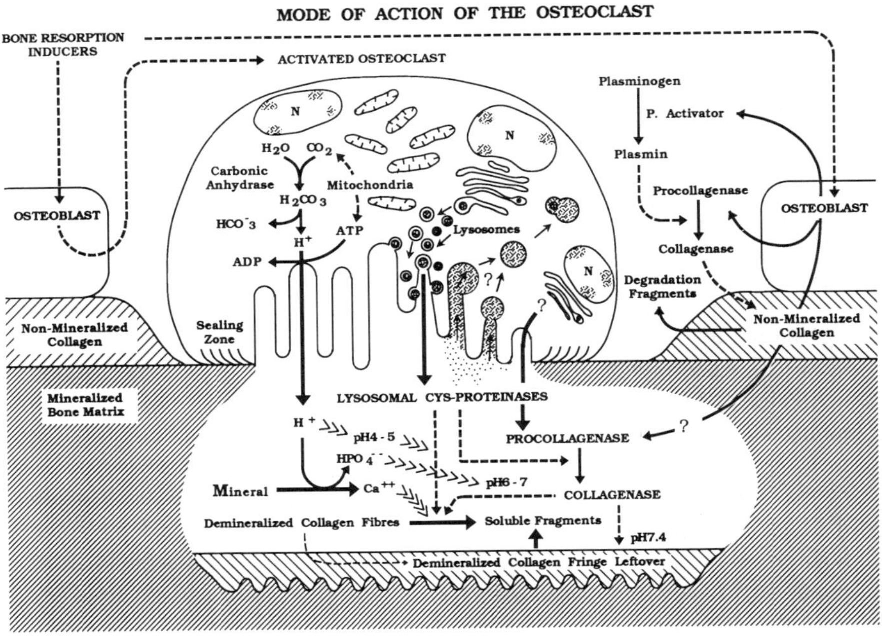

FIGURE 12. Hypothetical view of the mode of action of osteoclasts in the resorption of mineralized bone matrix. This action (center) is presumed to be prepared, whenever necessary, by osteoblast-secreted procollagenase (right) which, after its proper activation (achieved possibly through the plasminogen activator/plasmin cascade), may remove the nonmineralized collagen surface layer shielding the mineralized matrix, thus allowing attachment of the osteoclast to the mineralized bone. In the bone-resorptive lacuna established under the osteoclast, acid is secreted together with lysosomal cysteine proteinases; procollagenase is also present, but its origin is still unknown (the osteoclast itself or nearby osteoblast-type cells?) (see, however, the Note Added in Proof at Reference 134). Acid solubilizes the mineral, and the cysteine proteinases cooperate with collagenase in the removal of the demineralized collagenous matrix. The cysteine proteinases may activate procollagenase, and the subsequent concerted collagenolytic actions of these two types of enzymes are likely to result in a more efficient degradation of collagen than would be achieved by each enzyme acting independently. When the osteoclast detaches from the bone surface and moves away, the acidity of the resorptive lacuna is suddenly neutralized, inhibiting further actions of the cysteine proteinases. Collagenase, active at neutral pH, may then complete the resorbing process by removing the fringes of demineralized collagen left at the base of the lacuna. (The necessity of osteoblasts to allow the activation of osteoclasts by most inducers of bone resorption — see Reference 4 and Chapter 5 of this volume — is indicated at the left of the figure.) (Modified from Reference 4.)

Besides their direct collagenolytic action, cysteine proteinases could be responsible for the generation of active collagenase out of its zymogen, as cathepsin B is known to activate procollagenase.[69] Collagenase could then cooperate with cysteine proteinases in the degradation of the demineralized collagen. We have speculated elsewhere, with Baron et al.[112] and Everts et al.,[111] about possible modes of cooperation between the two enzyme systems. Conceivably, the collagenolytic action of the cysteine proteinases, optimal around pH 4 to 5.[74-78,89] could be exerted preferentially in the most acid portion of the bone-resorption lacuna, in the immediate vicinity of the ruffled border, while that of collagenase, effective between pH 6 and 7.5,[66] could be predominant deeper in the lacuna, at the interface between demineralized and mineralized matrix, where the pH is likely to be more neutral due to the buffering capacity of the dissolved bone salts. It is even likely that the concerted, sequential action of both types of collagenolytic enzymes may render collagen degradation much more efficient compared to the isolated action of each separately.[129] Also, the high Ca^{2+} concentrations, which are presumed to occur in the acid microenvironment of the resorption lacuna, should favor bone collagenolysis, since they increase the sensitivity of insoluble bone collagen toward proteolysis.[130,131] Besides these actions, collagenase could also be responsible for degradation of the fringes of yet undegraded, but already demineralized collagen likely to be left at the base of the resorption lacuna when the osteoclast detaches from the bone (see References 5 and 132, Figure 1), and thereby allow a sudden neutralization of the pH of the resorption lacuna. The acid-requiring cysteine proteinases would then no longer be active, so that the removal of these collagen fringes likely would require the action of neutral collagenase to allow completion of the resorbing process. Interestingly, the microcinematographic studies of Gaillard[133] had already shown, in 1957, that the process of matrix resorption often goes on even after the osteoclast has moved away from the bone surface it was eroding. This could represent the terminal, collagenase-dependent phase of the osteoclastic resorption process.

ACKNOWLEDGMENTS

The work done in our laboratory was supported by the Fund for Medical Scientific Research (Belgium) and by the Belgian State Prime Minister's Office, Science Policy Programming (interuniversity attraction poles, grant 7bis, and concerted actions, grant 88/93.122). We gratefully acknowledge the skillful secretarial assistance of Y. Marchand.

REFERENCES

1. **Kölliker, A.,** Die normale Resorption des Knochengewebes und ihre Bedeutung für die Entstehung der typischen Knochenformen, Vogel, Leipzig, 1873.
2. **Veis, A.,** Bones and teeth, in *Extracellular Matrix Biochemistry,* Piez, K. A. and Reddi, A. H., Eds., Elsevier, New York, 1984, 329.
3. **Neuman, W. F.,** Bone mineral and calcification mechanism, in *Fundamental and Clinical Bone Physiology,* Urist, M. R., Ed., Lippincott, Philadelphia, 1980, 83.

4. **Vaes, G.,** Cellular biology and biochemical mechanism of bone resorption. A review of recent developments on the formation, activation, and mode of action of osteoclasts, *Clin. Orthop. Relat. Res.*, 231, 239, 1988.
5. **Murrills, R. J., Shane, E., Lindsay, R., and Dempster, D. W.,** Bone resorption by isolated human osteoclasts *in vitro:* effects of calcitonin, *J. Bone Miner. Res.*, 4, 259, 1989.
6. **Vaes, G.,** La Résorption Osseuse et l'Hormone Parathyroïdienne, De Visscher, Bruxelles, and Maloine, Paris, 1966.
7. **Vaes, G.,** Lysosomes and the cellular physiology of bone resorption, in *Lysosomes in Biology and Pathology,* Vol. 1, Dingle, J. T. and Fell, H. B., Eds., Elsevier/North-Holland, Amsterdam, 1969, 217.
8. **Vaes, G.,** Collagenase, lysosomes and osteoclastic bone resorption, in *Collagenase in Normal and Pathological Connective Tissue,* Woolley, D. E. and Evanson, J. M., Eds., John Wiley & Sons, New York, 1980, 185.
9. **Neuman, W. F. and Neuman, M. W.,** *The Chemical Dynamics of Bone Mineral,* University of Chicago Press, Chicago, 1958, chap. 2.
10. **Firschein, H. E., Martin, G. R., Mulryan, B. J., Strates, B. S., and Neuman, W. F.,** Concerning the mechanism of action of parathyroid hormone. I. Ion-gradients, *J. Am. Chem. Soc.*, 80, 1619, 1958.
11. **Martin, G. R., Firschein, H. E., Mulryan, B. J., and Neuman, W. F.,** Concerning the mechanism of action of parathyroid hormone. II. Metabolic effects, *J. Am. Chem. Soc.*, 80, 6201, 1958.
12. **Mecca, C. E., Martin, G. R., and Goldhaber, P.,** Alteration of bone metabolism in tissue culture in response to parathyroid extract, *Proc. Soc. Exp. Biol. Med.*, 113, 538, 1963.
13. **Vaes, G.,** On the mechanism of bone resorption. The action of parathyroid hormone on the excretion and synthesis of lysosomal enzymes and on the extracellular release of acid by bone cells, *J. Cell Biol.*, 39, 676, 1968.
14. **Vaes, G. and Nichols, G., Jr.,** Metabolic studies of bone *in vitro*. III. Citric acid metabolism and bone mineral solubility, *J. Biol. Chem.*, 236, 3323, 1961.
15. **Vaes, G. and Nichols, G., Jr.,** Effect of a massive dose of parathyroid extract on bone metabolic pathways, *Endocrinology,* 70, 546, 1962.
16. **Baron, R., Neff, L., Louvard, D., and Courtoy, P. J.,** Cell-mediated extracellular acidification and bone resorption: evidence for a low pH in resorbing lacunae and localization of a 100-kD lysosomal membrane protein at the osteoclast ruffled border, *J. Cell Biol.*, 101, 2210, 1985.
17. **Green, P. M., Marshall, M. J., and Nisbet, N. W.,** A study of osteoclasts on calvaria of normal and osteopetrotic (mi/mi) mice by vital staining with acridine orange, *Br. J. Exp. Pathol.*, 67, 85, 1986.
18. **Anderson, R. E., Jee, W. S. S., and Woodbury, D. M.,** Stimulation of carbonic anhydrase in osteoclasts by parathyroid hormone, *Calcif. Tissue Int.*, 37, 646, 1985.
19. **Anderson, R. E., Woodbury, D. M., and Jee, W. S. S.,** Humoral and ionic regulation of osteoclast acidity, *Calcif. Tissue Int.*, 39, 252, 1986.
20. **Hunter, S. J., Schraer, H., and Gay, C. V.,** Characterization of isolated and cultured chick osteoclasts: the effects of acetazolamide, calcitonin, and parathyroid hormone on acid production, *J. Bone Miner. Res.*, 3, 297, 1988.
21. **Fallon, M. D.,** Bone resorbing fluid from osteoclasts is acidic. An *in vitro* micropuncture study, in *Endocrine Control of Bone and Calcium Metabolism,* Cohn, D. V., Fujita, T., Potts, J. T., Jr., and Talmage, R. V., Eds., Elsevier, Amsterdam, 1984, 144.
22. **Silver, I. A., Murrills, R. J., and Etherington, D. J.,** Microelectrode studies on the acid microenvironment beneath adherent macrophages and osteoclasts, *Exp. Cell Res.*, 175, 266, 1988.

23. **Tuukkanen, J. and Väänänen, H. K.,** Omeprazole, a specific inhibitor of H^+-K^+-ATPase, inhibits bone resorption *in vitro, Calcif. Tissue Int.,* 38, 123, 1986.
24. **Blair, H. C., Teitelbaum, S. L., Ghiselli, R., and Gluck, S.,** Osteoclastic bone resorption by a polarized vacuolar proton pump, *Science,* 245, 855, 1989.
25. **Bekker, P. J. and Gay, C. V.,** Biochemical characterization of an electrogenic vacuolar proton pump in purified chicken osteoclast plasma membrane vesicles, *J. Bone Miner. Res.,* 5, 569, 1990.
26. **Väänänen, H. K., Karhukorpi, E. K., Sundquist, K., Wallmark, B., Roininen, I., Hentunen, T., Tuukkanen, J., and Lakkakorpi, P.,** Evidence for the presence of a proton pump of the vacuolar H^+-ATPase type in the ruffled borders of osteoclasts, *J. Cell Biol.,* 111, 1305, 1990.
27. **Sundquist, K., Lakkakorpi, P., Wallmark, B., and Väänänen, K.,** Inhibition of osteoclast proton transport by bafilomycin Al abolishes bone resorption, *Biochem. Biophys. Res. Commun.,* 168, 309, 1990.
28. **Borle, A. B., Nichols, N., and Nichols, G., Jr.,** Metabolic studies of bone *in vitro.* II. The metabolic patterns of accretion and resorption, *J. Biol. Chem.,* 235, 1211, 1960.
29. **Johnston, C. C. and Deiss, W. P.,** An inhibitory effect of thyrocalcitonin on calcium release *in vivo* and bone metabolism *in vitro, Endocrinology,* 78, 1139, 1966.
30. **Heersche, J. N. M.,** The effect of thyrocalcitonin and parathyroid hormone on bone metabolism in tissue culture. Metabolism of bone in tissue culture: the effects of parathyroid hormone and thyrocalcitonin on the process of bone demineralization, *Proc. K. Ned. Akad. Wet.,* 72C, 594, 1969.
31. **Nisbet, J. A. and Nordin, B. E. C.,** Thyrocalcitonin inhibition of bone resorption in tissue culture, in *Calcitonin,* Taylor, S., Ed., Heinemann, London, 1968, 230.
32. **Forscher, B. K. and Cohn, D. V.,** *In vitro* carbohydrate metabolism of bone: effect of treatment of intact animal with parathyroid extract, in *Mechanism of Hard Tissue Destruction,* Sognnaes, R. F., Ed., American Association for the Advancement of Science, Washington, D.C., 1963, 577.
33. **Mahgoub, A. and Stern, P. H.,** Carbon dioxide and the effect of parathyroid hormone on bone *in vitro, Am. J. Physiol.,* 226, 1272, 1974.
34. **Siegmund, P. and Dulce, H. J.,** Zur Biochemie der Knochenauflösung. I. Einfluss des Carboanhydratase-Inhibitors 2-acetamino-2.3.4-thiodiazol-sulfonamid-(5) (Diamox) auf den Calciumstoffwechsel von Legehennen, *Hoppe Seyler's Z. Physiol. Chem.,* 320, 149, 1960.
35. **Dulce, H. J. and Siegmund, P.,** Zur Biochemie der Knochenauflösung. II. Der Einfluss von Diamox auf das Plasma-Calcium östron-behandelter Hähne, *Hoppe Seyler's Z. Physiol. Chem.,* 320, 160, 1960.
36. **Körber, F. and Siegmund, P.,** Versuche über Stoffwechselvorgange bei der Auflösung von Knochenmineral, in *Calcified Tissues,* Richelle, L. J. and Dallemagne, M. J., Eds., Collection des Colloques de l'Université, Liège, 1965, 207.
37. **Dulce, H. J., Siegmund, P., Körber, F., and Schütte, E.,** Zur Biochemie der Knochenauflösung. III. Ueber das Vorkommen von Carboanhydratase im Knochen, *Hoppe Seyler's Z. Physiol. Chem.,* 320, 163, 1960.
38. **Simasaki, M. and Yagi, T.,** Histochemistry of carbonic anhydrase with special reference to the osteoclast, *Dent. Bull. Osaka Univ.,* 1, 89, 1960.
39. **Gay, C. V. and Mueller, W. J.,** Carbonic anhydrase and osteoclasts: localization by labeled inhibitor autoradiography, *Science,* 183, 432, 1974.
40. **Gay, C. V., Ito, M. B., and Schraer, H.,** Carbonic anhydrase activity in isolated osteoclasts, *Metab. Bone Dis. Relat. Res.,* 5, 33, 1984.
41. **Anderson, R. E., Schraer, H., and Gay, C. V.,** Ultrastructural immunocytochemical localization of carbonic anhydrase in normal and calcitonin-treated chick osteoclasts, *Anat. Rec.,* 204, 9, 1982.

42. **Väänänen, H. K. and Parvinen, E. K.,** High active isoenzyme of carbonic anhydrase in rat calvaria osteoclasts. Immunohistochemical study, *Histochemistry,* 78, 481, 1983.
43. **Väänänen, H. K.,** Immunohistochemical localization of carbonic anhydrase isoenzymes I and II in human bone, cartilage and giant cell tumor, *Histochemistry,* 81, 485, 1984.
44. **Cao, H. and Gay, C. V.,** Effects of parathyroid hormone and calcitonin on carbonic anhydrase location in osteoclasts of cultured embryonic chick bone, *Experientia,* 41, 1472, 1985.
45. **Sundquist, K. T., Leppilampi, M., Jarvelin, K., Kumpulainen, T., and Väänänen, H. K.,** Carbonic anhydrase isoenzymes in isolated rat peripheral monocytes, tissue macrophages and osteoclasts, *Bone,* 8, 33, 1987.
46. **Sly, W. S., Hewett-Emmett, D., Whyte, M. P., Yu, Y. S. L., and Tashian, R. E.,** Carbonic anhydrase II deficiency identified as the primary defect in the autosomal recessive syndrome of osteopetrosis with renal tubular acidosis and cerebral calcification, *Proc. Natl. Acad. Sci. U.S.A.,* 80, 2752, 1983.
47. **Sly, W. S., Whyte, M. P., Sundaram, V., Tashian, R. E., Hewett-Emmett, D., Guibaud, P., Vainsel, M., Baluarte, H. J., Gruskin, A., Al-Mosawi, M., Sakati, N., and Ohlsson, A.,** Carbonic anhydrase II deficiency in 12 families with the autosomal recessive syndrome of osteopetrosis with renal tubular acidosis and cerebral calcification, *N. Engl. J. Med.,* 313, 139, 1985.
48. **Waite, L. C., Volkert, W. A., and Kenny, A. D.,** Inhibition of bone resorption by acetazolamide in the rat, *Endocrinology,* 87, 1129, 1970.
49. **Waite, L. C.,** Carbonic anhydrase inhibitors, parathyroid hormone and calcium metabolism, *Endocrinology,* 91, 1160, 1972.
50. **Conaway, H. H., Waite, L. C., and Kenny, A. D.,** Immobilization and bone mass in rats. Effects of parathyroidectomy and acetazolamide, *Calcif. Tissue Res.,* 11, 323, 1973.
51. **Kenny, A. D.,** Role of carbonic anhydrase in bone: partial inhibition of disuse atrophy of bone by parenteral acetazolamide, *Calcif. Tissue Int.,* 37, 126, 1985.
52. **Brown, G. M., Morris, C. A., Mitnick, M. A., and Insogna, K. L.,** Treatment of humoral hypercalcemia of malignancy in rats with inhibitors of carbonic anhydrase, *J. Bone Miner. Res.,* 5, 1037, 1990.
53. **Minkin, C. and Jennings, J. M.,** Carbonic anhydrase and bone remodeling: sulfonamide inhibition of bone resorption in organ culture, *Science,* 176, 1031, 1982.
54. **Hall, G. E. and Kenny, A. D.,** Role of carbonic anhydrase in bone resorption induced by 1,25-dihydroxyvitamin D_3 *in vitro, Calcif. Tissue Int.,* 37, 134, 1985.
55. **Hall, G. E. and Kenny, A. D.,** Role of carbonic anhydrase in bone resorption induced by prostaglandin E_2 *in vitro, Pharmacology,* 30, 339, 1985.
56. **Hall, G. E. and Kenny, A. D.,** Bone resorption induced by parathyroid hormone and dibutyryl cyclic AMP. Role of carbonic anhydrase, *J. Pharmacol. Exp. Ther.,* 238, 778, 1986.
57. **Hall, G. E. and Kenny, A. D.,** Role of carbonic anhydrase in bone resorption: effect of acetazolamide on basal and parathyroid hormone-induced bone metabolism, *Calcif. Tissue Int.,* 40, 212, 1987.
58. **Raisz, L. G., Simmons, H. A., Thompson, W. J., Shepard, K. L., Anderson, P. S., and Rodan, G. A.,** Effects of a potent carbonic anhydrase inhibitor on bone resorption in organ culture, *Endocrinology,* 122, 1083, 1988.
59. **Hunter, S. J., Rosen, C. J., and Gay, C. V.,** *In vitro* resorptive activity of isolated chick osteoclasts: effects of carbonic anhydrase inhibition, *J. Bone Miner. Res.,* 6, 61, 1991.
60. **Garrett, I. R., Boyce, B. F., Oreffo, R. O. C., Bonewald, L., Poser, J., and Mundy, G. R.,** Oxygen-derived free radicals stimulate osteoclastic bone resorption in rodent bone *in vitro* and *in vivo, J. Clin. Invest.,* 85, 632, 1990.
61. **Weiss, S. J., Peppin, G., Ortiz, X., Ragsdale, C., and Test, S. T.,** Oxidative autoactivation of latent collagenase by human neutrophils, *Science,* 227, 747, 1985.

62. **Peppin, G. J. and Weiss, S. J.,** Activation of the endogenous metalloproteinase, gelatinase, by triggered human neutrophils, *Proc. Natl. Acad. Sci. U.S.A.,* 83, 4322, 1986.
63. **Rivett, A. J. and Levine, R. L.,** Enhanced proteolytic susceptibility of oxidized proteins, *Biochem. Soc. Trans.,* 15, 816, 1987.
64. **Woolley, D. E. and Evanson, J. M., Eds.,** *Collagenase in Normal and Pathological Connective Tissues,* John Wiley & Sons, New York, 1980.
65. **Birkedal-Hansen, H., Ed.,** Matrix Metalloproteinases, *Proc. Matrix Metalloproteinase Conference, Destin, Florida,* Gustav Fischer, Stuttgart, 1992.
66. **Vaes, G.,** The release of collagenase as an inactive proenzyme by bone explants in culture, *Biochem. J.,* 126, 275, 1972.
67. **Vaes, G. and Eeckhout, Y.,** Procollagenase and its activation, in *Dynamics of Connective Tissue Macromolecules,* Burleigh, P. M. C. and Poole, A. R., Eds., Elsevier/North-Holland, Amsterdam, 1975, 129.
68. **Vaes, G. and Eeckhout, Y.,** The zymogen of bone collagenase and its activation, in *Protides of the Biological Fluids,* Vol. 22, Peeters, H., Ed., Pergamon Press, Oxford, 1975, 391.
69. **Eeckhout, Y. and Vaes, G.,** Further studies on the activation of procollagenase, the latent precursor of bone collagenase: effects of lysosomal cathepsin B, plasmin and kallikrein, and spontaneous activation, *Biochem. J.,* 166, 21, 1977.
70. **Springman, E. B., Angleton, E. L., Birkedal-Hansen, H., and Van Wart, H. E.,** Multiple modes of activation of latent human fibroblast collagenase: evidence for the role of a Cys^{73} active-site zinc complex in latency and a "cysteine switch" mechanism for activation, *Proc. Natl. Acad. Sci. U.S.A.,* 87, 364, 1990.
71. **Vaes, G.,** Multiple steps in the activation of the inactive precursor of bone collagenase by trypsin, *FEBS Lett.,* 28, 198, 1972.
72. **Murphy, G., Cockett, M. I., Stephens, P. E., Smith, B. J., and Docherty, A. J. P.,** Stromelysin is an activator of procollagenase. A study with natural and recombinant enzymes, *Biochem. J.,* 248, 265, 1987.
73. **He, C., Wilhelm, S. M., Pentland, A. P., Marmer, B. L., Grant, G. A., Eisen, A. Z., and Goldberg, G. I.,** Tissue cooperation in a proteolytic cascade activating human interstitial collagenase, *Proc. Natl. Acad. Sci. U.S.A.,* 86, 2632, 1989.
74. **Burleigh, M. C., Barrett, A. J., and Lazarus, G. S.,** Cathepsin B1. A lysosomal enzyme that degrades native collagen, *Biochem. J.,* 137, 387, 1974.
75. **Evans, P. and Etherington, D. J.,** Characterization of cathepsin B and collagenolytic cathepsin from human placenta, *Eur. J. Biochem.,* 83, 87, 1978.
76. **Kirschke, H., Kembhavi, A. A., Bohley, P., and Barrett, A. J.,** Action of rat liver cathepsin L on collagen and other substrates, *Biochem. J.,* 201, 367, 1982.
77. **Kirschke, H. and Barrett, A. J.,** Chemistry of lysosomal proteases, in *Lysosomes: Their Role in Protein Breakdown,* Glaumann, H. and Ballard, F. J., Eds., Academic Press, London, 1987, 193.
78. **Maciewicz, R. A. and Etherington, D. J.,** A comparison of four cathepsins (B, L, N and S) with collagenolytic activity from rabbit spleen, *Biochem. J.,* 256, 433, 1988.
79. **Hayashi, T. and Nagai, Y.,** Effect of pH on the stability of collagen molecule in solution, *J. Biochem.,* 73, 999, 1973.
80. **Parakkal, P. F.,** Involvement of macrophages in collagen resorption, *J. Cell Biol.,* 41, 345, 1969.
81. **Melcher, A. H. and Chan, J.,** Phagocytosis and digestion of collagen by gingival fibroblasts *in vivo:* a study of serial sections, *J. Ultrastruct. Res.,* 77, 1, 1981.
82. **Everts, V., Beertsen, W., and Tigchelaargutter, W.,** The digestion of phagocytosed collagen is inhibited by the proteinase inhibitors leupeptin and E-64, *Collagen Relat. Res.,* 5, 315, 1985.
83. **Okhuma, S. and Poole, B.,** Fluorescence probe measurement of intralysosomal pH in living cells and perturbation of pH by various agents, *Proc. Natl. Acad. Sci. U.S.A.,* 75, 3327, 1978.

84. **Reeves, J. P.,** The mechanism of lysosomal acidification, in *Lysosomes in Biology and Pathology,* Vol. 7, Dingle, J. T., Dean, R. T., and Sly, W., Eds., Elsevier, Amsterdam, 1984, 175.
85. **Maciewicz, R. A., Etherington, D. J., Kos, J., and Turk, V.,** Collagenolytic cathepsins of rabbit spleen: a kinetic analysis of collagen degradation and inhibition by chicken cystatin, *Collagen Relat. Res.,* 7, 295, 1987.
86. **Vaes, G.,** Excretion of acid and of lysosomal acid hydrolases during bone resorption induced in tissue culture by parathyroid extract, *Exp. Cell Res.,* 39, 470, 1965.
87. **Vaes, G.,** Inhibitory actions of calcitonin on resorbing bone explants in culture and on their release of lysosomal hydrolases, *J. Dent. Res.,* 51 (Suppl.), 362, 1972.
88. **Baron, R., Neff, L., Brown, W., Courtoy, P. J., Louvard, D., and Farquhar, M. G.,** Polarized secretion of lysosomal enzymes: co-distribution of cation-independent mannose-6-phosphate receptors and lysosomal enzymes along the osteoclast exocytic pathway, *J. Cell Biol.,* 106, 1863, 1988.
89. **Delaissé, J. M., Ledent, P., and Vaes, G.,** Collagenolytic cysteine proteinases of bone tissue: cathepsin B, (pro)cathepsin L and a cathepsin L-like 70 kDa proteinase, *Biochem. J.,* 279, 167, 1991.
90. **Delaissé, J. M., Eeckhout, Y., and Vaes, G.,** *In vivo* and *in vitro* evidence for the involvement of cysteine proteinases in bone resorption, *Biochem. Biophys. Res. Commun.,* 125, 441, 1984.
91. **Delaissé, J. M., Eeckhout, Y., and Vaes, G.,** Inhibition of bone resorption in culture by inhibitors of thiol proteinases, *Biochem. J.,* 192, 365, 1980.
92. **Delaissé, J. M., Ledent, P., Eeckhout, Y., and Vaes, G.,** Cysteine proteinases and bone resorption, in *Cysteine Proteinases and Their Inhibitors,* Turk, V., Ed., Walter de Gruyter, Berlin, 1986, 259.
93. **Delaissé, J. M., Eeckhout, Y., Sear, C., Galloway, A., McCullagh, K., and Vaes, G.,** A new synthetic inhibitor of mammalian tissue collagenase inhibits bone resorption in culture, *Biochem. Biophys. Res. Commun.,* 133, 483, 1985.
94. **Delaissé, J. M., Eeckhout, Y., and Vaes, G.,** Bone-resorbing agents affect the production and distribution of procollagenase as well as the activity of collagenase in bone tissue, *Endocrinology,* 123, 264, 1988.
95. **Lelièvre, Y., Bouboutou, R., Boiziau, J., Faucher, D., Achard, D., and Cartwright, T.,** Low molecular weight, sequence based, collagenase inhibitors selectively block the interaction between collagenase and TIMP (tissue inhibitor of metalloproteinases), *Matrix,* 10, 292, 1990.
96. **Barrett, A. J. and Salvesen, G., Eds.,** *Proteinase Inhibitors,* Elsevier, Amsterdam, 1986.
97. **Kirschke, H., Wikstrom, P., and Shaw, E.,** Active center differences between cathepsins L and B: the S_1 binding region, *FEBS Lett.,* 228, 128, 1988.
98. **Delaissé, J. M., Boyde, A., Maconnachie, E., Ali, N. N., Sear, C. H. J., Eeckhout, Y., Vaes, G., and Jones, S. J.,** The effects of inhibitors of cysteine-proteinases and collagenase on the resorptive activity of isolated osteoclasts, *Bone,* 8, 305, 1987.
99. **Everts, V., Beertsen, W., and Schroder, R.,** Effects of the proteinase inhibitors leupeptin and E-64 on osteoclastic bone resorption, *Calcif. Tissue Int.,* 43, 172, 1988.
100. **Jilka, R. L. and Hamilton, J. W.,** Evidence for two pathways for stimulation of collagenolysis in bone, *Calcif. Tissue Int.,* 37, 300, 1985.
101. **Lenaers-Claeys, G. and Vaes, G.,** Collagenase, procollagenase and bone resorption: effects of heparin, parathyroid hormone and calcitonin, *Biochim. Biophys. Acta,* 584, 375, 1979.
102. **François-Gillet, C., Delaissé, J. M., Eeckhout, Y., and Vaes, G.,** Immunoreactive collagenase and bone resorption, *Biochim. Biophys. Acta,* 673, 1, 1981.

103. **Eeckhout, Y., Delaissé, J. M., and Vaes, G.,** Direct extraction and assay of bone tissue collagenase and its relation to parathyroid hormone-induced bone resorption, *Biochem. J.*, 239, 739, 1986.
104. **Delaissé, J. M. and Van Ngoc, H.,** Induction of procollagenase in bone tissue under *in vivo* conditions, *Calcif. Tissue Int.*, 46(Suppl. 2), A17, 1990.
105. **Sakamoto, S. and Sakamoto, M.,** Biochemical and immunohistochemical studies on collagenase in resorbing bone in tissue culture. A novel hypothesis for the mechanism of bone resorption, *J. Periodont. Res.*, 17, 523, 1982.
106. **Heath, J. K., Atkinson, S. J., Meikle, M. C., and Reynolds, J. J.,** Mouse osteoblasts synthesize collagenase in response to bone resorbing agents, *Biochim. Biophys. Acta*, 802, 151, 1984.
107. **Otsuka, K., Sodek, J., and Limeback, H. F.,** Synthesis of collagenase and collagenase inhibitors by osteoblast-like cells in culture, *Eur. J. Biochem.*, 145, 123, 1984.
108. **Sakamoto, M. and Sakamoto, S.,** Immunocytochemical localization of collagenase in isolated mouse bone cells, *Biomed. Res.*, 5, 29, 1984.
109. **Partridge, N. C., Jeffrey, J. J., Ehlich, L. S., Teitelbaum, S. L., Fliszar, C., Welgus, H. G., and Kahn, A. J.,** Hormonal regulation of the production of collagenase and a collagenase inhibitor activity by rat osteogenic sarcoma cells, *Endocrinology*, 120, 1956, 1987.
110. **Delaissé, J. M., Terlain, B., Cartwright, T., Lefebvre, V., and Vaes, G.,** A collagenase inhibitor inhibits bone resorption *in vivo*, as evaluated by ^{3}H-tetracycline retention in bones of prelabelled mice, *Bone*, 12 (Abstr.), 289, 1991.
111. **Everts, V., Delaissé, J. M., Korper, W., Niehof, A., Vaes, G., and Beertsen, W.,** The degradation of collagen in the bone-resorbing compartment underlying the osteoclast involves the participation of both cysteine-proteinases and matrix metalloproteinases, *J. Cell. Physiol.* 150, 221, 1992.

112a. **Baron, R., Eeckhout, Y., Neff, L., François-Gillet, C., Henriet, P., Delaissé, J. M., and Vaes, G.,** Affinity purified antibodies reveal the presence of (pro)collagenase in the subosteoclastic bone resorbing compartment, *J. Bone Miner. Res.*, 5 (Suppl. 2) (Abstr.), 203, 1990.

112b. **Delaissé, J. M., Eeckhout, Y., Neff, L., François-Gillet, C., Henriet, P., Su, Y., Vaes, G., and Baron, R.,** (Pro)collagenase is present in rodent osteoclasts and in the underlying bone resorbing compartment, submitted.

113. **Shimizu, H., Sakamoto, M., and Sakamoto, S.,** Bone resorption by isolated osteoclasts in living versus devitalized bone: differences in mode and extent and the effects of human recombinant tissue inhibitor of metalloproteinases, *J. Bone Miner. Res.*, 5, 411, 1990.
114. **Blair, H. C., Kahn, A. J., Crouch, E. C., Jeffrey, J. J., and Teitelbaum, S. L.,** Isolated osteoclasts resorb the organic and inorganic components of bone, *J. Cell Biol.*, 102, 1164, 1986.
115. **Chambers, T. J., Darby, J. A., and Fuller, K.,** Mammalian collagenase predisposes bone surfaces to osteoclastic resorption, *Cell Tissue Res.*, 241, 671, 1985.
116. **Jilka, R. L.,** Procollagenase associated with the noncalcified matrix of bone and its regulation by parathyroid hormone, *Bone*, 10, 353, 1989.
117. **Hamilton, J. A., Lingelbach, S., Partridge, N. C., and Martin, T. J.,** Regulation of plasminogen activator production by bone-resorbing hormones in normal and malignant osteoblasts, *Endocrinology*, 116, 2186, 1985.
118. **Leloup, G., Peeters-Joris, Ch., Delaissé, J. M., Opdenakker, G., and Vaes, G.,** Tissue and urokinase type plasminogen activators in bone tissue and their regulation by parathyroid hormone, *J. Bone Miner. Res.*, 6, 1081, 1991.
119. **Vaes, G.,** Letter to the editor: concerning the presence or absence of an unmineralized matrix layer beneath the osteoblastic lining cells (reply to Dr. M. A. Andersen), *Clin. Orthop. Relat. Res.*, 241, 309, 1989.

120. **Blair, H. C., Dean, D. D., Howell, D. S., Teitelbaum, S. L., and Jeffrey, J. J.,** Hypertrophic chondrocytes produce immunoreactive collagenase *in vivo, Connect. Tissue Res.*, 23, 65, 1989.
121. **Laub, R., Huybrechts-Godin, G., Peeters-Joris, Ch., and Vaes, G.,** Degradation of collagen and proteoglycan by macrophages and fibroblasts: individual potentialities of each cell type and cooperative effects through the activation of fibroblasts by macrophages, *Biochim. Biophys. Acta,* 721, 425, 1982.
122. **Cury, J. D., Campbell, E. J., Lazarus, C. J., Albin, R. J., and Welgus, H. G.,** Selective up-regulation of human alveolar macrophage collagenase production by lipopolysaccharide and comparison to collagenase production by fibroblasts, *J. Immunol.*, 141, 4306, 1988.
123. **Shapiro, S. D., Campbell, E. J., Kobayashi, D. K., and Welgus, H. G.,** Immune modulation of metalloproteinase production in human macrophages. Selective pretranslational suppression of interstitial collagenase and stromelysin biosynthesis by interferon-gamma, *J. Clin. Invest.*, 86, 1204, 1990.
124. **Welgus, H. G., Campbell, E. J., Cury, J. D., Eisen, A. Z., Senior, R. M., Wilhelm, S. M., and Goldberg, G. I.,** Neutral metalloproteinases produced by human mononuclear phagocytes. Enzyme profile, regulation, and expression during cellular development, *J. Clin. Invest.*, 86, 1496, 1990.
125. **Case, J. P., Sano, H., Lafyatis, R., Remmers, E. F., Kumkumian, G. K., and Wilder, R. L.,** Transin stromelysin expression in the synovium of rats with experimental erosive arthritis. *In situ* localization and kinetics of expression of the transformation-associated metalloproteinase in euthymic and athymic Lewis rats, *J. Clin. Invest.*, 84, 1731, 1989.
126. **Sakamoto, S. and Sakamoto, M.,** Isolation and characterization of collagenase synthesized by mouse bone cells in culture, *Biomed. Res.*, 5, 39, 1984.
127. **Gillet, C., Eeckhout, Y., and Vaes, G.,** Purification of procollagenase and collagenase by affinity chromatography on Sepharose-collagen, *FEBS Lett.*, 74, 126, 1977.
128. **Mechanic, G. L., Binderman, I., and Harell, A.,** A novel hypothesis for bone resorption and remodeling, in *Current Advances in Skeletogenesis. Int. Congr. Ser.*, Vol. 589, Silberman, M. and Slavkin, H. C., Eds., Excerpta Medica, Amsterdam, 1982, 322.
129. **Danielsen, C. C.,** Age-related thermal stability and susceptibility to proteolysis of rat bone collagen, *Biochem. J.*, 272, 697, 1990.
130. **Etherington, D. J. and Birkedal-Hansen, H.,** The influence of dissolved calcium salts on the degradation of hard-tissue collagens by lysosomal cathepsins, *Collagen Relat. Res.*, 7, 185, 1987.
131. **Eeckhout, Y.,** Possible role and mechanism of action of dissolved calcium in the degradation of bone collagen by lysosomal cathepsins and collagenase, *Biochem. J.*, 272, 529, 1990.
132. **Ali, N. N., Boyde, A., and Jones, S. J.,** Mobility and resorption: osteoclastic activity *in vitro, Anat. Embryol.*, 170, 51, 1984.
133. **Gaillard, P. J.,** Parathyroid gland and bone *in vitro, Schweiz. Med. Wochenschr.*, 87, 217, 1957.
134. **Note Added in Proof (April 1992).** Our most recent studies[112b] have identified the presence of immunoreactive collagenase within the osteoclasts themselves, particularly in their perinuclear areas. This suggests that these cells are at the source of the collagenase found in the subosteoclastic resorption zone, a view which is further supported by the recent detection of collagenase mRNA in bovine odontoclasts (T. Okamura, *Jpn. J. Oral Biol.*, 34, 95, 1992).

Chapter 15

REACTIVE OXYGEN SPECIES AS POTENTIAL MEDIATORS OF OSTEOCLAST ACTION

Robert A. Greenwald and Barry R. Rifkin

TABLE OF CONTENTS

I. INTRODUCTION

Specialized cell types generally employ a variety of mediators and effector substances in order to carry out their ordained functions. The families of mediators used by such cells range from macromolecules of great molecular weight, to enzymes and proteins of modest size, to small peptides and cytokines, and, at the low end of the molecular weight chain, to a family of tiny (under 100 Da) but highly reactive substances derived from oxygen and possessing a number of interesting properties. Over the past 2 decades, there has been a sustained interest in the reactive species derived from oxygen as mediators of events in connective tissue biology. In light of the relationship between the osteoclast and the monocyte/macrophage lines — the latter, including phagocytic cells, well known for their ability to generate reactive oxygen species — and with the recent recognition that osteoclasts are perhaps capable of using such radicals for their bone-resorbing functions, the osteoclast joins the ranks of other cell types that appear to be able to convert oxygen into a species (or series of species) that can alter macromolecules, activate enzymes, affect matrix synthesis, induce chemotaxis, and perhaps even interact with bone mineral. These potential functions are the subject of this review.

Biomedical scientists are not accustomed to thinking of oxygen as a toxic substance, nor of contemplating molecules with half-lives of perhaps a nanosecond. Thus, superoxide has a certain mystical quality to it that captivates the interest of many researchers, and it is therefore not surprising that the literature on the role of oxygen radicals in disease is rather large. A few years ago, no less an authority than Linus Pauling determined that more papers are written each year about superoxide dismutase (SOD) than about any other enzyme. In this chapter, we address specifically those aspects that appear to relate directly to bone matrix in general and to osteoclasts in particular.

II. BIOLOGICALLY RELEVANT RADICALS AND THEIR REACTIONS

A full description of oxygen radical biochemisry and the possible role of such radicals in human diseases is far beyond the scope of this chapter. Several comprehensive reviews have appeared.[1-4] In brief, a free radical is defined as any atom, group of atoms, or molecule with an unpaired electron occupying an outer orbital. Molecular oxygen, O_2, is a biradical with two such unpaired electrons. A free radical need not be negatively charged, i.e., it may be neutral if the number of electrons equals the number of protons in the nucleus.

The biologically relevant activated oxygen species derived from oxygen are the superoxide anion, singlet oxygen, hydroxyl radical, and hydrogen peroxide. The one-electron reduction of oxygen, i.e., addition of one additional electron to an O_2 molecule, results in formation of the superoxide anion (superoxide radical), O_2^-. The two-electron reduction product of oxygen, when fully protonated, forms hydrogen peroxide. A third species of activated oxygen is recognized, called singlet oxygen, in which the number of electrons is not altered, but the spin of one electron and its orbital position is reversed. Finally, there is an additional reduction product of oxygen, called the hydroxyl radical, ·OH, which is the most highly reactive of all the oxy radicals. Although the phrase is somewhat cumbersome, the best global term which has been applied to all four oxygen-based species as a class is "reactive oxygen species, ROS". The term "oxy radicals" seems somewhat trivializing and belies their origin; the phrase "oxygen-derived free radicals" is close, but hydrogen peroxide is not a radical. Hence, ROS is used herein, except when a particular radical species is emphasized.

Free radicals might seem at first glance to be chemical curiosities of no relevance to mammalian biology. However, we now know that ROS can be generated via many naturally occurring biologic processes. For example, many cellular enzymes (including xanthine oxidase [XO], aldehyde oxidase, flavin dehydrogenase, and some peroxidases) generate superoxide as part of their action. Mitochondria and chloroplasts generate a steady flux of superoxide, and irradiation of tissue is a well-known source of such radicals. However the sources of greatest relevance to human disease are phagocytic cells and tissue ischemia/reperfusion. The latter mechanism has not yet been brought to bear on bone as an organ, but the monocyte/macrophage heritage of the osteoclast certainly makes the former seem relevant.

It was initially believed that the superoxide radical itself was directly toxic to target tissues, but direct evidence for this has not been forthcoming. It now appears that many tissue effects of O_2^- result from the secondary formation of other oxy radicals. Superoxide in aqueous media undergoes a spontaneous reaction with itself known as dismutation which yields one molecule each of hydrogen peroxide and molecular oxygen:

$$O_2^- + O_2^- + 2H^+ \rightarrow H_2O_2 + O_2 \tag{1}$$

Although this reaction proceeds spontaneously under physiologic conditions, it is catalyzed to take place at an even faster rate by the ubiquitously present intracellular enzyme called superoxide dismutase (abbreviated SOD and discussed in detail below). The steady-state intracellular level of O_2^- has been estimated to be 1×10^{-11} *M* in the presence of SOD, with a half-life on the order of 5 msec. The short half-life should not be misinterpreted as mitigating the potential reactivity of O_2^-, as the half-life is actually quite long in relation to the phenomenal diffusion coefficient of the radical. It should be noted that both the spontaneous dismutation of superoxide and that catalyzed by SOD yield hydrogen peroxide. A common error is to assume that SOD increases the local concentration of the latter, but in fact, the yield is the same; it just is faster when enzyme catalyzed than when spontaneous.

Under physiologic conditions, the superoxide radical itself is not as highly reactive as had been initially thought, and this species does not appear to be directly capable of attacking target tissues. H_2O_2, unlike O_2^-, can cross most biologic membranes and perhaps find an intracellular target to attack, but its reactivity is actually quite limited. The most toxic oxy radical is the hydroxyl radical, $\cdot OH$. It had been proposed a number of years ago that the hydroxyl radical could be produced from the interaction of superoxide and hydrogen peroxide via a chemical process known as the Haber-Weiss reaction:

$$O_2^- + H_2O_2 \rightarrow O_2 + \cdot OH + OH^- \tag{2}$$

However, detailed studies of the rate of this reaction have revealed that it could not take place under physiologic conditions with sufficient reaction parameters to account for hydroxyl radical generation.

An alternative explanation is now widely accepted whereby trace amounts of metal ions, primarily ferrous ion, react with hydrogen peroxide in what is known as the iron-catalyzed Fenton reaction to produce hydroxyl radical; ferrous ion is produced by the action of superoxide on ferric ion:

$$Fe^{3+} + O_2^- \rightarrow O_2 + Fe^{2+} \tag{3}$$

$$Fe^{2+} + H_2O_2 \rightarrow Fe^{3+} + OH^- + \cdot OH \tag{4}$$

Note that the sum of reactions 3 and 4 is equal to Reaction 2; the iron ions cancel from the summed equations. The oxygen produced by this reaction is partially in the activated state known as singlet. There is considerable debate as to whether protein-bound metal ions, e.g., lactoferrin, hemoglobin, etc., catalyze this reaction to any great degree. Sensitive measurements of the free (unbound) iron concentration in certain tissues, e.g., synovial fluid, have been reported to show a concentration sufficient to catalyze Reactions 3 and 4. It should be pointed out that other naturally occurring reducing substances, notably ascorbate (and perhaps reduced glutathione [GSH] or reduced nicotinamide-adenine dinucleotide [NADH]) can reduce ferric ion to ferrous, and thus the contribution of superoxide anion per se to the actual *in vivo* generation of hydroxyl radical remains plausible but unproven. Finally, recent attention has been drawn to what is called "site-specific" hydroxyl radical formation[5] wherein an iron ion bound to a macromolecule catalyzes $\cdot OH$ generation at the actual spot on the substrate where cleavage eventually ensues.

In summary, tissue toxicity from extracellular superoxide generation appears to be based on the dismutation of O_2^- to form hydrogen peroxide and the concomitant reduction of ferric ion to ferrous ion; reaction of these two products yields the highly toxic hydroxyl radical which can cleave covalent bonds in proteins and carbohydrates, cause lipid peroxidation, and destroy cell membranes. Evidence in support of such a mechanism comes from inhibitor studies where catalase, which scavenges hydrogen peroxide and/or chelators of iron salts, can lessen the effects of superoxide generation. In neutrophil-dependent systems, H_2O_2 also plays a role in reactions involving myeloperoxidase (MPO). Thus, there are three strategies which present themselves if one wishes to "detoxify" locally produced oxygen radicals: (1) deliver SOD or a SOD mimic to the area, (2) deliver catalase or a related peroxide scavenger, or (3) chelate (and thereby inactivate) the trace iron which catalyzes the reaction.

III. MECHANISMS OF RADICAL GENERATION

If ROS were test-tube novelties produced only by irradiation or chemical processes, they would obviously have little biologic relevance. Interest in ROS as mediators of inflammation comes from identification of two relevant sources: phagocytic cells, whose superoxide-generating capability was recognized in 1973, and the ischemia/reperfusion hypothesis, first suggested in 1983. From the osteoclast perspective, it is the phagocytic cell aspect which is of greatest relevance.

The biologic generation of superoxide was first brought to the world's attention by McCord and Fridovich in 1969,[6] but their initial observations were in enzymatic, cell-free systems. In 1973, Babior et al.,[7] reported that leukocytes incubated with latex particles produced a species that reduced cytochrome c and appeared to indeed be superoxide. A large literature has subsequently greatly expanded our knowledge in this area, especially with regard to the afferent aspect of the phenomena, i.e., how stimulus is coupled to response. A detailed review of this body of data is beyond the scope of this chapter, and can be found elsewhere.[8,9] The following emphasizes the efferent side of the equation.

In brief, phagocytic cells, when subjected to certain soluble stimuli or ingestion of particulate matter, undergo a "respiratory burst" characterized by a marked (up to 20-fold) increase in oxygen consumption associated with increased glucose metabolism via the hexosemonophosphate shunt. Simultaneously, these cells generate both superoxide anion and hydrogen peroxide. The cells capable of this response include the polymorphonuclear neutrophil (PMN), monocyte, eosinophil, peritoneal macrophage, alveolar macrophage, and osteoclast. The stimuli which can evoke the response include opsonized zymosan, C5a from the complement system, aggregated immunoglobulin G (IgG), opsonized bacteria, digitonin, the tumor promotor phorbol myristate acetate, and *N*-formyl-L-methionyl-L-leucyl-L-phenylalanine (FMLP). (Not all the listed stimuli work with all cell types from all species.) Many of these substances can be demonstrated to stimulate the cells when nanogram quantities are added to as few as a million cells in sensitive *in vitro* detection systems. Most of the oxygen consumption by stimulated cells can be accounted for as O_2^- production, and most of the H_2O_2 as the product of superoxide dismutation.

The source of superoxide generation in PMNs is a membrane-associated NADPH (nicotinamide-adenine dinucleotide phosphate, reduced) (and perhaps NADH) oxidase. This system appears to be on the cell surface, where it remains dormant until activated by ligand binding, which in turn initiates a complex process of signal transduction, second-messenger formation, protein kinase activation, etc.[9] The major components of the system include a low-potential b cytochrome and a flavoprotein. One portion of the enzyme system binds the pyridine nucleotide (perhaps on the cytoplasmic side of the plasma membrane), while the oxidase component (within the lipid bilayer) generates O_2^-.

$$2O_2 + NADPH \rightarrow 2O_2^- + NADP^+ + H^+ \quad (5)$$

When cells are treated with cytochalasin b, a substance that paralyzes ingestion of particles which would be otherwise phagocytosed, the activity of this enzyme system in intact cells is augmented. The hydrogen peroxide generated by dismutation of superoxide can also be acted upon by MPO and halide ions to form reactive chloramines and hypochlorous acid.

PMNs and macrophages are obvious components of most acute and chronic inflammatory states. Aggregated immunoglobulins, bacterial products, crystals, and complement components are also clearly identifiable in many such conditions, e.g., the rheumatoid joint. Thus, it is highly plausible to conjecture that phagocytic cells within inflamed areas are probably secreting ROS into their environment. Direct evidence to support such a concept is hard to obtain. Most data implicating superoxide in pathologic processes are indirect; an effect eliminated by a scavenger is imputed to have been caused by the radical species. There exists no marker which can be instilled into an inflamed area to undergo a stable, specific chemical change that could be quantitated upon recovery of the marker. Most *in vitro* work is done with equine ferricytochrome c, a protein which pleasantly changes color in the spectrophotometer when subjected to a flux of O_2^-, but if this material were placed into an area of inflammation, those molecules which did not wash away rapidly would be reduced and oxidized by an army of other agents — proving the action of ROS would be nearly impossible. Until and unless such a marker becomes available, our extrapolation of the *in vitro* data to *in vivo* situations will have to depend on how closely the experimenter attempts to reproduce the known physiology of the system under scrutiny.

IV. EFFECTS OF REACTIVE OXYGEN SPECIES ON CONNECTIVE TISSUE TARGET TISSUES

Regardless of the mechanism by which radicals are generated, their role in disease pathogenesis depends on the effect of the radicals on local targets. There is a large literature in which free radical fluxes have been applied to potential targets, either isolated molecular species, tissues, or organs (Table

TABLE 1
Molecules, Tissues, and Organs Reported to be Sensitive to a Flux of Reactive Oxygen Species

Tissue or organ	Effect seen
Bacteria	Decreased viability
Erythrocytes	Lysis
Neutrophils	Lysis, cell death
Lymphocytes	Decreased transformation, chromosome damage
Tumor cells	Lysis, cell death
Fibroblasts	Lysis, cell death
Endothelial cells	Lysis, cell death
Liposomes	Lysis
Mitochondria	Impaired respiration
Isolated, perfused lung	Decreased serotonin uptake
Hamster cheek pouch	Increased vascular permeability
Rat brain	Edema
Canine jejunum	Increased albumin clearance
Cultured cartilage	Decreased proteoglycan synthesis
Molecular species	**Effect seen**
Hyaluronic acid	Decreased viscosity, schism
Collagen	Impaired gelation, schism
DNA	Schism
Hemoglobin	Methemoglobin formation
Lipids	Peroxidation
Arachidonic acid	Oxidation
Papain	Inactivation
α1-Antiprotease	Decreased activity
NADH oxidase	Decreased activity
Ribonuclease	Inactivation

1). Almost without exception, the expected end product is death or destruction; very few salutary effects of free radicals are known. For bone biology considerations, the most relevant targets are the connective tissue macromolecules — collagen, hyaluronic acid, and proteoglycans.

A. HYALURONIC ACID

HA is the dominant macromolecule of synovial fluid, accounting for the viscosity (but not the lubricating function) of that fluid. HA is also an important constituent of proteoglycan aggregates within cartilage. The susceptibility of HA to degradation by free radicals was first observed over 30 years ago, well before oxygen radicals had been implicated in human disease. It was the seminal paper by McCord,[10] however, which really sparked the biomedical oxygen radical field, and after HA was chosen as the substrate for that study, it became the subject of many subsequent papers.

Degradation of HA can be produced in two ways: enzymatically by the specific class of enzymes known as hyaluronidases, or by chemical interaction of the macromolecule with a highly reactive chemical species such as a free radical. The enzymes which degrade HA reduce the substrate to oligosaccharides (or ultimately tetrasaccharides) of four to ten units in length, but such degradation pieces have never been found in joint fluid. There is no neutral hyaluronidase in polymorphonuclear leukocytes, and the initial observation of hyaluronidase activity in joint fluid by Bollet et al. in 1963[11] has not been (and probably cannot be) confirmed (see Reference 12 for a review). The overwhelming evidence now suggests that HA degradation *in vivo* in inflammatory joint fluid is radical mediated.

Free radical-mediated degradation of HA has been produced by (1) γ-irradiation (usually pulse radiolysis) from a ^{60}Co source,[13] (2) action of xanthine oxidase on hypoxanthine or xanthine,[10,14-16] (3) the action of ferrous ion, with or without addition of hydrogen peroxide, ascorbate, thiols, or penicillamine,[17-19] (4) action of singlet oxygen produced by exposure of a sensitizing dye to light,[20] (5) free radicals in cigarette smoke,[21] (6) neutrophils,[19,22] and (7) the myeloperoxidase system.[23] Iron chelation usually lessens the observed effects, subject to the limitation of iron contamination of laboratory reagents.

Methods for assessment of the degree of degradation and/or the biochemical changes wrought by free radicals include (1) depolymerization (dynamic viscosity loss as well as intrinsic viscosity), (2) water content, (3) reducing-group analysis, (4) loss of and/or change in the hexosamine or hexuronate ratio, (5) elemental analysis, (6) metachromasia, (7) impaired mucin clot, (8) circular dichroism, (9) gel chromatography, and (10) proteoglycan aggregation. HA degradation can be demonstrated for purified HA from several sources — bovine synovial fluid (SF), human SF, and even equine SF[24] — where SOD is often used therapeutically. Of interest is the fact that the end product of all of these purely chemical reactions appears to be a polymer with a molecular weight no smaller than 10^4. Enzymatic action by lysosomal carbohydrases in synovial fluid could possibly reduce the polymer length further[15] after initial steric hindrance had been obviated by free radical action, perhaps down to mono- or disaccharides (which would be rapidly cleared and undetectable). On the other hand, it has been reported that complexation of HA with certain proteins, especially α1-antitrypsin and haptoglobin, made the HA resistant to oxy radical degradation.[25] A detailed study of the chemistry, saccharide chain length, and end group analysis of HA depolymerized by ferrous ion has recently appeared.[26]

B. COLLAGENS

Collagens are the major structural proteins of connective tissues, accounting for the tensile strength of bone, skin, ligaments, tendon, etc. Collagen molecules in their native, helical, undenatured state are rather resistant to degradation and are not susceptible to action by serine, thiol, or other conventional proteases. Only metal-dependent, specific collagenases can make an initial "clip" in native collagen, after which other proteases can finish the job. Thus, it was of interest when reports indicated that oxygen radicals could also degrade collagen, albeit to a very limited degree.

The first experiments designed to alter collagen by radical action were attempts to strengthen the molecular structure rather than degrade it; workers in the leather industry reported that exposure of collagen to singlet oxygen reduced its viscosity, impeded fibril formation, and altered the primary amino acid composition.[27] In the 1979 study by Greenwald and Moy[28] collagen solutions exposed to superoxide failed to gel normally; molecular schism was implied by loss of dialyzable HyPro, but could not be confirmed at that time by sodium dodecyl sulfate-polyacrylamide gel electrophoresis SDS-PAGE. However, Monboisse and co-workers[29,30] were able to show low molecular weight bands on gel, along with confirmation of impeded gelation and loss of small peptides. They further showed that collagen could be attacked by any of three methods — pulse radiolysis from a particle accelerator, γ-radiation from ^{60}Co, or xanthine oxidase[31] — and that type I and III collagen were both susceptible. In all of the studies, however, the total percent of the collagen which can be so affected is always small, generally under 5%, suggesting that perhaps only that fraction which may have been artefactually altered during the experimental preparation is amenable to oxygen radical action.

Since collagen is a component of many tissues other than bone and joint, its degradation is of interest in lung disease, kidney pathology (basement membrane), skin, etc. Several reviews have appeared dealing with matrix degradation in pulmonary diseases, where oxidant action can be more directly implicated.[32,33] Collagen in solution is degraded by exposure to ozone or Fenton reagents,[34] or in lung slices exposed to hyperoxia.[35] Treatment of collagen with ultraviolet B(UVB) in a manner designed to generate singlet oxygen caused increased cross-linking and decreased solubility;[36] it had previously been hypothesized that perhaps superoxide was the ''cross-linkage'' which detached one collagen molecule from another, setting the scene for enzymatic action. In the latter regard, studies from Winterbourn's laboratory[37] showed that preexposure of glomerular basement membrane to a ROS system (based on myeloperoxidase) led to elastase susceptibility (solubilization of HyPro); hydrogen peroxide alone had no such effect. A review has appeared describing the effect of oxygen radicals as enhancers of protein susceptibility to enzymatic degradation; collagen is one such protein for which the mechanism may operate.[38] The PMN would appear therefore to have several methods at its disposal to remove collagen — a specific enzyme (collagenase) or a nonspecific combination of MPO/H_2O_2 and neutral proteases. Finally, exposure of proline peptides to a Fenton system (Cu(II)/peroxide) resulted in conversion of some proline residues into HyPro, along with formation of γ-aminobutyric acid.[39]

The collagen story is one of the few examples where not all ROS action is destructive. As is well known, the enzymatic hydroxylation of proline to HyPro requires molecular oxygen, ascorbate, Fe^{2+}, and α-ketoglutarate. How curious that the first three are also players in the ROS game! It was known in 1972 that nitro-blue tetrazolium, a free radical scavenger (since used as the substrate in a superoxide assay), would inhibit proline hydroxylation.[40] Both lysyl and prolyl hydroxylases were reported to be inhibited by several low molecular weight copper chelates with SOD-like activity.[41] Additional studies showed that paraquat, a good radical generator, enhanced collagen synthesis in cultured rat lungs;[42] direct generation of superoxide had a similar effect on hepatocytes.[43] Exposure of chondrocyte and fibroblast cultures to Fe^{2+}/ascorbate induced collagen synthesis.[44] On the other hand, culture of rat tibias in the presence of hydrogen peroxide decreased collagen production dramatically.[45] (As a general rule, whenever reagent-bottle H_2O_2 is added to cells or organ cultures *in vitro,* toxic effects are seen; it is hard to extrapolate such an experiment to the *in vivo* situation.)

In summary, ROS would appear to be agents capable of augmenting other systems for collagen breakdown and disposal by disrupting quaternary structure or making initial cleavages that lead to further enzymatic susceptibility. In connective tissues, especially bone, it is hard to imagine how ROS would get to the collagen without being intercepted by another substrate first. There has been, to our knowledge, no specific study of the susceptibility of bone type I collagen to free radicals, or any data on a protective (or enhancing) effect which might be attributable to the deposition of hydroxyapatite. In disorders where long-term, slow collagen degradation takes place (rheumatoid arthritis, osteoporosis, etc.), radical scavenging would probably not be preservative of collagen integrity.

C. PROTEOGLYCANS AND CONNECTIVE TISSUE MATRIX

One can study the action of ROS on HA and collagen with relative ease. Neither macromolecule is susceptible to degradation by garden-variety proteases, and a variety of analytic methods exist for assessment of the degradation. For cartilage (or bone) proteoglycans, however, the situation is much more difficult. Most proteoglycans are exquisitely sensitive to proteases, and there are few "physiologic" oxygen radical-generating systems that are protease free. Commercial XO is heavily contaminated[46] and must be cleaned up either chromatographically or mixed with a cocktail(s) of inhibitors; numerous controls must be run in parallel to distinguish ROS action from artefact. PMNs stimulated by zymosan, crystals, FMLP, etc. will secrete elastase and similar enzymes that readily degrade proteoglycan. There is also the problem of separating the intact from the degraded proteoglycan before doing a chemical assay, e.g., uronic acid, to determine percent degradation. Finally, the anionic side chains of chondroitin and keratan sulfate appear to be "sinks" that soak up the radicals before either the protein or carbohydrate backbones can be attacked. (Chondroitin sulfate is readily degraded by hydroxyl radicals generated with UV radiation.[47]) None of these problems has been satisfactorily resolved to date; much of the reported work involves "nonphysiologic" ferrous ion/EDTA and/or hydrogen peroxide-generating systems.

Such a system was used by Bartold et al.[48] to study degradation of gingival PGs. ROS degraded both gingival HA and gingival proteoglycan as assessed by viscosity and chromatographic profile; in addition, the metachromasia (with Alcian blue) of exposed tissue slices was decreased as well. In the same year, Bates et al. used purified XO to degrade CsCl-separated bovine nasal PG-HA aggregates,[49] but they observed only a mild effect and encountered substantial background degradation as well. Dean et al. measured release of ^{35}S from biosynthetically labeled cartilage after exposure to ^{60}Co or Fe^{2+}/-H_2O_2, as well as rediscovering the proteolytic contamination of XO.[50] A similar study with copper and H_2O_2 also showed "reagent-bottle" degradation.[51] When cartilage has been incubated with stimulated PMNs, the major effect has been that of elastase, and it has been difficult to demonstrate an additional effect due to radicals.[52] HOCl can also be shown to degrade proteoglycans.[53] Finally, Roberts et al.[54,55] have exposed purified proteoglycan aggregate to H_2O_2 and reported depolymerization, impaired HA interaction, and cleavage of link proteins; species variability exists as well, and the site of amino acid attack can be pinpointed in some cases.

ROS generated in the osteoclastic lacunae would have a difficult time finding a bone proteoglycan to attack except in the most superficial layer. The high reactivity of radicals with other molecules encountered first, combined with the apparent low susceptibility of PGs, would be expected to preclude any substantive effect in the matrix interior. In the relatively anoxic extracellular environment of the osteoclast, oxy radical generation is similarly expected to be sparse. Thus, there are conceptual problems in expecting a "chondroprotective" effect from an oxygen radical scavenger, should one become available for therapeutic installation.

D. CELLS

Hydrogen peroxide is a wonderful cellular toxin, especially when poured out of bottles, which is why it is sold in drugstores. A large number of studies have been reported in which application of H_2O_2 to cellular preparations has impeded their viability and/or function, and it has been known for over 15 years that PMNs in a superoxide-generating mode will eventually self-destruct from membrane damage.[56] Two cell types are of interest here: lymphocytes and matrix cells.

Niwa et al. reported in 1983[57] that when stimulated PMNs were cocultured with lymphocytes, the number of OKT4$^+$ and OKT8$^+$ cells was reduced, as was responsiveness to PHA and Concanavalin A (ConA); the authors discussed implications for lymphocyte reactivity within inflamed synovial fluid as a result of such a reaction. Similar results were noted using XO as the ROS source[58] or reagent-bottle H_2O_2.[59]

ROS generated by XO have been applied to rabbit chondrocyte cultures, leading to a dose-dependent decrease in cell growth; detailed study of the proliferation kinetics was reported.[60] Fibroblasts, on the other hand, appear to respond with a dualism: high fluxes caused visible damage, detachment, and morphologic changes, whereas low fluxes enhanced cell density and thymidine uptake.[61] The fibroblasts themselves were noted to generate O_2^- even under basal conditions, increasing with agitation of the culture flasks. Finally, exposure of cartilage to H_2O_2 in culture caused oxidative inactivation of glyceraldehyde-3-phosphate dehydrogenase, resulting in decreased chondrocyte ATP levels, with subsequent adverse effects on overall synthetic rates.[62] Once again, the extrapolation of these studies to the *in vivo* situation is conceptually difficult.

E. PROTEOGLYCAN SYNTHETIC REACTIONS

In a series of papers from the Australian group headed by Lowther,[63-65] the effect of ROS on proteoglycan synthesis by cultured bovine articular chondrocytes has been examined. Using XO as the source, they have shown inhibition of sulfate and acetate incorporation in glycosaminoglycans (GAGs), along with suppression of protein synthesis in general, with a similar effect on HA synthesis as well. Once again, H_2O_2 appears to be the mediator, and at high concentrations, a long-term effect is noted, whereas at low concentrations, the cells can recover after a few days in tissue culture. Cell viability was apparently not affected.

F. COLLAGENASE

Collagenase is an extracellular enzyme which mediates destruction of connective tissue matrix. In two imaginative papers, Weiss et al.[66,67] have shown that an oxidizing species produced by PMNs autoactivates the otherwise latent neutrophil collagenase. The reactive species which produces the activation depends on MPO, HOCl, and H_2O_2, and was characterized as a long-lived *N*-chloramine. If such a mechanism exists *in vivo,* then theoretically, radical scavenging might have a long-term beneficial effect in preventing tissue damage. In view of the apparent paradox regarding the possible role of collagenase in bone resorption, i.e., that osteoclasts do not appear to release large amounts of readily detectable collagenase whereas osteoblasts do, the possibility that the former use reactive oxygen species to activate the enzyme from the latter deserves pathophysiologic consideration.

A collagenase localized to the subosteoclastic resorption zone has been identified by Baron et al.,[68] Although collagenase has yet to be found in large quantities as a secretory product of osteoclasts, the monocyte/macrophage heritage of the osteoclast suggests that these cells may secrete such an enzyme which has so far gone undetected. Since osteoclasts lack substantial quantities of cathepsin B[69] (a lysosomal cysteine proteinase which activates latent collagenase), reactive oxygen species may be the actual activators *in vivo.*

Proteases such as collagenase are normally kept in check by protein inhibitors, one of which has also been used as a target for oxygen radical fluxes, viz., the serum elastase inhibitor, α1-proteinase inhibitor (α1-PI). The initial observation was made by Carp and Janoff in the context of the pathogenesis of emphysema using either XO or phagocytosing PMNs to attack the substrate;[70] the PMN reaction required MPO, suggesting that hypochlorous acid might have been the mediator,[71] a finding confirmed several years later.[72]

V. MACROPHAGE ROS PRODUCTION

Most work on ROS generation from phagocytic cells has been done with PMNs, since this cell type is readily accessible and is an obvious player in most inflammatory and/or degenerative conditions. Macrophages, which share an ontogenic background with osteoclasts,[73,74] have also been studied, primarily in the context of pulmonary disease. Since the granulocyte-macrophage progenitor cell gives rise to both monocyte and osteoclast, it stands to reason that cells of both lineages would be capable of utilizing similar ROS-generating systems to create end effects in tissues.

Macrophage ROS generation differs somewhat from that of PMNs. First, PMNs are terminal cells that do not grow in culture; macrophages can be sustained for several days and, in fact, their ROS generation is enhanced by such culture. Second, peripheral PMNs need not be "activated" — they merely require an appropriate stimulus such as a cell surface contact or phagocytosis. Dormant macrophages, e.g., those freshly isolated from peritoneum or alveolar lavage, yield low levels of ROS. "Activation" by pretreatment of the entire animal *in vivo* with substances such as bacille Calmetti-Guérin (BCG), muramyl dipeptide, bacterial lipopolysaccharide, parasites, or bacterial products greatly enhances the subsequent yields of ROS from isolated and stimulated cells.[75] Such treatment induces greater activity of the cell-membrane NADPH oxidase. The initial generation of ROS by macrophages is usually less than that of comparable PMNs isolated together, but macrophages can continue to generate for long time periods, well beyond the point where the PMN response falls off.[76] Finally, macrophages do not make HOCl, whereas PMNs, using MPO, do.

Lymphokines also enhance the capacity of monocytes to generate ROS, and several studies in this area may be of particular relevance to osteoclasts. One substance of interest is M-CSF, macrophage colony-stimulating factor, which is known to stimulate the formation of multinucleated osteoclast-like cells in long-term cultures of bone marrow cells.[73] M-CSF also enhances respiratory burst activity in peritoneal macrophages, increasing their production of O_2^- and H_2O_2.[77] In addition, γ-interferon enhanced both O_2^- and H_2O_2 production by monocytes as they were differentiating into macrophages *in vitro*.[78] Finally, there has been a report that bovine PTH, oxidized by treatment *in vitro* with H_2O_2, can stimulate the development of osteoclast ruffled borders in Japanese quail in the same manner as native PTH, suggesting that the products of ROS generation might be irrelevant in this aspect of osteoclast biology.[79]

VI. SPECIFIC ROS FEATURES OF OSTEOCLASTS

As of this writing, there have been only a handful of papers specifically dealing with osteoclastic generation of ROS. First are several abstracts presented at the 1986 to 1988 ASBMR meetings. Fallon et al.[80] reported that bone particles caused oxygen consumption by isolated osteoclasts (OCs), as well as cytochrome c reduction. Chemiluminescence was low, but the cells apparently contained a high endogenous level of SOD. Calcitonin, as expected, reversed the effects. The authors hypothesized a linkage between OC bone resorption and ROS generation.

Garret et al.[81] reported data from a bone-resorbing system stimulated with interleukin-1 (IL-1). Using nitro-blue tetrazolium (NBT) as a marker of O_2^- production, NBT staining (in the cytosol) was present only in osteoclasts which were in resorption lacunae. The staining was inhibited by calcitonin, and enhanced by exposure to vitamin A or to bone particles, stimuli which activate osteoclasts directly. The authors concluded that ROS were required for hormone-stimulated osteoclastic bone resorption. Key et al.[82] reported that H_2O_2 production could be localized to the osteoclast cell membrane by electron microscopic localization techniques.

Several of the aforementioned authors followed up their abstracts with full-length papers in 1990. Key et al.[83] used NBT to localize osteoclast superoxide production by both isolated rat cells and mouse calvaria. NBT is converted to an insoluble yellow compound, formazan, which remains localized at the site of its formation. Osteoclasts from both sources produced O_2^- at the ruffled border-bone interface, between the external osteoclastic membrane and the bone. The reaction was not inhibited by either cyanide or aminotriazole (the latter is an inhibitor of catalase), but recombinant human calcitonin, which of course inhibits OC bone resorption, also suppressed oxygen radical formation. The implication is that OCs generate such radicals into the interface region with resorbing bone, but the role of these radicals remains to be determined.

Garrett et al.[84,85] used the well-known xanthine oxidase system to measure the ability of ROS to resorb bone. Applying the reactants to fetal rat bones and neonatal mouse calvaria, the enzyme system caused a release of radiolabeled calcium which was partially inhibitable by SOD but not affected by catalase. The reaction was also suppressed by calcitonin. These authors also noted that the XO/X system also induced formation of increased numbers of osteoclasts and expanded the amount of resorption surface, effects once again reversed by SOD. Interestingly, the stimulation of OC formation/resorption in mouse calvaria induced by PTH, IL-1α, and TNF was also inhibited by SOD, and the authors detected formazan production in such tissues. Finally, they showed direct generation of superoxide by isolated avian OC using cytochrome c as a detector. This body of data once again, albeit circumstantially, suggests a role for ROS in the resorption of bone mediated by OC.

VII. OSTEOCLAST SUPEROXIDE DISMUTASE

SOD has been mentioned above several times. For many years, it was known that hepatic tissue contained a blue-green copper protein of fairly low molecular weight (32,000) called hemocuprein whose function was unknown until 1969, when McCord and Fridovich discovered that this molecule could catalyze the dismutation of the superoxide radical via Equation 1 above. Although the reaction proceeds spontaneously, its rate is accelerated several orders of magnitude by SOD, which functions enzymatically, i.e., SOD itself is unchanged during the reaction. Since that discovery, more attention has been paid to SOD (and more papers written each year) than for any other enzyme.

SOD is a ubiquitous, intracellular enzyme found in virtually all aerobic organisms. SOD, originally isolated from bovine erythrocytes, is widely distributed in many respiring cell types. The copper-zinc enzyme is found in the cytosol, as well as in nuclei, plasma membranes, and mitochondria. A manganese enzyme has been found primarily in the matrix space of mammalian mitochondria, as well as in the cell matrix of aerobic prokaryote.[86] (There is also an iron SOD, found in anaerobic bacteria and plants.) Virtually all aerobic bacteria contain SOD, and when suitable strains are exposed to increased oxygen tensions, higher levels of the enzyme are induced; similar induction can be shown in various mammalian cell lines and tissues. Exogenously added SOD is protective of oxy radical-mediated damage in a wide variety of test systems much too numerous to list here, and is identical to orgotein, an antiinflammatory protein previously isolated from liver as well. Oxy radicals have been implicated in many experimental systems by the demonstration that SOD added to the experiment abolishes the observed phenomenon.

In two related papers,[87,88] Oursler et al. have reported their findings concerning a glycoprotein isolated from osteoclast membrane which has homology to manganese SOD. This group was studying the cell membrane protein library of chick OCs using a series of monoclonal antibodies which they had raised against various surface markers. An antibody designated 121F was found to recognize a 150-kDa protein which was then isolated by an elaborate series of steps involving cell membrane preparation, immunoaffinity and gel permeation chromatography, SDS-polyacrylamide gels, etc. When the reactant protein was eventually isolated in purified form, amino acid sequence studies of N-terminal fragments revealed an unexpected but substantial degree of homology with manganese SOD from several species. The OC membrane protein so identified appears to be unique to OCs, as the antibody did not recognize other cell types.

In the companion paper, Oursler et al. used their antibodies to immunolocalize the protein, by both light and electron microscopy, to the OC plasma membrane, with little intracellular staining. The antibodies partially reacted with purified SODs from several species, including a SOD showing substantial homology to the OC membrane protein. The authors were unable to directly demonstrate SOD activity in their protein, due to an apparent destructive effect of the extraction conditions on the enzyme activity. They were, however, able to demonstrate in several different preparations a correlation between overall SOD activity on zymogram and reactivity to their antibody. This and several other types of circumstantial evidence allowed the authors to conclude that OCs do indeed possess a high molecular weight, plasma membrane protein which may have functional SOD activity. It remains to be determined if this protein serves an actual scavenging role for ROS *in vivo*.

VIII. SUMMARY

The field of ROS generation by bone cells and their possible pathophysiologic effects is just emerging, and few conclusions can be drawn. The following seem reasonable as of this writing:

1. Osteoclasts probably possess ROS-generating capacity, although the exact species formed are still uncertain, and it is possible that more than one cytokine and/or particulate stimulus might directly cause their production. Alternatively, it is likely that modulation of OC activity is based on osteoblasts; hormones (except calcitonin), cytokines, growth factors, etc. appear to modulate OC activity via interaction with osteoblasts, although the factor(s) produced by osteoblasts which activate osteoclasts have yet to be determined. Hormonal inhibition of ROS generation by PMNs and macrophages has attracted little attention; the phenomena noted with calcitonin deserves much further exploration. It must not be forgotten that ROS represent but one subset of a vast panorama of mediators of inflammation and connective tissue degradation, and their role remains to be defined.
2. ROS clearly can degrade HA, may have an effect on collagen and/or proteoglycans, and can perhaps alter the actions of some enzymes and/or their inhibitors. The role of such effects in disease remains far from certain. The effect of bone mineral on these actions has not been explored.
3. If OC-generated ROS can activate latent osteoblastic collagenase, an additional type of synergism between osteoblasts and osteoclasts in bone resorption may become apparent. If the observations of subosteoclastic collagenase are confirmed, then a self-contained mechanism for promoting collagen resorption might exist.
4. Immunological, biochemical, and functional data pertaining to an osteoclast plasma membrane manganese SOD have been reported. The role of this protein in OC function remains unknown.

The above notwithstanding, interest in ROS in many facets of mammalian biology is increasing, and their role in bone physiology, both normal and pathologic, merits further study.

REFERENCES

1. **Freeman, B. and Crapo, J. D.,** Free radicals and tissue injury, *Lab. Invest.,* 47, 412, 1982.
2. **Halliwell, B. and Gutteridge, J. M. C.,** Oxygen toxicity, oxygen radicals, transition metals, and disease, *Biochem. J.,* 219, 1, 1984.
3. **Fridovich, I.,** Biological effects of the superoxide radical, *Arch. Biochem. Biophys.,* 247, 1, 1986.
4. **Deby, C. and Goutier, R.,** New perspectives on the biochemistry of superoxide anion and the efficiency of superoxide dismutases, *Biochem. Pharmacol.,* 39, 399, 1990.
5. **Czapski, G., Aronovitch, J., Samuni, A., et al.,** The sensitization of the toxicity of superoxide and vitamin C by copper and iron: a site specific mechanism, in *Oxy Radicals and their Scavenger Systems,* Vol. 1, *Molecular Aspects,* Cohen, G. and Greenwald, R., Eds., Elsevier, New York, 1983, 111.
6. **McCord, J. M. and Fridovich, I.,** Superoxide dismutase: an enzymic function for erythrocuprein (hemocuprein), *J. Biol. Chem.,* 244, 6049, 1969.
7. **Babior, B. M., Kipnes, R. S., and Curnette, J. T.,** Biological defense mechanisms: the production by leukocytes of superoxide, a potential bactericidal agent, *J. Clin. Invest.,* 52, 741, 1973.
8. **Cohen, M. S., Britigan, B. E., Hassett, D. J., et al.,** Phagocytes, O2 reduction, and hydroxyl radical, *Rev. Infect. Dis.,* 10, 1088, 1988.
9. **Bellavite, P.,** The superoxide forming enzymatic system of phagocytes, *Free Radicals Biol. Med.,* 4, 225, 1988.
10. **McCord, J. M.,** Free radicals and inflammation: protection of synovial fluid by superoxide dismutase, *Science,* 185, 529, 1974.
11. **Bollet, A. J., Bonner, W. M., and Nance, J. L.,** The presence of hyaluronidase in various mammalian tissues, *J. Biol. Chem.,* 238, 3522, 1963.
12. **Greenwald, R. A.,** Oxygen radicals, inflammation, and arthritis: pathophysiological considerations and implications for treatment, *Semin. Arthritis Rheum.,* 20, 219, 1991.
13. **Myint, P., Deeble, D. J., Beaumont, P. C., et al.,** The reactivity of various free radicals with hyaluronic acid: steady state and pulse radiolysis studies, *Biochim. Biophys. Acta,* 925, 194, 1987.
14. **Greenwald, R. A. and Moy, W. W.,** Effect of oxygen derived free radicals on hyaluronic acid, *Arthritis Rheum.,* 23, 455, 1980.
15. **Betts, H. W., Cleland, L. G., Gee, D., et al.,** Effects of d-penicillamine on a model of oxygen derived free radical mediated tissue damage, *Agents Actions,* 14, 283, 1984.
16. **Hofmann, H. and Schmut, O.,** The inability of superoxide dismutase to inhibit the depolymerization of hyaluronic acid by ferrous ions and ascorbate, *Graefes Archiv. Klin. Ophthalmol.,* 214, 181, 1980.
17. **Wong, S. F., Halliwell, B., Richmond, R., et al.,** The role of superoxide and hydroxyl radicals in the degradation of hyaluronic acid induced by metal ions and by ascorbic acid, *J. Inorg. Biochem.,* 14, 127, 1981.
18. **Betts, W. H. and Cleland, L. G.,** Effect of metal chelators and antiinflammatory drugs on the degradation of hyaluronic acid, *Arthritis Rheum.,* 25, 1469, 1982.

19. **McNeil, J. D., Wiebkin, O. W., Betts, W. H., et al.,** Depolymerisation products of hyaluronic acid after exposure to oxygen derived free radicals, *Ann. Rheum. Dis.*, 44, 78, 1985.
20. **Andley, U. P. and Chakrabarti, B.,** Role of singlet oxygen in the degradation of hyaluronic acid, *Biochem. Biophys. Res. Commun.*, 115, 894, 1983.
21. **McDevitt, C. A., Beck, G. J., Ciunga, M. J., et al.,** Cigarette smoke degrades hyaluronic acid, *Lung,* 167, 237, 1989.
22. **Greenwald, R. A. and Moak, S. A.,** Degradation of hyaluronic acid by polymorphonuclear leukocytes, *Inflammation,* 10, 15, 1986.
23. **Baker, M. S., Green, S. P., and Lowther, D. A.,** Changes in the viscosity of hyaluronic acid after exposure to a myeloperoxidase derived oxidant, *Arthritis Rheum.*, 32, 461, 1989.
24. **Auer, D. E., Ng, J. C., and Seawright, A. A.,** Effect of Palosein (superoxide dismutase) and catalase upon oxygen derived free radical induced degradation of equine synovial fluid, *Equine Vet. J.*, 22, 13, 1990.
25. **Hutadilok, N., Ghosh, P., and Brooks, P. M.,** Binding of haptoglobin, inter-α-inhibitor, and α-1 proteinase inhibitor to synovial fluid hyaluronate and the influence of these proteins on its degradation by oxygen derived free radicals, *Ann. Rheum. Dis.*, 47, 377, 1988.
26. **Uchiyama, H., Dobashi, Y., Phkouchi, K., et al.,** Chemical changes involved in the oxidative reductive depolymerization of hyaluronic acid, *J. Biol. Chem.*, 265, 7753, 1990.
27. **Venkatasubramanian, K. and Joseph, K. T.,** Action of singlet oxygen on collagen, *Indian J. Biochem. Biophys.*, 14, 217, 1977.
28. **Greenwald, R. A. and Moy, W. W.,** Inhibition of collagen gelation by action of the superoxide radical, *Arthritis Rheum.*, 22, 251, 1979.
29. **Monboisse, J. C., Braquet, P., Randoux, A., et al.,** Non-enzymatic degradation of acid-soluble calf skin collagen by superoxide ion: protective effect of flavonoids, *Biochem. Pharmacol.*, 32, 53, 1983.
30. **Monboisse, J. C., Poulin, J., Braquet, P., et al.,** Effect of oxy radicals on several types of collagen, *Int. J. Tissue React.*, 6, 385, 1984.
31. **Monboisse, J. C., Gardes-Albert, M., Randoux, A., et al.,** Collagen degradation by superoxide anion in pulse and gamma radiolysis, *Biochim. Biophys. Acta,* 965, 29, 1988.
32. **Cantin, A. and Crystal, R. G.,** Oxidants, antioxidants and the pathogenesis of emphysema, *Eur. J. Respir. Dis.*, 66 (Suppl. 139), 7, 1985.
33. **Riley, D. J. and Kerr, J. S.,** Oxidant injury of the extracellular matrix: potential role in the pathogenesis of pulmonary emphysema, *Lung,* 163, 1, 1985.
34. **Curran, S. F., Amoruso, M. A., Goldstein, B. D., et al.,** Degradation of soluble collagen by ozone or hydroxyl radicals, *FEBS Lett.*, 176, 155, 1984.
35. **Kerr, J. S., Chae, C. U., Nagase, H., et al.,** Degradation of collagen in lung tissue slices exposed to hyperoxia, *Am. Rev. Respir. Dis.*, 135, 1334, 1987.
36. **Pathak, M. A., Carbonare, M. D., and Mobilio, S.,** The role of reactive oxygen in collagen crosslinking and photoaging, *J. Invest. Dermatol.*, 94, 564, 1990.
37. **Vissers, M. C. M. and Winterbourn, C. C.,** The effect of oxidants on neutrophil mediated degradation of glomerular basement membrane collagen, *Biochim. Biophys. Acta,* 889, 277, 1986.
38. **Naish-Byfield, S. and Dean, R. T.,** Antioxidants and the influence of free radical damage to proteins on proteolysis in and around mammalian cells, in *Current Trends in the Study of Intracellular Protein Degradation II,* Knecht, E. and Grisolia, S., Eds., Springer, Basel, 1990, 361.
39. **Uchida, K., Kato, Y., and Kawakishi, S.,** A novel mechanism for oxidative cleavage of prolyl peptides induced by the hydroxyl radical, *Biochem. Biophys. Res. Commun.*, 169, 265, 1990.

40. **Bhatnagar, R. S. and Liu, T. Z.,** Evidence for free radical involvement in the hydroxylation of proline: inhibition by nitroblue tetrazolium, *FEBS Lett.*, 26, 32, 1972.
41. **Myllyla, R., Schubotz, L. M., Weser, U., et al.,** Involvement of superoxide in the prolyl and lysyl hydroxylase reactions, *Biochem. Biophys. Res. Commun.*, 89, 98, 1979.
42. **Hussain, M. Z. and Bhatnagar, R. S.,** Involvement of superoxide in the paraquat induced enhancement of lung collagen synthesis in organ culture, *Biochem. Biophys. Res. Commun.*, 89, 71, 1979.
43. **Hussain, M. Z., Watson, J. A., and Bhatnagar, R. S.,** Increased prolyl hydroxylase activity and collagen synthesis in hepatocyte cultures exposed to superoxide, *Hepatology*, 7, 502, 1987.
44. **McManus, J., Hussain, M. Z., and Bhatnagar, R. S.,** The role of inflammation in wound healing: induction of collagen synthesis by superoxide anions, *J. Dent. Res.*, 55, 792, 1976.
45. **Ramp, W. K., Arnold, R. R., Russell, J. E., et al.,** Hydrogen peroxide inhibits glucose metabolism and collagen synthesis in bone, *J. Periodontol.*, 58, 340, 1987.
46. **Greenwald, R. A., Moak, S. A., Carsons, S., and Rush, S.,** Inhibition of the proteolytic contaminant in commercial xanthine oxidase by serum protein fractions, *J. Free Radicals Biol. Med.*, 1, 451, 1986.
47. **Khan, K. A. and Iqbal, M. N.,** Chemical effects of UV radiation on aqueous solutions of chondroitin sulfate, *Polym. Photochem.*, 4, 385, 1984.
48. **Bartold, P. M., Wiebkin, O. W., and Thonard, J. C.,** The effect of oxygen derived free radicals on gingival proteolycans and hyaluronic acid, *J. Peridodont. Res.*, 19, 390, 1984.
49. **Bates, E. J., Harper, G. S., Lowther, D. A., et al.,** Effect of oxygen derived reactive species on cartilage proteoglycan-hyaluronate aggregates, *Biochem. Int.*, 8, 629, 1984.
50. **Dean, R. T., Roberts, C. R., and Forni, L. G.,** Oxygen centered free radicals can efficiently degrade the polypeptide of proteoglycans in whole cartilage, *Biosci. Rep.*, 4, 1017, 1984.
51. **Chung, M. H., Kesner, L., and Chan, P. C.,** Degradation of articular cartilage by copper and hydrogen peroxide, *Agents Actions*, 15, 328, 1984.
52. **Burkhardt, H., Rehkopf, E., Kasten, M., et al.,** Interaction of polymorphonuclear leukocytes with cartilage *in vitro, Scand. J. Rheum.*, 17, 183, 1988.
53. **Kowanko, I. C., Bates, E. J., and Ferrante, A.,** Mechanisms of human neutrophil mediated cartilage damage *in vitro:* the role of lysosomal enzymes, hydrogen peroxide, and hypochlorous acid, *Immunol. Cell Biol.*, 67, 321, 1989.
54. **Roberts, C. R., Mort, J. S., and Roughley, P. J.,** Treatment of cartilage proteoglycan aggregate with hydrogen peroxide, *Biochem. J.*, 247, 349, 1987.
55. **Roberts, C. R., Roughley, P. J., and Mort, J. S.,** Degradation of human proteoglycan aggregate induced by hydrogen peroxide, *Biochem. J.*, 259, 805, 1989.
56. **Salin, M. L. and McCord, J. M.,** Free radicals and inflammation: protection of phagocytosing leukocytes by superoxide dismutase, *J. Clin. Invest.*, 56, 1319, 1975.
57. **Niwa, Y., Sakane, T., Shingu, M., et al.,** Effect of stimulated neutrophils from the synovial fluid of patients with rheumatoid arthritis on lymphocytes — a possible role of increased oxygen radicals generated by the neutrophils, *J. Clin. Immunol.*, 3, 228, 1983.
58. **Sagone, A. L., Kamps, S., and Campbell, R.,** The effect of oxidant injury on the lymphoblastic transformation on human lymphocytes, *Photochem. Photobiol.*, 28, 909, 1978.
59. **Zoschke, D. C. and Staite, N. D.,** Suppression of human lymphocyte proliferation by activated neutrophils or H2O2: surviving cells have an altered T helper/T suppressor ratio and an increased resistance to secondary oxidant exposure, *Clin. Immunol. Immunpathol.*, 42, 160, 1987.
60. **Vincent, F., Brun, H., Clain, E., et al.,** Effects of oxygen free radicals on proliferation kinetics of cultured rabbit articular chondrocytes, *J. Cell. Physiol.*, 141, 262, 1989.

61. **Murrell, G. A., Francis, M. J., and Bromley, L.,** Modulation of fibroblast proliferation by oxygen free radicals, *Biochem. J.*, 265, 659, 1990.
62. **Baker, M. S., Feigan, J., and Lowther, D. A.,** The mechanism of chondrocyte hydrogen peroxide damage: depletion of intracellular ATP due to supression of glycolysis caused by oxidation of glyceraldehyde-3-phosphate dehydrogenase, *J. Rheumatol.*, 16, 7, 1989.
63. **Bates, E. J., Lowther, D. A., and Handley, C. J.,** Oxygen free radicals mediate an inhibition of proteoglycan synthesis in cultured articular cartilage, *Ann. Rheum. Dis.*, 43, 462, 1984.
64. **Bates, E. J., Johnson, C. C., and Lowther, D. A.,** Inhibition of proteoglycan synthesis by hydrogen peroxide in cultured bovine articular cartilage, *Biochim. Biophys. Acta,* 838, 221, 1985.
65. **Bates, E. J., Lowther, D. A., and Johnson, C. C.,** Hyaluronic acid synthesis in articular cartilage: an inhibition by hydrogen peroxide, *Biochem. Biophys. Res. Commun.*, 132, 714, 1985.
66. **Weiss, S. J., Peppin, G., Ortiz, X., et al.,** Oxidative autoactivation of latent collagenase by human neutrophils, *Science,* 227, 747, 1985.
67. **Weiss, S. J. and Peppin, G. J.,** Collagenolytic metalloenzymes of the human neutrophil, *Biochem. Pharmacol.*, 35, 3189, 1986.
68. **Baron, R., Eeckhout, Y., Neff, L., Francois-Gillet, C., Henriet, P., Delaisse, J., and Vaes, G.,** Affinity purified antibodies reveal the presence of (pro)collagenase in the subosteoclastic bone resorbing compartment, *J. Bone Miner. Res.*, 5 (Suppl. 2), 5203, 1990.
69. **Rifkin, B. R., Vernillo, A. T., Kleckner, A. P., Auszmann, J. M., Rosenberg, L. R., and Zimmerman, M.,** Cathepsin B and L activities in isolated osteoclasts, *Biochem. Biophys. Res. Commun.*, 179, 63, 1991.
70. **Carp, H. and Janoff, A.,** *In vitro* suppression of serum elastase inhibitory capacity by reactive oxygen species generated by phagocytosing polymorphonuclear leukocytes, *J. Clin. Invest.*, 63, 793, 1979.
71. **Carp, H. and Janoff, A.,** Potential mediator of inflammation: phagocyte derived oxidants suppress the elastase inhibitory capacity of alpha 1-proteinase inhibitor *in vitro, J. Clin. Invest.*, 66, 987, 1980.
72. **Wasil, M., Halliwell, B., Hutchinson, D. C. S., et al.,** The antioxidant action of human extracellular fluids, *Biochem. J.*, 243, 219, 1987.
73. **Vaes, G.,** Cellular biology and biochemical mechanism of bone resorption, *Clin. Orthop. Relat. Res.*, 231, 239, 1988.
74. **Marks, S. C.,** Osteoclast biology: lessons from mammalian mutations, *Am. J. Med. Genet.*, 34, 43, 1989.
75. **Johnston, R. B., Jr.,** The production of superoxide by cultured macrophages, in *Handbook of Methods for Oxygen Radical Research,* Greenwald, R. A., Ed., CRC Press, Boca Raton, FL, 1988, 373.
76. **Bielefeldt, O. H. and Babiuk, L. A.,** *In vitro* generation of hydrogen peroxide and of superoxide anion by bovine polymorphonuclear neutrophilic granulocytes, blood monocytes, and alveolar macrophages, *Inflammation,* 8, 251, 1984.
77. **Wing, E. J., Ampel, N. M., Waheed, A., and Shadduck, R. K.,** Macrophage colony stimulating factor (M-CSF) enhances the capacity of murine macrophages to secrete oxygen reduction products, *J. Immunol.*, 135, 2052, 1985.
78. **Karhumaki, E. and Helin, H.,** Regulation of oxidative metabolism by interferon-gamma during human monocyte to macrophage differentiation, *Med. Biol.*, 65, 261, 1987.
79. **Miller, S. C. and Kenny, A. D.,** Activation of avian medullary bone osteoclasts by oxidized synthetic parathyroid hormone, *Proc. Soc. Exp. Biol. Med.*, 179, 38, 1985.
80. **Fallon, M., Silverton, S., Smith, P., Moskal, T., Constantinescu, C., Feldman, R., Golub, E., and Shapiro, I.,** The oxidative metabolism of isolated osteoclasts is regulated by calcitropic agents, *J. Bone Miner. Res.*, 1 (Suppl. 1), 1, 1986.

81. **Garrett, I. R., Oreffo, R. O. C., and Mundy, G. R.,** Production of oxygen-derived free radicals by bone resorbing osteoclasts, *J. Bone Miner. Res.,* 3 (Suppl. 1), 96, 1988.
82. **Key, L. L., Taylor, R. G., Burgess, M. C., and Ries, W. L.,** Oxidative metabolite production by plasma membranes at the osteoclast-bone interface, *J. Bone Miner. Res.,* 3 (Suppl. 1), 96, 1988.
83. **Key, L. L., Roes, W. L., Taylor, R. G., Hays, B. D., and Pitzer, B. L.,** Oxygen derived free radicals in osteoclasts: the specificity and localization of the nitroblue tetrazolium reaction, *Bone,* 11, 115, 1990.
84. **Garrett, L. R., Boyce, B. F., and Mundy, G. R.,** Osteoclastic bone resorption *in vitro* is dependent on oxygen derived free radical production, *J. Bone Miner. Res.,* 1 (Suppl. 1), 373, 1986.
85. **Garrett, I. R., Boyce, B. F., Oreffo, R. O. C., Bonewald, L., Poser, J., and Mundy, G. R.,** Oxygen derived free radicals stimulate osteoclast bone resorption in rodent bone *in vitro* and *in vivo, J. Clin. Invest.,* 85, 632, 1990.
86. **Petkau, A.,** Scientific basis for the clinical use of superoxide dismutase, *Cancer Treat. Rev.,* 13, 17, 1986.
87. **Oursler, M. J., Li, L., and Osdoby, P.,** Purification and characterization of an osteoclast membrane glycoprotein with homology to manganese superoxide dismutase, *J. Cell. Biochem.,* 46, 219, 1991.
88. **Oursler, M. J., Collin-Osdoby, P., Li, L., Schmitt, E., and Osdoby, P.,** Evidence for an immunological and functional relationship between superoxide dismutase and a high molecular weight osteoclast plasma membrane glycoprotein, *J. Cell. Biochem.,* 46, 331, 1991.

Chapter 16

OSTEOBLAST/OSTEOCLAST COUPLING

J. Edward Puzas and Motomi Ishibe

TABLE OF CONTENTS

I. INTRODUCTION

The skeleton of mammals is a dynamic and complex organ system responsible for two functions: the maintenance of blood levels of calcium and phosphate and mechanical support for the musculoskeletal system. Operationally, the metabolism of the skeleton can be divided into three phases: a formative phase, a maintenance phase, and an atrophic phase. The formative phase occurs during development and will ultimately determine the peak mass and density of the skeleton. Bone formation at these times involves endochondral formation, intramembranous formation, and a general consolidation of woven bone tissue into the more dense lamellar material which characterizes adulthood. Many excellent papers and texts have been published describing these processes, and readers are referred to these citations for a complete discussion of the topics. The present discourse deals only with the cellular and metabolic processes present in the maintenance and atrophic phases of the adult skeleton.

II. BONE CELLS

There are two types of bone cells: those that form bone and those that resorb bone, i.e., osteoblasts and osteoclasts. In order to better understand the interactions of these cells, a brief description of their functions are necessary. The development and functions of osteoblasts are presented here. Readings on osteoclast development and function can be found in other chapters of this volume.

A. THE OSTEOBLAST LINEAGE

Three forms of the osteoblast cell lineage are recognized. They are progenitor osteoblasts (preosteoblasts), mature osteoblasts, and osteocytes. From a phylogenetic point of view, preosteoblasts arise from cells in the condensing mesenchyme and, as such, are one form of connective tissue. Since mesenchymal cells can give rise to a number of tissue types, it is not until a cell is committed to the osteoblast lineage that it can be histologically differentiated as a bone-forming cell.

1. Preosteoblast

A committed progenitor cell destined to become an osteoblast has a number of distinguishing features. First, these cells are physically near bone-forming surfaces and are usually present where active, mature osteoblasts are synthesizing bone. Their appearance is that of an elongated cell with an elongated nucleus. Most often, they are found in a stratum type of configuration a few cell layers distant from active osteoblasts. Second, they have the capacity to divide, i.e., mitotic characters can be found in these cells. Third, these cells usually stain less intensely for alkaline phosphatase than mature osteoblasts, and they have no evidence of a developed rough endoplasmic reticulum, i.e., they have not yet acquired many of the differentiated characteristics and protein synthetic capabilities of fully developed osteoblasts.

2. Osteoblast

A mature osteoblast is derived from a preosteoblast and expresses all of the differentiated functions required to synthesize bone. The change from preosteoblast to osteoblast is not a quantal process. That is, there is a gradient of differentiation which becomes fully expressed when the mature form of the cell reaches the bone surface. Once the osteoblast has reached the surface, its function is to synthesize and secrete collagen and to ultimately mineralize it. Usually, active osteoblasts are found within a matrix that they themselves have synthesized. It is within this matrix at the mineralization front that the process of hydroxyapatite crystal growth occurs. The mineralization front is the advancing edge of calcification, and is usually 5 to 50 nm away from the osteoblast surface. The area between the osteoblast and the mineralizing front is often referred to as an ‘‘osteoid seam’’. The depth and character of the osteoid seam can be diagnostic for some forms of bone disease (such as osteomalacia and rickets).

3. Osteocyte

An osteocyte is an osteoblast which has become encased in calcified bone. During the process of bone formation, the osteoblast determines its own fate by calcifying itself into a lacuna. Approximately 10 to 20% of osteoblasts eventually become osteocytes, and although it can be said that not all osteoblasts survive as osteocytes, it is true that all osteocytes originated from osteoblasts. At the point of total encasement, the metabolic activity of the cell dramatically decreases due to the lack of nutrient diffusion. The only source of nutrients and gas exchange to which the osteocyte has access is that which can occur through small canals known as canaliculi. These canaliculi are actually the remnants of cellular processes which extended from the osteoblast during bone mineralization. The canaliculi form an extensive array of connecting tubules, and it has been speculated that these tubules form a communication as well as nutrient network.

4. Osteoblast Cell Models

A key to understanding the cell biology of any tissue is the ability to investigate its cells in an isolated, *in vitro* environment. In the case of bone, this presents a unique problem in that bone cells are embedded in a calcified matrix. Thus, the choice of the starting tissue for cell isolation dictates what methods can be used for the preparation of a cell model. Since mature, fully mineralized bone is relatively acellular and is impossible to demineralize without damage to the cells, most investigators have tended to use fetal or neonatal skeletal tissues for the isolation, preferentially employing the flat bones of the calvarium (Figure 1). Bone at this stage of development has a

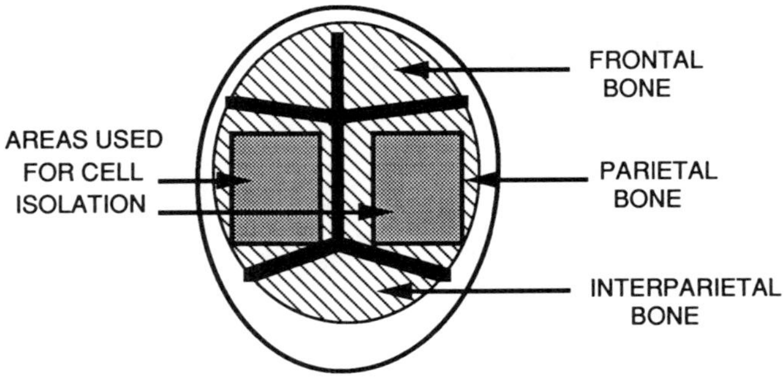

FIGURE 1. Representation of a neonatal rat calvarium. The suture lines are shown in black. A cartilage chondrocranium is found in the white border around the calvarium. By using only the segments of bone in the hatched areas for cell isolation, contamination of the cell preparation by cartilage and undifferentiated mesenchymal cells can be avoided.

high ratio of cells to tissue and is so lightly calcified that the cells can be released from the tissue with enzymatic digestions of only a few hours. Early work by Peck,[1,2] Wong and Cohn,[3] and Hefley and Brand[4] pioneered these isolation techniques which, for the most part, are still in use today. In fact, with the innovations employed by Hefley and Brand, it was possible to prepare highly enriched osteoblastic subpopulations from sequentially digested microdissected segments of parietal bone. Five cell fractions were routinely obtained: fraction 1, fibroblasts; fraction 2, fibroblasts plus osteoprogenitor cells; fraction 3, osteoprogenitor cells; fraction 4, osteoprogenitor cells plus osteoblasts; fraction 5, osteoblasts. Even though there is an unavoidable level of some cell contamination between the isolated fractions, fractions 1, 3, and 5 have proven to be very useful in sorting out many of the differentiation pathways down which progenitor cells can pass. Some of the characteristics of these cells are shown in Figures 2 to 4. These data indicate that fraction 5 cells (mature osteoblasts) have the highest level of alkaline phosphatase (Figure 2) and synthesize the most type I collagen (Figure 3). Fraction 3 cells (preosteoblasts) are the most responsive to vitamin D and parathyroid hormone (Figure 4). Fraction 1 cells (fibroblasts) express very few characteristics in the way of a bone phenotype and are frequently used as a control cell type for *in vitro* experiments.

Other models for studying osteoblastic cell activity are the transformed cell lines. Some of the most popular are ROS 17/2.8,[5] UMR 106,[6] and MC 3T3E1.[7] Although these cells are derived from normal bone cells and retain many of the characteristics of osteoblasts, a number of published reports have identified aberrations in their behavior. Thus, although these models can be convenient and easy to use, care in the interpretation of data as they relate to normal skeletal biology must be taken when working with such models.

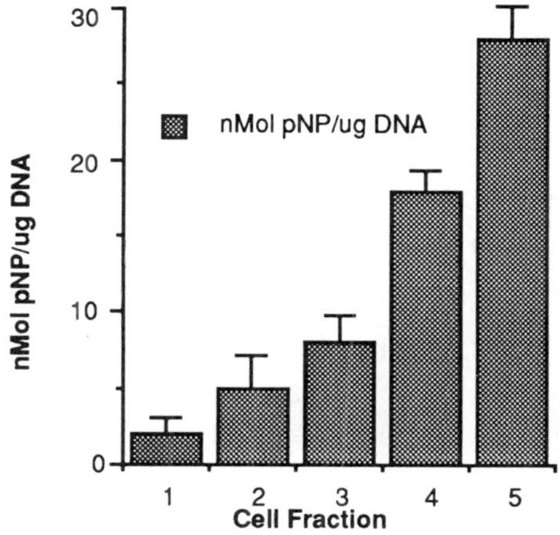

FIGURE 2. Alkaline phosphatase content in each of the cell fractions. Cells were plated at 20,000 cells per well on day 1 and assayed for hydrolysis of *p*-nitrophenol on day 2. The data are expressed as activity per μg DNA.

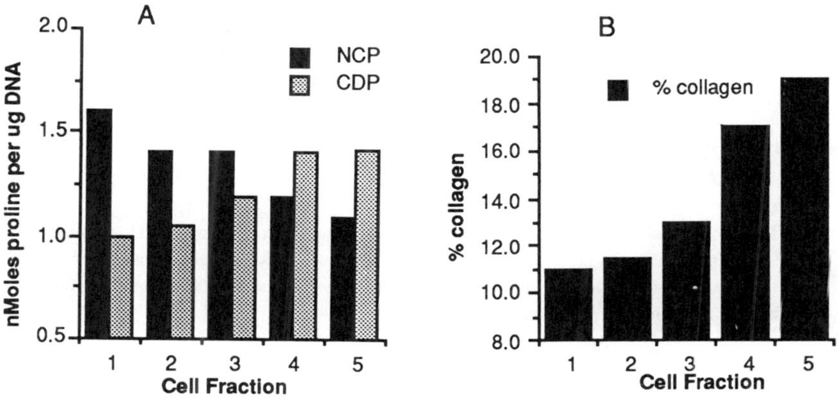

FIGURE 3. (A) Incorporation of a radiolabeled proline into noncollagen protein (NCP) and collagen (collagenase-digestible protein, CDP); (B) data recalculated to show the percent of total protein synthesized as collagen.

III. ORIGIN AND FATE OF OSTEOBLASTS

Many of the most enlightening studies demonstrating the origin and fate of preosteoblasts, osteoblasts, and osteocytes were performed a number of years ago.[8,9] These works utilized timed, radiolabeled thymidine exposure to monitor the progression of a cell through the osteoblast lineage. In principle, the method called for injection of a single dose of high-specific-activity tritium-labeled thymidine into a series of experimental animals. The animals were sacrificed at short (hours), moderate (days), and long (weeks) times

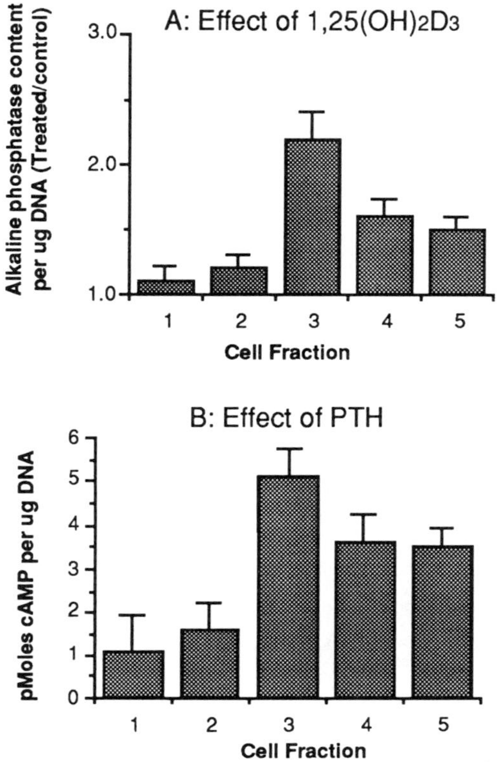

FIGURE 4. The effect of hormones on alkaline phosphatase content and PTH stimulation of cAMP. (A) Fraction 3 cells are the most sensitive to vitamin D metabolites after a 48-h exposure. Alkaline phosphatase activity is corrected for cell number. (B) Fractions 3, 4, and 5 are the most responsive to PTH for the stimulation of cAMP.

after administration of the thymidine, and the cells which incorporated the label were visualized with autoradiography. Since radioactive thymidine is incorporated only into newly forming DNA, the cells which were labeled after a short exposure were preosteoblasts in the process of cell division at the time of the exposure to the radiolabel. There was no label in osteoblasts or osteocytes. After a few days, the cells which demonstrated the label were osteoblasts, and after a few weeks the cells with the label were osteocytes. Because it is known that osteoblasts and osteocytes do not divide, it was deduced that the label which appeared in these cells originated in the preosteoblasts. Thus, the initial cells to be labeled had progressed through the maturation process, and their progression was monitored by visualizing the labeled DNA. In fact, the lifetime and differentiation time for osteoblasts could be calculated with these and other techniques, and it was shown that mature osteoblast appearance required not more than a few days and that they were active for up to 12 weeks.[10,11]

TABLE 1
Osteoblast Target Hormones and Growth Factors

Systemic hormones	Local growth factors
Vitamin D (1,25$(OH)_2D_3$)	Transforming growth factors (TGF-α and $\beta_{1,2,3}$)
Parathyroid hormone (PTH)	Basic fibroblast growth factor (bFGF)
Thyroid hormone (T_3, T_4)	Acidic fibroblast growth factor (aFGF)
Vitamin A (retinoic acid)	Platelet-derived growth factor (PDGF)
Estrogen (17β-estradiol)	Interleukins (IL-1α and -1β, IL-2, IL-6)
Insulin	Prostaglandins (PGE_1)
Glucocorticoids (cortisol)	Epidermal growth factor (EGF)
	Insulin-like growth factors (IGF-I and -II)
	Granulocyte/macrophage colony-stimulating factor (GM-CSF)
	Tumor necrosis factor (TNF-α)

IV. OSTEOBLAST SENSITIVITY

More modern studies have shown that osteoblasts possess the ability to respond to a number of regulatory factors and hormones.[12-14] In fact, it is now apparent that the most responsive cells in bone with regard to controlling both bone formation and bone resorption are cells in the osteoblast lineage. In bone, the activity and number of osteoblasts is influenced by a whole host of hormones and growth factors. A list of the most important, along with their standard abbreviations, is presented in Table 1.

By inference, it can be assumed that osteoblasts possess some type of receptor mechanism for all of these molecules. Thus, not only is the presence of a modulatory factor an important point of regulation in a cell's response, but so is the expression of the factor's receptor.

The actual cells in the osteoblast lineage which can respond to the list of factors is still under investigation. However, many investigators believe that the preosteoblast is the most likely target. It is generally believed that it is not possible to significantly enhance the activity of a fully mature, metabolically active osteoblast, nor is it possible to stimulate these cells to divide. Thus, both the differentiating effects and proliferative effects of growth factors and hormones are probably aimed at the progenitor form of osteoblasts. This is supported by direct evidence in the literature[15] (Figure 4).

V. THE OSTEOBLAST AS A PIVOTAL CELL IN ALL OF BONE METABOLISM

This area is one of the most active and important topics being investigated in bone research today. It is in this arena where molecular biological, immunological, and biochemical techniques have merged in an attempt to understand not only the disease processes, but also the normal processes of bone metabolism. It is probably from these lines of research that new therapies will emerge for diseases such as osteoporosis.

For many years, it was believed that the two most important bone target hormones were parathyroid hormone (PTH) and 1,25-dihydroxyvitamin D_3. Both of these hormones were known to be responsible for increasing the level of calcium in the blood, PTH by virtue of its stimulatory action on bone resorption, and 1,25-dihydroxyvitamin D_3 by virtue of its stimulatory action on intestinal calcium transport and bone resorption.[16,17] The direct effects of these hormones on osteoblasts were, at the early stages of our knowledge, controversial. Many investigators expected that the direct effect of these hormones might be to inhibit bone formation. However, in general, published reports in the literature supported an anabolic role for the skeleton for both hormones, a conclusion which was inconsistent with their role as stimulators of bone resorption.

In osteoblasts, PTH has been shown to mediate ion and amino acid transport, stimulate cAMP, regulate collagen synthesis, bind to a specific receptor, enhance the production of other growth factors, etc.[18-22] 1,25-dihydroxyvitamin D_3 was also shown to stimulate matrix and alkaline phosphatase synthesis, production of bone-specific proteins, bind to a receptor, etc.[23-26] The question was raised as to why hormones responsible for elevating serum calcium levels would have such marked and diverse effects on bone-forming cells. Many theories were put forward to explain some of these phenomena, but it was not until new cell-isolation techniques were developed and interest in bone growth factors was raised that a clearer picture became evident. It now appears that the osteoblast may be the central cell through which bone resorption as well as bone formation is mediated, i.e., many of the initial effects observed on osteoblasts (such as calcium flux changes, cAMP elevation, phospholipase C activation, etc.) may have been reflections of a signaling system which controls all of the cells in bone.

A. OSTEOBLASTS AND BONE RESORPTION

Perhaps because it was not an expected point of regulation, the role of the osteoblast in mediating bone resorption is just now being investigated. Through the work of Chambers et al., we now believe that primary stimulators of bone resorption have little if any effect directly on osteoclasts, i.e., molecules such as PTH, $1,25(OH)_2D_3$, IL-1, TNF-α, lymphotoxin, TGF-β, and the prostaglandins, which have stimulatory effects for bone resorption *in vitro* and *in vivo,* have virtually no effect on isolated, disaggregated osteoclasts. However, all of these primary resorbers stimulate osteoblasts to produce a soluble resorption stimulator which, in turn, can induce bone resorptive activity.[27] An active investigation is now proceeding to identify and purify this stimulator.

Moreover, osteoblasts may actually direct where the activity of osteoclast resorption occurs. One hypothesis which has gained favor suggests that all bone surfaces are covered by either an extended cellular membrane from a bone-lining osteoblast, or by a very thin and frequently imperceptible osteoid layer. Such bone surfaces are unavailable to osteoclastic activity. One of the actions of bone resorption stimulators may, therefore, cause the osteoblast to make these surfaces available. In the case of the lining cell, this would mean a drawing back of the membrane, and in the case of the osteoid layer, it would be through the action of a secreted collagenase. Both observations have been made in the literature,[28,29] again reinforcing that the initial target cell and mediator of bone resorption is the osteoblast.

VI. BALANCE BETWEEN BONE RESORPTION AND BONE FORMATION

From the early visions of Frost,[31] Parfitt,[30] and Jowsey,[32] it has come to be realized that the processes of osteoclastic bone resorption and osteoblastic bone formation occur in discreet, locally regulated units, i.e., osteoclasts become activated, select a site for resorption, excavate a lacuna, and at the completion of their activity, either disintegrate or migrate to another site. Subsequently, progenitor osteoblastic cells take up residence in the same lacuna, differentiate into mature osteoblasts, and reform the bone. This sequence of events occurs on both trabecular surfaces and the endosteal surfaces of cortical bone, the one distinction being that osteoclastic activity tends to tunnel deeper into cortical bone. A critical feature of these activities is that the events of bone resorption and bone formation appear to be both temporally and spatially linked to each other. In the adult skeleton, no bone formation occurs on endosteal or endocortical surfaces unless it is accompanied by a prior event of bone resorption. In fact, the magnitudes of the two processes are also linked to each other. This was most clearly shown in the work of Harris and Heaney.[33,34] With kinetic calcium analyses, they were able to measure the rates of both bone formation and bone resorption in patients with different bone diseases. They found a strong correlation between the skeletal-wide level of resorption and formation. They were the first to term this connection between the two processes as ''skeletal coupling'', i.e., the coupling of osteoclastic bone resorption to osteoblastic bone formation. This observation provided, for the first time, evidence that the skeleton had a type of intrinsic set point which could compensate for different rates of resorptive activity. The coupling mechanisms also explained a number of clinical and experimental findings. For example, it was known that patients with primary hyperparathyroidism had elevated rates of bone formation and lost very little bone in the early stages of the disease, even though they were subjected to a large stimulus to resorb bone.[35] This was probably due to a compensatory increase in coupled rates of bone formation. Another observation made many years ago by Parsons et al.[36] suggested that PTH actually stimulated bone formation as well as bone resorption. Although it is now known that PTH may have important effects on anabolic growth factors,[37,38] part of the original observation may have been related to the coupled activities of osteoclasts and osteoblasts. In experimental animal systems, it has been shown that administration of a primary stimulus for bone resorption (i.e., PTH) to young and presumably not tightly coupled animals can enhance the osteoblastic activity to such a point that the compensatory stimulus for formation exceeds that for resorption. This results in a decreased medulary canal and a large number of forming surfaces.[39] Perhaps one of the most important deductions to emerge from an understanding of the coupled activity of osteoclasts and osteoblasts was that bone loss which occurs in the aging skeleton is as much due to a defect in the formation of bone as to the resorption of bone. If individuals were able to exactly match the degree of bone resorption with an equal amount of bone formation over their lifetime, then they would remain at a plateau level for bone mass irrespective of the stimuli to resorb bone. However, in aging, the skeleton does lose mass, and it must be concluded that the coupled activity of osteoblasts cannot match the activity of osteoclasts.

VII. MECHANISMS OF OSTEOBLAST/OSTEOCLAST COUPLING

The coupling mechanisms were directly validated and attributed to actual cellular activity by Baylink et al.,[40] who also began to provide a mechanistic explanation for the sequence of events which occur in a remodeling site. During the course of these investigations, they described both a short-term and long-term type of coupled bone remodeling in an experimental animal model. Stress placed on the skeleton by depletion of calcium in the diet of rats led to marked and sustained increases in bone resorption and, consequently, a decrease in bone mass. This phenomenon was termed ''depletion''. The hallmarks of depletion were depressed levels of serum calcium, elevated levels of PTH, marked activity of osteoclasts, and large numbers of resorption lacunae.[41,42] Surprisingly an increased number of osteoblasts, compared to control treated animals, was also seen in depletion. However, the activity of the osteoblasts was near zero and, consequently, the rates of bone formation were very low. Nevertheless, it appeared as though the resorptive processes had somehow primed the osteoblast populations such that when the low calcium stress was removed, there would be a large increase in bone formation due to the activation of the considerable number of quiescent osteoblasts. Indeed, this did occur when calcium was restored to the diet.[43] The reformation phase of skeletal remodeling was termed ''repletion''. In a strict sense, the phenomenon of depletion/repletion does not exactly mimic the normal remodeling process in the skeleton because of the extended times involved. With the depletion/repletion model, it is possible to keep the compensatory bone formation rates depressed for extended periods of time (while the low-calcium diet is in place), whereas in normal remodeling, the sequence of events falls within relatively well-defined time frames. However, two important points can be concluded from these types of animal studies: (1) the activation mechanisms for increasing osteoblast number are probably similar in normal remodeling and dietary calcium depletion and (2) systemic factors,

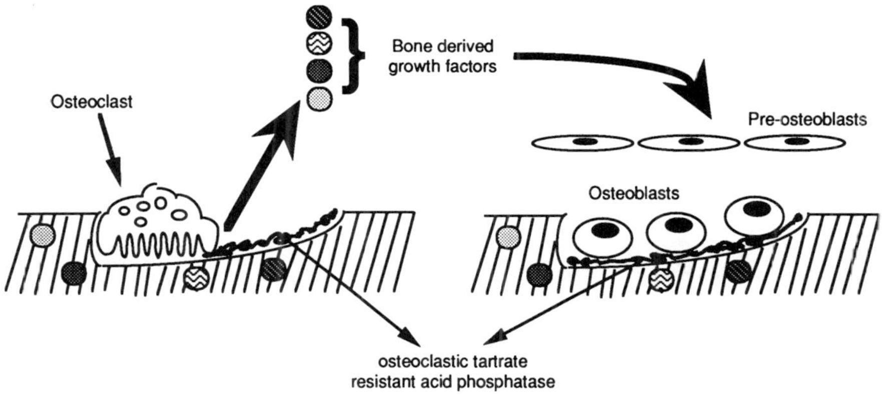

FIGURE 5. Diagrammatic representation of osteoclastic bone resorption followed by osteoblastic bone formation. During the resorption process, growth factors are released from the matrix which then activate osteoprogenitor cells. The osteoprogenitor cells mature into osteoblasts and ultimately replace the resorbed bone. The mechanism by which osteoblasts are directed to form bone only in the resorption lacunae may be due to the presence of molecules, such as tartrate-resistant acid phosphatase, which are left behind during osteoclastic activity.

or perhaps calcium itself, is a triggering mechanism to allow quiescent osteoblasts to begin to actively make bone.

In the late 1970s, the biochemical mechanism of coupling became a very active area of investigation. The work of Baylink, Howard, and Puzas et al.[44-46] demonstrated that (1) the coupling of osteoblastic activity to osteoclastic activity could not be explained by known systemic bone hormones, (2) coupling could be shown to occur in a bone-organ culture system *in vitro,* and (3) stimulation of bone resorption in culture by agents such as PTH led to an increased release of regulatory factors into the culture medium. These were some of the first definitive results indicating that the regulation of bone mass was mediated by locally released bone-formation mediators. As a result of these findings, the early discoveries of an extractable bone morphogenetic protein (BMP) by Urist and Strates came into clearer focus.[47] Presumably, either chemical extraction or osteoclastic activity could release these factors from bone matrix. Since most of these factors have been shown to be (1) present in bone matrix, (2) synthesized by osteoblasts, and (3) have bone formation regulatory properties, they have been termed bone autocrine growth factors. A diagrammatic representation of the remodeling/coupling process is shown in Figure 5.

One of the key molecules which plays a central role in the coupling mechanism is TGF-β. This factor is part of a family of proteins which has such a widespread function in all cell types that it may actually be considered a primordial growth factor. Virtually every tissue examined can either synthesize at least one form of TGF-β or respond to it. With regard to bone, osteoblasts have been shown to produce more than one form of the molecule and to have TGF-β receptors.[48] Moreover, bone tissue itself, due to its mass, is the largest repository for TGF-β in the body. Original work by Mundy, Seyedin, and later by industrial scientists[49-51] has shown that TGF-β has all the characteristics of the "ideal" coupling factor.

- It is produced in large quantities by osteoblasts.
- Its synthesis is regulated by known bone target hormones.
- It resides in the bone matrix.
- It has potent and diverse actions on preosteoblasts.
- It has an identifiable receptor on osteoblasts and preosteoblasts.
- It is released from the matrix by the activity of osteoclasts.

Interestingly, the form of TGF-β present in bone is that of an inactive, high molecular weight precursor which can be activated by exposure to an acid environment. Thus, the low pH of an active resorption site may be yet another regulatory point for the initiation of a coupling event.

VIII. SITE-DIRECTED OSTEOBLASTIC BONE FORMATION

Although the release of growth factors from bone may explain the temporal regulation of coupling, it cannot explain all aspects of the spatial regulation. For example, it is now accepted that the diffusion of soluble factors subsequent to bone resorption activates progenitor osteoblasts in a wide radius around a resorption site.[52] These cells proliferate and begin to differentiate into mature osteoblasts. True bone formation, however, only occurs on the actual resorption surface. There is little, if any, new bone deposited on surfaces adjacent to the resorption lacunae. This is even more striking when one considers that an active formation surface may only be tens of microns from an inactive surface. Why do osteoblasts reside and differentiate at specific sites of resorption?

The intuitive answer to this question is that the surface has been altered by the osteoclast such that differentiation only occurs at this site. However, it is not known what mediates this differentiation nor why it occurs only after a resorption event. A hypothesis has been put forward to explain this observation which states that one of the proteinaceous products of the resorptive process, type 5 tartrate-resistant osteoclast acid phosphatase (TRAP), can act as a regulatory molecule and participates in directing the sites of new bone formation (Figure 5). The evidence in support of this hypothesis is both direct and indirect.

The indirect evidence comes mostly from histochemical studies in which localization of osteoclastic acid phosphatase has demonstrated a high concentration of the enzyme at the resorption surface.[53] Never is there any of this enzyme observed on adjacent surfaces. Some reports have indicated that the acid phosphatase remains at the resorption site long after the osteoclast has left the lacunae or disintegrated.[54] In fact, detectable levels of acid phosphatase have been shown at the reversal line after bone formation has reformed the lacunae. Other observations from the oncology literature also support the notion that acid phosphatase can induce osteoblast differentiation. For example, one of the most common complications of prostatic carcinoma in humans is the metastasis of the tumor to bone. In fact, of all neoplasms which demonstrate skeletal metastases, prostate tumors have one of the highest incidences, approaching 70%.[55] In the skeleton, the presence of tumor cells is associated with high levels of host osteoblastic activity,[56] formation of dense sclerotic bone,[57] and localization of large amounts of the enzyme, acid phosphatase, at the metastatic skeletal sites.[58,59] Frequently, osteomalacic areas are noted at metastatic lesions. These have been attributed to a marked increase in collagen synthesis by osteoblasts without the corresponding ability to mineralize the osteoid.[60] The stimulatory effects of prostatic tumor cells on bone formation are further supported by clinical studies in patients with bone metastases. These studies demonstrate (1) elevated serum levels of osteocalcin and alkaline phosphatase[60] in affected patients, (2) a generalized increase in lumbar bone mineral content in patients with spinal metastases,[62] (3) an increase in iliac trabecular bone volume,[60] and (4) a decreased osteoblastic response upon successful treatment of the tumor.[63] Prostatic regulatory factors have been implicated in this stimulation of bone formation,[64,65] but for the most part, these molecules have been shown to be mitogenic for bone cells *in vitro*. They demonstrate low molecular weights (10 to 18 kDa) and heparin-binding affinity. On Western analysis, some of these proteins appear to resemble a bFGF-like molecule. However, none of these studies demonstrate the presence of a phenotype-enhancing molecule for bone cells.

The direct evidence demonstrates that TRAP and human prostatic acid phosphatase (hPAP) can directly enhance the expression of the osteoblast phenotype *in vitro*.[66] Both enzymes have been shown to dramatically stimulate alkaline phosphatase content and collagen synthesis in preosteoblasts. Little, if any, effect on proliferation was demonstrated by either enzyme. Moreover, the mechanism of the effect for TRAP and hPAP appears to be mediated by a receptor mechanism which, among other second messengers, causes a net influx of calcium ion into the cells.[67]

IX. MECHANISM OF ACID PHOSPHATASE STIMULATION OF BONE CELLS

At present, six different isoenzymes of acid phosphatase are known to exist in mammalian species. The type 5 isoenzyme is the most cationic of the group and is the only one which is insensitive to L(+)-tartrate inhibition. The type 5 enzyme has been identified in spleen, lung, liver, uterus, and bone. In most of these tissues, the amount of the enzyme is low; however, it can be significantly elevated in disease states such as Gaucher's disease, hairy cell leukemia, Hodgkin's disease, giant cell tumors, and Paget's disease.

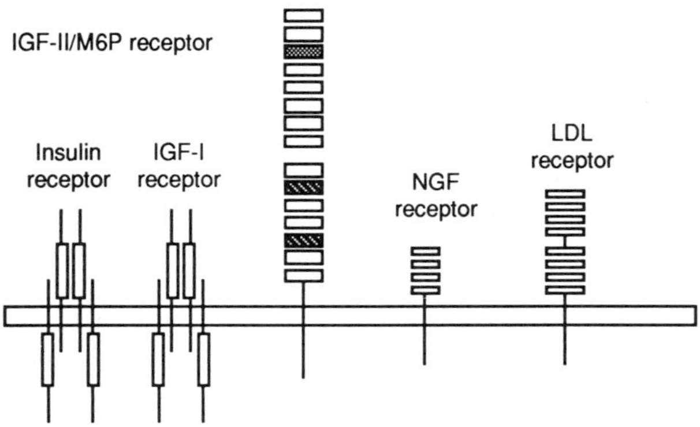

Dissociation Constant and Receptor Number for TRAP Binding to the IGF-II/Mannose-6-phosphate receptor in osteoblasts

Kd	Receptor Number
29.6 ± 9.7 nM	$1.9 \pm 0.9 \times 10^5$ sites/cell

FIGURE 6. The cation-independent IGF-II/mannose-6-phosphate receptor is a large molecular weight transmembrane protein. It has 15 subunits, two of which have been identified as mannose-6-phosphate binding sites and one of which has been identified as an IGF-II binding site. The IGF-II/mannose-6-phosphate receptor shares a structural similarity to the nerve growth factor receptor (NGF) and the LDL receptor, with less similarity to the insulin and IGF-I receptors. The binding of tartrate-resistant acid phosphatase (TRAP) to the osteoblastic IGF-II/mannose-6-phosphate receptor is of high affinity and of moderate capacity.

The osteoclastic isoenzyme is remarkably similar to the enzymes isolated from human placenta[68] and porcine uterus.[69] Homologies for known DNA sequences of the different type 5 enzymes approach 85%. All of the type 5 forms of acid phosphatase are glycosylated, with characteristic carbohydrate structures. One of the major carbohydrate residues is mannose-6-phosphate (M-6-P). The presence of M-6-P on the proteins directs the sequestration of these enzymes into lysosomal vacuoles in the cell by virtue of their binding to the M-6-P receptor on internal membranes.[70]

The M-6-P receptor, however, shares identity with the IGF-II receptor.[70-72] This binding molecule is an interesting, unique, and complex protein. The cationic form of the receptor is an approximately 250-kDa polypeptide with 15 transmembrane extracytoplasmic binding domains. The average length of the domains is 147 amino acids. All of the domains are contiguous and contain highly repetitive sequences except for one, domain 13.[70] Interestingly, in domain 13 there is a 43-residue sequence of amino acid repeats which resemble those in fibronectin.[73] Whether this area of the IGF-II/M-6-P is responsible for cell attachment in a way similar to that mediated by fibronectin remains to be demonstrated (Figure 6).

Each mole of receptor has one IGF-II binding domain and two M-6-P binding domains. However, the domains for M-6-P appear to show a variability in tertiary structure such that different configurations of ligand binding could occur. The role of the IGF-II binding site on this receptor presumably is to transduce the signal of this growth factor into the cell. The transduction signal is not an adenylate cyclase-mediated event,[74] but, rather, involves the production of inositol trisphosphate and an increase in intracellular calcium. These second messengers presumably trigger the effects of IGF-II. In many cases, as well as in bone, it appears that the IGF-II/M-6-P receptor controls differentiation.

The question which now should be asked is, can molecules other than authentic growth factors interact with a specific growth factor receptor and evoke a biological response? In other words, can molecules such as TRAP bind to the M-6-P site on the IGF-II/M-6-P receptor and stimulate bone cells like IGF-II? The evidence suggests that they can. It is not known how widespread such a mechanism might be in different tissues, but in bone it can explain the spatial recognition phenomenon in the coupling sequence. In fact, of all the growth factor receptors in bone, the IGF-II/M-6-P receptor is the most likely one to stimulate differentiation of progenitor osteoblasts. Thus, a likely sequence of events is (1) osteoclastic activity releases mitogenic factors from the matrix, (2) progenitor bone cells are stimulated to proliferate, (3) the progenitor cells preferentially adhere to the resorbed surface, and (4) the M-6-P-containing proteins on the surface stimulate the final differentiation pathways for the progenitor osteoblasts.

Even though a great body of literature has been written on IGF-II and M-6-P binding, the biological significance of consolidating the IGF-II binding site on the same protein as that for the M-6-P binding site remains uncertain. The IGF-II binding site was not always on the same protein as that for M-6-P. In fact, the combination of these sites appears to be an evolutionarily new event. Chicken and frog M-6-P receptors fail to bind IGF-II, whereas rat, human, and other mammalian M-6-P receptors bind both M-6-P and IGF-II.[75,76] These observations have been further documented with cross-linking studies, purification studies, and affinity chromatography studies. These data imply that the effects of IGF-II may be characteristic of regulatory processes in higher species. Thus, the emergence of a high-affinity binding site for IGF-II appears to have occurred after the divergence of mammals from other vertebrates, and may explain the differences in bone remodeling in these species.

X. SUMMARY

It is known that the osteoblast (or preosteoblast):

1. Can both produce and respond to locally released growth factors
2. Plays an important role in the recruitment and activity of osteoclasts
3. Utilizes site-directing signals for identifying the location for new bone formation

The conclusion which can be drawn from these observations is that the osteoblast is a critically important cell in bone metabolism in ways which go beyond its ability to simply form bone. This cell type is probably responsible for the initiating and coordinating actions of many aspects of the coupling mechanisms. In fact, by extension, it is probably safe to say that disease states of the underproduction and overproduction of bone are linked to biochemical defects in osteoblast functions as either a signaling cell or synthesizing cell. Thus, the sooner we understand the complete functions of this cell, the sooner we may be able to understand the skeleton as a whole organ system.

REFERENCES

1. **Peck, W. A., Carpenter, J., Messinger, K., and DeBra, D.,** Cyclic 3′5′-adenosine monophosphate in isolated bone cells. Response to low concentrations of parathyroid hormone, *Endocrinology,* 92, 692, 1973.
2. **Peck, W. A., Carpenter, J., and Messinger, K.,** Cyclic 3′,5′-adenosine monophosphate in isolated bone cells. II. Responses to adenosine and parathyroid hormone, *Endocrinology,* 94, 148, 1974.
3. **Wong, G. L. and Cohn, D. V.,** Target cells in bone for parathormone and calcitonin are different: enrichment for each cell type by sequential digestion of mouse calvaria and selective adhesion to polymeric surfaces, *Proc. Natl. Acad. Sci. U.S.A.,* 72, 3167, 1975.
4. **Hefley, T., Cushing, J., and Brand, J. S.,** Enzymatic isolation of cells from bone: cytotoxic enzymes of bacterial collagenase, *Am. J. Physiol.,* 240, C234, 1981.
5. **Majeska, R. J., Rodan, S. B., and Rodan, G. A.,** Parathyroid hormone-responsive clonal cell lines from rat osteosarcoma, *Endocrinology,* 107, 1494, 1980.
6. **Partridge, N. C., Alcorn, D., Michelangeli, V. P., Ryan, G., and Martin, T. J.,** Morphological and biochemical characterization of four clonal osteogenic sarcoma cell lines of rat origin, *Cancer Res.,* 43, 4308, 1983.
7. **Farley, J. R., Tarbaux, N. M., Hall, S. L., Linkhart, T. A., and Baylink, D. J.,** The anti-bone-resorptive agent calcitonin also acts *in vitro* to directly increase bone formation and bone cell proliferation, *Endocrinology,* 123, 159, 1988.
8. **Tonna, E. A. and Cronkite, E. P.,** The periosteum: autoradiographic studies on cellular proliferation and transformation utilizing tritiated thymindine, *Clin. Orthop. Relat. Res.,* 30, 218, 163.
9. **Tonna, E. A. and Cronkite, E. P.,** An autoradiographic study of periosteal cell proliferation with tritiated thymidine, *Lab. Invest.,* 11, 455, 1962.

10. **Kimmel, D. B. and Jee, W. S. S.,** Bone cell kinetics during longitudinal bone growth in rat, *Calcif. Tissue Int.,* 32, 123, 1980.
11. **Tran, V. P. T., Vignery, A., and Baron, R.,** Cellular kinetics of the bone remodelling sequence in the rat, *Anat. Rec.,* 202, 445, 1982.
12. **Jin, C. H., Miyaura, C., Ishimi, Y., Hong, M. H., Sato, T., Abe, E., and Suda, T.,** Interleukin 1 regulates the expression of osteopontin mRNA by osteoblasts, *Mol. Cell. Endocrinol.,* 74, 221, 1990.
13. **Gruber, H. E. and Baylink, D. J.,** The effects of fluoride on bone, *Clin. Orthop.,* June, 267, 264, 1991.
14. **Canalis, E., McCarthy, T. L., and Centrella, M.,** Growth factors and cytokines in bone cell metabolism, *Annu. Rev. Med.,* 42, 17, 1991.
15. **Puzas, J. E. and Jensen, J. A.,** Electrophoretically separated bone cell types from the foetal rat calvarium; a histochemical and biochemical study, *Histochem. J.,* 14, 561, 1982.
16. **Talmage, R. V. and Elliott, J. R.,** Removal of calcium from bone as influenced by the parathyroids, *Endocrinology,* 62, 717, 1958.
17. **Henry, H. L. and Norman, A. W.,** Vitamin D: metabolism and biological actions, *Annu. Rev. Nutr.,* 4, 493, 1984.
18. **Donahue, H. J., Fryer, M. J., Eriksen, E. F., and Heath, H.,** Differential effects of parathyroid hormone and its analogues on cytosolic calcium ion and cAMP levels in cultured rat osteoblastlike cells, *J. Biol. Chem.,* 263, 13522, 1988.
19. **Rosenbusch, J. P. and Nichols, G., Jr.,** Parathyroid hormone effects on amino acid transport into bone cells, *Endocrinology,* 81, 553, 1967.
20. **Lomri, A. and Marie, P. J.,** Effect of parathyroid hormone and forskolin on cytoskeletal protein synthesis in cultured mouse osteoblastic cells, *Biochim. Biophys. Acta,* 970, 333, 1988.
21. **Kream, B. E., Rowe, D., Smith, M. D., Maher, V., and Majeska, R.,** Hormonal regulation of collagen synthesis in a clonal rat osteosarcoma cell line, *Endocrinology,* 119, 1922, 1986.
22. **Hesch, R. D., Brabant, G., Rittinghaus, E. F., Atkinson, M. J., and Harms, H.,** Pulsatile secretion of parathyroid hormone and its action on a type I and type II PTH receptor: a hypothesis for understanding osteoporosis, *Calcif. Tissue Int.,* 42, 341, 1988.
23. **Harrison, J. R. and Clark, N. B.,** Avian medullary bone in organ culture: effects of vitamin D metabolites on collagen synthesis, *Calcif. Tissue Int.,* 39, 35, 1986.
24. **Fritsch, J., Grosse, B., Lieberherr, M., and Balsan, S.,** 1,25-dihydroxyvitamin D is required for growth-independent expression of alkaline phosphatase in cultured rat osteoblasts, *Calcif. Tissue Int.,* 37, 639, 1985.
25. **Price, P. A. and Baukol, S. A.,** 1,25-dihydroxyvitamin D increases synthesis of the vitamin K dependent bone protein by osteosarcoma cells, *J. Biol. Chem.,* 255, 11660, 1980.
26. **McDonnell, D. P., Pike, J. W., and O'Malley, B. W.,** The vitamin D receptor: a primitive steroid receptor related to thyroid hormone receptor, *J. Steroid Biochem.,* 30, 41, 1988.
27. **Chambers, R. J., McSheehy, P. M. J., Thomson, B. M., and Fuller, K.,** The effect of calcium regulating hormones and prostaglandins on bone resorption by osteoclasts disaggregated from neonatal rabbit bones, *Endocrinology,* 116, 234, 1985.
28. **Norimatsu, H., Vander Wiel, C. J., and Talmage, R. V.,** Morphologic support of a role for cell lining bone surfaces in maintenance of plasma calcium concentration, *Clin. Orthop. Relat. Res.,* 138, 254, 1979.
29. **Puzas, J. E. and Brand, J. S.,** Parathyroid hormone stimulation of collagenase secretion by isolated bone cells, *Endocrinology,* 104, 559, 1979.
30. **Parfitt, A. M.,** Quantum concept of bone remodeling and turnover: implications for the pathogenesis of osteoporosis, *Calcif. Tissue Int.,* 28, 1, 1979.

31. **Frost, H. M.,** Treatment of osteoporoses by manipulation of coherent bone cell populations, *Clin. Orthop. Relat. Res.,* 143, 227, 1979.
32. **Jowsey, J.,** Studies of haversian systems in man and some animals, *J. Anat.,* 100, 857, 1966.
33. **Harris, W. H. and Heaney, R. P.,** Skeletal renewal and metabolic bone disease, *N. Engl. J. Med.,* 280, 253, 1969.
34. **Harris, W. H. and Heaney, R. P.,** (1969): Skeletal renewal and metabolic bone disease, *N. Engl. J. Med.,* 280, 303, 1969.
35. **Tam, C. S., Bayley, A., Cross, E. G., Murray, T. M., and Harrison, J. E.,** Increased bone apposition in primary hyperparathyroidism: measurements based on short interval tetracycline labeling of bone, *Metabolism,* 31, 759, 1982.
36. **Tam, C. S., Heersche, J. N., Murray, T. M., and Parsons, J. A.,** Parathyroid hormone stimulates the bone apposition rate independently of its resorptive action: differential effects of intermittent and continuous administration, *Endocrinology,* 110, 506, 1982.
37. **McCarthy, T. L., Centrella, M., and Canalis, E.,** Parathyroid hormone enhances the transcript and polypeptide levels of insulin-like growth factor I in osteoblast-enriched cultures from fetal rat bone, *Endocrinology,* 124, 1247, 1989.
38. **Canalis, E., Centrella, M., Burch, W., and McCarthy, T. L.,** Insulin-like growth factor I mediates selective anabolic effects of parathyroid hormone in bone cultures, *J. Clin. Invest.,* 83, 60, 1989.
39. **Liu, C. C. and Kalu, D. N.,** Human parathyroid hormone-(1-34) prevents bone loss and augments bone formation in sexually mature ovariectomized rats, *J. Bone Miner. Res.,* 5, 973, 1990.
40. **Baylink, D. J., Farley, J. R., Howard, G. A., Drivdahl, R. H., Puzas, J. E., Masuda, T., Ivey, J. L., and Gruber, H.,** Coupling factor, *Adv. Exp. Biol. Med.,* 152, 409, 1982.
41. **Drivdahl, R. H., Liu, C. C., and Baylink, D. J.,** Regulation of bone repletion in rats subjected to varying low-calcium stress, *Am. J. Physiol.,* 246, R190, 1984.
42. **Baylink, D. J. and Liu, C. C.,** The regulation of endosteal bone volume, *J. Periodontol.,* 50, 43, 1979.
43. **Liu, C. C., Rader, J. I., Gruber, H., and Baylink, D. J.,** Acute reduction in osteoclast number during bone repletion, *Metab. Bone. Dis. Relat. Res.,* 4, 201, 1982.
44. **Howard, G. A., Carlson, C. A., and Baylink, D. J.,** Coupled bone metabolism *in vitro:* embryonic chick limbs in organ culture, *Prog. Clin. Biol. Res.,* 101, 259, 1982.
45. **Howard, G. A., Bottemiller, B. L., Turner, R. T., Rader, J. I., and Baylink, D. J.,** Parathyroid hormone stimulates bone formation and resorption in organ culture: evidence for a coupling mechanism, *Proc. Natl. Acad. Sci. U.S.A.,* 78, 3204, 1981.
46. **Drivdahl, R. H., Puzas, J. E., Howard, G. A., and Baylink, D. J.,** Regulation of DNA synthesis in chick calvarial cells by factors from bone organ culture, *Proc. Soc. Exp. Biol. Med.,* 168, 143, 1981.
47. **Urist, M. R. and Strates, B. S.,** Bone morphogenetic protein, *J. Dent. Res.,* 50, 1392, 1971.
48. **ten-Dijke, P., Iwata, K. K., Goddard, C., Pieler, C., Canalis, E., McCarthy, T. L., and Centrella, M.,** Recombinant transforming growth factor type beta 3: biological activities and receptor-binding properties in isolated bone cells, *Mol. Cell. Biol.,* 10, 4473, 1990.
49. **Mundy, G. R. and Bonewald, L. F.,** Role of TGF beta in bone remodeling, *Ann. N.Y. Acad. Sci.,* 593, 91, 1990.
50. **Bentz, H., Nathan, R. M., Rosen, D. M., Armstrong, R. M., Thompson, A. Y., Segarini, P. R., Mathews, M. C., Dasch, J. R., Piez, K. A., and Seyedin, S. M.,** Purification and characterization of a unique osteoinductive factor from bovine bone, *J. Biol. Chem.,* 264, 20805, 1989.

51. **Celeste, A. J., Iannazzi, J. A., Taylor, R. C., Hewick, R. M., Rosen, V., Wang, E. A., and Wozney, J. M.,** Identification of transforming growth factor beta family members present in bone-inductive protein purified from bovine bone, *Proc. Natl. Acad. Sci. U.S.A.*, 87, 9843, 1990.
52. **Vaughan, J. M.,** *The Physiology of Bone,* Clarendon Press, Oxford, 1970, chap 2.
53. **Miller, S. C.,** The rapid appearance of acid phosphatase activity at the developing ruffled border of parathyroid hormone activated medullary bone osteoclasts, *Calcif. Tissue Int.*, 37, 526, 1985.
54. **Liu, C. C., Sherrard, D. J., Maloney, N. A., and Howard, G. A.,** Reactivation of inhibited bone acid phosphatase and its significance in bone histomorphometry, *J. Histochem. Cytochem.*, 35, 1355, 1987.
55. **Harrington, K. D.,** in *Orthopaedic Treatment of Metastatic Bone Disease,* Harrington, K. D., Ed., C. V. Mosby, St. Louis, 1988, 15.
56. **Jacobs, S. C.,** Spread of prostatic cancer to bone, *Urology,* 21, 337, 1983.
57. **Resnick, D. and Niwayama, G.,** in *Diagnosis of Bone and Joint Disorders,* Resnick, D. and Niwayama, G., Eds., W. B. Saunders, Philadelphia, 1988, 3944.
58. **Ahmann, F. R., Woo, L., Hendrix, M., and Trent, J. M.,** Growth in semisolid agar of prostate cancer cells obtained from bone marrow aspirates, *Cancer Res.*, 46, 3560, 1986.
59. **Steffens, J., Friedmann, W., and Lobeck, H.,** Immunohistochemical diagnosis of the metastasizing prostatic carcinoma, *Eur. Urol.*, 11, 91, 1985.
60. **Charhon, S. A., Chapuy, M. C., Delvin, E. E., Valentin Opran, A., Edouard, C. M., and Meunier, P. J.,** Histomorphometric analysis of sclerotic bone metastases from prostatic carcinoma, with special reference to osteomalacia, *Cancer,* 51, 918, 1983.
61. **Coleman, R. E., Mashiter, G., Fogelman, I., Whitaker, K. D., Caleffi, M., Moss, D. W., and Rubens, R. D.,** Osteocalcin: a potential marker of metastatic bone disease and response to treatment, *Eur. J. Cancer Clin. Oncol.*, 24, 1211, 1988.
62. **Mobb, G. E., Moisey, C. U., and Ring, E. F.,** Bone mineral content in patients with carcinoma of the prostate, *Br. J. Urol.*, 62, 355, 1988.
63. **Dann, J., Castronovo, F. P., McKusick, K. A., Griffin, P. P., Strauss, H. W., and Prout, G. R.,** Total bone uptake in management of metastatic carcinoma of the prostate, *J. Urol.*, 137, 444, 1987.
64. **Jacobs, S. C. and Lawson, R. K.,** Mitogenic factor in human prostate extracts, *Urology,* 16, 488, 1980.
65. **Koutsilieris, M., Rabbani, S. A., and Goltzman, D.,** Selective osteoblast mitogens can be extracted from prostatic tissue, *Prostate,* 9, 109, 1986.
66. **Ishibe, M., Rosier, R. N., and Puzas, J. E.,** Human prostatic acid phosphatase directly stimulates collagen synthesis and alkaline phosphatase content of isolated bone cells, *J. Clin. Endocrinol. Metab.*, 73, 785, 1991.
67. **Ishibe, M., Zuscik, M., Rosier, R. N., and Puzas, J. E.,** Activation of IGF-II/cation independent mannose-6-phosphate receptor by acid phosphatase stimulates a rapid net influx of Ca^{+2} in BALB/c 3T3 fibroblasts, unpublished data currently under review.
68. **Ketcham, C. M., Roberts, R. M., Simmen, R. C. N., and Nick, H. S.,** Molecular cloning of the type 5, iron-containing tartrate-resistant acid phosphatase from human placenta, *J. Biol. Chem.*, 264, 557, 1989.
69. **Hunt, D. F., Yales, J. R., Shabanowitz, J., Zhu, H. Z., Zirino, T., Averill, B. A., Davarat-Larroque, S. T., Shewale, J. G., Roberts, R. M., and Brew, K. W.,** Sequence homology in the metalloproteins; purple acid phosphatase from beef spleen and uteroferrin from porcine uterus, *Biochem. Biophys. Res. Commun.*, 144, 1154, 1987.
70. **Dahms, N. M., Lobel, P., and Kornfeld, S.,** Mannose 6-phosphate receptors and lysosomal enzyme targeting, *J. Biol. Chem.*, 264, 12115, 1989.

71. **Morgan, D. O., Edman, J. C., Standring, D. N., Fried, V. A., Smith, M. C., Roth, R. A., and Rutter, W. J.,** Insulin-like growth factor II receptor as a multifunctional binding protein, *Nature (London),* 329, 301, 1987; published erratum, *Nature (London),* 20 (7), 442, 1988.
72. **Tong, P. Y., Tollefsen, S. E., and Kornfeld, S.,** The cation-independent mannose 6-phosphate receptor binds insulin-like growth factor II, *J. Biol. Chem.,* 263, 2585, 1988.
73. **Lobel, P., Dahms, N. M., and Kornfeld, S.,** Cloning and sequence analysis of the cation-independent mannose 6-phosphate receptor, *J. Biol. Chem.,* 263, 2563, 1988.
74. **Kojima, I., Hishimoto, I., Iiri, T., Ogata, E., and Rosenfeld, R.,** Evidence that type II insulin-like growth factor receptor is coupled to calcium gating system, *Biochem. Biophys. Res. Commun.,* 154, 9, 1988.
75. **Canfield, W. M. and Kornfeld, S.,** The chicken liver cation-independent mannose 6-phosphate receptor lacks the high affinity binding site for insulin-like growth factor II, *J. Biol. Chem.,* 264, 7100, 1989.
76. **Clairmont, K. B. and Czech, M. P.,** Chicken and Xenopus mannose 6-phosphate receptors fail to bind insulin-like growth factor II, *J. Biol. Chem.,* 264, 16390, 1989.

Chapter 17

EFFECTS OF PHARMACOLOGICAL AGENTS ON OSTEOCLASTS

Paula H. Stern and Peter Lakatos

TABLE OF CONTENTS

I. INTRODUCTION

Since pharmacology encompasses all interactions of chemicals with living systems, some arbitrary decisions were required to define the scope of this chapter. Clearly, it should include drugs used therapeutically in the treatment of bone disease and agents with important bone toxicities. However, it could also be expanded to include effects of agents used as tools to determine the mechanism of action of both the therapeutic agents and physiological factors regulating the activity of the tissue. Another complexity in constructing this chapter was that many of the candidate agents were the subjects of other chapters, e.g., those on "systemic factors", "local factors", "signal transduction", "diphosphonates", and others. We have elected to approach the topic from two directions. The first section will describe effects of a wide range of substances which have activity when added directly to osteoclasts, osteoclast-enriched systems, or cultures of putative osteoclast precursor cells. This section will be divided into the topics of systemic physiological factors, local physiological factors, other therapeutic agents, and pharmacologic tools. The second part will review effects of agents administered *in vivo* on osteoclast function. These will largely be agents used therapeutically.

II. EFFECTS OF PHARMACOLOGICAL AGENTS ON OSTEOCLASTS *IN VITRO*

A. SYSTEMIC FACTORS

1. Calcitonin

Calcitonin and the calcitonin gene-related peptides (CGRP) have direct effects in inhibiting the activity of osteoclasts. The actions include reduction of the area, pseudopodial ruffling, and motility of the cells, as well as decreased ability to resorb bone when the cells are incubated on devitalized bone surfaces.[1-6] An immunocytochemical study of the effects of calcitonin on chick osteoclasts revealed changes in cytoskeletal elements; as the cells assumed a more rounded shape, the density of the cytoskeletal elements increased, and the concentration of vimentin was elevated in the cell center.[7] Calcitonin effects on isolated rat osteoclasts are observed at concentrations as low as 0.1 pg/ml.[3] The CGRP is two to three orders of magnitude less potent than salmon calcitonin (sCT) in eliciting these responses in rat osteoclasts.[4] Rat and human α- and β-CGRP were equivalent in potency, whereas amino and carboxyterminal fragments were ineffective.[4] Differences in responsiveness to calcitonin have been observed. For example, human fetal osteoclasts exhibited antiresorptive effects in response to calcitonin, whereas osteoclasts obtained from transiliac crest biopsies were unresponsive.[5] In another study, osteoclasts from chicken bone were not responsive to either rat or chicken calcitonin, whereas rat osteoclasts responded to both.[3] The effects of calcitonin in decreasing osteoclast motility are rapid, occurring within minutes of the addition of the agent.[1] There is a subsequent recovery of cell activity, with increased motility observed after 1.5 h and increased resorption after 9.7 h.[6] This observation has been interpreted to suggest that the phenomenon of "escape" may be due to recovery of mature osteoclasts rather than to an increase in cell number. Calcitonin receptors have been demonstrated on isolated osteoclasts,[8-10] in multinucleated cells elicited by bone particle implants,[11] and in osteoclast-like, multinucleated tartrate-resistant acid phosphatase (TRAP)-positive cells formed in marrow cultures stimulated with calcitriol.[10] Photoactive [Arg,[11,18] Lys[14]] sCT-labeled 88,000- and 71,000-Da proteins in rat osteoclasts.[9] Calcitonin increased cyclic AMP in osteoclasts.[8,10,12] The effect was enhanced by forskolin, but was unaffected by pertussis toxin, suggesting that the effect is independent of N_i, the inhibitory G-protein regulating adenylate cyclase.[13] Calcitonin increased intracellular free calcium in osteoclasts, an effect that was mimicked by 8Br-cAMP.[14] This response was enhanced in the presence of added calcium. It has been proposed that the increased calcium activates the cytoskeletal system, leading to cell retraction and the attenuation of bone resorption. This model would imply a single signal transduction pathway through elevated cyclic AMP. Another model visualizes two separate G-protein-activated pathways, the effects of calcitonin on cell motility being mediated through a cyclic AMP pathway and the effects on cell retraction through a calcium pathway.[15] Evidence for this were the observations that dibutyryl cyclic AMP and forskolin mimicked the effects of calcitonin on motility, whereas ionomycin and elevated calcium mimicked the latter.[15] Pertussis toxin likewise stimulated cell retraction. Since pertussis toxin affects several G-proteins, interpretation of the pertussis toxin effect is complex.

2. Glucocorticoids

Glucocorticoids have direct effects on bone, inhibiting resorption. They have been found to affect cell spreading and bone-resorbing activity of mature osteoclasts[16] and dentinoclasts.[17] Dexamethasone and hydrocortisone decreased the appearance of mature cells with osteoclastic characteristics in marrow cultures and in cultures of U937 cells,[18,19] suggesting that glucocorticoids can inhibit osteoclast development as well as the activity of mature osteoclasts. Dexamethasone was one to two orders of magnitude more potent than hydrocortisone in inhibiting osteoclast activity.[16] Both the dose-response characteristics and the time course are consistent with the direct effects on osteoclasts being responsible for the inhibition of bone resorption observed in organ cultures. The concentration ranges are also consistent with those expected to be achieved when glucocorticoids are used therapeutically, such as in the treatment of hypercalcemia. At high concentrations, both agents elicited toxicity, which was detectable at 6 h and greater at 12 h.[16] The cytotoxic effects were reversed by progesterone, suggesting that they were receptor mediated.

3. Estrogens

Estrogens are effective antiresorptive agents *in vivo* (see Section III). However, they have displayed only minimal effects on bone *in vitro*. Radioimmunoassays for estrogen receptors demonstrated their presence in human osteoclasts.[2] Estrogens have been found to have dual actions on isolated osteoclasts.[21] Over a narrow concentration range (1 m*M*, not 0.1 or 10 m*M*), 17β-estradiol inhibited osteoclast bone-resorbing effects when calvarial cells, osteoblastic osteosarcoma cells (UMR or HOS), were added.[21] This action was specific and not mimicked by dihydrotestosterone or progesterone. It is conceivable that the long-term *in vivo* effects of estrogens in inhibiting bone resorption could reflect this direct inhibitory effect on osteoclasts. A dose-dependent stimulatory effect was also elicited by estradiol.[21] The difficulties that have been experienced in attempting to elicit direct effects of estrogens in bone organ cultures might be a consequence of these dual antagonistic effects. The antiestrogen tamoxifen, which inhibits bone resorption both *in vivo* and *in vitro*, prevented the stimulatory effect of estrogen on osteoclasts.[21]

4. Parathyroid Hormone

Parathyroid hormone (PTH) actions on osteoclasts have been considered to be indirectly mediated through receptors on osteoblastic cells. Evidence for this has included the lack of responses of enriched populations of osteoclasts to PTH unless osteoblasts were added.[1,2,22] Consistent with this, receptors have been described on osteoblasts and their precursor cells in bone.[23,24] However, PTH receptors have been detected on isolated mature osteoclasts and osteoclast precursors.[25,26] (See Chapter 6 of this volume.) PTH can promote fusion of marrow cells, leading to the formation of multinucleated cells with an osteoclast phenotype.[27] PTH stimulated the appearance of TRAP-positive cells in densely plated cultures of cells obtained late in digestion from neonatal mouse calvaria.[28] A direct effect of PTH on mature osteoclasts leading to enhanced bone resorption still appears unlikely.

5. Vitamin D

1,25-dihydroxyvitamin D_3, like PTH, lacks direct effects on mature osteoclasts, even at high concentrations of the hormone.[1,2] 1,25-dihydroxyvitamin D_3 promotes the fusion and differentiation of candidate osteoclast precursor cells in bone marrow.[10,27,29-33] TRAP activity, calcitonin binding, and increased cAMP in response to calcitonin were increased in mouse marrow cells exposed to 10 n*M* 1,25-dihydroxyvitamin D_3.[10] Cultures from young animals have been shown to be more responsive to calcitriol than those from more mature animals.[31] Although other steroids in the vitamin D pathway have not been shown to have differentiating activity, the 1,25-dihydroxyvitamin D_3 analog calcipotriol (MC903), which is a potent differentiating agent and is weak or inactive as a hypercalcemic agent, had activity similar to calcitriol in increasing osteoclast number and acid phosphatase activity.[32] The 23S, 25R, 23R, and 25S 1α-$25(OH)_2D_3$-26,23 lactones, which have been reported to antagonize bone resorption, antagonized the effects of 1,25-dihydroxyvitamin D_3 on osteoclast precursor fusion.[33] Both transforming growth factor-β (TGF-β) and prostaglandins appear to be involved in the effect of calcitriol on osteoclasts, since both antibodies to TGF-β and the prostaglandin synthesis inhibitor indomethacin inhibited the effect of 1,25-dihydroxyvitamin D_3.[34]

6. Vitamin A

Retinoic acid has been shown to promote proliferation of osteoclast precursors and the fusion of precursor cells in bone organ cultures and bone marrow culture.[29] Older bones did not show the proliferative response, suggesting that the effect is on an early, less differentiated cell.[29] Retinol and retinoic acid, added to isolated mature chicken osteoclasts, enhanced the parameters of activity, including acid phosphatase and release of proline from labeled bone particles.[35] Effects were elicited over a 10-m*M* to 10-μ*M* concentration range. Microtubule depolarization and increases in the number of podosomes were observed.[35] A specific retinol binding protein has been demonstrated in isolated osteoclasts.[36]

B. LOCAL FACTORS

Prostaglandins, cytokines, and growth factors have diverse direct effects on bone resorption and formation. Studies with osteoclast precursor populations as well as mature cells suggest that some of these effects reflect changes in osteoclast generation and activity.

1. Prostaglandins

Prostaglandin Es, added directly to enriched osteoclast populations on bone surfaces, have a rapid effect in inhibiting motility and resorption.[37-39] Inhibition of resorption, assessed by the area of excavated pits, was followed by a stimulatory effect measurable after 48 h.[39] The delayed stimulatory effect was attributed to increased proliferation of stromal cells, since prostaglandins lacked direct stimulatory effects on osteoclasts.[37] Prostaglandin E_2 inhibited the formation of osteoclastic cells in marrow cultures.[40] Evidence that thromboxane A_2 may play a role in bone resorption was obtained in studies with a stable analog of thromboxane A_2, 9,11-epithio-11,12-methanothromboxane A_2.[41] This compound had *in vitro* bone-resorbing activity and elicited increased formation of multinucleated cells in mouse marrow cultures.[41] The bone-resorbing effects of the analog were prevented by an antagonist of thromboxane A_2.

2. Cytokines and Growth Factors

Osteoclast formation and activity are affected by a number of cytokines and growth factors, and it is conceivable that these agents could regulate their activity *in vivo*. α- and γ-Interferon generally have inhibitory effects on osteoclast fusion, consistent with their inhibitory effects on bone resorption.[42-44] Proliferation does not seem to be inhibited, and in one study the number of osteoclasts was found to be increased by γ-INF.[42]

Macrophage colony-stimulating factor (M-CSF), as well as granulocyte-monocyte colony-stimulating factor (GM-CSF), inhibited proliferation of precursor cells and osteoclast formation in mouse marrow cultures.[30,45] Human M-CSF inhibited resorption by isolated rat osteoclasts.[46] In baboon marrow culture, recombinant human CSF-GM and CSF-1 stimulated formation of multinucleated cells, CSF-GM being the more effective agent.[47] A role of endogenous GM-CSF was suggested by the observation that neutralizing antibodies reversed the inhibitory effects of high cell density on marrow cell fusion.[30]

α- and β-Interleukin-1 and tumor necrosis factor-α and -β, at low concentrations, increase differentiation of marrow cells to form osteoclastic cells.[48] The resorptive effects probably also involve activation of receptors on osteoblasts and subsequent activation of osteoclasts, since these cytokines had no effects on isolated osteoclasts,[2] but in the presence of osteoblastic cells, stimulated resorption at concentrations less than 1 m*M*.[49] Interleukin-2 stimulated acid production and resorptive activity of isolated osteoclasts cultured on calvarial surfaces.[50] Interleukin-3 promoted osteoclast generation, activating pluripotent hemotopoetic stem cells.[51,52]

Transforming growth factor-β (TGF-β) had both prostaglandin-dependent and -independent effects on osteoclasts, consistent with the dual effects it exhibits in bone organ cultures. Osteoclast formation was decreased,[34,53-55] but at low TGF-β concentrations, there was an increase in osteoclast activity, which was inhibited by indomethacin, suggesting prostaglandin mediation.[34] Inhibitory effects of TGF-β on cell fusion in marrow cultures were synergistic with those of the osteoinductive factor (OIF).[54] TGF-β increased CFU-GM and the proportion of granulocytic colonies in marrow cultures.[53] It was postulated that the enhanced production of granulocytes may decrease the precursor pool for osteoclasts.[53] The inhibitory effects of TGF-β on osteoclast formation do not appear to be mediated through prostaglandin production.[29] Isolated osteoclasts can release active TGF-β from a latent complex in bone.[55]

Two other stimulators of resorption, epidermal growth factor and TGF-α, which may act through similar receptors, failed to increase the number of multinucleated cells in a human marrow cell culture, but markedly promoted the actions of calcitriol.[56] Effects of sequential addition as well as autoradiographic examination supported the conclusion that the effects of the growth factors were on proliferation, whereas those of calcitriol were on precursor cell fusion in the human marrow cell cultures.[56]

C. PHARMACOLOGICAL THERAPEUTIC AGENTS

1. Diphosphonates

Diphosphonates have a high affinity for bone mineral and are likely to produce their selective effects on osteoclasts due to incorporation into the matrix. Demethylation of amino-bisphosphonate (APD) increased its basicity and inhibitory potency, possibly by increasing the affinity for sites on the matrix.[57] Effects of diphosphonates on resorption by mature osteoclasts displayed a delay, which may reflect the time required for a process by which the agents becomes incorporated into the matrix.[58,59] Inhibitory effects on isolated rat and chicken osteoclasts are elicited at bisphosphonate concentrations in the middle- to high-nanomolar range.[58,60] In studies of the effects of diphosphonates on isolated rat osteoclasts, 4-amino-1-hydroxybutylidene-1,1-bisphosphonic acid (ABP) was more potent than 3-amino-1,1-hydroxypropane-1, 1-diphosphonic acid (APD) or 1-hydroxyethane-1,1-diphonic acid (EHDP) in decreasing the number of resorption pits and the resorption area.[60] Inhibition of proton pump activity and protein synthesis is observed, suggesting that the agents act as metabolic inhibitors.[58] Stimulatory effects on isolated osteoclasts were observed at low-nanomolar concentrations.[60] In addition to having effects on the activity of mature osteoclasts, the diphosphonates have also been shown to affect differentiation of bone marrow cells into resorbing cells with osteoclast characteristics.[61,62]

2. Fluoride

Effects of fluoride in increasing bone mass may be mediated through actions on the osteoclast as well as the osteoblast. Resorptive activities of rabbit and chicken osteoclasts were inhibited by fluoride at 0.5 to 2 m*M*, concentrations that also inhibited bone resorption *in vitro*.[63,64] A decrease in pit depth and a concomitant peripheral expansion of the region of contact between the cells and a dentine surface also was observed.[65]

3. Gallium Nitrate

Gallium nitrate is another antiresorptive agent which, in bone organ cultures, exhibits a delayed effect.[66,67] Bone slices exposed to the compound and subsequently washed were resistant to resorption by added osteoclasts.[68] It has been proposed that gallium nitrate is incorporated into the calcified matrix and subsequently produces an inhibition of resorption.[66,68] Gallium nitrate, at inhibitory concentrations, did not affect osteoclast morphology.[68] Based upon evidence that the cultures did not contain osteoblasts (lack of stimulation by added PTH), it was concluded that the effect of gallium nitrate in inhibiting resorption is a direct effect on osteoclasts.[68]

4. Cyclosporines

Immunosuppressive cyclosporines have inhibitory effects on *in vitro* bone resorption elicited by a number of agents.[69,70] Although immunosuppressive cyclosporines can stimulate bone loss *in vivo* (see Section III), their actions on osteoclasts are consistent with what is observed in bone organ cultures. In bone marrow and cord blood monocyte cultures, cyclosporine A inhibited cell fusion.[71] There is less effect on proliferation, consistent with the lack of significant effects on thymidine incorporation.[70] At the present time, it is not established whether the effects of immunosuppressive cyclosporines on bone are mediated through actions on cytokine production.

D. PHARMACOLOGIC TOOLS

A number of pharmacologic agents affect cellular processes in a sufficiently specific manner that they have been useful in elucidating the mechanisms by which osteoclasts act to resorb bone. Several of these are also useful or potentially useful therapeutic agents for the treatment of disorders of calcium metabolism.

1. Carbonic Anhydrase Inhibitors

Carbonic anhydrase inhibitors decrease bone resorption *in vivo* and *in vitro*. The *in vitro* effect is more pronounced at low carbon dioxide concentrations, consistent with the enzyme inhibition being the mechanism of their effect.[72] In isolated osteoclast populations, cell acidity, as measured by acridine orange, is decreased.[73] Ethoxzolamide prevented the parathyroid hormone-stimulated increase in carbonic anhydrase activity in a population of predominantly osteoclastic cells.[74] The effectiveness of carbonic anhydrase inhibitors in preventing bone resorption is consistent with a role of this enzyme in the mechanism of bone resorption. (See Chapters 6 and 14 of this volume.)

2. Agents Affecting H^+ Secretion

Several agents that inhibit hydrogen ion secretion, including amiloride, an inhibitor of Na^+-H^+ exchange, and omeprazole, an inhibitor of K^+,H^+-ATPase, inhibited bone resorption *in vitro*.[75,76] When tested as inhibitors of the activity of isolated osteoclasts, inhibitors of the Na^+-H^+ exchange were more effective than omeprazole.[77] These findings also support the importance of increased acid production in the stimulation of bone resorption. Mg-dependent ATP-hydrolyzing activity in osteoclast membranes was inhibited by several agents that inhibited proton transport, including *N,N'*-dicyclohexylcarbodiimide, *N*-ethylmaleamide, nolinium bromide, and 4,4'-diisothiocyano-2,2'-stilbene sulfonic acid.[78]

3. Proteolytic Enzyme Inhibitors

Lysosomal enzyme release probably plays a role in osteoclastic bone resorption. Several proteinase inhibitors, including leupeptin and E-64, prevented the action of osteoclasts in resorbing bone matrix without affecting demineralization.[79] Agents that are specific inhibitors of cysteine proteinases decreased the numbers, depths, and volumes of the resorption pits produced by isolated chicken osteoclasts on dentin.[80] This effect was not elicited by the collagenase inhibitor Cl-1.[80]

4. Inhibitors of Osteoclast Attachment

The likely importance of osteoclast adhesion structures in the activity of the osteoclast is indicated by the inhibitory effects of echistatin.[81] This protein, which has an Arg-Gly-Asp (RGD) sequence, bound to a receptor characterized as an integrin.[81] Echistatin inhibited the resorption of bone by isolated rat osteoclasts. Replacement of the arginine with an alanine in the echistatin molecule resulted in a compound that was inactive.[81]

5. Agents Affecting Second-Messenger Pathways

The use of pharmacologic agents that affect second-messenger pathways has provided information regarding their role in the action of hormones, local factors, and other pharmacologic agents on bone. Calcitonin increases cyclic AMP production in osteoclasts.[8,10,12] Forskolin, which increases adenylate cyclase activity, had activities much like those of calcitonin on isolated osteoclasts.[82,83] Cell spreading was inhibited by 10 to 30 μM forskolin.[83] Lower concentrations of forskolin potentiated the effects of calcitonin.[13,84] When the effects on pit formation were quantified, a biphasic effect of forskolin was detected.[85] Low concentrations caused stimulation associated with increased osteoclast number. Higher concentrations of forskolin, as well as isobutylmethylxanthine and dibutyryl cAMP, decreased pit formation. Forskolin and dibutyryl cAMP treatment resulted in shape changes in osteoclasts.[85]

The possible role of cyclic GMP in bone resorption has been less extensively examined than that of cAMP. Intense staining of osteoclasts for cyclic GMP was revealed by immunohistochemical techniques.[86] Nitric oxide, which activates guanylate cyclase, and sodium nitroprusside and 3-morpholinosydnonimine, two agents that release nitric acid, inhibited the motility of isolated osteoclasts, a response similar to that elicited by calcitonin.[87] However, dibutyryl cGMP and 8BR-cGMP did not affect osteoclast motility, making the role of this second-messenger pathway in calcitonin action uncertain.[87]

Lithium inhibited the effects of calcitonin on osteoclastic activity, possibly by blocking an effect of calcitonin on G-protein coupling.[88] Since lithium has multiple effects in cells, other mechanisms are also possible.

6. Agents Affecting Ion Channels

Electrophysiological studies and responses to pharmacologic agents indicate the presence and importance of calcium channels in osteoblasts. KCl depolarization transiently increased intracellular calcium in osteoclasts, and the effect was blocked by several classes of calcium channel blockers.[89,90] A dihydropyridine blocker of L-type calcium channels was the most effective inhibitor. High calcium induced an increase in intracellular calcium which was shown to derive from both intracellular and extracellular sources.[91] This was associated with a decrease in podosomes.[92] However, there were no effects of nifedipine, BAY K 8644, or altered membrane voltage on the inhibitory influence of elevated calcium on osteoclast activity.[93] High concentrations ($\geq 300\ \mu M$) of verapamil, a less specific calcium channel blocker, increased intracellular calcium in osteoclasts and decreased osteoclastic activity, mimicking the effect of high calcium.[94] Lower verapamil concentrations were ineffective.

The presence of sodium channels in osteoclast was suggested by electrophysiological studies showing inhibitory effects of the selective sodium channel blocker tetrodotoxin.[90]

II. EFFECTS OF PHARMACOLOGICAL AGENTS ON OSTEOCLASTS *IN VIVO*

In this section, we concentrate on agents that are either used for the treatment of bone metabolic disorders or administered in other diseases, but may affect osteoclastic activity *in vivo*. The agents discussed here are either stimulators or inhibitors of osteoclasts. The overall effects on bone metabolism *in vivo* of some of these compounds may differ considerably from what are seen *in vitro,* since systemic factors are present that can modify their effects on calcium metabolism. Therefore, this section has been divided into two parts: stimulators and inhibitors of bone resorption. The effects of some agents on osteoclasts have not yet been established well *in vivo;* thus, their classification is somewhat arbitrary. However, all of the listed agents share the common feature of having effects on osteoclastic activity in the concentrations found in body fluids after administration. We have also included agents that are not as widely used in clinical practice currently, but may be of importance in the near future. Finally, some compounds are discussed that are environmental factors and may have an impact on osteoclastic activity as well. Diphosphonates are not discussed in this section, since the following chapter (Chapter 18) has been devoted to those drugs.

A. STIMULATORS OF BONE RESORPTION

1. Parathyroid Hormone (PTH)

The hypercalcemic feature of PTH was recognized by MacCallum and Voegtlin[95] at the beginning of this century. Later, it was recognized that in rats, parathyroidectomy causes increased calcium excretion and lower serum calcium.[96] PTH may also enhance the production of 1,25-dihydroxyvitamin D_3, thus stimulating intestinal calcium absorption.[97] As a consequence, PTH plays a major role in the maintenance of calcium homeostasis.[98]

PTH stimulates bone resorption *in vivo* at concentrations ranging from 1 p*M* to 0.1 n*M*, about 1000-fold less than concentrations required to elicit resorption *in vitro*. Several factors could be responsible for this discrepancy. PTH might be accumulated in bone. Atkinson et al.[99] reported increased PTH levels in bone marrow. Martin et al.[100] showed that only the 1-34 fragment is retained in higher concentration in bone. Retention of active PTH fragments may lead to levels that are close to those required for stimulation of bone resorption *in vitro*. The other possible explanation of the difference between *in vivo* and *in vitro* bone-resorbing levels of PTH is that PTH-stimulated 1,25-dihydroxyvitamin D_3 production may contribute to bone resorption *in vivo*. This is supported by studies showing decreased responsiveness to PTH in the presence of 1,25-dihydroxyvitamin D_3.[101,102] Finally, the antisera utilized in various immunoassays for the measurement of PTH react with different parts of the molecule, and active fragments may be present in lower concentrations than the intact hormone.

Injection of PTH promotes the development of new osteoclasts and increases the number of osteoclasts in various species *in vivo*.[103-105] The latter fact is the major histological finding following the exposure of bone to PTH. An increase in metabolic activity of both osteoclast precursors and osteoclasts is seen following PTH injection into rabbits.[106] In another study, PTH increased the number, but not the activity, of osteoclasts in dogs.[107] Mears[108] has shown that PTH alters the membrane potential of osteoclasts in preparations from rabbit bone. Studies on the time course of PTH action have shown a correlation between its effect on osteoclast number and increases in serum calcium levels *in vivo*.[109] As a result of these events, osteoclastic bone resorption is increased *in vivo*.[110-112]

2. Vitamin D

Osteoclastic bone resorption is increased in response to 1,25-dihydroxyvitamin D_3.[113] The effects of the active vitamin D metabolite on resorption *in vivo* are produced by effects on both the number and activity of osteoclasts.[114] In mice, 1,25-dihydroxyvitamin D_3 increases bone turnover.[115] In normal rabbits, exposure to low doses of 1,25-dihydroxyvitamin D_3 increased bone turnover and the frequency of TRAP-positive osteoclasts without a significant change in the number of cells.[114] At higher doses, osteoclast number increased significantly without a change in the percentage of heavily stained cells. At high doses of 1,25-dihydroxyvitamin D_3, normal rabbits get progressive osteopenic.[116] These data suggest that osteoclasts increase the production of acid hydrolases in response to 1,25-dihydroxyvitamin D_3 before increasing the number of cells. The persistence of lightly staining osteoclasts with increasing dose may represent a maturation of enzymatic capacity in developing osteoclasts. Osteopetrotic rabbits responded to exogenous calcitriol by increasing the number and acid hydrolase content of osteoclasts; however, rabbits treated in this manner do not increase bone resorption and show no radiographic or histologic changes in bone density.[116] In dogs, administration of 1,25-dihydroxyvitamin D_3 increased the activity, but not the number, of osteoclasts.[107] This suggests that 1,25-dihydroxyvitamin D_3 represents a trophic hormone for differentiation of osteoclasts. Analysis of bone implants from vitamin D-deficient rats showed that the total number of osteoclasts, osteoclast-mediated resorption, and remodeling of newly synthesized bone are either suppressed or reduced in magnitude.[117] On the other hand, 1,25-dihydroxyvitamin D_3 has a protective effect on the glucocorticoid-induced decrease in bone turnover by increasing both resorption and formation.[118]

3. Glucocorticoids

Despite the fact that glucocorticoids can inhibit osteoclastic activity *in vitro,* their endogenous excess or systemic administration results in bone loss[119-126] *in vivo*. However, when administered locally, corticosteroids may prevent resorption.[17] In these cases, they probably act via the inhibition of local prostaglandin synthesis and thus cause the inhibition of osteoclast activity.[127] Subcutaneous injection of cortisone acetate may decrease bone resorption and formation as well.[118] Systemic administration of steroids in hypercalcemia of malignancy may inhibit osteoclastic resorption,[128] but it may be mediated through inhibition of the production of hypercalcemic factor(s) and not through a direct effect.

Steroid-induced bone loss occurs primarily in trabecular bone.[129,130] Histologically, there is a reduction of the mean wall thickness of trabecular bone, with a consequent decrease in total bone volume and increase in eroded surface.[131] Osteoclast-covered surface as well as osteoclast number are elevated,[132] and bone resorption is increased.[133] Urinary hydroxyproline excretion, a marker of increased osteoclastic activity, is increased in patients who have developed glucocorticoid-induced osteoporosis.[134] To resolve the inconsistency between *in vitro* and *in vivo* data, the concept of a biphasic effect of glucocorticoids on osteoclasts has been suggested.[135] According to this view, physiologic concentrations are required for the late stages of differentiation and function of osteoclasts, whereas the generation of new osteoclasts that involves cell replication is inhibited by high doses and prolonged exposure.

Other mechanisms may also play an important role in the contrasting effects of glucocorticoids on bone *in vitro* and *in vivo.* Decreased gastrointestinal absorption[136] and increased renal loss[137] of calcium result in secondary hyperparathyroidism.[138] The sensitivity of osteoblasts to parathormone is enhanced by glucocorticoids,[139] and thus it might have implications for resorption because of osteoblastic control of the initiation of the resorption process. *In vivo,* the enhanced resorption is prevented by parathyroidectomy.[140,141] Glucocorticoids inhibit ovarian, testicular, and adrenal gonadal hormone secretion,[142-144] and this may lead to increased resorptive activity as well.[145] It should also be noted that a glucocorticoid-induced decrease in bone formation[146] may contribute to bone loss *in vivo.* In summary, the direct effect of steroids on osteoclasts may be inhibition, whereas *in vivo,* systemic administration may cause other secondary effects that can compete with this inhibition and result in a net bone loss.

4. Thyroid Hormones

Thyroid hormones stimulate osteoclastic bone resorption *in vitro* and *in vivo.*[147,148] This was first suggested by calcium kinetic studies[149] and was confirmed in animal experiments.[150] Increased osteoclastic bone resorption and fibroblastic proliferation resembling osteitis fibrosa cystica were described in histologic sections of patients with hyperthyroidism.[151] Later, it was noted that osteoclastic resorption was not as greatly increased in hyperthyroidism as in hyperparathyroidism.[152]

The direct stimulation of osteoclastic bone resorption by thyroid hormones explains many of the effects observed *in vivo.*[153] Increased release of calcium from bone raises the serum calcium concentration, and this leads to compensatory decreases in PTH secretion. Decreased PTH results in decreased tubular reabsorption of calcium, producing hypercalciuria. PTH also increases the tubular reabsorption of phosphate which, together with increased bone resorption, produce hyperphosphatemia, leading to decreased vitamin D activation and impaired intestinal calcium absorption. Calcium absorption is further decreased by the increased intestinal motility. Decreased intestinal calcium absorption and hypercalciuria prevent further increases in serum calcium. Urinary excretions of hydroxyproline,[154] bone-derived pyridinium cross-links of pyridinoline and deoxypyridinoline,[155] are increased, reflecting increased osteoclastic activity.

In Graves' disease, thyroid antigens can stimulate the release of lymphokines.[156] Thus, osteoclast activating-factor production might also be increased. It has been suggested that increased resorption in hyperthyroidism might be due to the decreased circulating levels of calcitonin. However, calcitonin concentrations are not decreased in thyrotoxicosis.[157] The role of PTH in thyroid hormone-stimulated resorption is still not clear. It has been reported that thyroid hormones sensitize bone to actions of circulating PTH.[158]

The above data explain why hyperthyroidism is often (in approximately 20% of cases) accompanied by overt hypercalcemia,[159,160] elevated levels of alkaline phosphatase,[161] and osteoporosis.[162-164] Hypercalcemia can be reduced by propranolol treatment.[165] Mild hyperthyroidism may enhance bone loss in women after menopause.[166] Reduction in thyroid hormone levels results in reduced bone turnover.[167] Significant bone loss was also reported during replacement therapy of hypothyroidism in postmenopausal women.[168,169] Furthermore, bone loss in the forearm[170,171] and spine[172] was demonstrated in thyroid hormone-treated euthyroid patients.

5. Cyclosporines

While there is no doubt that immunosuppressive cyclosporines reduce osteoclastic activity and prevent calcemic hormone-stimulated bone loss *in vitro* (see Section II), *in vivo* data show contrasting results. High-turnover osteopenia that was dose and time dependent was reported in several studies in rats after the administration of cyclosporine A.[173-176] One study in rats[177] showed that cyclosporine A, at a dose of 7 mg/kg/d for 14 d, inhibited bone resorption and reduced urinary hydroxyproline, plasma acid phosphatase, and the number of osteoclasts in caudal vertebrae. Several reports noted increased bone remodeling in human patients given cyclosporine A. Bone-specific alkaline phosphatase was found to be elevated in about one third of patients receiving cyclosporine A.[178,179] In kidney transplant patients given cyclosporine A, histomorphometric indices showed increased bone turnover with high osteoclastic activity in contrast to patients receiving azathioprine.[180] It is of interest that in one study, a subset of patients that did not show elevated alkaline phosphatase levels exhibited an early fall in serum calcium.[178] This fact may represent an early transient phase in the action of the compound on bone. Lower bone-mineral content was found in heart[181,182] and liver[183] transplant patients; however, these patients received other drugs also known to affect osteoclastic activity.

Several previous studies have used cyclosporine A for the treatment of autoimmune (type II collagen) arthritis in rats with variable results, dependent upon the time of cyclosporine A administration.[184-186] One study has demonstrated that rats with adjuvant arthritis showed regression of bone loss in a dose-responsive manner after treatment with cyclosporine A.[187] However, that study did not address any mechanisms of action. Also, an improvement in the condition of patients with rheumatoid arthritis was seen with low-dose cyclosporine A.[188]

The apparent inconsistency between the *in vitro* and *in vivo* data might be resolved by the recent finding that cyclosporine A stimulates the production of 1,25-dihydroxyvitamin D in rats and mice.[189] This could explain the increased osteoclastic activity produced by cyclosporine A *in vivo*. In adjuvant arthritis, lower initial serum 1,25-dihydroxyvitamin D levels were found.[190] The beneficial effect of cyclosporine A in this model could also be explained by its stimulatory effect on the production of the active vitamin D metabolite. In this case, the originally low-serum 1,25-dihydroxyvitamin D may be increased to normal level by the cyclosporine A treatment. Nevertheless, further studies are required to elucidate the precise mechanism of action of cyclosporines on bone.

6. Transforming Growth Factor-β

In addition to its stimulatory effects on bone formation, TGF-β may have effects on resorption as well. There are only a limited number of observations available so far that assess the effects of TGF-β on osteoclasts *in vivo*. After five daily injections of 1 μg TGF-β_1 or TGF-β_2 administered s.c. over the frontal and parietal bones of 7-week-old mice, osteoclast numbers in sections of bone were significantly elevated on days 8 and 19 (3 and 14 d after the last injection, respectively).[191] A stimulatory effect on osteoblasts preceded that on osteoclasts. There was no difference between the effects of TGF-β_1 and TGF-β_2. Indomethacin did not inhibit these effects. The authors concluded that TGF-β did not directly influence bone resorption *in vivo*, but osteoclasts were recruited in order to remodel TGF-β-induced woven bone.

The results of Marcelli et al.[192] are somewhat in contrast to those mentioned above. They injected 2.5 to 5 μg human recombinant TGF-β_1 into the subcutaneous tissue overlying the calvaria of normal adult mice for 2 to 5 d TGF-β_1 treatment had no effect on plasma calcium concentration. Beside stimulating bone formation processes locally, TGF-β_1 increased the number of osteoclasts in marrow spaces, and the osteoclasts formed were unusually large. When TGF-β_1 was given together with indomethacin, no increase in osteoclast number was seen, while stimulation of bone formation was only partially inhibited. These observations suggest that TGF-β_1 did not stimulate bone resorption directly, but, rather, enhanced endogenous prostaglandin production, which in turn stimulated bone resorption. These results are similar to those obtained *in vitro* (see Section II).

In both studies, a net gain in bone mass was observed despite the increase in osteoclast number; however, there were considerable differences in the findings. The reason for the discrepancies between these two studies might be due to differences in the doses used, ages of animals, or other factors. The relevance of these data to the physiologic role of TGF-β remains to be established. However, TGF-β may have important therapeutic implications in the future.

7. Colony-Stimulating Factors

The role of these factors in osteoclastic activity and resorption *in vivo* is controversial and not well established. The injection of recombinant human macrophage colony-stimulating factor (rhM-CSF) into osteopetrotic (a disease characterized by the lack of bone resorption) mice cured completely the deficiencies in the mutant mice.[193] Before treatment, the number of osteoclasts in osteopetrotic mice was decreased, and the osteoclasts were reported to have an abnormal cytoplasmic distribution of acid phosphatase, toluidine blue-positive and electron-dense cytoplasmic inclusions, and hypertrophy of clear zones and ruffled borders.[194] After treatment with rhM-CSF, increased numbers of osteoclasts were observed on the surface of bone trabeculae.[193] No obvious abnormality was found except that these cells were smaller in size and contained fewer nuclei than osteoclasts found in normal mice. In another study,[195] TRAP-positive mononucleated cells were found in the bone marrow cavity of osteopetrotic mice implanted with diffusion chambers containing M-CSF-producing L929 cells.

Failure of tooth eruption is due to reduced bone resorption in these mutant mice. After the 14-d treatment with rhM-CSF, the mucous membrane covering the molars was broken down and the crowns of these teeth emerged through the perforations.[193] Osteoclasts were detected on the surface of alveolar bone. These observations indicate that osteoclasts appear in response to rhM-CSF, show a distribution similar to that observed in normal mice, and restore normal resorption processes.

It has been suggested that M-CSF plays an essential role in the differentiation of osteoclasts.[193] Contrasting data also are available. Hattersley et al. found that rhM-CSF inhibited bone resorption by osteoclasts disaggregated from rat bone.[46] It has also been proposed that CSFs may serve as possible mediators of PTH effect.[196] However, there are differences between CSF- and PTH-stimulated bone resorption in terms of the time required for stimulation.[196] The mechanisms of action are presumably different as well. These differences in resorptive activity make CSF an unlikely mediator for PTH action.

B. INHIBITORS OF BONE RESORPTION

1. Calcitonin Gene Family

In young, growing rats, calcitonin has an acute inhibitory effect on osteoclasts that is reflected in a significant hypocalcemia, reaching its lowest value at 50 min after intraperitoneal administration.[197]

Calcitonin has been used in the treatment of metabolic bone diseases where the resorbing activity of osteoclasts is higher.[198] In postmenopausal osteoporosis, it is one of the most effective antiresorption agents.[199] Its inhibitory effect on osteoclasts is relatively rapid. In one study,[200] when 100 IU salmon calcitonin was administered intramuscularly to postmenopausal women, a significant fall of urinary hydroxyproline excretion was observed that peaked at 8 to 16 h postcalcitonin. A significant diminution of the urinary excretion of calcium was found 8 to 24 h postcalcitonin. The effect was more pronounced in patients with greater basal excretion of hydroxyproline, i.e., presumably with higher osteoclastic activity, and in those with more significant bone loss. When the effect of calcitonin in postmenopausal women was compared to that in patients with Paget's disease, the fall in hydroxyproline excretion was significantly higher in the postmenopausal women in terms of absolute values, but the percent diminution was not significantly different.

After a 3-month treatment of bone lesions of multiple myeloma with nasal spray calcitonin, the osteoclast resorption surface was found to be reduced significantly, with a consequent increase in trabecular bone volume, cortical thickness, and osteoid volume.[201] There was also a decline in the corrected serum calcium and urinary hydroxyproline:creatinine ratio, reflecting the diminished bone resorption. In patients with idiopathic hypercalciuria, calcitonin caused a significant drop in urine calcium, phosphorus, and hydroxyproline excretion and a concomitant increase in serum parathormone levels.[202] Treatment with calcitonin in malignant hypercalcemia can successfully reduce osteoclastic activity.[128] In medullary carcinoma of the thyroid, when endogenous production of calcitonin is increased, no sign of an inhibitory effect on resorption, i.e., increased bone mass, was found.[203] On the other hand, low serum levels of calcitonin after total thyroidectomy due to thyroid neoplasm may result in significant bone loss as a consequence of increased bone turnover.[204] Although calcitonin can prevent thyroid hormone-stimulated bone resorption *in vitro,* there is as yet no strong evidence of the same effect *in vivo*.

Beside calcitonin, the calcitonin gene complex encodes a family of novel peptides, including CGRP, amylin, and katacalcin. The physiological role of CGRP still remains unclear. CGRP inhibits bone resorption by active rat osteoclast and thus lowers plasma calcium when injected into young rats.[205] CGRP was found to be 100- to 1000-fold less potent than calcitonin. The effects of CGRP and calcitonin were additive. Human CGRP caused a marked dose-dependent elevation of bone cyclic AMP levels in mice that was similar to that caused by calcitonin. It has been suggested that CGRP is weakly agonistic for the calcitonin receptor on the osteoclast to produce effects on bone resorption and plasma calcium. It is still not clear whether the elevation of bone cyclic AMP represents an osteoclastic effect or originates from an effect on osteoblasts. Amylin has been shown to be a powerful inhibitor of osteoclastic resorption and to lower plasma calcium in rats and rabbits.[206] It is quite conceivable that these peptides may play a role in the systemic or local regulation of bone metabolism.

2. Estrogens and Antiestrogens

High-affinity nuclear binding sites for estrogens have been recently found in bone cells;[207] however, the mechanism by which estrogens exert their effects on bone *in vivo* is still unknown. Ovariectomy causes a high-turnover bone loss in animals and humans as well.[208,209] In rats, a 50% loss in trabecular volume and a two- to ninefold increase in osteoclast number, relative resorption surface, osteoblast number, and relative osteoid surface were found.[210] The local infusion of 17β-estradiol for 8 d into the limbs of ovariectomized rats restored the trabecular bone volume dose-dependently to 75 and 85% of control (nonovariectomized) levels. Infusion of 17α-estradiol had no effect on bone volume and osteoclast number. It was concluded from these findings that 17β-estrogen delivered directly to the bone *in vivo* acted locally to inhibit bone resorption. Cultures of bone marrow cells from ovariectomized animals formed significantly more TRAP-positive cells than those from sham-operated controls.[211] This finding supports the hypothesis that ovarian hormone deficiency promotes the expansion of a pool of marrow-derived progenitor cells that differentiate into bone-resorbing osteoclasts under the influence of osteotropic hormones. Estrogens and antiestrogens (tamoxifen) may act synergistically in preserving bone mass. These agents can inhibit osteoclasts on the endosteal surface.[212] They prevented the increases in bone formation rate and medullary area in ovariectomized rats.[213] Tamoxifen reduced the length of resorbing surface and osteoclast number to values observed in control animals.

In experimental animals[214] and humans,[215-217] the major effect of estrogens is to decrease bone resorption. This antiresorptive effect is followed by a delayed decreased in bone formation, possibly secondary to the coupling between the two phases of bone remodeling.[218] Although estrogen receptors on osteoclasts have been described (see Section II), their functional significance has not been established. With the exception of calcitonin, other hormonal regulators of bone metabolism (PTH, 1,25-dihydroxyvitamin D, interleukin-1, TNF-α, and prostaglandin E_2) have their receptors on osteoblasts rather than on osteoclasts.[219,220] The effects of these factors on bone resorption appear to be achieved by stimulating the release of osteoblast-derived regulators of osteoclastic activity. The antiresorptive effects of estrogens and antiestrogens may then be exerted through altering the response of osteoblasts to stimulators of bone resorption.

3. Progestagens

In several different *in vivo* systems, progesterone appears to act as an antiglucocorticoid agent and thus to interfere with glucocorticoid-related increased bone resorption.[221] Studies on animals and humans suggest that progesterone acts in a manner similar to estrogen to decrease the markers of osteoclastic bone resorption.[222-225] Different progestagenic compounds were used and the doses given and routes of administration varied, but responses to treatments showed a consistent pattern. Serum calcium may decrease or stay the same, but never increases. Urinary calcium excretion always decreases, whereas hydroxyproline excretion may remain the same or decrease.

Ovariectomized rats placed on ethynodiol therapy had an increase in femoral width and ashed mineral for the length of the femur which was not different from that in similarly growing intact animals.[226,227] In beagle dogs receiving progestagen treatment, resorption on the corticoendosteal surface decreased.[228] In another study in which dogs were treated with medroxyprogesterone, trabecular bone showed an increased percentage of surface covered by osteoid.[229] There was a 30 to 50% increase in the number of osteoclasts and osteoblasts. Although medroxyprogesterone treatment was associated with an increased depth of resorptive cavities, there was no change in wall thickness. This suggests that progestagen treatment activates all aspects of bone remodeling, with a net balance of increased bone formation.

The addition of 10 day/month of low-dose norethisterone to the estrogen treatment of postmenopausal bone loss prevented the rapid bone turnover that occurred when estrogen was withdrawn.[230] In three separate metabolic bone disorders — glucocorticoid-induced osteoporosis,[231] primary hyperparathyroidism,[232] and heparin-induced osteoporosis[233] — progestagen treatment also reduced bone loss. These experimental and clinical data indicate that progestagens are active on bone by promoting bone formation and increasing bone turnover. It is possible that they play a role in the coupling of bone resorption with bone formation.

4. Gallium Nitrate

Radioactive [^{67}Ga] has been used for several years to localize malignant tumors, especially lymphomas, and abscesses.[234,235] Also, gallium nitrate has been used in treating cancer-related hypercalcemia.[236-238] Warrell et al. observed a transient hypocalcemia not due to the increased urinary excretion of calcium in patients during gallium nitrate therapy, and proposed that inhibition of bone resorption could explain the hypocalcemic effect of gallium nitrate observed *in vivo*.[239,240] Gallium has been shown to be incorporated into the hydroxyapatite crystals, and these crystals are larger and less soluble than normal hydroxyapatite crystals and are significantly more resistant to osteoclasts.[241] In gallium-treated rats, the number of osteoclasts, acid phosphatase content in bone, and serum immunoreactive PTH were increased.[66] The calcemic response to PTH (i.v., 50 μg/kg) was decreased 2 h after injection of the hormone. It was suggested that gallium increases osteoclast number *in vivo* through stimulation of parathyroid activity induced by a fall in serum calcium. Recently, it has been reported that bone particles from gallium-treated rats are resistant to resorption *in vivo* when implanted in nontreated rats.[242] The number of osteoclast-like cells and TRAP activity in the gallium-containing bone particle implants were increased at 21 d. These data indicate that gallium incorporation into bone matrix confers resistance to osteoclastic resorption.

Matkovic et al. have reported favorable responses to intravenous gallium nitrate in patients with Paget's disease of bone.[243] Urinary hydroxyproline excretion decreased earlier than serum alkaline phosphatase, indicating that gallium nitrate primarily inhibits the activity of osteoclasts. Warrell et al.[244] reported similar results with the gallium nitrate treatment of pagetic patients.

5. Carbonic Anhydrase Inhibitors and Thiazides

The hydration of CO_2 into H_2CO_3, catalyzed by carbonic anhydrase, appears to be required for the proper functioning of the osteoclast proton pump. The presence of carbonic anhydrase in bone tissue has been demonstrated,[245,246] and it has been suggested that the activity of this enzyme was necessary for the osteoclastic secretion of hydrogen ions that was presumably required for bone demineralization. The osteoclast carbonic anhydrase corresponds to one of the four isoenzymes known to exist in mammals, isoenzyme II.[247] Deficiency of the enzyme has been identified in an autosomal recessive syndrome of osteopetrosis occurring in association with renal tubular acidosis.[248,249] This finding provides genetic evidence that carbonic anhydrase II is involved in osteoclast function and bone resorption *in vivo*. As early as the 1960s, acetazolamide, an inhibitor of carbonic anhydrase, was shown to inhibit bone resorption *in vivo*.[250,251] The inhibition of bone resorption by acetazolamide and other inhibitors of carbonic anhydrase was observed under various *in vivo* conditions.[252,253] The hypercalcemic response to parathyroid hormone or 1,25-dihydroxyvitamin D_3 administered to parathyroidectomized-nephrectomized rats is blunted or prevented by acetazolamide.[254,255] Inactive analogs of acetazolamide do not exhibit any of the inhibitory effects on resorption *in vivo*.[254]

In a rat model of humoral hypercalcemia of malignancy, where intense osteoclastic bone resorption appears to be largely responsible for the hypercalcemia,[256] treatment with acetazolamide and a more potent carbonic anhydrase inhibitor, 5-(3-hydroxybenzoyl)-2-thiophenesulfonamide, reduced serum calcium significantly.[257] A retrospective study by Pierce et al.[258] has shown the bone-sparing effect of acetazolamide and methazolamide in postmenopausal women during long-term (>4 years) therapy. In another study, Gram et al.[259] found no effect of acetazolamide on bone metabolism; however, the duration of drug treatment was very short (28 d). Prospective studies are necessary to determine the therapeutic value of specific carbonic anhydrase inhibitor treatment in metabolic bone diseases. Extraosseous side effects of these drugs might be of major concern in such a future treatment. A possible solution, bone targeting of carbonic anhydrase inhibitors, has recently been suggested.[260,261]

The thiazides have positive effects on calcium balance, and it has been demonstrated that thiazide use is associated with a bone-sparing effect in humans.[262] This effect is unrelated to hypertension or sodium balance. Lehmann et al.[263] reported that hydrochlorothiazide use inhibits bone resorption in men. Thiazide diuretics, as aromatic sulfonamides, are weak inhibitors of carbonic anhydrase. This activity and the renal effects of thiazides may result in a positive effect on bone mass, including reduction in fracture rate. Human studies seem to support this possibility.[264,265]

6. Prostaglandin Synthesis Inhibitors

It is believed that prostaglandins, particularly the E series, are involved in various steps of the activation phase of the bone remodeling sequence.[266,267] Tran Van et al.[268] developed a model of synchronized bone remodeling in the rat. After extraction of the upper molar teeth, antagonist mandibular molars egress and, concomitantly, a sequence of remodeling is initiated along the periosteal aspect of the buccal mandibular cortex, where it rarely occurs. Osteoclastic bone resorption reaches a peak 4 d after activation of this sequence. In this model, Richard and Saffar[269] observed that a 4-d treatment with indomethacin inhibited the resorption phase of this *de novo* remodeling sequence. This treatment throughout the entire activation period (4 d in this model) inhibited osteoclastic resorption completely. By modifying the treatment procedure,[270] it appeared that prostaglandins were operative mainly between the second and third day of action. Prostaglandins did not seem to act on precursor recruitment; rather, they might be involved in osteoclast activity, as the mean interface between osteoclasts and bone surface was reduced in the treated groups. However, indomethacin did not inhibit the remodeling sequence persistently, since a 6-d treatment resulted in a high profile of resorption. In hamster periodontitis,[271] high-dose (6 mg/kg/d) indomethacin has been reported to reduce the number of osteoclasts, and the extent of bone formation was markedly increased at the same time. In the same study, indomethacin had no influence on femoral periosteal remodeling and epiphyseal trabecular density, nor on growth plate thickness. These results emphasize the role of prostaglandins in uncoupling the remodeling sequence. Indomethacin also reduced the migration of polymorphonuclear cells.

Immobilization-related osteopenia was prevented by the use of indomethacin in rats.[272] No significant differences were noted in osteoclast number, resorption surfaces, and trabecular volume of the immobilized tibiae, compared to the nonimmobilized limbs. Osteoclast number and resorption surface increased, and trabecular volume decreased, in the nontreated group. Bone formation parameters remained reduced in the immobilized legs of both the indomethacin- and vehicle-treated groups compared to the control legs. This way, indomethacin inhibited bone resorption, but did not prevent the decrease in the bone formation produced by immobilization. A similar effect was seen in paraplegic rats treated with indomethacin.[273]

Flurbiprofen inhibits endosteal resorption by reducing the activity and recruitment of osteoclasts in rats, thus increasing cancellous bone mass by depressed trabecular bone resorption without bone formation.[274,275] Similar results were found with the use of etodolac.[276] In ovariectomized rats, naproxen, at a serum level of 4 μg/ml, inhibited bone resorption by decreasing osteoclastic activity.[277] In contrast, naproxen reduced bone formation only at serum levels >20 μg/ml.

The involvement of prostaglandins in bone resorption has been demonstrated *in vivo* in malignant osteolysis,[278] dental and other bone cysts,[279,280] and periodontitis.[281] Malignant hypercalcemia was successfully treated in mice by indomethacin.[278] Indomethacin has also been used in the treatment of hypercalcemia due to malignancy in humans.[282] Flurbiprofen has been shown to inhibit human alveolar bone loss.[283]

7. Fluoride

Since the first use of fluoride in the treatment of postmenopausal osteoporosis in 1961,[284] it has been used and studied extensively. Although it affects mainly the osteoblast population, it has marked effects on osteoclasts as well. Fluoride treatment was found to induce secondary hyperparathyroidism in some studies,[285-288] while others reported normal parathyroid function[289-292] in various models. The effects of fluoride on bone resorption may be partly due to secondary hyperparathyroidism; however, some of the effects appear to be independent of parathyroid activity. In rats, periosteal resorption increased after 15, 35, and 113 d of fluoride treatment,[285] and the same changes were found in parathyroidectomized and sham-operated animals. Fluoride-containing bone resorbed slower than normal bone.[286] An increase in eroded surface was reported in pigs.[293] This enhancement of resorption was associated with a decrease in osteoclast number in the beagle.[294] It was concluded that fluoride inhibited osteoclast differentiation, but that each osteoclast was more efficient.

Chavassieux et al.[288] found that cancellous-eroded perimeter and osteoclast number increased after 45 d of fluoride treatment, but between 45 and 90 d, both of these parameters tended to return to their initial value. Histomorphometric analysis of osteoclasts resorbing bone above the erupting first rat molar tooth germ showed that single-dose fluoride treatment reduced the number of osteoclasts at 2 and 6 h after its administration, after which a gradual increase to control levels was observed.[295] An increased number of inactive osteoclasts, not attached to the surface, accompanied the reduced number of active osteoclasts. Marie and Hott[296] reported an absence of changes in resorption parameters after fluoride treatment in mice. Despite these data, it is conceivable that changes in resorption during fluoride administration are secondary to other effects of the drug on bone formation and mineralization. These possibilities are discussed in a recent review.[297] However, at present, fluoride therapy is questionable in the treatment of osteoporosis.[298]

8. Mithramycin

Mithramycin is an antitumor drug which inhibits osteoclastic bone resorption and induces hypocalcemia in normocalcemic human patients and normocalcemia in hypercalcemic patients.[299] The mechanism by which mithramycin lowers serum calcium has not yet been completely elucidated.[300] In tumor-bearing nude mice, a single dose of mithramycin (8 mg/kg) decreased hypercalcemia and urine calcium levels.[301] The tumor volume was reduced and the tumor tissue had areas of hemorrhage and necrosis. Mithramycin did not reduce the number of TRAP-positive osteoclasts in bone; however, there were ultrastructural changes in osteoclasts. The osteoclasts were smaller, with a diminished ruffled border area, and the clear zone occupied a greater percentage of osteoclasts. The villi of the ruffled border were thickened and fused to form a more prominent clear zone.

Presently, it is thought that mithramycin can lower serum calcium without reducing the number of osteoclasts. The ultrastructural changes in osteoclasts induced by mithramycin may diminish their capability to mobilize calcium from bone. The antitumor effect of mithramycin[302,303] may also have an impact on serum calcium.

9. Strontium

Strontium at low dosage has also been proposed to be of therapeutic value in patients with osteoporosis.[304,305] There are reports that long-term treatment with strontium stimulates bone formation without altering bone resorption in rats.[306] However, short-term administration of strontium to mice had no significant effects on bone formation parameters except a more than 10% increase in osteoid surface.[296] By contrast, the number of active osteoclasts fell by more than 11%. Despite this effect, the calcified bone volume remained similar to normal values.

10. Lead

It is well known that lead is incorporated in bone matrix during calcification and stays there until bone is resorbed by the osteoclasts.[307] Despite this incorporation, the structure of the metaphyseal bones did not appear greatly changed in lead-intoxicated rats.[308] The only apparent abnormality was an initial condition of osteosclerosis. Osteoclasts were more numerous than in controls, and most of them contained a lot of shrunken, condensed nuclei. These osteoclasts proved to be abnormal. They contained dense intranuclear and cytoplasmic granules collected in small clusters. Under electron microscope, bone-attached osteoclasts showed normal resorption activity despite the presence of inclusion bodies. These inclusions contained a glycoprotein component and lead. In semithin sections, some osteoclasts were detached from the bone matrix, dislocated in the central zones of the medullary spaces, and appeared to be no longer engaged in bone resorption. These processes may represent the ultimate fate of lead-intoxicated osteoclasts. However, no data have yet been reported on changes in bone density after long-term lead exposure.

REFERENCES

1. **Chambers, T. J., Chambers, J. C., Symonds, J., and Darby, J. A.,** The effect of human calcitonin on the cytoplasmic spreading of rat osteoclasts, *J. Clin. Endocrinol. Metab.*, 63, 1080, 1986.
2. **de Vernejoul, M. C., Horowitz, M., Demignon, J., Neff, L., and Baron, R.,** Bone resorption by isolated chick osteoclasts in culture is stimulated by murine spleen cell supernatant fluids (osteoclast-activating factor) and inhibited by calcitonin and prostaglandin E2, *J. Bone Miner. Res.*, 3, 69, 1988.

3. **Dempster, D. W., Murrills, R. J., Horbert, W. R., and Arnett, T. R.,** Biological activity of chicken calcitonin: effects on neonatal rat and embryonic chick osteoclasts, *J. Bone Miner. Res.*, 2, 443, 1987.
4. **Zaidi, M., Chambers, T. J., Gaines Das, R. E., Morris, H. R., and MacIntyre, I.,** A direct action of human calcitonin gene-related peptide on isolated osteoclasts, *J. Endocrinol.*, 115, 511, 1987.
5. **Murrills, R. J., Shane, E., Lindsay, R., and Dempster, D. W.,** Bone resorption by isolated human osteoclasts *in vitro:* effects of calcitonin, *J. Bone Miner. Res.*, 4, 259, 1989.
6. **Kanehisa, J.,** Time course of "escape" from calcitonin-induced inhibition of motility and resorption of disaggregated osteoclasts, *Bone,* 10, 125, 1989.
7. **Hunter, S. J., Schraer, H., and Gay, C. V.,** Characterization of the cytoskeleton of isolated chick osteoclasts: effect of calcitonin, *J. Histochem. Cytochem.*, 37, 1529, 1989.
8. **Nicholson, G. C., Moseley, J. M., Sexton, P. M., Mendelsohn, F. A. O., and Martin, T. J.,** Abundant calcitonin receptors in isolated rat osteoclasts, *J. Clin. Invest.*, 78, 355, 1986.
9. **D'Santos, C. S., Nicholson, G. C., Moseley, J. M., Evans, T., Martin, T. J., and Kemp, B. E.,** Biologically active, derivatizable salmon calcitonin analog: design, synthesis, and applications, *J. Endocrinol.*, 123, 1483, 1988.
10. **Takahashi, N., Akatsu, T., Sasaki, T., Nicholson, G. C., Moseley, J. M., Martin, T. J., and Suda, T.,** Induction of calcitonin receptors by 1 alpha, 25-dihydroxyvitamin D3 in osteoclast-like multinucleated cells formed from mouse bone marrow cells, *Endocrinology,* 123, 1504, 1988.
11. **Goldring, S. R., Roelke, M., and Glowacki, J.,** Multinucleated cells elicited in response to implants of devitalized bone particles possess receptors for calcitonin, *J. Bone Miner. Res.*, 3, 117, 1988.
12. **Rifkin, B. R., Auszmann, J. M., Kleckner, A. P., Vernillo, A. T., and Fine, A. S.,** Calcitonin stimulates cAMP accumulation in chicken osteoclasts, *Life Sci.*, 42, 799, 1988.
13. **Nicholson, G. C., Moseley, J. M., Yates, A. J., and Martin, T. J.,** Control of cyclic adenosine 3′,5′-monophosphate production in osteoclasts: calcitonin-induced persistent activation and homologous desensitization of adenylate cyclase, *Endocrinology,* 120, 1902, 1987.
14. **Malgaroli, A., Meldosi, J., Zambonin Zallone, A., and Teti, A.,** Control of cytosolic free calcium in rat and chicken osteoclasts: the role of extracellular calcium and calcitonin, *J. Biol. Chem.*, 264, 14342, 1989.
15. **Zaidi, M., Datta, H. K., Moonga, B. S., and MacIntyre, I.,** Evidence that the action of calcitonin on rat osteoclasts is mediated by two G proteins acting via separate post-receptor pathways, *J. Endocrinol.*, 126, 473, 1990.
16. **Tobias, J. and Chambers, T. J.,** Glucocorticoids impair bone resorptive activity and viability of osteoclasts disaggregated from neonatal rat long bones, *Endocrinology,* 125, 1290, 1989.
17. **Pierce, A. M. and Lindskog, S.,** Early responses by osteoclasts *in vivo* and dentinoclasts *in vitro* to corticosteroids, *J. Submicro. Sc. Cytol. Pathol.*, 21, 501, 1989.
18. **Pharoah, M. J. and Heersche, J. N.,** Dexamethasone inhibits formation of osteoclast-like cells in bone-marrow cultures, *J. Dent. Res.*, 65, 1006, 1986.
19. **Dodd, R. C. and Winslow, B. T.,** Corticosteroids inhibit the differentiation of U937 cells induced by the combination of lymphocyte-conditioned media and calcitriol, *Bone Miner.*, 2, 281, 1987.
20. **Pensler, J. A., Radosevich, J. A., Higbee, R., and Langman, C.,** Osteoclasts isolated from membranous bone in children exhibit nuclear estrogen and progesterone receptors, *J. Bone Miner. Res.*, 5, 797, 1990.
21. **Tobias, J. H. and Chambers, T. J.,** The effect of sex hormones on bone resorption by rat osteoclasts, *Acta Endocrinol.*, 124, 121, 1991.

22. **McSheehy, P. M. and Chambers, T. J.,** Osteoblastic cells mediate osteoclastic responsiveness to parathyroid hormone, *Endocrinology,* 118, 824, 1986.
23. **Silve, C. M., Hradek, G. T., Jones, A. L., and Arnaud, C. D.,** Parathyroid hormone receptor in intact embryonic chicken bone: characterization and cellular localization, *J. Cell Biol.,* 94, 379, 1982.
24. **Rouleau, M. F., Mitchell, J., and Goltzman, D.,** *In vivo* distribution of parathyroid hormone receptors in bone: evidence that a predominant osseous target cell is not the mature osteoblast, *Endocrinology,* 123, 187, 1988.
25. **Teti, A., Rizzoli, R., and Zambonin Zallone, A.,** Parathyroid hormone binding to cultured avian osteoclasts, *Biochem. Biophys. Res. Commun.,* 174, 1217, 1991.
26. **Hakeda, Y., Hiura, K., Sato, T., Okazaki, R., Matsumoto, T., Ogata, E., Ishitani, R., and Kumegawa, M.,** Existence of parathyroid hormone binding sites on murine hemopoietic blast cells, *Biochem. Biophys. Res. Commun.,* 163, 481, 1989.
27. **Teti, A., Volleth, G., Carano, Z., and Zambonin Zallone, A.,** The effects of parathyroid hormone or 1,25-dihydroxyvitamin D3 on monocyte-osteoclast fusion, *Calcif. Tissue Int.,* 42, 302, 1988.
28. **Jilka, R. L.,** Parathyroid hormone-stimulated development of osteoclasts in cultures of cells from neonatal murine calvaria, *Bone,* 7, 29, 1986.
29. **Scheven, B. A. and Hamilton, N. J.,** Retinoic acid and 1,25-dihydroxyvitamin D3 stimulate osteoclast formation by different mechanisms, *Bone,* 11, 53, 1990.
30. **Shinar, D. M., Sato, M., and Rodan, G. A.,** The effect of hemopoietic growth factors on the generation of osteoclast-like cells in mouse bone marrow culture, *Endocrinology,* 126, 1728, 1990.
31. **Takahashi, N., Mundy, G. R., Kuehl, T. J., and Roodman, G. D.,** Osteoclast-like cell formation in fetal and newborn long-term baboon marrow cultures is more sensitive to 1,25-dihydroxyvitamin D3 than adult long-term marrow cultures, *J. Bone Miner. Res.,* 2, 311, 1987.
32. **Thavarajah, M., Evans, D. B., Binderup, L., and Kanis, J. A.,** 1,25(OH)2D3 and calcipotriol (MC903) have similar effects on the induction of osteoclast-like cell formation in human bone marrow cultures, *Biochem. Biophys. Res. Commun.,* 171, 1056, 1990.
33. **Ishizuka, S., Kurihara, N., Hakeda, S., Maeda, N., Ikeda, K., Kumegawa, M., and Norman, A. W.,** 1 alpha,25-Dihydroxyvitamin D3[1 alpha,25-(OH)2D3]-26,23-lactone inhibits 1,25-(OH)2D3-mediated fusion of mouse bone marrow mononuclear cells, *Endocrinology,* 123, 781, 1988.
34. **Shinar, D. M. and Rodan, G. A.,** Biphasic effects of transforming growth factor-beta on the production of osteoclast-like cells in mouse bone marrow cultures: the role of prostaglandins in the generation of these cells, *Endocrinology,* 126, 3153, 1990.
35. **Oreffo, R. O., Teti, A., Triffitt, J. T., Francis, M. J., Carano, A., and Zallone, A. Z.,** Effect of vitamin A on bone resorption: evidence for direct stimulation of isolated chicken osteoclasts by retinol and retinoic acid, *J. Bone Miner. Res.,* 3, 203, 1988.
36. **Teti, A., Oreffo, R. O., Zambonin Zallone, A., Triffitt, J. T., and Francis, M. J.,** The effect of retinol on osteoclasts is mediated by a specific cytosolic binding protein, *Boll. Soc. Ital. Biol. Sper.,* 62, 1315, 1986.
37. **Chambers, T. J., McSheehy, P. M. J., Thomson, B. M., and Fuller, K.,** The effect of calcium-regulating hormones and prostaglandins on bone resorption by osteoclasts disaggregated from neonatal rabbit bones, *Endocrinology,* 116, 234, 1985.
38. **Arnett, T. R. and Dempster, D. W.,** A comparative study of disaggregated chick and rat osteoclasts *in vitro:* effects of calcitonin and prostaglandins, *Endocrinology,* 120, 602, 1987.
39. **Okuda, A., Taylor, L. M., and Heersche, J. N.,** Prostaglandin E2 initially inhibits and then stimulates bone resorption in isolated rabbit osteoclast cultures, *Bone Miner.,* 7, 255, 1989.

40. **Chenu, C., Kurihara, N., Mundy, G. R., and Roodman, G. D.,** Prostaglandin E2 inhibits formation of osteoclastlike cells in long-term human marrow cultures but is not a mediator of the inhibitory effects of transforming growth factor beta, *J. Bone Miner. Res.*, 5, 677, 1990.
41. **Saito, S., Yamasaki, K., Yamada, S., Matsumoto, A., Akatsu, T., Takahashi, N., Shibasaki, Y., Suda, T., and Fukuhara, T.,** A stable analogue of thromboxane A2, 9,11-epithio-11,12-methano-thromboxane A2, stimulates bone resorption *in vitro* and osteoclast-like cell formation in mouse marrow culture, *Bone Miner.*, 12, 15, 1991.
42. **Vignery, A., Niven-Fairchild, T., and Shepard, M. H.,** Recombinant murine interferon-gamma inhibits the fusion of mouse alveolar macrophages *in vitro* but stimulates the formation of osteoclastlike cells on implanted syngeneic bone particles in mice *in vivo, J. Bone Miner. Res.*, 5, 637, 1990.
43. **Takahashi, N., Mundy, G. R., and Roodman, G. D.,** Recombinant human interferon-gamma inhibits formation of human osteoclast-like cells, *J. Immunol.*, 137, 3544, 1986.
44. **Kurihara, N. and Roodman, G. D.,** Interferons-alpha and -gamma inhibit interleukin-1 beta-stimulated osteoclast-like cell formation in long-term human marrow cultures, *J. Interferon Res.*, 10, 541, 1990.
45. **van de Wijngaert, F. P., Tas, M. C., van der Meer, J. W., and Burger, E. H.,** Growth of osteoclast precursor-like cells from whole mouse bone marrow: inhibitory effect of CSF-1, *Bone Miner.*, 3, 97, 1987.
46. **Hattersley, G., Dorey, E., Horton, M. A., and Chambers, T. J.,** Human macrophage colony-stimulating factor inhibits bone resorption by osteoclasts disaggregated from rat bone, *J. Cell. Physiol.*, 137, 199, 1988.
47. **MacDonald, B. R., Mundy, G. R., Clark, S., Wang, E. A., Kuehl, T. J., Stanley, E. R., and Roodman, G. D.,** Effects of human recombinant CSF-GM and highly purified CSF-1 on the formation of multinucleated cells with osteoclast characteristics in long-term bone marrow cultures, *J. Bone Miner. Res.*, 1, 227, 1986.
48. **Pfeilschifter, J., Chenu, C., Bird, A., Mundy, G. R., and Roodman, G. D.,** Interleukin-1 and tumor necrosis factor stimulate the formation of human osteoclastlike cells *in vitro, J. Bone Miner. Res.*, 4, 113, 1989.
49. **Thomson, B. M., Mundy, G. R., and Chambers, T. J.,** Tumor necrosis factors alpha and beta induce osteoblastic cells to stimulate osteoclastic bone resorption, *J. Immunol.*, 138, 775, 1987.
50. **Ries, W. L., Seeds, M. C., and Key, L. L.,** Interleukin-2 stimulates osteoclastic activity: increased acid production and radioactive calcium release, *J. Periodont. Res.*, 24, 242, 1989.
51. **Hagenaars, C. E., van der Kraan, A. A., Kawilarang-de Haas, E. W., Visser, J. W., and Nijweide, P. J.,** Osteoclast formation from cloned pluripotent hemopoietic stem cells, *Bone Miner.*, 6, 179, 1989.
52. **Scheven, B. A., Visser, J. W., and Nijweide, P. J.,** *In vitro* osteoclast generation from different bone marrow fractions, including a highly enriched haematopoietic stem cell population, *Nature (London)*, 321, 79, 1986.
53. **Chenu, C., Pfeilschifter, J., Mundy, G. R., and Roodman, G. D.,** Transforming growth factor beta inhibits formation of osteoclast-like cells in long-term human marrow cultures, *Proc. Natl. Acad. Sci. U.S.A.*, 85, 5683, 1988.
54. **Kukita, A., Bonewald, L., Rosen, D., Seyedin, S., Mundy, G. R., and Roodman, G. D.,** Osteoinductive factor inhibits formation of human osteoclast-like cells, *Proc. Natl. Acad. Sci. U.S.A.*, 87, 3023, 1990.
55. **Oreffo, R. O., Mundy, G. R., Seyedin, S. M., and Bonewald, L. F.,** Activation of the bone-derived latent TGF beta complex by isolated osteoclasts, *Biochem. Biophys. Res. Commun.*, 158, 817, 1989.
56. **Takahashi, N., MacDonald, B. R., Hon, J., Winkler, M. E., Derynck, R., Mundy, G. R., and Roodman, G. D.,** Recombinant human transforming growth factor-alpha stimulates the formation of osteoclast-like cells in long-term human marrow cultures, *J. Clin. Invest.*, 78, 894, 1986.

57. **Boonekamp, P. M., Lowik, C. W., van der Wee-Pals, L. J., van Wijk-van Lennep, M. L., and Bijvoet, O. L.,** Enhancement of the inhibitory action of APD on the transformation of osteoclast precursors into resorbing cells after dimethylation of the amino group, *Bone Miner.*, 2, 29, 1987.
58. **Carano, A., Teitelbaum, S. L., Konsek, J. D., Schlesinger, P. H., and Blair, H. C.,** Bisphosphonates directly inhibit the bone resorption activity of isolated avian osteoclasts *in vitro, J. Clin. Invest.*, 85, 456, 1990.
59. **Flanagan, A. M. and Chambers, T. J.,** Dichloromethylene-bisphosphonate (Cl2MBP) inhibits bone resorption through injury to osteoclasts that resorb Cl2MBP-coated bone, *Bone Miner.*, 6, 33, 1989.
60. **Sato, M. and Grasser, W.,** Effects of bisphosphonates on isolated rat osteoclasts as examined by reflected light microscopy, *J. Bone Miner. Res.*, 5, 31, 1990.
61. **Hughes, D. E., MacDonald, B. R., Russell, R. G., and Gowen, M.,** Inhibition of osteoclast-like cell formation by bisphosphonates in long-term cultures of human bone marrow, *J. Clin. Invest.*, 83, 1930, 1989.
62. **Lowik, C. W., van der Pluijm, G., van der Wee-Pals, L. J., van Treslong-De Groot, H. B., and Bijvoet, O. L.,** Migration and phenotypic transformation of osteoclast precursors into mature osteoclasts: the effect of a bisphosphonate, *J. Bone Miner. Res.*, 3, 85, 1988.
63. **Okuda, A., Kanehisa, J., and Heersche, J. N.,** The effects of sodium fluoride on the resorptive activity of isolated osteoclasts, *J. Bone Miner. Res.*, 5, S115, 1990.
64. **Taylor, M. L., Maconnachie, E., Frank, K., Boyde, A., and Jones, S. J.,** The effect of fluoride on the resorption of dentine by osteoclasts *in vitro, J. Bone Miner. Res.*, 5, S121, 1990.
65. **Taylor, M. L., Boyde, A., and Jones, S. J.,** The effect of fluoride on the patterns of adherence of osteoclasts cultured on and resorbing dentine: a 3-D assessment of vinculin-labelled cells using confocal optical microscopy, *Anat. Embryol.*, 180, 427, 1989.
66. **Warrell, R. P., Bockman, R. S., Coonley, C. J., Isaacs, M., and Staszewski, H.,** Gallium nitrate inhibits calcium resorption from bone and is effective treatment for cancer-related hypercalcemia, *J. Clin. Invest.*, 73, 1487, 1984.
67. **Lakatos, P. and Stern, P. H.,** Gallium nitrate inhibits bone resorption and collagen synthesis in neonatal mouse calvariae, *J. Bone Miner. Res.*, 6, 1121, 1991.
68. **Hall, T. J. and Chambers, T. J.,** Gallium inhibits bone resorption by a direct effect on osteoclasts, *Bone Miner.*, 8, 211, 1990.
69. **Skjodt, H., Crawford, A., Elford, P. R., Ihrie, E., Wood, D. D., and Russell, R. G. G.,** Cyclosporine A modulates interleukin-1 activity on bone *in vitro, J. Rheumatol.*, 24 (Suppl. 1), 165, 1985.
70. **Stewart, P. J., Green, O. C., and Stern, P. H.,** Cyclosporine A inhibits calcemic hormone-induced bone resorption *in vitro, J. Bone Miner. Res.*, 1, 285, 1986.
71. **Orcel, P., Denne, M. A., and de Vernejoul, M. C.,** Cyclosporine-A *in vitro* decreases bone resorption, osteoclast formation, and the fusion of cells of the monocyte-macrophage lineage, *Endocrinology,* 128, 1638, 1991.
72. **Mahgoub, A. and Stern, P. H.,** Carbon dioxide and the effect of parathyroid hormone on bone *in vitro, Am. J. Physiol.*, 226, 1272, 1974.
73. **Hunter, S. J., Schraer, H., and Gay, C. V.,** Characterization of isolated and cultured chick osteoclasts: the effects of acetazolamide, calcitonin, and parathyroid hormone on acid production, *J. Bone Miner. Res.*, 3, 297, 1988.
74. **Silverton, S. F., Dodgson, S. J., Fallon, M. D., and Forster, R. E., II,** Carbonic anhydrase activity of chick osteoclasts is increased by parathyroid hormone, *Am. J. Physiol.*, 253, E670, 1987.
75. **Kim, S. G. and Krieger, N. S.,** Dichlorobenzamil inhibits stimulated bone resorption *in vitro, Endocrinology,* 122, 415, 1988.
76. **Tuukkanen, J. and Vaananen, H. K.,** Omeprazole, a specific inhibitor of H+-K+-ATPase, inhibits bone resorption *in vitro, Calcif. Tissue Int.*, 38, 123, 1986.

77. **Hall, T. J. and Chambers, T. J.,** Na+/H+ antiporter is the primary proton transport system used by osteoclasts during bone resorption, *J. Cell. Physiol.*, 142, 420, 1990.
78. **Bekker, P. J. and Gay, C. V.,** Biochemical characterization of an electrogenic vacuolar proton pump in purified chicken osteoclast plasma membrane vesicles, *J. Bone Miner. Res.*, 5, 569, 1990.
79. **Everts, V., Beertsen, W., and Schroder, R.,** Effects of the proteinase inhibitors leupeptin and E-64 on osteoclastic bone resorption, *Calcif. Tissue Int.*, 43, 172, 1988.
80. **Delaisse, J. M., Boyde, A., Maconnachie, E., Ali, N. N., Sear, C. H., Eeckhout, Y., Vaes, G., and Jones, S. J.,** The effects of inhibitors of cysteine-proteinases and collagenase on the resorptive activity of isolated osteoclasts, *Bone*, 8, 305, 1987.
81. **Sato, M., Sardana, M. K., Grasser, W. A., Garsky, V. M., Murray, J. M., and Gould, R. J.,** Echistatin is a potent inhibitor of bone resorption in culture, *J. Cell. Biol.*, 111, 1713, 1990.
82. **Lerner, U. H., Ransjo, M., Klaushofer, K., Horandner, H., Hoffmann, O., Czerwenka, E., Koller, K., and Peterlik, M.,** Comparison between the effects of forskolin and calcitonin on bone resorption and osteoclast morphology *in vitro*, *Bone*, 10, 377, 1989.
83. **Ransjo, M., Lerner, U. H., and Heersche, J. N.,** Calcitonin-like effects of forskolin and choleratoxin on surface area and motility of isolated rabbit osteoclasts, *J. Bone Miner. Res.*, 3, 611, 1988.
84. **Nicholson, G. C., Yumita, S., Moseley, J. M., Yates, A. J., and Martin, T. J.,** Forskolin augments the effects of calcitonin on cytoplasmic spreading of isolated rat osteoclasts and plasma calcium levels in the rat, *J. Bone Miner. Res.*, 3, 181, 1988.
85. **Murrills, R. J. and Dempster, D. W.,** The effects of stimulators of intracellular cyclic AMP on rat and chick osteoclasts *in vitro:* validation of a simplified light microscope assay of bone resorption, *Bone*, 11, 333, 1990.
86. **Davidovitch, Z., Montgomery, P. C., and Shanfield, J. L.,** Cellular localization and concentration of bone cyclic nucleotides in response to acute PTE administration, *Calc. Tissue Res.*, 24, 81, 1977.
87. **MacIntyre, I., Zaidi, M., Towhidul, A. S. M., Datta, H. K., Moonga, B. S., Lidbury, P. S., Hecker, M., and Vane, J.,** Osteoclastic inhibition: an action of nitric oxide not mediated by cyclic AMP, *Proc. Natl. Acad. Sci. U.S.A.*, 88, 2936, 1991.
88. **Zaidi, M., Patchell, A., Datta, H. K., and MacIntyre, I.,** Uncoupling of receptor-mediated cellular responses by ionic lithium, *J. Endocrinol.*, 123, R5, 1989.
89. **Teti, A., Grano, M., Colucci, S., Argentino, L., Barattolo, R., Miyauchi, A., Teitelbaum, S. L., Hruska, K. A., and Zambonin Zallone, A.,** Voltage dependent calcium channel expression in isolated osteoclasts, *Boll. Soc. Ital. Biol. Sper.*, 65, 1115, 1989.
90. **Chesnoy-Marchais, D. and Fritsch, J.,** Voltage-gated sodium and calcium currents in rat osteoblasts, *J. Physiol.*, 398, 291, 1988.
91. **Zaidi, M., Datta, H. K., Patchell, A., Moonga, B., and MacIntyre, I.,** "Calcium-activated" intracellular calcium elevation: a novel mechanism of osteoclast regulation, *Biochem. Biophys. Res. Commun.*, 163, 1461, 1989.
92. **Teti, A., Barattolo, R., Grano, M., Colucci, S., Argentino, L., Teitelbaum, S. L., Hruska, K. A., Santacroce, G., and Zambonin Zallone, A.,** Podosome expression in osteoclasts: influence of high extracellular calcium concentration, *Boll. Soc. Ital. Biol. Sper.*, 65, 1039, 1989.
93. **Datta, H. K., MacIntyre, I., and Zaidi, M.,** Intracellular calcium in the control of osteoclast function. I. Voltage-insensitivity and lack of effects of nifedipine, BAYK8644 and diltiazem, *Biochem. Biophys. Res. Commun.*, 167, 183, 1990.
94. **Zaidi, M., MacIntyre, I., and Datta, H.,** Intracellular calcium in the control of osteoclast function. II. Paradoxical elevation of cytosolic free calcium by verapamil, *Biochem. Biophys. Res. Commun.*, 167, 807, 1990.
95. **MacCallum, W. G. and Voegtlin, C.,** On the relation of tetany to the parathyroid glands and to calcium metabolism, *J. Exp. Med.*, 11, 118, 1909.

96. **Talmage, R. V., Kraintz, F. W., and Buchanan, G. D.,** Effect of parathyroid extract and phosphate salts on renal calcium and phosphate excretion after parathyroidectomy, *Proc. Soc. Exp. Biol. Med.*, 88, 600, 1955.
97. **Garabedian, M., Tanaka, Y., Holick, M. F., and DeLuca, H. F.,** Response of intestinal calcium transport and bone calcium mobilization to 1,25-dihydroxyvitamin D_3 in thyroparathyroidectomized rats, *Endocrinology,* 94, 1022, 1974.
98. **Parfitt, A. M.,** The actions of parathyroid hormone on bone: relation to bone remodeling and turnover, calcium homeostasis, and metabolic bone disease. Mechanisms of calcium transfers between blood and bone and their cellular basis: morphological and kinetic approaches to bone turnover, *Metabolism,* 25, 809, 1976.
99. **Atkinson, M. J., Bodenstein, H., and Hesch, R. D.,** Parathyroid hormone concentration gradient across the human bone marrow, *Calcif. Tissue Int.*, 35, 591, 1983.
100. **Martin, K. J., Freitag, J., Conrades, M. B., Hruska, K., Klahr, S., and Slatopolsky, E.,** Selective uptake of the synthetic amino terminal fragment of parathyroid hormone by isolated perfused bone, *J. Clin. Invest.*, 62, 656, 1978.
101. **Au, W. Y. W. and Raisz, L. G.,** Restoration of parathyroid hormone responsiveness in vitamin D-deficient rats by parenteral calcium or dietary lactose, *J. Clin. Invest.*, 46, 1572, 1967.
102. **Morii, H. and DeLuca, H. F.,** Relationship between vitamin D deficiency, thyrocalcitonin, and parathyroid hormone, *Am. J. Physiol.*, 213, 358, 1967.
103. **Barnicot, N. A.,** The local action of the parathyroid and other tissues on bone in intracerebral grafts, *J. Anat.*, 82, 233, 1948.
104. **Tatevossian, A.,** Effect of parathyroid extract on blood calcium and osteoclast count in mice, *Calcif. Tissue Res.*, 11, 251, 1973.
105. **Takano-yamamoto, T. and Rodan, G. A.,** A model for investigating the local action of bone-acting agents *in vivo:* effects of hPTH(1-34) on the secondary spongiosa in the rat, *Calcif. Tissue Int.*, 47, 158, 1990.
106. **Bingham, P. J., Brazell, I. A., and Owen, M.,** The effect of parathyroid extract on cellular activity and plasma calcium levels *in vitro, J. Endocrinol.*, 45, 387, 1969.
107. **Malluche, H. H., Matthews, C., Faugere, M., Fanti, P., Endres, D. B., and Friedler, R. M.,** 1,25-dihydroxyvitamin D maintains bone cell activity and parathyroid hormone modulates bone cell number in dogs, *Endocrinology,* 119, 1298, 1986.
108. **Mears, D. C.,** Effects of parathyroid hormone and thyrocalcitonin on the membrane potential of osteoclasts, *Endocrinology,* 88, 1021, 1971.
109. **Holtrop, M. E., King, G. J., Cox, H. A., and Reit, B.,** Time-related changes in the ultrastructure of osteoclasts after injection of parathyroid hormone in young rats, *Calcif. Tissue Int.*, 27, 129, 1979.
110. **Heller, M., McLean, F. C., and Bloom, W.,** Cellular transformation in mammalian bones induced by parathyroid extract, *Am. J. Anat.*, 87, 315, 1950.
111. **Walker, D. G.,** The induction of osteopetrotic changes in hypophysectomized, thyroparathyroidectomized and intact rats of various ages, *Endocrinology,* 89, 1389, 1971.
112. **McGuire, J. L. and Marks, S. C.,** The effects of parathyroid hormone on bone cell structure and function, *Clin. Orthop. Relat. Res.*, 100, 392, 1974.
113. **Chambers, T. J., McSheehy, P. M. J., Thomson, B. M., and Fuller, K.,** The effect of calcium-regulating hormones and prostaglandins on bone resorption by osteoclasts disaggregated from neonatal rabbit bones, *Endocrinology,* 116, 234, 1985.
114. **Grise, M. A., Marks, S. C., MacKay, C. A., and Popoff, S. N.,** Effects of 1,25-dihydroxyvitamin D on osteoclast number and cytochemistry in normal and osteopetrotic (os) rabbits, *Am. J. Anat.*, 189, 261, 1990.
115. **Marie, P. J. and Travers, R.,** Continuous infusion of 1,25-dihydroxyvitamin D_3 stimulates bone turnover in the normal young mouse, *Calcif. Tissue Int.*, 35, 418, 1983.
116. **Popoff, S. N., McGuire, J. L., Zerwekh, J. E., and Marks, S. C.,** Treatment of congenital osteopetrosis with high-dose 1,25-dihydroxyvitamin D, *J. Bone Miner. Res.*, 4, 57, 1989.

117. **Vandersteenhoven, J. J., DeLustro, F. A., Bell, N. H., and Turner, R. T.,** Osteoinduction by implants of demineralized allogenic bone matrix is diminished in vitamin D-deficient rats, *Calcif. Tissue Int.*, 42, 39, 1988.
118. **Jowell, P. S., Epstein, S., Fallon, M. D., Reinhardt, T. A., and Ismail, F.,** 1,25-dihydroxyvitamin D_3 modulates glucocorticoid-induced alteration in serum bone Gla protein and bone histomorphometry, *Endocrinology,* 120, 531, 1987.
119. **Cushing, H.,** Basophile adenomas, *J. Nerv. Ment. Dis.*, 76, 50, 1932.
120. **Adinoff, A. D. and Hollister, J. R.,** Steroid-induced fractures and bone loss in patients with asthma, *N. Engl. J. Med.*, 309, 265, 1983.
121. **Ruegsegger, P., Medici, T. C., and Anliker, M.,** Corticosteroid-induced bone loss. A longitudinal study of alternate day therapy in patients with bronchial asthma using quantitative computed tomography, *Eur. J. Clin. Pharmacol.*, 25, 615, 1983.
122. **Gennari, C.,** Glucocorticoids and bone, in *Bone and Mineral Research/3,* Peck, W. A., Ed., Elsevier, Amsterdam, 1985, 213.
123. **Gluck, O. S., Murphy, W. A., Hahn, T. J., and Hahn, B.,** Bone loss in adults receiving alternate day glucocorticoid therapy. A comparison with daily therapy, *Arthritis Rheum.*, 24, 892, 1985.
124. **Bressot, C., Meunier, P. J., Chapuy, M. C., Lejeune, E., Edourd, C., and Darby, A. J.,** Histomorphometric profile, pathophysiology and reversibility of corticosteroid-induced osteoporosis, *Metab. Bone Dis. Relat. Res.*, 1, 303, 1979.
125. **de Deuxchaisnes, C. N., Devogelaer, J. P., Esselinckx, W., et al.,** The effect of low dosage glucocorticoids on bone mass in rheumatoid arthritis: a cross-sectional study and a longitudinal study using single photon absorptiometry, *Adv. Exp. Med. Biol.*, 171, 210, 1984.
126. **Versraeten, A. and Dequeker, J.,** Vertebral and peripheral bone mineral content and fracture incidence in postmenopausal patients with rheumatoid arthritis: effect of low-dose corticosteroids, *Ann. Rheum. Dis.*, 45, 852, 1986.
127. **Shindell, R., Huurman, W. W., Lipiello, L., and Connolly, J. F.,** Prostaglandin levels in unicameral bone cysts treated by intralesional steroid injection, *J. Pediatr. Orthop.*, 9, 516, 1989.
128. **Heath, D.,** The treatment of hypercalcemia of malignancy, *Clin. Endocrinol.*, 34, 155, 1991.
129. **Seeman, E., Wagner, H. W., Offord, K. P., Kumar, R., Johnson, W. J., and Riggs, B. L.,** Differential effects of endocrine dysfunction on the axial and appendicular skeleton, *J. Clin. Invest.*, 69, 1302, 1982.
130. **Gallagher, J. C., Aaron, J., Horsman, A., Wilkinson, R., and Nordin, B. E.,** Corticosteroid osteoporosis, *Clin. Endocrinol. Metab.*, 2, 355, 1973.
131. **Dempster, D. W., Arlot, M. A., and Meunier, P. J.,** Mean wall thickness and formation periods of trabecural bone packets in corticosteroid-induced osteoporosis, *Calcif. Tissue Int.*, 35, 410, 1983.
132. **Hahn, T. J., Halstead, L. R., Teitelbaum, S. L., and Hahn, B. H.,** Altered mineral metabolism in glucocorticoid-induced osteopenia. Effect of 25-hydroxyvitamin D administration, *J. Clin. Invest.*, 64, 655, 1979.
133. **Wajchenberg, B. L., Pereira, V. G., Kieffer, J., and Ursic, S.,** Effect of dexamethasone on calcium metabolism and ^{47}Ca kinetics in normal subjects, *Acta Endocrinol.*, 61, 173, 1983.
134. **Need, A. G., Philcox, J. C., Hartley, T. F., and Nordin, B. E.,** Calcium metabolism and osteoporosis in corticosteroid-treated postmenopausal women, *Aust. N.Z. J. Med.*, 16, 341, 1986.
135. **Wong, G. L.,** Basal activities and hormone responsiveness of osteoclast-like and osteoblast-like bone cell are regulated by glucocorticoids, *J. Biol. Chem.*, 254, 6337, 1979.
136. **Colette, C., Monnier, L., Pares Herbute, N., Blotman, F., and Mirouze, J.,** Calcium absorption in corticoid treated subjects: effects of a single oral dose of calcitriol, *Horm. Metab. Res.*, 19, 335, 1987.

137. **Nielsen, H. K., Thomsen, K., Eriksen, E. F., Charles, P., Storm, T., and Mosekilde, L.,** The effect of high-dose glucocorticoid administration on serum bone gamma carboxyglutamic acid-containing protein, serum alkaline phosphatase and vitamin D metabolites in normal subjects, *Bone Miner.*, 4, 105, 1988.
138. **Suzuki, Y., Ichikawa, Y., Saito, E., and Homma, M.,** Importance of increased urinary calcium excretion in the development of secondary hyperparathyroidism of patients under glucocorticoid therapy, *Metabolism,* 32, 151, 1983.
139. **Catherwood, B. D.,** 1,25-Dihydrocholecalciferol and glucocorticoid regulation of adenylate cyclase in an osteoblast-like cell line, *J. Biol. Chem.*, 160, 736, 1985.
140. **Stern, P. H.,** Inhibition by steroids of parathyroid hormone-induced ^{45}Ca release from embryonic rat bone *in vitro, J. Pharmacol. Exp. Ther.*, 168, 211, 1969.
141. **Kukreja, S. C., Bowser, E. N., Hargis, G. K., Henderson, W. J., and Williams, G. A.,** Mechanism of glucocorticoid-induced osteopenia: role of parathyroid glands, *Proc. Soc. Exp. Biol. Med.*, 152, 358, 1976.
142. **Luton, J. P., Thieblot, P., Valcke, J. C., Mahoudeau, J. A., and Bricaire, H.,** Reversible gonadotropin deficiency in male Cushing's disease, *J. Clin. Endocrinol. Metab.*, 32, 639, 1977.
143. **Sakakura, M., Takebe, K., and Nakagawa, S.,** Inhibition of luteinizing hormone secretion induced by synthetic LRH by long-term treatment with glucocorticoids in human subjects, *J. Clin. Endocrinol. Metab.*, 40, 774, 1975.
144. **Crilly, R. G., Cawood, M., Marshall, D. H., and Nordin, B. E.,** Hormonal status in normal, osteoporotic and corticosteroid-treated postmenopausal women, *J. R. Soc. Med.*, 71, 733, 1978.
145. **Goulding, A. and Gold, E.,** Effects of chronic prednisolone treatment on bone resorption and bone composition in intact and ovariectomized rats and in ovariectomized rats receiving β-estradiol, *Endocrinology,* 122, 482, 1988.
146. **Lukert, B. P. and Raisz, L. G.,** Glucocorticoid-induced osteoporosis: pathogenesis and management, *Ann. Intern. Med.*, 112, 352, 1990.
147. **Mundy, G. R., Shapiro, J. L., Bandelin, J. G., Canalis, E. M., and Raisz, L. G.,** Direct stimulation of bone resorption by thyroid hormones, *J. Clin. Invest.*, 58, 529, 1976.
148. **Mundy, G. R. and Raisz, L. G.,** Thyrotoxicosis and calcium metabolism, *Miner. Electrolyte Metab.*, 2, 285, 1979.
149. **Krane, S. M., Brownell, G. L., Skanbury, J. B., and Corrigon, H.,** The effect of thyroid disease on calcium metabolism in man, *J. Clin. Invest.*, 35, 874, 1956.
150. **Adams, P. H., Jowsey, J., Kelly, P. J., Riggs, B. L., Kinney, V. R., and Jones, J. D.,** Effects of hyperthyroidism on bone and mineral metabolism in man, *Q. J. Med.*, 36, 1, 1967.
151. **Follis, R. H.,** Skeletal changes associated with hyperthyroidism, *Bull. Johns Hopkins Hosp.*, 92, 405, 1953.
152. **Meunier, P. J., Bianchi, G. G. S., Edouard, C. M., Bernard, J. C., Courpron, P., and Vignon, G. E.,** Bony manifestations of thyrotoxicosis, *Orthop. Clin. North Am.*, 3, 745, 1972.
153. **Mosekilde, L., Eriksen, E. F., and Charles, P.,** Effects of thyroid hormones on bone and mineral metabolism, *Endocrinol. Metab. Clin. North Am.*, 19, 35, 1990.
154. **Charles, P., Poser, J. W., Mosekilde, L., et al.,** Estimation of bone turnover evaluated by ^{47}Ca-kinetics. Efficiency of serum bone gamma-carboxyglutamic acid-containing protein, serum alkaline phosphatase, and urinary hydroxyproline excretion, *J. Clin. Invest.*, 76, 2254, 1985.
155. **Harvey, R. D., McHardy, K. C., Reid, I. W., Paterson, F., Bewsher, P. D., Duncan, A., and Robins, S. P.,** Measurement of bone collagen degradation in hyperthyroidism and during thyroxine replacement therapy using pyridinium crosslinks as specific urinary markers, *J. Clin. Endocrinol. Metab.*, 72, 1189, 1991.

156. **Lamki, L., Row, V. V., and Volpe, R.,** Cell-mediated immunity in Graves' disease and in Hashimoto's thyroiditis as shown by demonstration of migration inhibition factor (MIF), *J. Clin. Endocrinol. Metab.,* 36, 358, 1973.
157. **Fraser, S. A. and Wilson, G. M.,** Plasma-calcitonin in disorders of thyroid function, *Lancet,* 1, 725, 1971.
158. **Castro, J. H., Genuth, S. M., and Klein, L.,** Comparative response to parathyroid hormone in hyperthyroidism and hypothyroidism, *Metabolism,* 24, 839, 1975.
159. **Krane, S. M.,** Skeletal system, neuromuscular system, emotions and mentation, in *Thyroid Diseases,* Ingbar, S. H. and Werner, S. C., Eds., Harper & Row, New York, 1971, 598.
160. **Allwerx, J. and Bouillon, R.,** Mineral and bone metabolism in thyroid disease: a review, *Q. J. Med.,* 232, 737, 1986.
161. **Michie, W., Duncan, T., Hamer-Hodges, D. W., Bewsher, P. D., Hems, G., Skowers, J. M., Pegg, C. A. S., and Hedley, A. J.,** Mechanism of hypocalcemia after thyroidectomy for thyrotoxicosis, *Lancet,* 1, 508, 1971.
162. **Fraser, S. A., Anderson, J. B., Smith, D. A., and Wilson, G. M.,** Osteoporosis and fractures following thyrotoxicosis, *Lancet,* 1, 981, 1971.
163. **Krolner, B., Jorgensen, J. V., and Nielsen, S. P.,** Spinal bone mineral content in myxoedema and thyrotoxicosis, effects of thryoid hormone(s) and antithyroid treatment, *Clin. Endocrinol.,* 18, 439, 1983.
164. **Toh, S. H., Claunch, B. C., and Brown, P. H.,** Effects of hyperthyroidism and its treatment on bone mineral content, *Arch. Int. Med.,* 145, 883, 1985.
165. **Rude, R. K., Oldham, S. B., Singer, F. R., and Nicoloff, J. T.,** Treatment of thyrotoxic hypercalcemia with propranolol, *N. Engl. J. Med.,* 294, 431, 1976.
166. **Lakatos, P., Hollo, I., and Horvath, C.,** Severe postmenopausal osteoporosis and thyroid hormones, *Arch. Intern. Med.,* 146, 1859, 1986.
167. **Kragstrup, J., Melsen, F., and Mosekilde, L.,** Effects of thyroid hormone(s) on mean wall thickness of trabecular bone pockets, *Metab. Bone Dis. Relat. Res.,* 3, 181, 1981.
168. **Fallon, M. D., Perry, H. M., III, Bergfeld, M., Droke, D., Teitelbaum, S. L., and Avioli, L. V.,** Exogenous hyperthyroidism with osteoporosis, *Arch. Intern. Med.,* 143, 442, 1983.
169. **Baran, D. T. and Braverman, L. E.,** Editorial: thyroid hormones and bone mass, *J. Clin. Endocrinol. Metab.,* 72, 1182, 1991.
170. **Ross, D. S., Neer, R. M., Ridgway, E. C., and Daniels, G. H.,** Subclinical hyperthyroidism and reduced bone density as a possible result of prolonged suppression of the pituitary-thyroid axis with L-thyroxine, *Am. J. Med.,* 82, 1167, 1987.
171. **Taelman, P., Kaufman, J. M., Janssens, X., Vandecauter, H., and Vermeulen, A.,** Reduced forearm bone mineral content and biochemical evidence of increased bone turnover in women with euthyroid goiter treated with thyroid hormone, *Clin. Endocrinol.,* 33, 107, 1990.
172. **Lakatos, P., Tarjan, G., Foldes, J., Krasznai, I., and Hollo, I.,** The effect of thyroid treatment on bone mineral content in patients with hypothyroidism and euthyroid benign adenoma, in *Proceedings of the International Symposium on Osteoporosis,* Vol. 1, Christiansen, C., Johansen, J. S., and Riis, B. J., Eds., Osteopress Aps, Copenhagen, Denmark, 1987, 452.
173. **Movsowitz, C., Epstein, S., Fallon, M., Ismail, F., and Thomas, S.,** The bisphosphonate 2-PEB inhibits cyclosporin A induced high-turnover osteopenia in the rat, *J. Lab. Clin. Med.,* 115, 62, 1990.
174. **Movsowitz, C., Epstein, S., Ismail, F., Fallon, M., and Thomas, S.,** Cyclosporin A in the oophorectomized rats: unexpected severe bone resorption, *J. Bone Miner. Res.,* 4, 393, 1989.
175. **Schlosberg, M., Movsowitz, C., Epstein, S., Ismail, F., Fallon, M., and Thomas, S.,** The effect of cyclosporin A administration and its withdrawal on bone mineral metabolism in the rat, *Endocrinology,* 124, 2179, 1989.

176. **Movsowitz, C., Epstein, S., Fallon, M., Ismail, E., and Thomas, S.,** Cyclosporin A *in vivo* produces severe osteopenia in the rat: effect of dose and duration of administration, *Endocrinology,* 123, 2571, 1988.
177. **Orcel, P., Bielakoff, J., Modrowski, D., Miravet, L., and de Vernejoul, M. C.,** Cyclosporin A induces *in vivo* inhibition of resorption and stimulation of formation in rat bone, *J. Bone Miner. Res.,* 4, 387, 1989.
178. **Loertscher, R., Thiel, G., Hardner, F., and Brunner, F. P.,** Persistent elevation of alkaline phosphatase in cyclosporine-treated renal transplant patients, *Transplantation,* 36, 115, 1983.
179. **Briner, V., Brunner, F. P., Landmann, J., Ritz, E., Dambacher, M. A., and Thiel, G.,** Raised bone alkaline phosphatase levels in patients (PTS) on cyclosporine A (CsA): bone disease or repair, *Kidney Int.,* 32, 431, 1987.
180. **Aubia, J., Masramon, J., Serrano, S., Lloveras, J., and Marinoso, L.,** Bone histology in renal transplant patients receiving cyclosporin, *Lancet,* 1, 1048, 1988.
181. **Rich, G. M., Mudge, G., and LeBoff, M. S.,** Cyclosporin A-associated osteoporosis in cardiac transplant patients, *J. Bone Miner. Res.,* 5 (Suppl. 2), 439, 1990.
182. **Muchmore, J. S., Cooper, D. K. C., Ye, Y., Schlegel, V. T., and Zuhdi, N.,** Loss of vertebral bone density in heart transplant patients, *Transplant. Proc.,* 23, 1184, 1991.
183. **Porayko, M. K., Wiesner, R. H., Hay, J. E., Krom, R. A. F., Dickson, E. R., Beaver, S., and Schwerman, L.,** Bone disease in liver transplant recipients: incidence, timing, and risk factors, *Transplant. Proc.,* 23, 1462, 1991.
184. **Henderson, B., Staines, N. A., Burrai, I., and Cox, J. H.,** The anti-arthritic and immunosuppressive effects of cyclosporine on arthritis induced in the rat by type II collagen, *Clin. Exp. Immunol.,* 57, 51, 1984.
185. **Kaibara, N., Hotokebuchi, T., Takagishi, K., and Katsuki, I.,** Paradoxical effects of cyclosporin A on collagen arthritis in rats, *J. Exp. Med.,* 158, 2007, 1983.
186. **Hom, J. T., Butler, L. D., Riedl, P. E., and Bendele, A. M.,** The progression of the inflammation in established collagen-induced arthritis can be altered by treatments with immunological or pharmacological agents which inhibit T cell activities, *Eur. J. Immunol.,* 18, 881, 1988.
187. **del Pozo, E., Graeber, M., Elford, P., and Payne, T.,** Regression of bone and cartilage loss in adjuvant arthritic rats after treatment with cyclosporin A, *Arthritis Rheum.,* 33, 247, 1990.
188. **Tugwell, P., Bombardier, C., Gent, M., Bennett, K. J., Bensen, W. G., Carette, S., Chalmers, A., Esdaile, J. M., Klinkhoff, A. V., Kraag, G. R., Ludwin, D., and Roberts, R. S.,** Low-dose cyclosporin versus placebo in patients with rheumatoid arthritis, *Lancet,* 335, 1051, 1990.
189. **Stein, B., Halloran, B. P., Reinhardt, T., Engstrom, G. W., Bales, C. W., Drezner, M. K., Currie, K. L., Takizawa, M., Adams, J. S., and Epstein, S.,** Cyclosporin-A increases synthesis of 1,25-dihydroxyvitamin D_3 in the rat and mouse, *Endocrinology,* 128, 1369, 1991.
190. **Langman, C. B., Ford, K. K., Pachman, L. M., and Glorieux, F.,** Vitamin D metabolism in rats with adjuvant-induced arthritis, *J. Bone Miner. Res.,* 5, 905, 1990.
191. **Mackie, E. J. and Trechsel, U.,** Stimulation of bone formation *in vivo* by transforming growth factor-β: remodeling of woven bone and lack of inhibition by indomethacin, *Bone,* 11, 295, 1990.
192. **Marcelli, C., Yates, A. J., and Mundy, G. R.,** *In vivo* effects of human recombinant transforming growth factor β on bone turnover in normal mice, *J. Bone Miner. Res.,* 5, 1087, 1990.
193. **Kodama, H., Yamasaki, A., Nose, M., Niida, S., Ohgame, Y., Abe, M., Kumegawa, M., and Suad, T.,** Congenital osteoclast deficiency in osteopetrotic (op/op) mice is cured by injections of macrophage colony-stimulating factor, *J. Exp. Med.,* 173, 269, 1991.
194. **Marks, S. C.,** Morphological evidence of reduced bone resorption in osteopetrotic (op) mice, *Am. J. Anat.,* 163, 157, 1982.

195. **Wiktor-Jedrzejczak, W., Bartocci, A., Ferrante, A. W., Jr., Ahmed-Ansari, A., Sell, K. W., Pollard, J. W., and Stanley, E. R.,** Total absence of colony-stimulating factor 1 in the macrophage-deficient osteopetrotic (op/op) mouse, *Proc. Natl. Acad. Sci. U.S.A.,* 87, 4828, 1990.
196. **MacDonald, B. R.,** Parathyroid hormone, prostaglandins and bone resorption, *World Rev. Nutr. Diet.,* 47, 163, 1986.
197. **Boross, M. and Hollo, I.,** Effect of aminoglutethimide on calcitonin sensitivity, *Akt. Gerontol.,* 7, 471, 1977.
198. **Civitelli, R., Gonnelli, S., Zacchei, F., Bigazzi, S., Vattimo, A., Avioli, L., and Gennari, C.,** Bone turnover in postmenopausal osteoporosis, *J. Clin. Invest.,* 82, 1268, 1988.
199. **MacIntyre, I., Stevenson, J., Whitehead, M., Wimalawansa, S., Banks, L., and Healy, M.,** Calcitonin for prevention of postmenopausal bone loss, *Lancet,* 1, 990, 1988.
200. **Gonzalez, D., Ghiringhelli, G., and Mautalen, C.,** Acute antiosteoclastic effect of salmon calcitonin in osteoporotic women, *Calcif. Tissue Int.,* 38, 71, 1986.
201. **Rico, H., Hernandez, E. R., Diaz-Mediavilla, J., Alvarez, A., Martinez, R., and Espinos, D.,** Treatment of multiple myeloma with nasal spray calcitonin: a histomorphometric and biochemical study, *Bone Miner.,* 8, 231, 1990.
202. **Friipponi, P., Mannarelli, C., Gubbiotti, G., Blass, A., Moretti, I., Tini, S., Giuseppetti, N., Ballanti, S., and Morucci, P.,** The effects of calcitonin on idiopathic nephrolithisasis. Evidence of bone involvement in fasting hypercalciuria, *J. Endocrinol. Invest.,* 11, 509, 1988.
203. **Hurley, D. L., Tiegs, R. D., Wahner, H. W., and Heath, H., III,** Axial and appendicular bone mineral density in patients with long-term deficiency or excess of calcitonin, *N. Engl. J. Med.,* 317, 537, 1987.
204. **McDermott, M. T., Kidd, G. S., Blue, P., Ghaed, V., and Hofeldt, F. D.,** Reduced bone mineral content in totally thyroidectomized patients: possible effect of calcitonin deficiency, *J. Clin. Endocrinol. Metab.,* 56, 936, 1983.
205. **Zaidi, M., Chambers, T. J., Bevis, P. J., Beachem, J. L., Gaines Das, R. E., and MacIntyre, I.,** Effects of peptides from the calcitonin genes on bone and bone cells, *O. J. Exp. Physiol.,* 73, 471, 1988.
206. **Datta, H. K., Zaidi, M., Wimalawansa, S. J., Ghatei, M. A., Baecham, J. L., Bloom, S. R., and MacIntyre, I.,** *In vivo* and *in vitro* effects of amylin and amylinamide on calcium metabolism in the rat and rabbit, *Biochem. Biophys. Res. Commun.,* 162, 876, 1989.
207. **Eriksen, E. F., Colvard, D. S., Berg, N. J., Graham, M. L., Mann, K. G., Spelsberg, T. C., and Riggs, B. L.,** Evidence of estrogen receptors in normal human osteoblast-like cells, *Science,* 241, 84, 1988.
208. **Turner, R. T., Hannon, K. S., Demers, L. M., Buchanan, J., and Bell, N. H.,** Differential effects of gonadal function on bone histomorphometry in male and female rats, *J. Bone Miner. Res.,* 4, 557, 1989.
209. **Hreshchyshyn, M. M., Hopkins, A., Zylstra, S., and Anbar, M.,** Effects of natural menopause, hysterectomy, and oophorectomy on lumbar spine and femoral neck bone densities, *Obstet. Gynecol.,* 72, 631, 1988.
210. **Takano-Yamamoto, T. and Rodan, G. A.,** Direct effects of 17 beta-stradiol on trabecular bone in ovariectomized rats, *Proc. Natl. Acad. Sci. U.S.A.,* 87, 2172, 1990.
211. **Kalu, D. N.,** Proliferation of tartarate-resistant acid phosphatase-positive multinucleate cells in ovariectomized animals, *Proc. Soc. Exp. Biol. Med.,* 195, 70, 1990.
212. **Ohashi, T., Kusuhara, S., and Ishida, K.,** Effects of oestrogen and anti-oestrogen on the bone cells of the endosteal surface of male Japanese quail, *Br. Poult. Sci.,* 28, 727, 1987.
213. **Turner, R. T., Wakley, G. K., Hannon, K. S., and Bell, N. H.,** Tamoxifen prevents the skeletal effects of ovarian hormone deficiency in rats, *J. Bone Miner. Res.,* 2, 449, 1987.

214. **Turner, R. T., Vandersteenhoven, J. J., and Bell, N. H.,** The effects of ovariectomy and 17β-estradiol on cortical bone histomorphometry in growing rats, *J. Bone Miner. Res.,* 2, 115, 1987.
215. **Riggs, B. L., Jowsey, J., Goldsmith, R. S., Kelly, P. J., Hoffman, D. L., and Arnaud, C. D.,** Short- and long-term effects of estrogen and synthetic anabolic hormone in postmenopausal osteoporosis, *J. Clin. Invest.,* 51, 1659, 1972.
216. **Heaney, R. P.,** A unified concept of osteoporosis, *Am. J. Med.,* 39, 877, 1965.
217. **Lindsay, R., Hart, D. M., Forrest, C., and Baird, C.,** Prevention of spinal osteoporosis in oophorectomized women, *Lancet,* 2, 1151, 1980.
218. **Parfitt, A. M.,** Treatment of osteoporosis: theoretical possibilities, *Clin. Invest. Med.,* 5, 181, 1982.
219. **Kahn, A. J. and Partridge, N. C.,** New concepts in bone remodeling: an expanding role for the osteoblast, *Am. J. Otolaryngol.,* 8, 258, 1987.
220. **McSheehy, P. M. J. and Chambers, T. J.,** Osteoblastic cells mediate osteoclastic responsiveness to parathyroid hormone, *Endocrinology,* 118, 824, 1986.
221. **Prior, J. C.,** Progesterone as a bone-trophic hormone, *Endocrinol. Rev.,* 11, 386, 1990.
222. **Lindsay, R., Hart, D. M., Aitken, J. M., and Purdee, D.,** The effects of ovarian sex steroids on bone mineral status in the oophorectomized rat and in the human, *Postgrad. Med.,* 54S, 50, 1978.
223. **Erlick, Y., Meldrum, D. R., Lagassee, L. D., and Judd, H. L.,** Effect of megestrol acetate on flushing and bone metabolism in postmenopausal women, *Maturitas,* 3, 167, 1981.
224. **Deeny, M. F., Farish, E. T., Dagen, M., Hart, D. M., and Fletcher, C. D.,** Changes in the bone and liver isoenzymes of alkaline phosphatase in postmenopausal women being treated with norethisterone, *Clin. Chim. Acta,* 171, 103, 1988.
225. **Christiansen, C. and Riis, B. J.,** Five years with continuous combined oestrogen/progestogen therapy. Effects on calcium metabolism, lipoproteins and bleeding pattern, *Br. J. Obstet. Gynecol.,* 97, 1087, 1990.
226. **Aitken, J. M., Armstrong, E., and Anderson, J. B.,** Osteoporosis after oophorectomy in the mature female rat and the effect of oestrogen and progestogen replacement therapy in its prevention, *J. Endocrinol.,* 55, 79, 1972.
227. **Barbagallo, M., Carbogani, A., Palummeri, E., Chiavaini, M., Pedrazzoni, M., Bracchi, P. G., and Passeri, M.,** The comparative effect of ovarian hormone administration on bone mineral status in oophorectomized rats, *Bone,* 10, 113, 1989.
228. **Karambolova, K. K., Snow, G. R., and Anderson, C.,** Surface activity on the periosteal and corticoendosteal envelopes following continuous progestogen supplementation in spayed beagles, *Calcif. Tissue Int.,* 38, 239, 1986.
229. **Snow, G. R. and Anderson, C.,** The effect of 17 beta estradiol and progestagen on trabecular bone remodeling in oophorectomized dogs, *Calcif. Tissue Int.,* 39, 198, 1986.
230. **Christiansen, C., Christiansen, M. S., and Transbol, I.,** Bone mass in postmenopausal women after withdrawal of oestrogen/gestagen replacement therapy, *Lancet,* 1, 459, 1981.
231. **Grecu, E. D., Weinshelbaum, A., and Simmons, R.,** Effective therapy of glucocorticoid-induced osteoporosis with medroxyprogesterone acetate, *Calcif. Tissue Int.,* 46, 294, 1990.
232. **Horowitz, M., Wishart, J., Need, A. G., Morris, H., Philcox, J., and Nordin, B. E. C.,** Treatment of postmenopausal hyperparathyroidism with norethindrone, *Arch. Intern. Med.,* 147, 681, 1987.
233. **Robin, J. C., Ambrus, J. L., and Ambrus, C. M.,** Studies on osteoporosis. X. Effect of estrogen-progestin combination on heparin-induced osteoporosis, *Steroids,* 42, 669, 1983.
234. **Foster, B. J., Clagett-Carr, K., Hoth, D., and Leyland-Jones, B.,** Gallium nitrate: the second metal with clinical activity, *Cancer Treat. Rep.,* 70, 1311, 1986.
235. **Larson, S. M.,** Mechanism of localization of gallium-67 in tumours, *Semin. Nucl. Med.,* 8, 192, 1978.

236. **Krakoff, I. H., Newman, R. A., and Goldberg, R. S.,** Clinical toxicological and pharmacologic studies of gallium nitrate, *Cancer,* 44, 1722, 1979.
237. **Warrell, R. P., Jr., Skelos, A., Alcock, N. W., and Bockman, R. S.,** Gallium nitrate for acute treatment of cancer-related hypercalcemia: clinicopharmacological and dose response analysis, *Cancer Res.,* 46, 4208, 1986.
238. **Warrell, R. P., Jr., Isaacs, M., Alcock, N., and Bockman, R. S.,** Gallium nitrate for treatment of refractory hypercalcemia from parathyroid carcinoma, *Ann. Intern. Med.,* 107, 683, 1987.
239. **Warrell, R. P., Jr., Coonley, C. J., Straus, D. J., and Young, C. W.,** Treatment of patients with advanced malignant lymphoma using gallium nitrate administered as a seven day continuous infusion, *Cancer,* 51, 1982, 1983.
240. **Warrell, R. P., Jr., Bockman, R. S., Coonley, C. J., Isaacs, M., and Staszewski, H.,** Gallium nitrate inhibits bone resorption and is effective treatment for cancer-related hypercalcemia, *J. Clin. Invest.,* 73, 1487, 1984.
241. **Bockman, R. S., Bostey, A., Bullough, P., Alcock, N., and Warrell, R. P.,** Gallium increases bone calcium and crystallite perfection of hydroxyapatite, *Calcif. Tissue Int.,* 9, 3376, 1986.
242. **Donnelly, R., Bockman, R. S., Doty, S. B., and Boskey, A. L.,** Bone particles from gallium-treated rats are resistant to resorption in vivo, *Bone Miner.,* 12, 167, 1991.
243. **Matkovic, V., Apseloff, G., Shepard, D. R., and Gerber, N.,** Use of gallium to treat Paget's disease of bone: a pilot study, *Lancet,* 335, 72, 1990.
244. **Warrell, R. P., Jr., Bosco, B., Weinerman, S., Levine, B., Lane, J., and Bockman, R. S.,** Gallium nitrate for advanced Paget disease of bone: effectiveness and dose-response analysis, *Ann. Intern. Med.,* 113, 847, 1990.
245. **Dulce, H. J., Siegmund, P., Körber, F., and Schütte, E.,** Zur Biochemie der Knochenauflösung. III. Über das Vorkommen von Carboanhydratase im Knochen, *Hoppe Seyler's Z. Physiol. Chem.,* 320, 163, 1960.
246. **Väänänen, H. K.,** Immunohistochemical localization of carbonic anhydrase isoenzymes I and II in human bone, cartilage and giant cell tumor, *Histochemistry,* 81, 485, 1984.
247. **Sundquist, K. T., Leppilampi, M., Jarvelin, K., Kumpulainen, T., and Väänänen, H. K.,** Carbonic anhydrase isoenzymes in isolated rat peripheral monocytes, tissue macrophages and osteoclasts, *Bone,* 8, 33, 1987.
248. **Sly, W. S., Hewett-Emett, D., Whyte, P. M., Yu, Y. S. L., and Tashian, R. E.,** Carbonic anhydrase II deficiency identified as the primary defect in the autosomal recessive syndrome of osteopetrosis with renal tubular acidosis and cerebral calcification, *Proc. Natl. Acad. Sci. U.S.A.,* 80, 2752, 1983.
249. **Sly, W. S., Whyte, P. M., Sundaram, V., Tashian, R. E., Hewett-Emett, D., Gilbaud, P., Vainsel, M., Baluarte, H. J., Gruskin, A., Al-Mosawi, M., Sakati, N., and Ohlsson, A.,** Carbonic anhydrase II deficiency in 12 families with the autosomal recessive syndrome of osteopetrosis with renal tubular acidosis and cerebral calcification, *N. Engl. J. Med.,* 313, 139, 1985.
250. **Dulce, H. J. and Siegmund, P.,** Zur Biochemie der Knochenauflösung. II. Der Einfluss von Diamox auf das plasma-Calcium östron-behandelter Hähne, *Hoppe Seyler's Z. Physiol. Chem.,* 320, 160, 1960.
251. **Körber, F. and Siegmund, P.,** Versuche über Stoffwechselvorgange bei der Auflösung von Knochenmineral, in *Calcified Tissues,* Richelle, L. J. and Dallemagne, M. J., Eds., Collection des Colloques de l'Universtité, Liége, Belgium, 1965, 207.
252. **Kenny, A. D.,** Role of carbonic anhydrase in bone: partial inhibition of disuse atrophy of bone by parenteral acetazolamide, *Calcif. Tissue Int.,* 37, 126, 1985.
253. **Waite, L. C., Volkert, W. A., and Kenny, A. D.,** Inhibition of bone resorption by acetazolamide in the rat, *Endocrinology,* 87, 1129, 1971.
254. **Waite, L. C.,** Carbonic anhydrase inhibitors, parathyroid hormone and calcium metabolism, *Endocrinology,* 91, 1160, 1972.

255. **Pierce, W., Lineberry, M., and Waite, L.,** Effect of sulfonamides on the hypercalcemic response to vitamin D, *Horm. Metab. Res.*, 14, 607, 1982.
256. **Insogna, K. and Broadus, A.,** Hypercalcemia of malignancy, *Annu. Rev. Med.*, 38, 241, 1987.
257. **Brown, G. M., Morris, C. A., Mitnick, M. A., and Insogna, K. L.,** Treatment of humoral hypercalcemia of malignancy in rats with inhibitors of carbonic anhydrase, *J. Bone Miner. Res.*, 5, 1037, 1990.
258. **Pierce, W. M., Jr., Nardin, G. F., Fuqua, M. F., Sabah-Maren, E., and Stern, S. H.,** Effect of chronic carbonic anhydrase inhibitor therapy on bone mineral density in white women, *J. Bone Miner. Res.*, 6, 347, 1991.
259. **Gram, J., Bollerslev, J., Nielsen, H. K., Larsen, H. F., and Mosekilde, L.,** The effect of carbonic anhydrase inhibition on calcium and bone homeostasis in healthy postmenopausal women, *J. Intern. Med.*, 228, 367, 1990.
260. **Pierce, W. M., Jr. and Waite, L. C.,** Bone-targeted carbonic anhydrase inhibitors: effect of a proinhibitor on bone resorption *in vitro, Proc. Soc. Exp. Biol. Med.*, 186, 96, 1987.
261. **Pierce, W. M., Jr., Kaysinger, K. K., Waite, L. C., and Waite, K. J.,** Bone targeted carbonic anhydrase inhibitors: prodrugs and active configurations, *FASEB J.*, 4, A476, 1990.
262. **Wasnich, R. D., Benfante, R. J., Yano, K., Heilbrun, L., and Vogel, J. M.,** Thiazide effect on the mineral content of bone, *N. Engl. J. Med.*, 309, 344, 1983.
263. **Lehmann, J., Gray, R. W., Maierhofer, W. J., and Cheung, H. S.,** Hydrochlorothiazide inhibits bone resorption in men despite experimentally elevated serum 1,25-dihydroxyvitamin D concentrations, *Kidney Int.*, 28, 951, 1985.
264. **Ray, W. A., Griffin, M. R., Downey, W., and Melton, L. J., III,** Long-term use of thiazide diuretics and risk of hip fracture, *Lancet,* 1, 687, 1989.
265. **LaCroix, A. Z., Wienpahl, J., White, L. R., Wallace, R. B., Scherr, P. A., George, L. K., Cornoni-Huntley, J., and Ostfeld, A. M.,** Thiazide diuretic agents and the incidence of hip fracture, *N. Engl. J. Med.*, 322, 286, 1990.
266. **Rodan, G. A. and Martin, T. J.,** Role of osteoblasts in hormonal control of bone resorption. A hypothesis, *Calcif. Tissue Int.*, 33, 349, 1981.
267. **Hamilton, J. A., Lingelbach, S., Partridge, N. C., and Martin, T. J.,** Regulation of plasminogen activator production by bone-resorbing hormones in normal and malignant osteoblasts, *Endocrinology,* 116, 2186, 1985.
268. **Tran Van, P., Vignery, A., and Baron, R.,** Cellular kinetics of the bone remodeling sequence in the rat, *Anat. Rec.*, 202, 445, 1982.
269. **Richard, A. and Saffar, J. L.,** Effect of indomethacin on the periosteal remodeling induced by antagonist tooth extraction, *J. Dent. Res.*, 64, 709, 1985.
270. **Saffar, J. L. and Leroux, P.,** Role of prostaglandins in bone resorption in a synchronized remodeling in the rat, *Bone,* 9, 141, 1988.
271. **Carte-Bartlett, P., Dersott, J. M., and Saffar, J. L.,** Periodontal and femoral bone status in periodontitis-affected hamsters receiving a high dose of indomethacin treatment, *J. Biol. Buccale,* 17, 93, 1989.
272. **Thompson, D. D. and Rodan, G. A.,** Indomethacin inhibition of tenotomy-induced bone resorption in rats, *J. Bone Miner. Res.*, 3, 409, 1988.
273. **Shoutens, A., Verhas, M., Dourov, N., Bergmann, P., Caulin, F., Verschaeren, A., Mone, M., and Heilporn, A.,** Bone loss and bone blood flow in paraplegic rats treated with calcitonin, diphosphonate and indomethacin, *Calcif. Tissue Int.*, 42, 136, 1988.
274. **Li, X. J., Jee, W. S., and Li, Y. L.,** Flurbiprofen enhances growth and cancellous and cortical bone accumulation in rapidly growing long bones, *Bone,* 10, 35, 1989.
275. **Jee, W. S., Li, X. J., and Li, Y. L.,** Flurbiprofen-induced stimulation of periosteal bone formation and inhibition of bone resorption in older rats, *Bone,* 9, 381, 1988.
276. **Hayward, M. A., Howard, G. A., Neuman, R. G., Wood, D. D., Weichman, B. M., and Van Sickle, D. C.,** Prostaglandins in inflammatory bone pathology: mechanism and therapeutic benefit of etodolac, *Agents Actions,* 26, 310, 1989.

277. **Lane, N., Coble, T., and Kimmel, D. B.,** Effect of naproxen on cancellous bone in ovariectomized rats, *J. Bone Miner. Res.*, 5, 1029, 1990.
278. **Tashjian, A. H., Voelkel, E. F., Goldhaber, P., and Levine, L.,** Successful treatment of hypercalcemia by indomethacin in mice bearing a prostaglandin-producing fibrosarcoma, *Prostaglandins*, 3, 515, 1973.
279. **Harvey, W. and Meghji, S.,** Prostaglandins in pathological bone resorption, *Agents Actions*, 19, 73, 1986.
280. **Shindell, R., Huurman, W. W., Lipiello, L., and Connolly, J. F.,** Prostaglandin levels in unicameral cysts treated by intralesional steroid injection, *J. Pediatr. Orthop.*, 9, 516, 1989.
281. **Saffar, J. L. and Lasfargues, J. J.,** A histometric study of the effect of indomethacin and calcitonin on bone remodeling in hamster periodontitis, *Arch. Oral Biol.*, 29, 555, 1984.
282. **Gibbs, J., Dillon, M. J., Lang, S., Meghji, S., and Pritchard, J.,** Indomethacin responsive hypercalcemia associated with a renal sarcoma, *Arch. Dis. Child.*, 65, 1168, 1990.
283. **Williams, R. C., Jeffcoat, M. K., Howell, T. H., Rolla, A., Stubbs, D., Teoh, K. W., Reddy, M. S., and Goldhaber, P.,** Altering the progression of human alveolar bone loss with non-steroidal anti-inflammatory drug flurbiprofen, *J. Periodont.*, 60, 485, 1989.
284. **Rich, C. and Ensinck, F.,** Effect of sodium fluoride on calcium metabolism of human beings, *Nature (London)*, 191, 184, 1961.
285. **Baylink, D., Wergedal, J., Stauffer, M., and Rich, C.,** Effects of fluoride on bone formation, mineralization, and resorption in the rat, in *Fluorine in Medicine*, Vischer, T. L., Ed., Hans Huber, Bern, 1970, 37.
286. **Faccini, J. M.,** Inhibition of bone resorption in the rabbit by fluoride, *Nature (London)*, 214, 1269, 1967.
287. **Milhaud, G., Riviere, F., and Enriquez, B.,** Etude experimentale de la fluorose de l'agneau de boucherie, *Ann. Rech. Vet.*, 16, 369, 1985.
288. **Chavassieux, P., Pastoureau, P., Boivin, G., Charhon, S. A., Chapuy, M. C., Delmas, P. D., and Meunier, P. J.,** Effects of sodium fluoride on bone remodelling in ewes, *J. Bone Miner. Res.*, 2 (Suppl. 1), 359, 1988.
289. **Raisz, L. G. and Taves, D. R.,** The effect of fluoride on parathyroid function and responsiveness in the cat, *Calcif. Tissue Res.*, 1, 219, 1967.
290. **Liu, C. C. and Baylink, D. J.,** Stimulation of bone formation and resorption by fluoride in thyroparathyroidectomized rats, *J. Dent. Res.*, 56, 304, 1977.
291. **Rosenquist, J.,** Effects of supply and withdrawal of fluoride. Experimental studies on growing and adult rabbits. IV. Serum alkaline phosphatase isoenzymes, *Acta Pathol. Microbiol. Scand. Sect. A*, 81, 645, 1973.
292. **Andersen, L., Richards, A., Care, A. D., Kerzel Andersen, H. M., Kragstrup, J., and Fejerskov, O.,** Parathyroid glands, calcium and vitamin D in experimental fluorosis in pig, *Calcif. Tissue Int.*, 38, 222, 1986.
293. **Kragstrup, J., Richars, A., and Fejerskov, O.,** Experimental osteofluorosis in domestic pigs: a histomorphometric study of vertebral trabecular bone, *J. Dent. Res.*, 63, 885, 1984.
294. **Snow, G. R. and Anderson, C.,** Short-term chronic fluoride administration and trabecular bone remodeling in beagles: a pilot study, *Calcif. Tissue Int.*, 38, 217, 1986.
295. **Lindskog, S., Flores, M. E., Lilja, E., and Hammarstrom, L.,** Effect of a high dose of fluoride on resorbing osteoclasts *in vivo*, *Scand. J. Dent. Res.*, 97, 483, 1989.
296. **Marie, P. J. and Hott, M.,** Short-term effects of fluoride and strontium on bone formation and resorption in mouse, *Metabolism*, 35, 547, 1986.
297. **Chavassieux, P.,** Bone effects of fluoride in animal models *in vivo*. A review and a recent study, *J. Bone Miner. Res.*, 5 (Suppl. 1), 95, 1990.

298. **Inkovaara, J. A.,** Is fluoride treatment justified today?, *Calcif. Tissue Int.*, 49 (Suppl.), 68, 1991.
299. **Sleeboom, H. P. and Bijvoet, L. M.,** Treatment of tumor-induced hypercalcemia, in *Bone Resorption, Metastasis, and Diphosphonates,* Garattini, S., Ed., Raven Press, New York, 1985, 59.
300. **Kiang, D. T., Loken, M. K., and Kennedy, B. J.,** Mechanism of hypocalcemic effect of mithramycin, *J. Clin. Endocrinol. Metab.*, 48, 341, 1979.
301. **Rosol, T. J. and Capen, C. C.,** The effect of low calcium diet, mithramycin, and dichlorodimethylene bisphosphonate on humoral hypercalcemia of malignancy in nude mice transplanted with the canine adenocarcinoma tumor line (CAC-8), *J. Bone Miner. Res.*, 2, 395, 1987.
302. **Asanuma, K.,** Antitumor activity and pharmacokinetics of 7-n-(p-hydroxy-phenyl) mitomycin C in human tumor xenografts transplanted in nude mice, *J. Antibiot.*, 38, 401, 1985.
303. **Nishikawa, S., Rosol, T. J., and Capen, C. C.,** Effects of mithramycin in transplantable canine perianal gland carcinoma (CAC-9) in nude mice: biochemical, histomorphometric, and ultrastructural investigations, *Vet. Pathol.*, 23, 698, 1986.
304. **Shorr, E. and Carter, A. C.,** The usefulness of strontium as an adjuvant to calcium in the remineralization of the skeleton in osteoporosis in man, *Bull. Hosp. Jt. Dis. Orthop. Inst.*, 13, 59, 1952.
305. **Janes, M. and McCaslin, F.,** The effect of strontium lactate in the treatment of osteoporosis, *Proc. Mayo Clin.*, 34, 329, 1959.
306. **Marie, P. J., Garba, M. T., Hott, M., et al.,** Effect of low doses of stable strontium on bone metabolism in rats, *Miner. Electrolyte Metab.*, 11, 5, 1985.
307. **Scheiman-tagger, E. and Brodie, A. G.,** Lead acetate as a marker of growing calcified tissues. A modified method, *Anat. Rec.*, 150, 435, 1964.
308. **Bonucci, E. and Silvestrini, G.,** Ultrastructural changes in experimental lead intoxication, *Contrib. Nephrol.*, 64, 93, 1988.

Chapter 18

BISPHOSPHONATES

Paul H. Schlesinger and Harry C. Blair

TABLE OF CONTENTS

I. INTRODUCTION

Geminal bisphosphonates (bisphosphonates*, diphosphonates) are compounds containing a carbon bound to two phosphorus molecules. The subject of this chapter is their rational application to the treatment of skeletal diseases. Because extensive reviews of the older literature are available,[1-3] we will concentrate on new concepts and observations regarding bisphosphonates, with particular focus on specific interactions of bisphosphonates with bone matrix and bone cells. Similarly, elements of fundamental bone metabolism bearing on bisphosphonate activity, and the relationship of bone diseases to abnormal rates of bone turnover potentially treatable with bisphosphonates, are described elsewhere in this volume and will not be considered in detail.

Bisphosphonates have offered the first artificial, end organ-specific drugs for bone disease. Although they do not always provide the specific effects desired, they have the potential to be improved by chemical substitution at the 1-carbon position with a variety of chemical groups (rather like production of a class of antibiotics from a prototype). The major impediment to effective improvement of bisphosphonate therapy has been a lack of understanding as to how they act, and what functions side-group replacement may subserve. That information would facilitate development of new bisphosphonates with enhanced properties. Thus far, the only absolute chemical requirement appears to be the nonbiologically hydrolyzable phosphate-carbon-phosphate (P-C-P) group, which is analogous to pyrophosphate (see Figure 1).

II. THE IMPORTANCE OF PHARMACOLOGIC MANIPULATION OF BONE MINERAL METABOLISM

During a lifetime, the skeleton is remodeled many times, and the process is continuous. This activity is required for growth, to maintain a strong skeleton, and to fulfill the skeleton's metabolic roles as a repository of calcium, base equivalents, and phosphate. Because of its many roles, an abnormality in bone resorption ultimately leads to major disruption in orthopedic or metabolic functions, and the need for therapeutic intervention.

With the discovery of the roles of vitamin D and parathyroid hormone in the regulation of calcium metabolism, we have become more proficient at the treatment of many hard tissue-related disease states, including (classical) rickets and hyperparathyroidism. However, as powerful as these hormones are in their effect on growth and calcium metabolism, they are less useful with many diseases affecting bone mass or quality, but not [Ca^{2+}].

* Abbreviations: HEBP (also called EHDP and by the trade name Didronel), 1-hydroxy-1,1-ethylidenebisphosphonate; Cl_3MBP, dichloromethylidenebisphosphonate; AHPrBP (also called APD), 3-amino-1-hydrohypropylidenebisphosphonate; ACHBP, azacycloheptylidene-2,2-bisphosphonate. Variants of these names substituting the organic-based carbon names (ethane for ethylidene), phosphonic acid for phosphonate, and the organic-based *di*phosphonate for the inorganic-type *bis*pyrophosphate should be regarded as the same compounds.

The skeleton is the repository of almost all of the body's calcium, and the only significant, accessible source of manipulable calcium to fasting, land-living vertebrates. Intracellular calcium activities are carefully regulated at 10^{-7} to 10^{-9} *M*, because changes in calcium activity in this range are commonly used for cellular regulation and control. Maintenance of extracellular calcium activity (i.e., for practical purposes, the concentration of ionized calcium) at ~1.5 m*M* is, in turn, necessary for the maintenance of intracellular calcium activities. Even when developing gradually, the body has only limited ability to survive hyper- or hypocalcemia. Evolution of calcium as a major mediator of intracellular signals has been, for the higher vertebrates, particularly advantageous because of extensive extracellular reservoirs of this crucial ion. On the other hand, this requires the maintenance of intracellular solutions, having calcium concentrations four orders of magnitude below ambient solution concentrations, adjacent to solid calcium salt reservoirs. Furthermore, there must be complex regulation, since calcium in its role as a signaling ion, is traded off against the physical integrity of the skeleton. Therefore, any attempt to treat metabolic bone disease by influencing calcium metabolism through mass action or by dramatic hormone shifts would certainly be difficult and potentially dangerous.

III. RATIONALE FOR DEVELOPING END ORGAN-SPECIFIC THERAPY

The ability to regulate bone mass and quality without significantly altering nonbone calcium and phosphate regulation is the goal of treatment of most types of chronic metabolic bone diseases. Complete achievement of this goal is probably not possible because of the intimate integration between bone and [Ca^{2+}]. However, a means to selectively affect either bone formation or dissolution would perhaps improve on existing therapy in that the nonbone regulatory mechanisms would not be altered. In this case, the natural feedback mechanisms of serum ion homeostasis would compensate for the altered rate of bone deposition or formation, so that the only noticeable effect might be on the skeleton. Underlying our lack of specific bone-directed treatments is an incomplete understanding of the basic biochemical and cellular mechanisms of bone formation and dissolution. However, recent advances in bone physiology suggest that certain compounds selectively binding to hydroxyapatite, which is essentially unique to bone, may also inhibit specific biochemical processes of bone cells, thus achieving the goal of selective therapy. The bisphosphonates are a leading hope for development of this type of compound.

IV. DISEASES FOR WHICH END-ORGAN THERAPY OF BONE IS REQUIRED

A. OSTEOPOROSIS AND OTHER DISORDERS INVOLVING BONE LOSS

Of the diseases for which development and specific end-organ therapy is needed, undoubtedly the chief is osteoporosis, which may affect as much as half of our increasingly elderly population.[4,5] This alone would certainly be sufficient justification for research targeted toward specifically reducing bone loss. There are, however, additional conditions that could benefit, including the secondary hyperparathyroidism commonly associated with end-stage renal disease, and general or localized bone loss in a variety of rheumatic and neoplastic diseases. Of these, hypercalcemia resulting from neoplastic invasion of bone is one of the most critical, and has provided dramatic evidence of the potential benefits of bisphosphonate therapy.[6]

B. OTHER MINERALIZATION DISORDERS

Ultimately, with safe and effective tools for the manipulation of bone metabolism, a large number of conditions may be favorably affected. Presently, only a few disorders have been carefully studied. Paget's disease of bone (adult type), characterized by a locally excessive rate of bone turnover, has been treated with bisphosphonates, where the ability of these drugs to inhibit mineralization has been documented. On the other hand, use of bisphosphonates has also been proposed for disorders where ectopic deposition of calcium salts is observed.

V. DEVELOPMENT OF BISPHOSPHONATES AS EFFECTORS OF MINERAL METABOLISM

A. CELLULAR VS. PHYSICOCHEMICAL EFFECTS

Since there are real limitations to intervention in bone disease by manipulating global calcium metabolism, much recent research has focused on the molecular mechanisms of bone metabolism. In this regard, the discovery of bisphosphonates, which affect bone mineral metabolism at the end-organ level, has provided both a potential mechanism of intervention and the means to refine several hypotheses concerning the physiologic mechanism of mineral formation and dissolution.

The processes of mineral deposition and removal involve chemical crystallization and dissolution as directed by cellular activity. Various hypotheses have emphasized either the chemical or cellular mechanisms at the expense of the other.[7-9] However, recently developments in bone biology, especially concerning the HCl secretion mechanism of bone dissolution by osteoclasts, have demonstrated that bone cells play a crucial role of adjusting the environment at the bone solution interface. Bone formation remains less well understood, but it would seem that the physicochemical aspects of crystal nucleation and growth dominate that process.[1] That being the case, it is possible that bone deposition or removal could be independently and specifically affected by pharmacologic agents modifying crystal nucleation or acid transport, respectively. This would not be true if both processes were controlled by passive hydrozyapatite crystal growth and dissolution.

B. THEORIES OF THE MOLECULAR MECHANISM OF BONE FORMATION RELATED TO BISPHOSPHONATE ACTION

By the late 1950s, it was apparent that bone formation and metabolism could not be explained on the basis of a simple solubility constant, the product of serum $[Ca^{2++}]$ and $[PO_4^{-3}]$.[7] Some of the reasons for this are now evident, including the (obvious in retrospect) fact that in going from solid OH^- and PO_4^{-3} to aqueous solution, a protonation must occur, and it has been amply demonstrated that production of a local low pH is the mechanism of osteoclastic bone dissolution.[10-12] To date, however, the complex behavior of calcium phosphate solubility in biological solutions has no accurate theoretical model.

A second observation leading to discovery of the bisphosphonates was the effect of serum on hydroxyapatite formation: the presence of serum reduces calcium salt precipitation when compared to solubility in simple electrolyte solutions.[13] In fact, the concentrations of calcium and phosphate required for precipitation in serum exceeded the physiological calcium phosphate product! Since serum inhibited the crystallization of hydroxyapatite, it was hypothesized that, for crystallization to occur in physiological solutions, collagen is required to provide nucleation sites for the formation of calcium phosphate crystals.[9,14] In this model, crystallization is proposed to proceed from a supersaturated, but metastable, solution. This proposal was supported by *in vitro* studies of crystallization of calcium phosphate in the presence of collagen.[15] On the other hand, this model does not explain why mineralization is restricted to specific sites and selected tissues, even though type I collagen, apparently identical to that of bone, is the most abundant extracellular protein.

C. INHIBITORS OF BONE FORMATION

In order to explain the lack of generalized hydroxyapatite crystallization throughout the body, the existence of endogenous inhibitors was proposed, which would prevent mineral precipitation except at specific sites. To initiate mineralization where required, the inhibitors would be eliminated or reduced.[9,16-18]

Because it is abundant in bone-forming cells, alkaline phosphatase was felt to play a role in bone formation. This has been confirmed by the description of the disease produced by lack of bone alkaline phosphatase activity.[19] These findings clearly point to an active cellular process involving phosphate cleavage controlling hydroxyapatite formation.[20]

From the opposite point of view, the knowledge that alkaline phosphatase is essential to bone formation suggested that complex phosphates are important mediators of bone mineral deposition. Indeed, it was soon apparent that pyrophosphate (Figure 1), which builds up to high levels in hypophosphatasia, but was thought to be absent in normal people, inhibits hydroxyapatite crystallization *in vivo*.[18,21] Pyrophosphate was known to inhibit the precipitation of calcium salts, and had been used as a water softener.[22] Interestingly, this inhibition is not due to chelation of calcium ions (i.e., reducing the calcium activity but not its concentration), with a resulting reduction of the ion product, and precipitation, by mass action. Rather, it results from binding of small quantities of the pyrophosphate to microscopic crystals of hydroxyapatite, disrupting the further growth of the crystals.[23] Therefore, the ability of pyrophosphate to prevent crystallization by this mechanism would be closely linked with its ability to bind to crystallized calcium salts. Furthermore, it would not be necessary to reduce $[Ca^{2+}]_{free}$ in order to influence the crystallization of calcium.

D. PYROPHOSPHATE AND BONE MINERAL METABOLISM

Pyrophosphate is, surprisingly, present in extracellular physiologic fluids,[17,20,21] and at concentrations ranging upward from 2 to 3 μM. This made it a reasonable candidate to be a functional inhibitor of hydroxyapatite crystallization. It was subsequently shown that pyrophosphate is tightly bound to hydroxyapatite crystals.[24] As a result of binding to hydroxyapatite, pyrophosphate inhibited further crystallization.[25] Pyrophosphate also inhibits dissolution of hydroxyapatite.[26]

Subsequently, it was shown that parentally administered pyrophosphate inhibited ectopic calcification in a number of tissues, including skin,[27,28] kidney, and aorta.[29,30] In these experiments, calcification had been experimentally induced by increasing the local ion product ($[Ca^{2+}]\cdot[PO_4^{-3}]$), such as via administration of vitamin D_3. Successful inhibition of calcification in the face of increased calcium phosphate product indicated that pyrophosphate had access to the crystallization site, and that the ion product could, after all, play an important role in initiating crystallization.

Disappointingly, parenterally administered pyrophosphate did not measurably affect bone metabolism.[30] This lack of effect was attributed to the rapid metabolism of pyrophosphate by endogenous phosphatases, which had been previously described.[17,18] Therefore, as a pharmacological agent, it was apparent that pyrophosphate was not suitable for clinical use, and interest in exploring the mechanism of crystallization inhibition waned. The search for compounds having similar activity on bone metabolism included phosphonates, because the P-C-P bond is resistant to phosphatases and pyrophosphatases.[31]

E. BISPHOSPHONATES AS PYROPHOSPHATE ANALOGS

The mechanism of pyrophosphate action was determined to be physicochemical, resulting from its binding to hydroxyapatite and consequent crystal poisoning. Therefore, when it was appreciated that the bisphosphonates also inhibited the precipitation of calcium salts, they were regarded as potentially clinically useful analogs of pyrophosphate.

Methanebisphosphonate may be regarded as substituted pyrophosphate, the diester oxygen being replaced with carbon. The two additional covalent bonding positions of that carbon are available for substitution, leading to the family of geminal bisphosphonates (Figure 1). The geminal bisphosphonates are completely resistant to the action of pyrophosphatases and phosphatases, and are therefore as stable *in vivo* as in solutions lacking phosphatases.[1] However, like pyrophosphate, bisphosphonates inhibit crystallization, crystal maturation, and crystal dissolution of hydroxyapatite *in vitro*.[25,26,29,32-35] Hence, these compounds showed obvious promise in inhibiting bone metabolism as nonhydrolyzable pyrophosphate analogs. This promise was soon borne out, as the ethane derivative was demonstrated to inhibit ectopic calcification, reduce bone resorption, and was, very importantly, effective orally.[36]

F. BISPHOSPHONATE ACTIVITY ON BONE MINERAL

Of the phosphonates, only the geminal bisphosphonates (having two phosphine bonds to the same carbon, and referred to, here, simply as bisphosphonates) have useful activity upon bone metabolism.[1] Monophosphonates are known to occur naturally, but do not appear to function directly in bone metabolism.[37] Vicinal (phosphine groups bound to adjacent carbons) and longer-chain polyphosphonates are not known to be active on bone metabolism.[1,37] Bisphosphonates form polynuclear complexes with calcium in solution at neutral pH.[38] Bidentate complexes are formed between bisphosphonates and calcium in hydroxyapatite. If the bisphosphonyl carbon is hydroxyl substituted (as in 1-hydroxy-1,1-ethanebisphosphonate, also called ethylene-1-hydroxy-1,1-diphosphonate, HEBP, or Didronel), the bisphosphonate forms a more stable tridentate complex with calcium[38] (Figure 1). The increased binding of 1-OH-1,1-bisphosphonates to hydroxyapatite correlates with their activity as crystal poisons *in vitro* and with their ability to inhibit mineralization *in vitro* and *in vivo*.[29,39,40] In general, there is a good correlation between the inhibition of calcium phosphate precipitation and the inhibition of ectopic calcification or normal mineralization *in vivo*.[29,40-42] In sum, these observations strongly support the physicochemical mechanism of action by bisphosphonates on mineralization. One exception to this rule is dichloromethanebisphosphonate (dichloromethylenediphosphonate, Cl_2MBP; Figure 1), which is an excellent inhibitor of crystallization in solution, but a poor inhibitor of mineralization.[42] The mechanistic reason for this inconsistency is not clear at present, although Cl_2MBP is more effective and better tolerated than EHBP for treatment of the hypercalcemia of malignancy *(vide infra)*.

G. RELATIONSHIP OF CHEMICAL STRUCTURE TO BISPHOSPHONATE ACTIVITY ON BONE RESORPTION

In a wide variety of *in vitro* or cell culture experiments,[26,34,43-54] and in trials *in vivo*,[26,34,42,50,55] bisphosphonates have proven to be effective inhibitors of bone resorption. However, the mechanism of this inhibition appears to be more complex than anticipated from earlier solubility studies.

In the test tube, bisphosphonates had been shown to inhibit the dissolution of hydroxyapatite crystals.[26,34,56] This occurs via binding of bisphosphonates to hydroxyapatite crystals and alteration of their stability.[39,57] However, in contrast to bisphosphonate effects on mineralization, there is no discernable relationship between *in vitro* inhibition of dissolution and *in vivo*, osteoclast-mediated inhibition of bone dissolution.[1,40] In spite of this lack of correlation, there is good circumstantial evidence that binding of bisphosphonates is required to inhibit bone resorption, and that this bisphosphonate binding occurs at low concentrations which are not generally toxic to cells.[51]

Consequently, several structural features that promote binding to hydroxyapatite have been employed to increase the inhibition of bone resorption by bisphosphonates. Two substitutions have been particularly useful. Simple unsubstituted bisphosphonates form bidentate complexes with calcium, with either one or two moles of calcium per mole of bisphosphonate. This occurs either in solution or on the growing surface of hydroxyapatite. The addition of a hydroxyl on the 1-carbon of a geminal bisphosphonate allows the formation of tridentate complexes (Figure 1; HEBP, AHPrBP) with calcium in hydroxyapatite,[38] increasing the stability of the complex (greater affinity binding to hydroxyapatite). Similarly, the addition of an amino group to an aliphatic side chain of the bisphosphonate carbon increases binding to hydroxyapatite (Figure 1, AHPrBP).[51] These two modifications provide large increases in the inhibition of bone resorption, producing the very active bisphosphonates.[40] Increasing the side-chain length up to nine carbons (or four carbons if there is amino substitution) further increases activity. Increases in size beyond these limits result in decreased activity, suggesting a general steric limitation.

The combination of optimal carbon-chain length and amino or hydroxy substitution leads to an increase in activity from a minimum effective dose of 15 to 50 μg/kg/d for the short-chain-length aliphatic bisphosphonates to 1 μg/kg/d for the most active compounds.[40,50] Recently, cyclic geminal bisphosphonates have been shown to have similar activities, ~1 μg/kg/d (Figure 1).[58]

In summary, the structural features important to bisphosphonate inhibition of bone resorption are the P-C-P bond, required for activity on bone resorption *in vivo*, and substitutions that increase interaction with calcium or reduce steric constraints to binding to hydroxyapatite. The ideal size appears to be four to nine carbons, depending on substitutions. It should be noted, however, that the ideal compound will be determined not so much by its minimum effective dose, but by the adverse effects at that effective dose, or by how close an effective dose is to the lethal dose (therapeutic margin).

Pyrophosphate

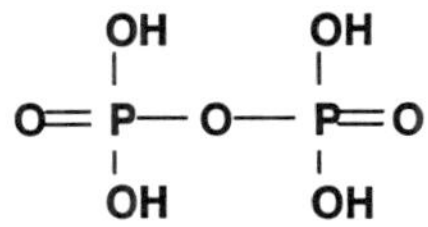

1-Hydroxyethylidene-1, 1-bisphosphonic Acid (HEBP)

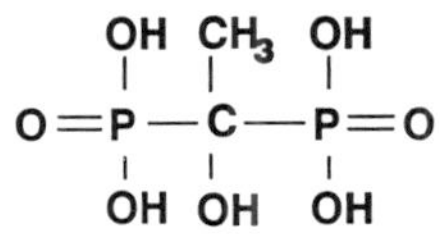

Dichloromethylenebisphosphonic Acid (Cl_2MBP)

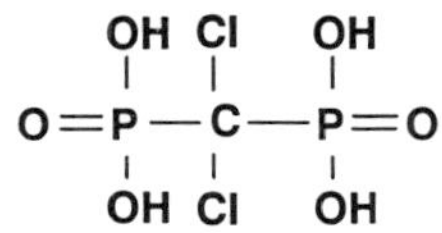

3-Amino-1-hydroxypropylidene-1, 1-bisphosphonic Acid (AHPrBP)

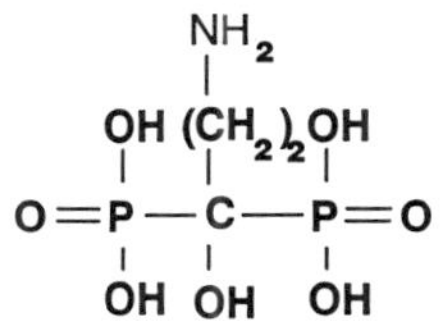

Azacycloheptylidene-2, 2-bisphosphonic Acid

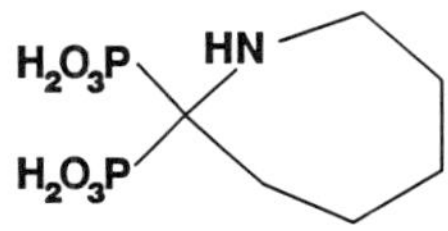

FIGURE 1. Molecular structures of pyrophosphate and four clinically used bisphosphonates discussed in the text.

H. CELLULAR ACTIVITY OF BISPHOSPHONATES

The inhibition of mineralization by bisphosphonates is clearly dominated by its physicochemical effect upon the crystallization of hydroxyapatite, so that these compounds will prevent the deposition of calcium salts in atherosclerotic lesions.[59,60] Bisphosphonates also inhibit the crystallization of other calcium salts *in vivo*.[61] In both of these situations, the bisphosphonates are presumed to have no effect upon the cellular processes associated with mineralization, since the effects seen *in vivo* are entirely consistent with the *in vitro* observation that they function to increase the solubility product of calcium phosphate.

In the case of bone resorption, the situation is very different, and is abundantly clear that the cellular effects of the bisphosphonates are very important. Morphological changes observed in cultured osteoclasts and *in vivo* established that bisphosphonates have direct cellular activity.[42,62-65] Since osteoclasts are known to be central to bone resorption, and the bisphosphonates are particularly effective in reducing bone resorption, it is not surprising that bisphosphonates should act on these cells. It was not clear, on the other hand, how bisphosphonates might specifically target to osteoclasts, and what the biochemical mechanism of their cellular effects is.

Studies of osteoclast biochemical and metabolic responses to bisphosphonates yielded several effects that could be attributed to the drugs, but did not provide a uniform, general mechanism to explain the activity of these compounds on bone resorption. Frequently, the direction or magnitude of the biochemical response depended on the compound, or the manner in which it was studied. Dichloromethylenebisphosphonate and 1-hydroxyethylidene-1,1-bisphosphonate reduce lactate production in osteoclasts,[44] isolated bone, and cartilage.[66] On the other hand, compounds with longer aliphatic chains, or the same compounds studied in different tissue preparations, resulted in increased lactate production.[40] From these studies, it seemed clear that whatever the bisphosphonate effect on glycolysis might be, it would probably not account for the inhibition of osteoclast activity.

It has also been observed that bisphosphonates increase fatty acid oxidation,[67] increase amino acid oxidation, stimulate the citric acid cycle,[1] and increase cellular glycogen.[68] Other studies showed that bisphosphonates reduce the rate of calcium release from renal mitochondria *in vitro,*[69] but increase mitochondrial calcium *in vivo*.[70] It is very difficult to propose a metabolic mechanism whereby bisphosphonates would selectively affect osteoclasts from these observations. Furthermore the effects reported would affect the general metabolism of the body rather than bone specifically. For the most part, the effects observed were not dramatic, indicating that they may be secondary actions of these drugs.

The activity of lysosomal enzymes has been clearly implicated in osteoclast-mediated bone degradation.[52] It has been observed that, at millimolar concentrations, bisphosphonates inhibit lysosomal enzymes in subcellular fractions from liver.[31] Further, on calvaria stimulated with parathyroid hormone, bisphosphonates inhibit the secretion of lysosomal enzymes.[49] The first inhibition might reflect the ability of bisphosphonates as chelators or as competitive nonhydrolyzable substrates. The inhibition of lysosomal enzyme secretion *in vivo* may be the consequence of cellular toxicity.

Additional effects upon macromolecular metabolism have been noted. These include increased production of alkaline phosphatase,[71] increased synthesis of bone and cartilage collagen,[72,73] and impaired collagenolysis.[74] Regarding further skeletal elements, bisphosphonates impair dentin and cementum formation,[75] and increase the synthesis of proteoglycans *in vitro* but cause a decrease *in vivo*.[76,77] Although the ability of bisphosphonates to bind metal cations could inhibit many of the steps in macromolecular synthesis and processing, it is also possible, as above, that these varied effects represent irrelevent secondary phenomena or toxic side effects.

Of the cellular regulatory mechanisms that have been studied, there are reports of contradictory effects upon cAMP production,[78,79] inhibition of prostaglandin synthesis by bone cells or calvaria *in vitro* or *in vivo,*[80,81] decreased cellular multiplication,[66,82] inhibition of macrophage multiplication,[83] and inhibition of migration.[84] In addition, there are several reports of T-cell-dependent macrophage effects, perhaps indicating immunologic consequenses of bisphosphonate administration.[85,86]

Studies finding cellular effects have in several cases focused on macrophages, suggesting that macrophages may be more sensitive to bisphosphonates than other cell types. This may indicate a commonality between macrophages and osteoclasts. This would not be surprising, considering the ontogeny of osteoclasts, which have a circulating precursor of macrophage lineage. On the other hand, as we will discuss shortly, this might also represent the endocytic and phagocytic activity of macrophages, which could simply be providing an increased exposure of these cells to the drugs.

In summary, it is easily seen that the data to date do not support a single definitive biochemical mechanism for bisphosphonate activity, but rather, that these compounds act upon osteoclasts at the concentrations used in therapy. Since the bisphosphonates are felt to enter cells and are known to be excellent chelators of divalent cations,[66] including Ca^{2+}, it seems plausible that this would play a role in their toxicity. It has been noted that hypocalcemia can be induced during administration of these drugs intravenously.[87] However, in experimental systems, it has not been possible to eliminate the effects of bisphosphonates by providing calcium to compensate for drug binding.[84] The other obvious possibility stems from the bisphosphonates' structural resemblance to pyrophosphate, which suggests that bisphosphonates, as inert analogs of the important phosphate esters, could inhibit any of a wide variety of enzymes. Therefore, the activity of bisphosphonates upon cells to which it has access requires further clarification.

I. MECHANISM OF OSTEOCLAST SPECIFICITY

The inhibition of bone resorption by osteoclasts correlates strongly with the binding of bisphosphonates to hydroxyapatite. This is in spite of the observation that the physicochemical effects of specific bisphosphonates upon hydroxyapatite crystallization do not parallel their inhibition of bone dissolution.[1,40] These apparently paradoxical findings are explained by evidence that these drugs act, in matrix dissolution, via effects on the osteoclast. The dissolution of bone takes place via vectorial HCl transport by the osteoclast, so it is not surprising that the solubility effects of bisphosphonates at neutral pH are not important to mineral solubilization. On the other hand, binding of bisphosphonates to mineral assures that, at a locally acidified site, a very high concentration of bisphosphonate would develop with sudden dissolution of mineral matrix that otherwise holds the drug out of solution. Such high local concentrations of bisphosphonates appear to affect bone-degrading cells, while other cells, exposed to much lower serum activities of the drugs, remain largely unaffected. The experimental findings leading to this conclusion began to emerge when it became possible to do detailed studies on osteoclasts *in vitro*, as outlined below.

The manner in which bisphosphonates achieve their rather dramatic specificity in reducing bone degradation was for a long time enigmatic. Morphological changes observed in osteoclasts during bisphosphonate exposure *in vitro* indicated cellular toxicity at concentrations similar to and greater than those used to inhibit bone resorption *in vivo*.[42,62-65] Furthermore, in animals treated with bisphosphonates, there is a decrease in the numbers of osteoclasts.[88] Although it has been appreciated for some time that bisphosphonates absorbed to bone were primarily responsible for osteoclast inhibition *in vivo*, recently it was proposed that osteoclasts are exposed to bisphosphonates in the act of resorbing bone.[51,53]

It had been demonstrated that macrophages demonstrated a similar sensitivity to bisphosphonates or bisphosphonate-treated bone.[46,83] These cells had different sensitivities to the drugs if they were not absorbed to bone. This, in turn, led to the interesting hypothesis that the unexpected sensitivity that bisphosphonates exhibit for osteoclasts resides in the route of delivery. In addition, it had been demonstrated that, to be affected by bone-bound drug, the cells must be degrading the bone (as both macrophages and osteoclasts do under these conditions).

It was under conditions where bone-bound bisphosphonate was present that greater sensitivity of the osteoclasts to bisphosphonates was noted. Fibroblasts cultured in the presence of bisphosphonates had the same sensitivity to these drugs as osteoclasts or macrophages without bone present. However, if bone was added to the fibroblast cultures, the sensitivity to bisphosphonates decreased almost to zero, because the drug was bound by the bone and, thus, prevented from interacting with the cells.[54]

Since bisphosphonates were known to bind strongly to bone, and their effectiveness was generally correlated to bone binding, it can be proposed that cells resorbing bone exposed to bisphosphonates would obtain an especially high exposure to these drugs as the bone was solubilized. Therefore, if an independent method of targeting the drug to cells could be employed, the mechanism of cellular inhibition could be studied independently of bone concentration effects. This has been accomplished using liposomes loaded with bisphosphonates, which demonstrated that bisphosphonates at high local concentration are potent general metabolic inhibitors. It is quite possible to eliminate phagocytic macrophages *in vitro* or *in vivo* without influencing osteoclasts. In other words, when bone-bound drug was not available, the osteoclast was not affected, whereas cells phagocytosing and degrading bisphosphonate-loaded liposomes were intoxicated.[89,90] In this way, the bisphosphonates were shown to be cytotoxic if provided in sufficient concentrations. Binding to bone is unnecessary for cell toxicity, but it provides a powerful vehicle for therapeutic targeting to osteoclasts.

Both the macrophage and the osteoclast, sensitive to bisphosphonates, employ a vesicular proton pump in their major cellular activities. Hence, the possibility existed that the bisphosphonates might be inhibitors of this proton pump. The proton pump requires MgATP, has a chemically reactive sulfhydryl group, and is located in the endosomal-phagosomal compartment of these cells. This was recently studied in cultured avian osteoclasts resorbing ^{3}H-proline-labeled bone. Under culture conditions, the bisphosphonates EHBP, Cl_2MBP, and AHPrBP bound to bone inhibited the osteoclast's ability to acidify its bone resorption site.[54] Direct examination of the proton pump activity in purified and reconstituted inside-out membrane vesicles containing the osteoclast proton pump, however, demonstrated that EHBP had no effect upon the osteoclast H^+-ATPase at concentrations ~20-fold the dose effective in cell culture. Therefore, it was concluded that bisphosphonates affect the osteoclasts as a general metabolic inhibitor. The bisphosphonate is delivered to the osteoclast as it solubilizes the bone to which the bisphosphonate is absorbed, and the high concentrations of the drug thus received are sufficient to inhibit metabolic processes of the cell.

Thus, the selective effect of bisphosphonates on osteoclasts reflects the avidity with which this metabolic inhibitor binds to bone mineral, and mobilization of these drugs from bone by the osteoclast itself, a sort of auto-inhibiting activity. The total effect upon bone resorption thus results from the combination of binding affinity for bone and inhibition by the bisphosphonates of as yet unknown, and possibly diverse, cellular metabolic reactions. This mechanism explains the observed lack of correlation between bisphosphonate inhibition of calcium phosphate crystal dissolution and bone resorption. Both processes are affected by bisphosphonate binding to hydroxyapatite, but crystal dissolution in acid is not, as a physicochemical process, significantly dependent upon that binding. Rather, inhibition of bone resorption is largely due to the antimetabolic activity of the bisphosphonate released at high concentrations from the bone surface. Therefore, specific inhibition of bone resorption depends on the fortuitous combination of at least two chemical properties of the bisphosphonates, their affinity for hydroxyapatite and their metabolic inhibition of macrophage-family cells.

VI. CLINICAL OVERVIEW OF BISPHOSPHONATES

Bisphosphonates have primarily been tested and applied in the two situations relevant to calcium metabolism: ectopic calcification and increased bone destruction. These, with basic bisphosphonate pharmacology, are considered here. In addition, ^{99m}Tc derivatives are used as skeletal markers; these uses are not further considered here.

A. ECTOPIC CALCIFICATION AND OSSIFICATION

A major disorder for which it was hoped bisphosphonates would provide a dramatic effect was urolithiasis, but they have not been useful for this purpose. The doses required to inhibit crystal growth in urine are high and cause adverse skeletal effects; the clinical benefit is questionable.[91,92] Bisphosphonates have also been used in an attempt to prevent traumatic or surgically induced calcifications, but with limited success.[93] A bisphosphonate, the cyclic compound azacycloheptylidene-2,2-bisphosphonate (Figure 1), is employed in the prevention of dental calculus by the addition to toothpaste.[94,95]

Fibrodysplasia ossificans progressiva has been treated with bisphosphonates, but the effectiveness of this therapy is unclear. Treatment of children with these drugs can produce rickets, so it must be done for short periods, further complicating the interpretation of results, and, as the disease is rare, a controlled prospective study would seem to be impractical.[96]

B. INCREASED BONE RESORPTION

The major clinical uses of bisphosphonates, to date, fall into the class of disorders with bone destruction. These compounds are useful in conditions of increased bone resorption because they inhibit osteoclastic bone resorption. However, it is important to remember that the processes of bone formation and resorption are closely linked, so it is frequently seen that mineralization will also decrease. It should also be noted that bisphosphonates may affect both processes, although neither equally nor predictably.

Paget's disease of bone is a common indication for bisphosphonates. The drugs are felt to be helpful in reducing symptoms, as well as abnormal bone remodeling with its consequent skeletal deformations. The three drugs studied — HEBP, Cl_2MBP and AHPrBP — have been shown to be effective. Interestingly, the bisphosphonates maintain their action for months after administration ceases.[97-99] This has not been formally studied, but probably is related to the slow turnover of these compounds, linked directly to the turnover of bone.

Hyperparathyroidism is not benefited by HEBP, although Cl_2MBP may be beneficial,[100,101] especially over the short term. However, since secondary hyperparathyroidism generally results from high $[PO_4^{-3}]$ or low ionized calcium, the effect of bisphosphonate treatment, which reduces the mobilization of bone calcium, could be to further stimulate PTH release. Recently, increased secretion of PTH following the inhibition of bone resorption with Cl_2MBP has been noted.[102] If this is confirmed, treatment of secondary hyperparathyroidism by inhibiting bone resorption with bisphosphonates would seem to be inappropriate in many cases, particularly given their persistence.

Hypercalcemia of malignancy, or tumoral bone destruction with or without hypercalcemia, are caused by various types of lesions. This may be mediated either via humoral osteolytic cytokines or by local destruction of bone tumors *in situ*.[103] Both situations respond to bisphosphonates, but they require aggressive parenteral administration of HEBP.[104,105] Cl_2MBP has been successfully administered orally, and the more potent AHPrBP has provided additional bisphosphonate effectiveness. However, bone resorption appears to preside over the hypercalcemia in most cases.[109]

It is clear from the experimental observations presented above that the action of bisphosphonates reflects their action as cellular toxins, targeted to bone cells by binding to the hydroxyapatite that these cells dissolve. Therefore, it is not surprising that bisphosphonates would inhibit tumor-induced dissolution of bone, whether caused by stimulation of osteoclasts or through the direct action of tumor cells as they dissolved the bone. Consequently, the promising report that bisphosphonates may inhibit the spread of tumor metastases is not surprising.[110] The bisphosphonates have benefited patients suffering from metastatic cancer, with a low incidence of serious side effects even in patients with renal difficulties.[111]

Studies of bisphosphonate use in osteoporosis have not shown a useful improvement in calcium balance.[112] This could be due to a number of factors; many patients with osteoporosis do not show excessive osteoclastic activity, for instance. Furthermore, the dose of bisphosphonate may have been either ineffectively low or high enough to affect mineralization as well as bone degradation. Continued studies using Cl_2MBP, which is less active on mineralization, and AHPrBP, which is more active, will be important to determine if it is possible to effectively treat bone resorption without inhibiting mineralization. Clearly, this would be an area where these drugs could improve the lives of millions of people, but where the possible side effects of long-term treatment require careful evaluation.

C. PHARMACOLOGY

Bisphosphonates are synthetic compounds that are not biologically degraded at the phosphine bonds; they are excreted unaltered by animals.[113] When given orally, 1 to 10% of the dose is absorbed, depending upon the compound.[114,115] Absorption is inhibited by calcium, so bisphosphonates should not be taken with meals.[116] The compounds with longer side chains are not absorbed from the gastrointestinal tract.

The serum half-life of the bisphosphonates is short; uptake into bone is similar to that of calcium and phosphate.[115,117,118] It is crucial to note that the fraction which is absorbed to bone is released very slowly, as the bone turns over, so that toxicity reversal is slow. The possibility of a toxic effect that disappears gradually, if at all, is a serious concern. From parenteral administration in large doses, the major problem is hypocalcemia.[87] However, the life-threatening toxicity of these compounds generally has not proved to be a problem, although bone and pain and an osteitis fibrosa-like biopsy appearance may develop with longer-term administration. With large chronic doses, fractions begin to appear. It is difficult to assess with certainty whether fractures after bisphosphonate therapy are due to inhibition of bone turnover, and consequent inability of the skeleton to renew and repair itself, or caused by inhibition of mineralization.[119]

ACKNOWLEDGMENT

We wish to thank Margaret Klemm for her invaluable assistance in assembling the manuscript.

REFERENCES

1. **Fleisch, H.,** Bisphosphonates: a new class of drugs in diseases of bone and calcium metabolism, in *Handbook of Experimental Pharmacology, Calcium in Drug Action,* Vol. 83, Baker, P. F., Ed., Springer-Verlag, New York, 1988, 441.

2. **Fleisch, H.,** Bisphosphonates — history and experimental basis, *Bone,* 8 (Suppl. 1), 23, 1987.
3. **Fleisch, H.,** Bisphosphonates: mechanisms of action and clinical applications, *Bone Miner. Res.,* 1, 319, 1983.
4. **Peck, W. A. and Avioli, L. V.,** *Osteoporosis — The Silent Thief,* Scott, Foresman, Glenview, IL, 1988.
5. **Francis, R. M.,** *Osteoporosis: Pathogenesis and Management,* Kluwer Academic, Boston, 1990.
6. **Mundy, G. R.,** Pathophysiology of cancer-associated hypercalcemia, *Semin. Oncol.,* 17 (Supp. 5), 10, 1990.
7. **Robison, R.,** *The Significance of Phosphoric Esters in Metabolism,* New York University Press, New York, 1932.
8. **Neuman, W. F. and Neuman, M. W.,** *The Chemical Dynamics of Bone Mineral,* University of Chicago Press, Chicago, 1958.
9. **Glimcher, M. J.,** Molecular biology of mineralized tissues with particular reference to bone, *Rev. Mod. Phys.,* 31, 359, 1959.
10. **Blair, H. C., Teitelbaum, S. L., Ghiselli, R., and Gluck, S.,** Osteoclast bone resorption by a polarized vacuolar proton pump, *Science,* 245, 855, 1989.
11. **Bekker, P. J. and Gay, C. V.,** Biochemical characterization of an electrogenic vacuolar proton pump in purified chicken osteoclast plasma membrane vesicles, *J. Bone Miner. Res.,* 5, 569, 1990.
12. **Blair, H. C., Teitelbaum, S. L., Tan, H.-L., Koziol, C. M., and Schlesinger, P. H.,** Passive chloride permeability charge-coupled to the H^+-ATPase of avian osteoclast ruffled membrane, *Am. J. Physiol.,* in press.
13. **Neuman, W. F.,** *Metabolic Interrelations,* Josiah Macy, Jr. Foundation, New York, 1950, 187.
14. **Freudenberg, F. and Gyorgi, P.,** Uber kalkbindung durch tierische gewebe. III, *Biochem. Z.,* 118, 50, 1921.
15. **Gutman, A. B. and Yu, T. F.,** *Metabolic Interrelations,* Josiah Macy, Jr. Foundation, New York, 1951, 205.
16. **Fleisch, H. and Neuman, W. F.,** On the role of phosphatase in the nucleation of calcium phosphate by collagen, *J. Am. Chem. Soc.,* 82, 3783, 1960.
17. **Fleisch, H. and Neuman, W. F.,** Mechanisms of calcification: role of collagen, polyphosphates, and phosphatase, *Am. J. Physiol.,* 200, 1296, 1961.
18. **Fleisch, H., Russel, R. G. G., and Straumann, F.,** Effect of pyrophosphate on hydroxyapatite and its implications in calcium homeostasis, *Nature (London),* 212, 901, 1966.
19. **Fraser, D.,** Hypophosphatasia, *Am. J. Med.,* 22, 1757, 1957.
20. **Russell, R. G. G., Bisaz, S., Donath, A., Morgan, D. B., and Fleisch, H.,** Inorganic pyrophosphate in plasma in normal persons and in patients with hypophosphatasia, osteogenesis imperfects and other disorders of bone, *J. Clin. Invest.,* 50, 961, 1971.
21. **Fleisch, H. and Bisaz, S.,** Isolation from urine of pyrophosphate, a calcification inhibition, *Am. J. Physiol.,* 203, 671, 1962.
22. **Reitemeyer, R. and Buehrer, T.,** The inhibiting action of minute amounts of sodium hexametaphosphate on the precipitation of calcium carbonate from ammoniacal solutions, *J. Phys. Chem.,* 44, 535, 1940.
23. **Buehrer, T. and Reitemeyer, R.,** The inhibiting action of minute amounts of sodium hexametaphosphate on the precipitation of calcium carbonate from ammoniacal solutions, *J. Phys. Chem.,* 44, 552, 1940.
24. **Krane, S. M. and Glimcher, M. J.,** Transphosphorylation from nucleoside di- and triphosphates by apatite, *J. Biol. Chem.,* 237, 2991, 1962.
25. **Francis, M. D., Russell, R. G. G., and Fleisch, H.,** Diphosphonates inhibit formation of calcium phosphate crystals *in vitro* and pathological calcification *in vivo, Science,* 165, 1264, 1969.

26. **Fleisch, H., Russell, R. G. G., and Francis, M. D.,** Diphosphonates inhibit hydroxyapatite dissolution *in vitro* and bone resorption in tissue culture and *in vivo, Science,* 165, 1262, 1969.
27. **Schibler, D. and Fleisch, H.,** Inhibition of skin calcification (calciphylaxis) by polyphosphates, *Experientia,* 22, 367, 1966.
28. **Gabbiani, G.,** Effect of phosphates upon experimental skin calcinosis, *Can. J. Physiol. Pharmacol.,* 44, 203, 1966.
29. **Fleisch, H., Russell, R. G. G., Bisaz, S., Muhlbauer, R. C., and Williams, D. A.,** The inhibitory effects of phosphonates on the formation of calcium phosphate crystals *in vitro* and on aortic and kidney calcification *in vivo, Eur. J. Clin. Invest.,* 1, 12, 1970.
30. **Schibler, D., Russell, R. G. G., and Fleisch, H.,** Inhibition of pyrophosphate and polyphosphate of aortic calcification induced by vitamin D(3) in rats, *Clin. Sci.,* 35, 363, 1968.
31. **Felix, R., Russell, R. G. G., and Fleisch, H.,** The effect of several diphosphonates on acid phosphohydrolases and other lysosomal enzymes, *Biochim. Biophys. Acta,* 429, 429, 1976.
32. **Francis, M. D.,** The inhibition of calcium hydroxyapatite crystal growth by polyphosphates, *Calcif. Tissue Res.,* 3, 151, 1969.
33. **Hansen, N. M., Felix, R., Bisaz, S., and Fleisch, H.,** Aggregation of hydroxyapatite crystals, *Biochim. Biophys. Acta,* 451, 549, 1976.
34. **Russell, R. G. G., Mulhbauer, R. C., Bisaz, S., Williams, D. A., and Fleisch, H.,** The influence of pyrophosphate, condensed phosphates, phosphonates and other phosphate compounds on the dissolution of hydroxyapatite *in vitro* and on bone resorption induced by parathyroid hormone in tissue culture and in thyroparathyroidectomized rats, *Calcif. Tissue Res.,* 6, 183, 1970.
35. **Meyer, J. L. and Nancollas, G. H.,** The influence of multidentate organic phosphonates on the crystal growth of hydroxyapatite, *Calcif. Tissue Res.,* 13, 205, 1973.
36. **Fleisch, H., Russell, R. G. G., Bisaz, S., Termine, J. D., and Posner, A. S.,** Influence of pyrophosphate on the transformation of amorphous to crystalline calcium phosphate, *Calcif. Tissue Res.,* 2, 49, 1968.
37. **Hilderbrand, R. L.,** Phosphonic acids in nature, in *The Role of Phosphonates in Living Systems,* Hilderbrand, R. L., Ed., CRC Press, Boca Raton, FL, 1983, 5.
38. **Barnett, B. L. and Strickland, L. C.,** Structure of disodium dihydrogen 1-hydroxyethylidene-diphosphonate tetrahydrate: a bone growth regulator, *Acta Crystallogr., Sect. B,* 35, 1212, 1979.
39. **Jung, A., Bisaz, S., and Fleisch, H.,** The binding of pyrophosphate and two diphosphonates on hydroxyapatite crystals, *Calcif. Tissue Res.,* 11, 269, 1973.
40. **Shinoda, H., Felix, R., and Fleisch, H.,** Structure-activity relationships of various biphosphonates, *Calcif. Tissue Int.,* 35, 87, 1983.
41. **Treschsel, U., Schenk, R., Bonjour, J. P., Russell, R. G. G., and Fleisch, H.,** Relation between bone mineralization, Ca absorption, and plasma Ca in phosphonate-treated rats, *Am. J. Physiol.,* 232, E298, 1977.
42. **Schenk, R., Merz, W. A., Muhlbauer, R., Russell, R. G. G., and Fleisch, H.,** Effect of ethane-1-hydroxy-1,1-diphosphonate (EHDP) and dichloromethylene diphosphonate (CL2MDP) on the calcification and resorption of cartilage and bone in the tibial epiphysis and metaphysis of rats, *Calcif. Tissue Res.,* 11, 196, 1973.
43. **Reynolds, J. J., Minkin, C., Morgan, D. B., Spycher, D., and Fleisch, H.,** The effect of two diphosphonates on the resorption of mouse calvaria *in vitro, Calcif. Tissue Res.,* 10, 302, 1972.
44. **Morgan, D. B., Monod, A., Russell, R. G. G., and Fleisch, H.,** Influence of dichloromethylene diphosphonate (Cl1MDP) and calcitonin on bone resorption, lactate production and phosphatase and pyrophosphatase content of mouse calvaria treated with parathyroid hormone *in vitro, Calcif. Tissue Res.,* 13, 287, 1973.

45. **Minkin, C., Rabadjija, L., and Goldhaber, P.,** Bone remodeling *in vitro*: the effect of two diphosphonates in osteoid synthesis and bone resorption in mouse calvaria, *Calcif. Tissue Res.*, 14, 161, 1974.
46. **Chambers, T. J.,** Diphosphonates inhibit bone resorption by macrophages *in vitro, J. Pathol.*, 132, 255, 1980.
47. **Gebauer, U. and Fleisch, H.,** The effect of two diphosphonates on the resorption of mouse calvaria induced by 1,25-dihydroxychloecalciferol and prostaglandin E1, in culture, *Miner. Electrolyte Metab.*, 4, 172, 1980.
48. **Reitsma, P. H., Teitelbaum, S. L., Verlinden-Ooms, H., and van der Wee-Pals, L. J. A.,** Differential action of bisphosphonates (3-amino-1-hydroxyprolidene)-1,1-bisphosphonate (APD) and disodium dichloromethylidene bisphosphonate (Cl2MMDP) on rat macrophage-mediated bone resorption *in vitro, J. Clin. Invest.*, 70, 927, 1982.
49. **Delaisse, J.-M., Eeckhout, Y., and Vaes, G.,** Bisphosphonates and bone resorption: effects on collagenase and lysosomal enzyme excretion, *Life Sci.*, 37, 2291, 1985.
50. **Schenk, R., Eggli, P., Felix, R., Fleisch, H., and Rosini, S.,** Quantitative morphometric evaluation of the inhibitory activity of new aminobisphosphonates on bone resorption in the rat, *Calcif. Tissue Res.*, 38, 342, 1986.
51. **Boonekamp, P. M., van der Wee-Pals, J. A., van Wijk-van Leenep, M. M. L., Thesing, C. W., and Bijvoet, O. L. M.,** Two modes of action of bisphosphonates on osteoclastic resorption of mineralized matrix, *Bone Miner.*, 1, 27, 1986.
52. **Vaes, G.,** Cell biology and biochemical mechanism of bone resorption, *Clin. Orthop. Relat. Res.*, 231, 239, 1988.
53. **Flannagan, A. M. and Chambers, T. J.,** Dichloromethylenebisphosphonate (Cl(2)MBP) inhibits bone resorption through injury to osteoclasts that resorb Cl(2)MBP-coated bone, *Bone Miner.*, 6, 33, 1989.
54. **Carano, A., Teitelbaum, S. L., Konsek, J. D., Schlesinger, P. H., and Blair, H. C.,** Bisphosphonates directly inhibit bone resorption activity of isolated osteoclasts *in vitro, J. Clin. Invest.*, 85, 456, 1990.
55. **Nilsson, O. S., Bauer, H. C. F., Brosjo, O., and Tornkvist, H.,** Bone resorption in orthoptic and heterotopic bone of dichloromethylene bisphosphonate-treated rats, *J. Orthop. Res.*, 8, 213, 1990.
56. **Bisaz, S., Felix, R., Hensen, N. M., and Fleisch, H.,** Disaggregation of hydroxyapatite crystals, *Biochim. Biophys. Acta,* 451, 560, 1976.
57. **Boonekamp, P. M., Lowik, C. W. G. M., van der Wee-Pals, L. J. A., van Wijk-van Lennep, M. L. L., and Bijvoet, O. L. M.,** Enhancement of the inhibitory action of APD on the transformation of osteoclast precursors into resorbing cells after dimethylation of the amino group, *Bone Miner.*, 2, 29, 1987.
58. **Bevan, J. A., Johnson, K. Y., Slough, C., Benedict, J., Fleisch, H., and Black, J.,** Skeletal effects of 2-(2-pyridinyl)-ethylidenebis(phosphonate) in acute and subchronic rat studies, *Calcif. Tissue Res.*, 38 (Suppl.), 31, 1985.
59. **Hollander, W., Prusty, S., Nagraj, S., Kirkpatrick, B., Paddock, J., and Colombo, M.,** Comparative effects of cetaben (PHB) and dichloromethylene diphosphonate (12MBP) on the development of atherosclerosis in the cynomolgus monkey, *Atherosclerosis,* 31, 307, 1978.
60. **Kramsch, D. M. and Chan, C. T.,** The effect of agents interfering with soft tissue calcification and cell proliferation on calcific fibrous plaques in rabbits, *Circ. Res.*, 42, 562, 1978.
61. **Fraser, D., Russell, R. G. G., Pohler, O., Robertson, W. G., and Fleisch, H.,** The influence of disodium enthane-1-hydroxy-1,1-diphosphonate (EHDP) on the development of experimentally induced urinary stones in rats, *Clin. Sci.*, 42, 197, 1972.
62. **Rowe, D. J. and Hausmann, E.,** The alteration of osteoclast morphology by diphosphonate in bone organ culture, *Calcif. Tissue Res.*, 10, 53, 1976.
63. **Miller, S. C. and Jee, W. S. S.,** Ethane-1-hydroxy-1,1-diphosphonate (EHDP): effects on growth and modeling of the rat tibia, *Calcif. Tissue Res.*, 18, 215, 1975.

64. **Miller, S. C., Jee, W. S. S., and Woodbury, L.,** Ethane-1-hydroxy-1,1-diphosphonate (EHDP) effects on incorporation and accumulation of osteoclast nuclei, *Calcif. Tissue Res.,* 22, 243, 1977.
65. **Rowe, D. J. and Housemann, E.,** The effects of calcitonin and colchicine on the cellular response to diphosphonate, *Br. J. Exp. Pathol.,* 61, 303, 1980.
66. **Fast, D. K., Felix, R., Dowse, C., Neuman, W. F., and Fleisch, H.,** The effects of diphosphonates on the growth and glycolysis of connective tissue cells, *Biochem. J.,* 172, 97, 1978.
67. **Felix, R. and Fleisch, H.,** Increase in fatty acid oxidation in calvaria cells cultured with diphosphonates, *Biochem. J.,* 196, 237, 1981.
68. **Felix, R., Fast, D. K., Sallis, J. D., and Fleisch, H.,** Effect of diphosphonate on glycogen content of rabbit ear cartilage cell in culture, *Calcif. Tissue Res.,* 30, 163, 1980.
69. **Guilland, D. F., Sallis, J. D., and Fleisch, H.,** The effect of two diphosphonates on the handling of calcium by rat kidney mitochondria *in vitro, Calcif. Tissue Res.,* 15, 303, 1974.
70. **Plasmans, C. M. T., Jap, P. H. K., Kujipers, W., and Sloof, T. J. J. H.,** Influence of diphosphonate on the cellular aspect of young bone tissue, *Calcif. Tissue Res.,* 32, 247, 1980.
71. **Felix, R. and Fleisch, H.,** Increase in alkaline phosphatase activity in calvaria cells cultured with diphosphonates, *Biochem. J.,* 183, 73, 1979.
72. **Guenther, H. L. and Guenther, H. E.,** The effects of 1-hydroxyethane-1,1-diphosphonate and dichloromethanediphosphonate on collagen synthesis by rabbit articular chondrocytes and rat bone cells, *Biochem. J.,* 196, 293, 1981.
73. **Guenther, H. L., Guenther, H. E., and Fleisch, H.,** The influence of 1-hydroxyethane-1,1-diphosphonate and dichloromethanediphosphonate on lysine hydroxylation and cross link formation in rat bone, cartilage and skin collagen, *Biochem. J.,* 196, 303, 1981.
74. **Gallagher, J. A., Guenther, H. L., and Fleisch, H.,** Rapid intracellular degradation of newly synthesized collagen by bone cells. Effect of dichloromethylenebisphosphonate, *Biochim. Biophys. Acta,* 719, 349, 1982.
75. **Beertsen, W., Niehof, A., and Everts, V.,** Effects of 1-hydroxyethylidene-1,1-bisphosphonate (HEBP) on the formation of dentin and periodontal attachment apparatus in the mouse, *Am. J. Anat.,* 174, 83, 1985.
76. **Guenther, H. L., Guenther, H. E., and Fleisch, H.,** Effects of 1-hydroxyethane-1,1-diphosphonate and dichloromethane-diphosphonate on rabbit articular chondrocytes in culture, *Biochem. J.,* 184, 203, 1979.
77. **Larsson, A.,** The metabolic heterogeneity of glycosaminoglycans of the different zones of epiphyseal growth plate and the effect of ethane-1-hydroxy-1,1-diphosphonate (EHDP) upon glycosaminoglycan synthesis *in vivo, Calcif. Tissue Res.,* 21, 67, 1976.
78. **Pilcyzk, R., Sutcliffe, H., and Martin, T. J.,** Effects of pyrophosphate and diphosphonates on parathyroid hormone- and fluoride-stimulated adenylate cyclase activity, *FEBS Lett.,* 24, 225, 1972.
79. **Gebauer, U., Russell, R. G. G., Touabi, M., and Fleisch, H.,** Effect of diphosphonates on adenosine 3′:5′ cyclic monophosphate in mouse calvaria after stimulation by parathyroid hormone *in vitro, Clin. Sci. Mol. Med.,* 50, 473, 1976.
80. **Felix, R., Bettex, J. D., and Fleisch, H.,** Effects of diphosphonates on the synthesis of prostaglandins in cultured calvarial cells, *Calcif. Tissue Res.,* 33, 549, 1981.
81. **Ohya, K., Yamada, S., Felix, R., and Fleisch, H.,** Effect of bisphosphonates on prostaglandin synthesis by rat bone cells and mouse calvaria in culture, *Clin. Sci.,* 69, 403, 1985.
82. **Evequoz, V., Trechsel, U., and Fleisch, H.,** Effect of bisphosphonates on the production of interleukin 1-like activity by macrophages and its effect on rabbit chrondrocytes, *Bone,* 6, 439, 1985.

83. **Cecchini, M. G., Felix, R., Fleisch, H., and Cooper, P. H.,** Effect of bisphosphonates on proliferation and viability of mouse bone marrow-derived macrophages, *J. Bone Miner. Res.,* 2, 135, 1987.
84. **Stevenson, P. H. and Stevenson, J. R.,** Cytotoxic and migration inhibitory effects of bisphosphonates on macrophages, *Calcif. Tissue Res.,* 38, 227, 1986.
85. **Milhaud, G., Labat, M.-L., and Moricad, Y.,** (Dichloromethene) diphosphonate-induced impairment of T-lymphocyte function, *Proc. Natl. Acad. Sci. U.S.A.,* 80, 4469, 1983.
86. **Labat, M. L., Tzehoval, E., Moricad, Y., Feldmann, M., and Milhaud, G.,** Lack of a T-cell dependent subpopulation of macrophages in (dichloromethylene) diphosphonate-treated mice, *Biomed. Pharmacol.,* 37, 270, 1983.
87. **Francis, M. D. and Slough, C. L.,** Acute intravenous infusion of disodium dihydrogen (1-hydroxyethylidene)diphosphonate: mechanism of acute toxicity, *J. Pharm. Sci.,* 73, 1097, 1984.
88. **Marshall, M. J., Wilson, A. S., and Davie, M. W. J.,** Effects of (3-amino-1-hydroxypropylidene)-1,1-diphosphonate on mouse osteoclast, *J. Bone Miner. Res.,* 5, 955, 1990.
89. **van Rooijen, N. and van Nieuwmegen, R.,** Elimination of phagocytic cells in the spleen after intravenous injection of liposome-encapsulated dichloromethylene diphosphonate, *Cell Tissue Res.,* 238, 355, 1984.
90. **van Rooijen, N., van Nieuwmegen, R., and Kamperdijk, E. W. A.,** Elimination of phagocytic cells in the spleen after intravenous injection of nephrolithiasis (calcium stones) with diphosphonate, *Metabolism,* 23, 1167, 1974.
91. **Ohata, M. and Pak, C. Y.,** Preliminary study on the treatment of nephrolithiasis (calcium stones) with diphosphonate, *Metabolism,* 23, 1167, 1974.
92. **Baumann, J. M., Bisaz, S., Fleisch, H., and Wacker, M.,** Biochemical and clinical effects of ethane-1-hydroxy-1,1-diphosphonate on calcium nephrolithasis, *Clin. Sci. Mol. Med.,* 54, 509, 1974.
93. **Thomas, B. J. and Amstutz, H. C.,** Results of the administration of diphosphonate for the prevention of heterotrophic ossification after total hip arthroplasty, *J. Bone J. Surg.,* 67A, 400, 1985.
94. **Muhlemann, H. R., Bowles, D., Schatt, A., and Bernimoulin, J. P.,** Effect of diphosphonate on human supragingival calculus, *Helv. Odonotol. Acta,* 14, 31, 1970.
95. **Sturzenberger, O. P., Swancar, J. R., and Reiter, G.,** Reduction of dental calculus in humans through the use of a dentifrice containing a crystal-growth inhibitor, *J. Periodontol.,* 42, 416, 1971.
96. **Reiner, M., Sautter, V., Olath, A., Bossi, A., Largiader, U., and Fleisch, H.,** Diphosphonate treatment in myositis ossificans progressiva, in *Etidronate,* Caniggia, A., Ed., Gentili, Pisa, 1980, 237.
97. **Russell, R. C. G., Smith, R., Preston, C., Walton, R. J., and Woods, C. G.,** Diphosphonates in Paget's disease, *Lancet,* 1, 894, 1974.
98. **Siris, E. S., Canfield, R. E., Jacobs, T. P., and Baquiran, D. C.,** Long-term therapy of Paget's disease of bone with EHDP, *Arthritis Rheum.,* 23, 1177, 1980.
99. **Chapuy, M. C., Charhon, S. A., and Meunier, P. J.,** Sustained biochemical effects of short treatment of Paget's disease of bone with dichloromethylene diphosphonate, *Metab. Bone Dis. Relat. Res.,* 4, 325, 1983.
100. **Douglas, D. L., Russell, R. G. G., Preston, C. J., Prenton, M. A., Duckworth, T., Kanis, J. A., Preston, F. E., and Woodhead, J. S.,** Effect of dichloromethylene diphosphonate in Paget's disease of bone and in hypercalcemia due to primary hyperparathyroidism or malignant disease, *Lancet,* 1, 1043, 1980.
101. **Shane, E., Baquiran, D. C., and Bilezikian, J. P.,** Effect of dichloromethylene diphosphonate on serum and urinary calcium in primary hyperthyroidism, *Ann. Intern. Med.,* 95, 23, 1981.

102. **Adami, S., Bertoldo, F., Rossini, M., Jayawerra, P., O'Riordan, J. L. H., and Lo Cascio, V.,** Regulation of calcium-parathyroid hormone feedback in primary hyperparathyroidism: effect of bisphosphonate treatment, *Clin. Endocrinol.,* 33, 391, 1990.
103. **Mundy, G. R.,** Pathophysiology of cancer-associated hypercalcemia, *Semin. Oncol.,* 17 (Suppl. 5), 10, 1990.
104. **Jung, A.,** Comparison of two parenteral diphosphonates in hypercalcemia of malignancy, *Am. J. Med.,* 72, 221, 1982.
105. **Ryzen, E., Martodam, R. R., Troxell, M., Benson, A., Paterson, A., Shepard, K., and Hicks, R.,** Intravenous etidronate in the management of malignant hypercalcemia, *Arch. Intern. Med.,* 145, 449, 1985.
106. **Chapuy, M. C., Meunier, P. J., Alexandre, C. M., and Vignon, E. P.,** Effects of disodium dichloromethylene diphosphonate on hypercalcemia produced by bone metastases, *J. Clin. Invest.,* 65, 1243, 1980.
107. **Siris, E. S., Sherman, W. H., Baquiran, D. C., Schlatterer, J. P., Osserman, E. F., and Canfield, R. E.,** Effects of dichloromethylene diphosphonate on skeletal mobilization of calcium in multiple myeloma, *N. Engl. J. Med.,* 302, 310, 1980.
108. **Siris, E. S., Hyman, G. A., and Canfield, R.,** Effects of dichloromethylene diphosphonate in women with breast carcinoma metastatic to the skeleton, *Am. J. Med.,* 74, 401, 1983.
109. **Thiebaud, D., Jaeger, P., Jacquet, A. F., and Burckhardt, P.,** A single-day treatment of tumor induced hypercalcemia by intravenous amino-hydroxy-propylidene bisphosphonate, *J. Bone Miner. Res.,* 1, 555, 1986.
110. **Elomaa, I., Blomqvist, C., Grohn, P., Porkka, L., Kairento, A. L., Selander, K., Lamberg-Allardt, C., and Holmstrom, T.,** Long-term controlled trial with diphosphonate in patients with osteolytic bone metastases, *Lancet,* 1, 146, 1983.
111. **Singer, F. R.,** Role of the bisphosphonate etidronate in the therapy of cancer-related hypercalcemia, *Semin. Oncol.,* 17, 34, 1990.
112. **Heaney, R. P. and Saville, P. D.,** Etidronate disodium in postmenopausal osteoporosis, *Clin. Pharmacol. Ther.,* 20, 593, 1976.
113. **Steber, J. and Wierich, P.,** Properties of hydroxyethane diphosphonate affecting its environmental fate: degradability, sludge adsorption, mobility in soils, and bioconcentrations, *Chemosphere,* 15, 929, 1986.
114. **Fogelman, I., Smith, L., Mazess, R., Wilson, M. A., and Bevan, J. A.,** Absorption of oral diphosphonate in normal subjects, *Clin. Endocrinol.,* 24, 57, 1986.
115. **Yakatan, G. J., Poynor, W. J., Talbert, R. L., Floyd, B. F., Slough, C. L., Ampulski, R. S., and Benedict, J. J.,** Clondronate kinetics and bioavailability, *Clin. Pharmacol. Ther.,* 31, 402, 1982.
116. **Francis, M. D. and Martodam, R. R.,** Chemical, biochemical, and medicinal properties of the diphosphonates, in *The Role of Phosphonates in Living Systems,* Hilderbrand, R. L., Ed., CRC Press, Boca Raton, FL, 1983, 56.
117. **Bisaz, S., Jung, A., and Fleish, H.,** Uptake of bone of pyrophosphate, diphosphonates and their technetium derivatives, *Clin. Sci. Mol. Med.,* 54, 265, 1978.
118. **Conrad, K. A. and Lee, S. M.,** Clondronate kinetics and dynamics, *Clin. Pharmacol.,* 30, 114, 1981.
119. **Flora, L., Hassing, G. S., Parfitt, A. M., and Villanueva, A. R.,** Comparative skeletal effects of two diphosphonates in dogs, *Metab. Bone Dis. Relat. Res.,* 2 (Suppl.), 389, 1980.
120. **Fleish, H.,** Diphosphonates: history and mechanisms of action, *Metab. Bone Dis. Relat. Res.,* 4/5, 279, 1981.

Chapter 19

MECHANICAL EFFECTS ON OSTEOCLAST FUNCTION

Elisabeth H. Burger

TABLE OF CONTENTS

I. FUNCTIONAL ADAPTATION OF BONE

The major physiological function of osteoclasts is to resorb bone. Osteoclasts also resorb calcified cartilage during endochondral ossification and dentin during eruption of the secondary dentition. Resorption of bone and calcified cartilage occur during modeling of the skeleton, when bone organs reach a functionally significant architecture, which includes their size, shape, mass, and distribution of bone tissue.[21] Remodeling is the process by which existing bone tissue is replaced by new bone in the course of tissue renewal, using basic multicellular units (BMU) as the cellular effector mechanism.[20,22,29,31] The function of the skeleton is to support the body tissues and facilitate movement, and also to protect vital organs from accidental injury. In addition, bone tissue plays a role in mineral metabolism as a store of calcium and phosphate. In fulfilling these tasks, bone, cartilage, and associated fibrous tissues are subjected to repeated stresses and strains during millions of cycles of mechanical loading throughout life. The repeated loadings over a specific period of time constitute the local loading histology of the tissue.[13]

The importance of tissue loading or stress histories in skeletal biology has been well appreciated by several investigators, such as Wolff,[66] D'Arcy Thompson,[14] Frost,[21-24] Carter,[10] and Lanyon,[37] to name a few. Although it is now generally acknowledged that mechanical loading history is an important regulator of bone biology, its specific role and mechanism of action are not understood. This is partly due to a general lack of knowledge as to how bone cells react to mechanical stress, in terms of cell proliferation, specific differentiation, type and rate of matrix synthesis, and the production of local regulatory factors. In the absence of cell biological data, skeletal adaptation to mechanical usage is necessarily dealt with at the anatomical level, in terms which consider bone and cartilage as biomaterials with certain specific biomechanical properties[1] rather than as collections of specific cells with their pertinent extracellular matrices. Recently, however, some *in vitro* experimental evidence has been accumulating on the reactions of isolated osteoblasts and chondrocytes[3,5,26,59,62a,70] and bone tissue and organ cultures[19,28,34-36] to mechanical signals. Also, careful *in vivo* studies[32,40,63] indicate that osteoclasts as well as osteoblasts are involved in the adaptive response to altered loading regimes. Last, some *in vitro* studies have begun to shed some light on the responsiveness of osteoclasts to mechanical loading *in vitro*.[28,34,35,53]

In the following text, these studies are reviewed, and an attempt is made to link their conclusions to the literature on bone adaptation at the anatomical level.

II. ANIMAL STUDIES OF THE ROLE OF OSTEOCLASTS IN THE FUNCTIONAL ADAPTATION OF BONE

Many animal studies have shown that bone mass in long bones decreases when the normal mechanical loading is withdrawn by experimental conditions leading to immobilization and disuse. Sometimes, these conditions involved surgical intervention, such as denervation, tenotomy, and amputation. Others obtained immobilization by casting one of the animal's limbs, or by suspending the animal (rat) by its tail. Eliminating the use of one leg induces overloading of the other, and so both disuse and hyperuse can be inflicted at the same time. By such methods, it has been convincingly shown that bone mass decreases under conditions of disuse and increases when extra mechanical loading is applied.[58] However, careful histomorphometric studies of the changes with time at the tissue level due to under- or overloading are scarce.[30,32,40,61,63] Best documented are the reactions in the rat tibial metaphyseal spongiosa.

In a study of short- and long-term effects of mobilization by casting, the right hindlimb of 9-month-old female rats was immobilized against the abdomen by elastic bandage tapes. The unloaded right and overloaded left tibiae were compared with normal tibiae of age-matched controls.[32,40] Extensive histomorphometry after double labeling with bone markers, as well as single-photon absorptiometry, were performed after 0, 2, 10, 18, and 26 weeks. Unloading caused a negative bone balance from a combination of elevated bone resorption and depressed bone formation. After 2 weeks, the percent of eroded bone surface was elevated 48% compared with age-matched controls, indicating an enhanced osteoclast activity. Enhanced osteoclast activity persisted at 10 weeks (43% increase of percent eroded surface) and fell to control values at 18 weeks.[40] A similar rapid bone loss between 0 and 12 weeks has been described[30] in response to one-legged immobilization by plaster casting in young adult dogs. In addition, in the overloaded (left) tibia of the rat study,[32] the percent of eroded surface was persistently lower over the 26-week time period, indicating that osteoclast activity was reduced. Osteo*blast* activity was persistently enhanced by overloading[32] and reduced by unloading.[30,40] Vico et al.[63] studied the effects of 5 d of weightlessness during space flight on bone mass and osteoclast numbers in 12-week-old pregnant rats. Osteoclasts were counted after staining for tartrate-resistant acid phosphatase (TRAP), which considerably improves their identification in histological sections.[11,62] The only detectable effect of the short exposure to weightlessness was an enhanced number of osteoclasts per square millimeter of trabecular bone surface. This parameter was increased 50 to 90% in the vertebrae of spaceflight rats compared with flight control animals which had been subjected to the stress conditions (noises, vibrations) associated with launch, flight, and reentry on earth, but without actual spaceflight. In the tibial metaphyseal spongiosa, the percentage of eroded surface was more than doubled after spaceflight, but the difference was not significant due to large standard deviations. Unfortunately, the staining regime used for the tibia sections did not allow osteoclast identification, and no data on osteoclast numbers could be provided. Earlier studies had shown no increase of osteoclastic parameters[9,33] after a spaceflight of long duration (20 d).

A study of the effects of 7 to 40 d of tail suspension of adult rats[64,65] showed that osteoclastic parameters (number of osteoclasts per square millimeter of trabecular surface, percentage of eroded surface) were doubled after 1 week, and had almost returned to control values after 2 weeks. Suspension of rats by their tail has been devised to mimick the effect of weightlessness during spaceflight,[69] but the kinetics of bone cell response seem to differ from the response to weightlessness.[65]

Increased osteoclast numbers were observed as early as 30 h after immobilization on the tibia by tenotomy in young adult male rats.[61] In this study, most of the resorptive effect occurred in the first 3 d, while inhibition of prostaglandin synthesis by treatment with indomethacin prevented increased osteoclast activity.[61]

These studies show that sudden experimental disuse of a bone organ leads to an early and transient increase of osteoclastic bone resorption, which is accompanied by a more persistent decrease of osteoblastic activity. However, interpretation of the data requires caution. Surgical intervention and trauma, such as inflicted by tenotomy, generally stimulate local cellular activity,[24] possibly as a result of locally increased levels of cytokines and growth factors shortly after injury. Cytokines such as interleukin-1 (IL-1) and tumor necrosis factors (TNF), and growth factors such as TGF-β have been shown to stimulate bone resorption.[2,25] The phenomenon of injury-induced bone loss must be distinguished from the possible effects of unloading per se, and this problem frustrates most, if not all, *in vivo* studies of the early responses to experimental unloading. Still, a direct response of osteoclasts to a sudden change of the mechanical loading regime of bone tissue cannot be excluded by the present evidence. The careful study of Li et al.[40] is especially convincing, because no surgical trauma was inflicted and because the osteoclastic response persisted for at least 10 weeks. However, closer study of mechanical regulation of osteoclastic activity obviously calls for *in vitro* approaches.

III. *IN VITRO* STUDIES OF OSTEOCLASTIC RESPONSE TO MECHANICAL STRESS

The *in vitro* evidence indicating that osteoclasts may be directly involved in the reaction of bone tissue to mechanical stress is limited. Only a few recent papers have addressed this issue, and many questions are still unanswered. However, some interesting observations have been reported.

Klein Nulend et al.[34] submitted fetal mouse calvaria cultures to intermittent (0.3 Hz, 1 s pressure, 2 s relaxation) hydrostatic compressive force

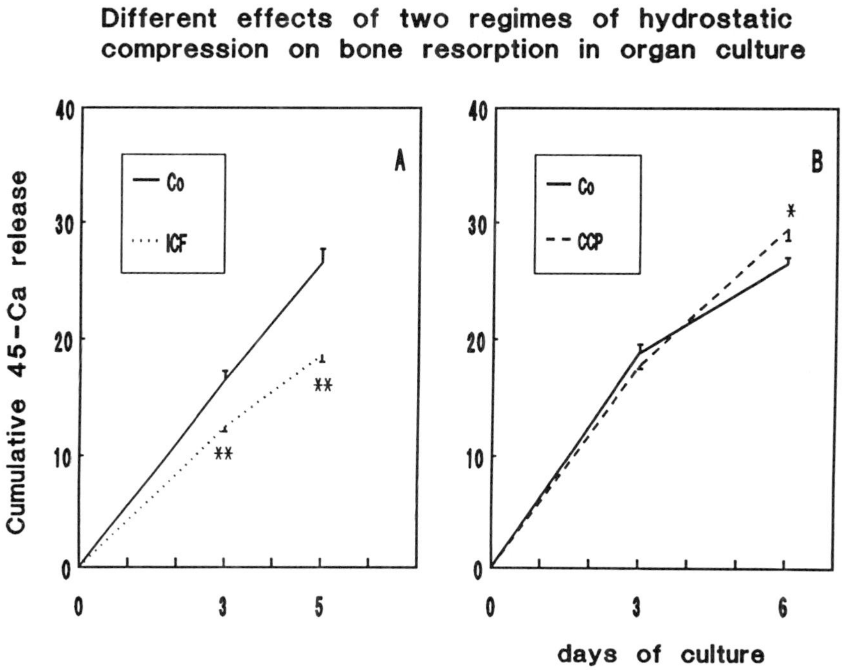

FIGURE 1. Radiolabeled half calvaria of 17-d-old mouse fetuses were subjected to (A) an intermittent (0.3 Hz) compressive hydrostatic force of 13 KPa (ICF) or (B) continuous hydrostatic compression of 100 KPa (CCP), and resorption monitored as the percent label released to the culture medium. 13-KPa ICF inhibited bone resorption after 3 and 5 d, while 100 KPa stimulated resorption after 6 d of culture. Values are means ± SEM; n > 6. * $p < 0.01$; ** $p < 0.001$, Student's *t*-test. (Data in A derived from Reference 34; B, from Reference 28.)

(ICF) of low magnitude (13 KPa or 130 mbar above ambient). This regimen caused a 20 to 30% inhibition of mineral resorption in the calvaria after 3 and 5 d *in vitro* (Figure 1A). At the same time, alkaline phosphatase activity, a parameter of osteoblastic activity, was about doubled.[34] The authors concluded that osteoclasts as well as osteoblasts responded to the mechanical stress, but in opposite directions. By contrast, application of 100 KPa (1000 mbar above ambient) continuous hydrostatic pressure (CCP) to similar mouse calvaria cultures[28] resulted in an *increased* mineral resorption after 4 to 6 d, but not after 2 to 3 d of culture[28] (Figure 1B). CCP enhanced the production of prostaglandins in the first 3 d of culture, which probably led to the increased resorption in the last days of culture.[28]

In both studies, ICF and CCP mechanical pressure was applied by compressing the gas phase of the culture system, leading to hydrostatic pressure in the fluid culture medium. Intermittent compression of 13 KPa also reduced mineral resorption in long-bone organ cultures of fetal mice[6,35] and stimulated mineralization in such bones.[36] The magnitude of the compressive force (13

KPa) was near the calculated maximal physiologic pressure of the fetal long-bone rudiments as a result of muscle contraction.[36] The effects on mineralization due to ICF could not be ascribed to the (small) changes in pO_2 and pCO_2 which occurred during gas phase compression,[36] and were thus ascribed to the mechanical pressure. In the CCP studies, care was taken to maintain the pO_2 and pCO_2 at ordinary tissue culture levels by using gas phases of different O_2, CO_2, and N_2 composition.[44]

The effects of 100 KPa continuous pressure and 13 KPa intermittent pressure also differed in other assays. In 8-d cultures of adult mouse bone marrow, 100 KPa CCP strongly *stimulated* the appearance of mononuclear cells with osteoclast characteristics such as TRAP and calcitonin-binding activity.[28] This effect was biphasic, 100 KPa having much more effect than 50 or 200 KPa, and also appeared to be mediated by increased prostaglandin E_2 (PGE_2) synthesis.[28] Again, 13 KPa ICF had the opposite effect. In 7-d cultures of mouse bone marrow, the number of mononuclear TRAP cells is *decreased* by some 75%. In both studies, multinucleation of TRAP cells was scant.

Intermittent (1-h interval, or 0.003 Hz) hydrostatic pressure of 3 KPa had no direct effect on bone resorption in organ cultures of neonatal mouse calvaria.[53] When applied to mouse bone marrow cultures during the first 4 d of an 8-d culture, 3 KPa at 0.0003 Hz increased the (low) number of multinucleated TRAP-positive cells and also increased their staining intensity.[53]

The inhibiting effect of 13 KPa ICF on bone mineral resorption was strengthened rather than inhibited by treatment with indomethacin,[6] while similar treatment completely inhibited the stimulating effect of 100 KPa CCP on TRAP cell formation and PGE_2 production.[28]

These studies suggest that hydrostatic compression of bone organ or bone marrow cultures may have opposite effects, depending on the magnitude of the pressure and/or whether static or intermittent loading is applied.

At first glance, application of hydrostatic pressure to skeletal tissue *in vitro* seems unphysiological, as stresses *in vivo* are always at least partially composed of shear stress which leads to distortion of the material. Hydrostatic (dilatational) stress causes a change in the volume of compressible materials, but cartilage and bone are nearly incompressible. However, in organ cultures containing mineralized tissue as well as nonmineralized periosteum or cartilage, the material mismatch across a calcified/noncalcified tissue interface causes local shear stresses and strain even when ambient hydrostatic pressure is applied.[67] It seems possible that local shear stress may also be generated at the fluid phase/plastic substrate interface of a monolayer cell culture, but this assumption needs further theoretical analysis.[54]

Differences between the effects of continuous vs. intermittently applied stress are reported in several studies. Sandy et al.[54] found that mechanical deformation of the culture dish bottom increased the production of bone-resorbing factors by mouse calvarial osteoblasts grown as monolayers. The culture dishes were deformed over a convex surface, producing an increase in surface area of approximately 5%. Intermittent deformation (0.2 Hz or every 5 s) was far more effective than continuous deformation. Continuous application of 13 KPa stimulated mineralization of long-bone organ cultures, but only up to one third of the effect of intermittent 13-KPa pressure.[36] In a study of bone remodeling in functionally deprived avian ulna preparations, a static load of physiological magnitude failed to protect the bone tissue against bone loss resulting from disuse, while a similar load applied intermittently (100 cycles per day) not only prevented disuse osteoporosis, but even increased the total bone mass of the preparation.[39] In addition, Meade et al.[42a] reported that a compressive force continuously applied *in vivo* to the diaphysis of adult mongrel dogs strikingly suppressed bone maturation. In a study of the effect of compressive force on organ cultures of cartilaginous mandibular condyles of neonatal rats, continuously applied force *reduced* the synthesis of sulfated glycosaminoglycans and collagen.[12] However, a *stimulatory* effect was observed by intermittently applied (0.7 Hz) compression, but only within the range of 0.5 to 1.0 g. This same force regimen increased alkaline phosphatase activity in the cultures, while continuously applied force decreased it.

Apart from the difference between static and dynamic application of force, the magnitude of the force seems to be important. Murray and Rushton[43] devised a method for subjecting calvarial cells to cyclic strains *in vitro* by stretching the plastic substrate on which they grew, and studied PGE_2 production in response to stretching. Their method of force application, cyclic stretching of the bottom of the culture well, allowed characterization of the applied stress by the deformation or strain which it accomplished. This allowed the cells to be subjected to known strains of physiological as well as pathological magnitude. PGE_2 production was not enhanced by a physiological strain of 3 millistrain,* but did increase when the cells were stretched to 7 and 10 millistrain,[43] in accordance with earlier studies.[59] In weight-bearing bones of a range of animals, the peak strains during strenuous activity were between 2 and 3 millistrain,[52] and the physiological range of strains, associated with normal activity but not microstructural bone damage, extends up to 5 millistrain. So, increased PGE_2 synthesis after 7 or more millistrain seems to be a cellular response to pathological rather than physiological loading. In the study of Sandy et al.,[54] mechanical deformation produced an increase in surface area of approximately 5%, which means a deformation in one dimension of about 2.5%, or 25 millistrain. This strain increased the production of bone-resorbing factors of high and low molecular weight. Although hydrostatic compression may induce shear stress in organ culture, the amount of local distortion or strain in the tissue is unknown. This has to do with the complex situation in a bone organ culture which contains various tissues (bone,

* $$1 \text{ millistrain} = \frac{\text{change in length}}{\text{initial length}} \times 10^{-3}.$$

connective tissue, and cartilage) and matrices (mineralized/nonmineralized) of different material properties. Based on studies of muscle force in humans, 13 KPa was calculated as being near the maximal pull in the mouse fetus.[36] If this assumption also holds for the calvarial anlagen in the same fetus, then 100 KPa, as used in the CCP study, would seem to mimick the effect of pathological loading rather than physiologic stress. In addition, 3 KPa, as used in the study of Saito et al.,[53] may have been too low to evoke a response in the tissue. It is of interest that Murray and Rushton[43] found a biphasic dose-response relationship between strain magnitude and PGE_2 release by calvarial cells. At 7 and 28 millistrain, the PGE_2 release was significantly enhanced (twofold), but not at 3 and 14 millistrain. A similar biphasic effect was found in the CCP study: 1 atmosphere above ambient, or 100 KPa, strongly (13-fold) increased the number of TRAP cells in bone marrow cultures, as well as PGE_2 synthesis, but 50 and 200 KPa had no significant effect.

In conclusion, although the data are still scant and no firm conclusions may be drawn, the available evidence indicates that the effects of static vs. dynamic stress are different. Prostaglandin-mediated catabolic responses are associated with stress of pathologic rather than physiologic magnitude, and may be related to repair processes in response to bone damage. Intermittent stress of physiological magnitude seems to be able to evoke an anabolic response in functionally deprived bone tissue such as the isolated avian ulna and organ cultures of fetal and neonatal bone rudiments. Osteoclasts seem to partake directly in both the catabolic and the anabolic responses of the tissue, by enhanced formation and activity in the catabolic response and decreased formation and activity in the anabolic response.

IV. TRANSDUCTION OF MECHANICAL SIGNALS TO OSTEOCLASTS

Nothing definitive is known about the mechanism of transduction of mechanical signals. In principle, two pathways may be considered. In the first, osteoclastic or preosteoclastic cells are directly stimulated by mechanical perturbation of the environment, which means that the cells must have some sort of mechanoreceptors. In the other pathway, osteoclasts react indirectly to mechanical stress, via signals derived from the bone matrix or from osteoblasts, osteocytes, or calcifying chondrocytes.[49] Osteoblasts are highly sensitive to fluid flow induced by mechanical stress.[49] Mechanosensitive ion channels have recently been described in osteoblast-like cells.[15,18] In addition, voltage-activated K^+ channels have been described in freshly isolated embryonic chicken osteoclasts,[48] but it is unknown if these channels are stretch activated, in which case they might act as mechanoreceptors. Application of 5 millistrain to cancellous bone core biopsies of adult dog femora induced an increase in RNA synthesis in osteocytes after 6 h and a rise in intracellular

G6PD activity in lining cells immediately after loading.[19] Increased RNA synthesis in osteocytes was also described 24 h after osteogenic loading *in vivo*.[45] In addition, within 5 min after loading, a strain magnitude-related increase in enzyme activity was shown in osteocytes *in vivo*.[57] This evidence indicates that osteocytes, which by their location within the mineralized matrix are ideally situated to perceive local effects of stress, do indeed react to mechanical loading.

The few *in vitro* studies of mechanical stress and bone resorption provide some evidence that the resorptive response is mediated by soluble factors, which may have derived from nonosteoclastic cells. Mouse calvarial osteoblasts in monolayer culture constitutively produced bone-resorbing factors; mechanical deformation stimulated this activity of the osteoblasts.[54] Fractionation of the culture medium resolved three peaks of bone-resorbing activity, two of which were stimulated by mechanical deformation.[54] The inhibiting effect of 13 KPa ICF on resorption of long-bone rudiments[35] could be transferred via the conditioned medium also inhibited the appearance of TRAP cells in cultures of bone marrow. Continuously applied 100 KPa CCP markedly increased PGE_2 (but not IL-1) in the culture medium of mouse calvaria.[28] CCP seemed to stimulate bone resorption by generating new osteoclasts through a mechanism involving PGE_2 production. These studies provide the first evidence for transduction of mechanical signals to osteoclasts, and suggest that the osteoclastic response is at least partly a reaction to signals generated by skeletal tissue under stress. However, proof that any of the two pathways is involved in functional adaptation of bone to stress is lacking, and any conclusion in this regard remains speculative.

V. MECHANICAL VS. HORMONAL REGULATION OF OSTEOCLASTS

Bone metabolism is regulated by systemic circulating hormones and by local growth factors and cytokines.[8,17,42,60] The mechanism by which these chemical factors influence osteoclasts is far better documented than the mechanism of mechanical modulation. This undoubtedly has to do with the difficulty of applying mechanical perturbation to osteoclasts *in vitro*, compared to adding chemical regulators to a tissue culture medium.

Several *in vivo* studies have combined the effects of mechanical and hormonal signals. Some have evaluated the course of bone loss in thyroidectomized and parathyroidectomized animals during immobilization.[7,41] In dogs, immobilization of a limb *increased* the effectiveness of parathyroid hormone (PTH) in inducing bone resorption. However, parathyroidectomy and thyroidectomy prevented development of osteopenia subsequent to limb immobilization.[6] In addition, bone loss of the immobilized femur and tibia of thyroparathyroidectomized rats was less pronounced than in immobilized intact rats.[41] Calcium accretion rate was also reduced in the nonimmobilized limb of the parathyroidectomized rats. From this, Lindgren[41] concluded that the hormone deprivation generally slowed down bone turnover, which led to a slower bone loss after immobilization.

In a study of local differences in tissue response to generalized increases of serum vitamin D_3 and parathyroid hormone in calcium-deficient rats, Liu and Baylink[42] showed that osteoclastic responses to calcium stress may vary, depending on local mechanical demands. After 12 d of calcium deprivation, a five- to sixfold increase was observed in osteoclast numbers at the endosteal surface of alveolar bone of the mandible. However, osteoclast numbers in the bony socket of the teeth (incisor as well as first molar) did not change, although their number was originally higher than in the alveolar endosteum. A similar differential response in alveolar bone to a systemic hormonal change has been described by Baron and Saffar.[4] Long-term calcitonin treatment decreased the number of osteoclasts in the alveolar wall of the tooth socket, but not at the endosteal surface of the alveolar bone. The difference in response is in agreement with the larger supportive role of the bone of the tooth sockets, compared to the trabecular bone of the mandible. This protected the bony socket against osteoclastic attack. Lanyon et al.[38] used the externally loadable ulna preparation of hen turkeys to study the degree of cortical bone loss associated with calcium insufficiency alone, calcium insufficiency together with disuse, and calcium insufficiency modified by a single, short period of external loading per day. The loss of bone attributable to calcium insufficiency and that attributable to disuse were additive, suggesting that even under conditions leading to systemic bone loss, functional activity provides a protective influence on bone mass. In addition, a single, short, daily period of loading, using physiological strains and strain rates, partially reduced the resorptive effect of calcium insufficiency plus disuse.

These studies provide good evidence that bone-resorbing hormones interact with local conditions of use or disuse in determining the extent of osteoclastic bone resorption. The cellular effector mechanism involved in this interplay is as yet unknown, although interaction of osteoblasts and osteoclasts seems likely.

VI. CONCLUSIONS

The recent reports on osteoclastic response to mechanical stress *in vitro* strongly support the conclusion that osteoclasts are able to respond to mechanical signals. It is still unclear if osteoclasts or their progenitors possess mechanoreceptors, but ion channels have recently been demonstrated in isolated chicken osteoclasts. Several studies show that mechanical perturbation of osteoblasts, or other skeletal cells such as periodontal fibroblasts or fetal cartilage, may produce factors affecting osteoclast function. This indicates that mechanical signals are transmitted from osteogenic to osteoclastic cells, similar to hormonal signals such as PTH and vitamin D.[27,46,68]

In vivo and *in vitro* studies agree that intermittent stimulation with a stress of physiological magnitude inhibits osteoclastic bone resorption, while higher stress of pathological magnitude stimulates resorption. Below a certain threshold, mechanical signals have no effect. Intermittent and continuous signals gave different results in all assays used, both *in vitro* and *in vivo*. Exploration of the dose-response relationship between stress magnitude, strain rates, and cycle type on the one hand, and osteoclast generation and function on the other, is an important issue for future research. In view of the increasing evidence that mechanical training may prevent osteoporosis associated with aging,[55,56] knowledge of the cellular basis of mechanical adaptation of the skeleton is urgently needed.

REFERENCES

1. **Ascenzi, A., Improta, S., Portigliatti Barbos, M., Carando, S., and Boyde, A.,** Distribution of lamellae in humoral shafts deformed by bending with inferences on mechanical properties, *Bone,* 8, 319, 1987.
2. **Bertolini, D. R., Nedwin, G. E., Bringman, T. S., Smith, D. D., and Mundy, G. R.,** Stimulation of bone resorption and inhibition of bone formation *in vitro* by human tumor necrosis factors, *Nature (London),* 319, 516, 1986.
3. **Binderman, I., Zor, U., Kaye, A. M., Shisushoni, Z., Harell, A., and Somjen, D.,** The transduction of mechanical force into biochemical events in bone cells may involve activation of phospholipase A_2, *Calcif. Tissue Int.*, 42, 261, 1988.
4. **Baron, R. and Saffar, J. L.,** A quantitative study on the effects of prolonged calcitonin treatment on alveolar bone remodeling in the gold hamster, *Calcif. Tissue Res.*, 22, 265, 1977.
5. **Buckley, M. J., Banes, A. J., Levin, L. G., Sumpio, B. E., Sato, M., Jordan, R., Filbert, J., Link, G. W., and Tran Son Tay,, R.,** Osteoblasts increase their rate of division and align in response to cyclic, mechanical tension *in vitro, Bone Miner.*, 4, 225, 1988.
6. **Burger, E. H., Klein Nurlend, J., and Veldhuijzen, J. P.,** Modulation of osteogenesis in fetal bone rudiments by mechanical stress *in vitro, J. Biomech.*, 245, 101, 1991.
7. **Burkhart, J. M. and Jowsey, J.,** Parathyroid and thyroid hormones in development of immobilization and osteoporosis, *Endocrinology,* 81, 1053, 1967.
8. **Canalis, E., McCarthy, T., and Centrella, M.,** The regulation of bone formation by local growth factors, in *Bone and Mineral,* Vol. 6, Peck, W. A., Ed., 1989, 27.
9. **Cann, E. C. and Adachi, R. R.,** Bone resorption and mineral excretion in rats during space flight, *Am. J. Physiol.*, 13, 327, 1983.
10. **Carter, D. R.,** Mechanical loading history and skeletal biology, *J. Biomech.*, 20, 1095, 1987.
11. **Chappard, D., Alexandre, C., and Riffat, A.,** Histochemical identification of osteoclasts. Review of current methods and reappraisal of a simple procedure for routine diagnosis on undecalcified human iliac bone biopsies, *Basic Appl. Histochem.*, 27, 75, 1983.
12. **Coprey, J. C. V. M., Jansen, H. W. B., and Duterloo, H. S.,** Effects of compressive forces on proliferation and matrix synthesis in mandibular condylar cartilage of the rat *in vitro, Arch. Oral Biol.*, 30, 305, 1984.

13. **Cowin, S. C.,** *Bone Mechanics,* CRC Press, Boca Raton, FL, 1988.
14. **D'Arcy Thompson, W.,** *On Growth and Form,* University Press, Cambridge, 1942.
15. **Davidson, R. M., Tatakis, D. W., and Auerbach, A. L.,** Multiple forms of mechanosensitive ion channels in osteoblast-like cells, *Pflügers Arch.,* 416, 646, 1990.
16. **Delling, G., Schäfer, A., Schleicher, H. J., and Ziegler, R.,** The effect of calcitonin on disuse atrophy of bone in the rat, *Calcif. Tissue Res.,* 6, 143, 1970.
17. **Dieudonné, S., Foo, P., Van Zoelen, J., and Burger, E. H.,** Inhibiting and stimulating effects of TGF-beta1 on osteoclastic bone resorption in fetal mouse bone organ cultures, *J. Bone Miner. Res.,* 6, 479, 1991.
18. **Duncan, R. and Misler, S.,** Voltage-activated and stretch-activated Ba^{2+} conducting channels in an osteoblast-like cell line (UMR 106), *FEBS Lett.,* 251, 17, 1989.
19. **El Haj, A. J., Minter, S. L., Rawkinson, S. C. F., Suswillo, R., and Lanyon, L. E.,** Cellular responses to mechanical loading *in vitro, J. Bone Miner. Res.,* 5, 923, 1990.
20. **Frost, H. M.,** *Mathematical Elements of Lamellar Bone Remodeling,* Charles C Thomas, Springfield, IL, 1964.
21. **Frost, H. M.,** Skeletal structural adaptations to mechanical usage (SATMU). I. Redefining Wolff's law: the bone modeling problem, *Anat. Rec.,* 226, 403, 1990.
22. **Frost, H. M.,** Skeletal structural adaptations to mechanical usage (SATMU). II. Redefining Wolff's law: the remodeling problem, *Anat. Rec.,* 226, 414, 1990.
23. **Frost, H. M.,** Vital biomechanics: some general concepts for skeletal adaptations to mechanical usage, *Calcif. Tissue Int.,* 42, 145, 1987.
24. **Frost, H. M.,** *Intermediary Organization of the Skeleton,* Vol. 1, 2, CRC Press, Boca Raton, FL, 1986.
25. **Gowen, M., Wood, D. D., Ihrie, E. J., McGuire, M. K. B., and Russell, R. G. G.,** An interleukin 1-like factor stimulates bone resorption *in vitro, Nature (London),* 306, 378, 1983.
26. **Hasegawa, S., Sato, S., Saito, S., Suzuki, Y., and Brunette, D. M.,** Mechanical stretching increases the number of cultured bone cells synthesizing DNA and alters their pattern of protein synthesis, *Calcif. Tissue Int.,* 37, 431, 1985.
27. **Heath, J. K., Atkinson, S. J., Meikle, M. C., and Reynolds, J. J.,** Mouse osteoblasts synthesize collagenase in response to bone resorbing agents, *Biochim. Biophys. Acta,* 802, 151, 1984.
28. **Imamura, K., Ozawa, H., Hiraide, T., Takahashi, N., Shibasaki, Y., Fukuhara, T., and Suda, T.,** Continuously applied compressive pressure induces bone resorption by a mechanism involving prostaglandin E_2 synthesis, *J. Cell Physiol.,* 144, 222, 1990.
29. **Jaworski, Z. F. G.,** Lamellar bone turnover system and its effector organ, *Calcif. Tissue Int.,* 36, S46, 1984.
30. **Jaworski, Z. F. G. and Uhthoff, H. K.,** Disuse osteoporosis: current status and problems, in *Current Concepts of Bone Fragility,* Uhthoff, H. K., Ed., Springer-Verlag, Berlin, 1986, 181.
31. **Jee, W. S. S.,** The skeletal tissues, in *Cell and Tissue Biology: A Textbook of Histology,* 6th ed., Weiss, L., Ed., Urban and Schwarzenberg, Baltimore, 1988, 211.
32. **Jee, W. S. S. and Li, X. J.,** Adaptation of cancellous bone to overloading in the adult rat: a single photon absorptiometry and histomorphometry study, *Anat. Rec.,* 227, 418, 1990.
33. **Jee, W. S. S., Wronski, T. J., Morey, E. R., and Kimmel, D. B.,** Effects of space flight on trabecular bone in rats, *Am. J. Physiol.,* 13, 310, 1983.
34. **Klein Nulend, J., Veldhuijzen, J. P., de Jong, M., and Burger, E. H.,** Increased bone formation and decreased bone resorption in fetal mouse calvaria as a result of intermittent compressive force *in vitro, Bone Miner.,* 2, 441, 1987.
35. **Klein Nulend, J., Veldhuijzen, J. P., van Strien, M. E., de Jong, M., and Burger, E. H.,** Inhibition of osteoclastic bone resorption by mechanical stimulation *in vitro, Arthritis Rheum.,* 33, 66, 1990.

36. **Klein Nulend, J., Veldhuijzen, J. P., and Burger, E. H.,** Increased calcification of growth plate cartilage as a result of compressive force *in vitro, Arthritis Rheum.*, 29, 1002, 1986.
37. **Lanyon, L. E.,** Functional strain as a determinant for bone remodeling, *Calcif. Tissue Int.*, 36, S56, 1984.
38. **Lanyon, L. E., Rubin, C. T., and Baust, G.,** Modulation of bone loss during calcium insufficiency by controlled dynamic loading, *Calcif. Tissue Int.*, 38, 209, 1986.
39. **Lanyon, L. E. and Rubin, C. T.,** Static versus dynamic loads as an influence on bone remodeling, *J. Biomech.*, 17, 897, 1984.
40. **Li, X. J., Jee, W. S. S., Chow, S.-Y., and Woodbury, D. M.,** Adaptation of cancellous bone to aging and immobilization in the rat: a single photon absorpitometry and histomorphometry study, *Anat. Rec.*, 227, 12, 1990.
41. **Lindgren, J. U.,** Studies of the calcium accretion rate of bone during immobilization in intact and thyroparathyroidectomized adult rats, *Calcif. Tissue Res.*, 22, 41, 1976.
42. **Liu, C. C. and Baylink, D. J.,** Differential response in alveolar bone osteoclasts residing at two different bone sites, *Calcif. Tissue Int.*, 36, 182, 1984.
42a. **Meade, J. B., Cowin, S. E., Klawitter, J. J., van Buskirk, W. C., and Skinner, H. B.,** Bone remodeling due to continuously applied loads, *Calcif. Tissue Int.*, 36, S25, 1984.
43. **Murray, D. W. and Rushton, N.,** The effect of strain on bone cell prostaglandin E_2 release: a new experimental model, *Calcif. Tissue Int.*, 47, 35, 1990.
44. **Ozawa, H., Imamura, K., Abe, E., Takahashi, N., Hiraide, T., Shibasaki, Y., Fukuhara, T., and Suda, T.,** Effect of a continuously applied compressive pressure on mouse osteoblast-like cells (MC3T3-E1) *in vitro, J. Cell. Physiol.*, 142, 177, 1990.
45. **Pead, M. J., Suswillo, R., Skerry, T. M., Vedi, S., and Lanyon, L. E.,** Increased ^3H uridine levels on osteocytes following a single short period of dynamic loading *in vivo, Calcif. Tissue Int.*, 43, 92, 1988.
46. **Perry, H. M., Skogen, W., Chappel, J. C., Wilner, G. D., Kahn, A. J., and Teitelbaum, S. L.,** Conditioned medium from osteoblast-like cells mediates parathyroid hormone induced bone resorption, *Calcif. Tissue Int.*, 40, 298, 1987.
47. **Raisz, L. G. and Lorenzo, J. A.,** Interaction of hormones, ions and drugs in the regulation of osteoclastic bone resorption, *Adv. Exp. Med. Biol.*, 128, 579, 1980.
48. **Ravesloot, J. H., Ypey, D. L., Vrijheid-Lammers, T., and Nijweide, P. J.,** Voltage-activated K^+ conductances in freshly isolated embryonic chicken osteoclasts, *Proc. Natl. Acad. Sci. U.S.A.*, 86, 6821, 1989.
49. **Reich, K. M., Gay, C. V., and Frangos, J. A.,** Fluid shear stress as a mediator of osteoblast cyclic adenosine monophosphate production, *J. Cell. Physiol.*, 143, 100, 1990.
50. **Rubin, C. T.,** Skeletal strain and the functional significance of bone architecture, *Calcif. Tissue Int.*, 36, S11, 1984.
51. **Rubin, C. T. and Lanyon, L. E.,** Regulation of bone formation by applied dynamic loads, *J. Bone J. Surg. Am. Vol.*, 66, 397, 1984.
52. **Rubin, C. T. and Lanyon, L. E.,** Regulation of bone mass by mechanical strain magnitude, *Calcif. Tissue Int.*, 37, 411, 1985.
53. **Saito, S., Ngan, P., Rosol, T., Saito, M., Shimizu, H., Shinjo, N., Shanfeld, J., and Davidovitch, Z.,** Involvement of PGE synthesis in the effect of intermittent pressure and Interleukin-1 beta on bone resorption, *J. Dent. Res.*, 70, 27, 1991.
54. **Sandy, J. R., Meghji, S., Scutt, A. M., Harvey, W., Harris, M., and Meikle, M. C.,** Murine osteoblasts release bone-resorbing factors of high and low molecular weights: stimulation by mechanical deformation, *Bone Miner.*, 5, 115, 1989.
55. **Silbermann, M., Bar-Shira-Maymon, B., Coleman, R., Reznick, A., Weisman, Y., Steinhagen-Thiessen, E., von der Mark, H., and von der Mark, K.,** Long-term physical exercise retards trabecular bone loss in lumbar vertebrae of aging female mice, *Calcif. Tissue Int.*, 46, 80, 1990.

56. **Simkin, A., Ayalon, J., and Leichter, I.,** Increased trabecular bone density due to bone-loading exercises in postmenopausal osteopenic women, *Calcif. Tissue Int.*, 40, 59, 1987.
57. **Skerry, T. M., Bitensky, L., Chayen, J., and Lanyon, L. E.,** Early strain related changes in enzyme activity in osteocytes following bone loading *in vivo, J. Bone Miner. Res.*, 4, 783, 1989.
58. **Smith, E. L. and Gilligan, C.,** Mechanical forces and bone, in *Bone and Mineral Research,* Vol. 6, Peck, W. A., Ed., Elsevier, Amsterdam, 1989, 139.
59. **Somjen, D., Binderman, I., Berger, E., and Harell, A.,** Bone remodeling induced by physical stress is prostaglandin E_2 mediated, *Biochim. Biophys. Acta,* 627, 91, 1980.
60. **Tashian, A. H., Voelkel, E. F., Lazarro, M., Singer, F. R., Roberts, A. B., Derynck, R., Winkler, M. E., and Levine, L.,** Alpha and beta human transforming growth factors stimulate prostaglandin production and bone resorption in cultures of mouse calvaria, *Proc. Natl. Acad. Sci. U.S.A.*, 82, 4535, 1985.
61. **Thompson, D. D. and Rodan, G. A.,** Indomethacin inhibition of tenotomy-induced bone resorption in rats, *J. Bone Miner. Res.*, 3, 409, 1988.
62. **Van de Wijngaert, F. P. and Burger, E. H.,** Demonstration of tartrate resistant acid phosphatase in undecalcified glycolmethacrylate embedded mouse bone: a possible marker for (pre)osteoclast identification, *J. Histochem. Cytochem.*, 34, 1317, 1986.

62a. **Van Kampen, G. P. J., Veldhuijzen, J. P., Kuijer, R., Van der Stadt, R. J., and Schipper, C. A.,** Cartilage response to mechanical force in high-density chondrocyte cultures, *Arthritis Rheum.*, 28, 419, 1985.

63. **Vico, L., Chappard, D., Alexandre, C., Palle, S., Minaire, P., Riffat, G., Novikov, V. E., and Bakulin, A. V.,** Effects of weightlessness on bone mass and osteoclast number in pregnant rats after a five-day spaceflight (Cosmos 1514), *Bone,* 8, 95, 1987.
64. **Vico, L. and Alexandre, C.,** Normalization of bone cellular responses occurs between 7 and 14 days of simulated weightlessness in rats, *Physiologist,* 32S, 25, 1989.
65. **Vico, L., Bourrin, S., Very, J. M., Chappard, D., and Alexandre, C.,** Bone adaptation to real and simulated microgravity, *ESA-SP,* 307, 359, 1990.
66. **Wolff, J.,** *The Law of Bone Remodeling* (transl.), Springer-Verlag, Berlin, 1986.
67. **Wong, M. and Carter, D. R.,** Theoretical stress analysis of organ culture osteogenesis, *Bone,* 11, 127, 1990.
68. **Wong, G. L.,** Paracine interactions in bone-secreted products of osteoblasts permit osteoclasts to response to parathyroid hormone, *J. Biol. Chem.*, 259, 4019, 1984.
69. **Wronski, T. J. and Morey-Holton, E. R.,** Skeletal responses to stimulated weightlessness: a comparison of suspension techniques, *Aviat. Space Environ. Med.*, 58, 63, 1987.
70. **Yeh, C. and Rodan, G. A.,** Tensile forces enhance prostaglandin E synthesis in osteoblasts grown on collagen ribbons, *Calcif. Tissue Int.*, 36, S67, 1984.

Chapter 20

OSTEOPETROSIS

Lois K. Osier and Sandy C. Marks, Jr.

TABLE OF CONTENTS

I. INTRODUCTION

This chapter focuses on congenital mammalian osteopetroses as illustrations of compromised osteoclast development and function. After a brief introduction to mammalian osteopetroses and to osteoclast development and function, we discuss osteoclast ontogeny, activation, and function from the perspectives provided by experimental investigations of osteopetrotic mutations. These observations are summarized in a brief, tentative review of each mammalian mutation's position within the cascade of osteoclast biology. Finally, we discuss a few unexplored frontiers in the biology of osteopetrosis and, thus, osteoclasts. Of particular interest is that the intricate details of osteoclast biology are interrupted at different sites by different mutations. We emphasize the unique perspectives which each mutation provides, some of which are known and others remain to be discovered. The more we learn about osteoclasts, the more we will value these vignettes provided by the osteopetroses. Reviews of osteopetrosis with different emphases have appeared,[72,74,81] and the interested reader is referred to them.

II. OVERVIEW OF OSTEOPETROSIS

Osteopetrosis is broadly defined as a condition of skeletal sclerosis. It can be induced in birds by a viral infection of osteoblasts,[131] and in mammals by treatment with diphosphonates,[91] low-dose daily injections of parathyroid hormone,[127] and by marrow ablation followed by transplantation of hemopoietic cells from some osteopetrotic animals.[69,155] These acquired osteoscleroses develop gradually and are of different etiologies. The avian and parathyroid hormone (PTH)-induced forms are caused by marked elevations in bone formation,[86,113] while those that appear after diphosphonate treatment and transplantation result from reduced osteoclastic activity.[23,69,155] Congenital osteopetrosis in mammals is a condition of generalized skeletal sclerosis inherited in animals as an autosomal recessive trait and in human beings in an autosomal recessive or autosomal dominant manner.[72] These congenital mammalian osteopetroses are the focus of this chapter because each appears to be a different illustration of intercepted osteoclast development or function.

Congenital osteopetrosis in mammals exhibits pathognomonic radiographic features (Figure 1) which include dense bones without marrow spaces and delayed appearance of some ossification centers. Microscopically (Figure 2), the absence of marrow cavities in long bones is characteristic, and is produced by failure to resorb the primary bone of endochondral origin. Upon examination at higher magnification, osteoclasts may or may not be obvious, and in any case, they fail to resorb bone in a manner that is timely or quantitatively adequate.

Additional similarities among osteopetrotic mutations in animals include abnormal size and shape of bones, susceptibility to pathologic fracture, re-

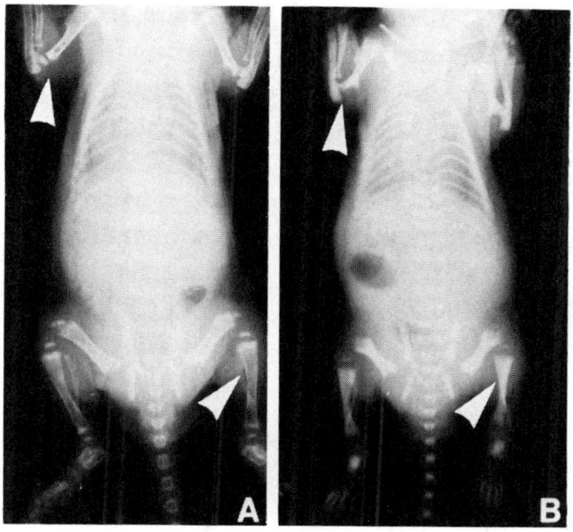

FIGURE 1. Osteopetrotic mutations exhibit a characteristic, generalized skeletal sclerosis radiographically, as seen in these X-rays of a 1-week-old toothless (tl) rat (B). Compared to its normal littermate (A), the tl skeleton (B) exhibits a generalized radioopacity best seen in the long bones of each limb (arrowheads). Reproduced actual size.

sistance to the hypercalcemic effects of PTH[70,150,151] and 1,25-dihydroxyvitamin D (1,25(OH)$_2$D),[104] reduced bone resorption,[74] and elevated blood levels of 1,25(OH)$_2$D.[162]

Mutations differ with respect to lifespan (some mutations are lethal within weeks of birth), blood calcium and phosphorus levels, bone formation,[81] presence and degree of development of the ruffled border is osteoclasts,[74] and numbers of osteoclasts in several skeletal sites.[84] Some hypocalcemic mutations also have rickets.[11] Finally, mutations differ in their response to the most significant experimental manipulation of osteopetroses, bone marrow transplantaion.[152-156] Only four of the eight mutations in rats, mice, and rabbits are cured by stem cell transplantation from a normal littermate.[73,74] These observations, considered together, underscore the remarkable heterogeneity among osteopetroses. As we shall demonstrate, a pathogenetic heterogeneity accounts for much of these variations in biochemical and cellular phenotype.

Human osteopetrosis was first described in an adult male by the German radiologist Albers-Schönberg in 1904. There have been several hundred reports in the clinical literature since then and a few clinical studies. Most of these, by necessity, have been descriptive, since little clinical experimentation can be justified without data from other sources. The discovery of osteopetrotic mutations in mice and rats 40 years later provided replicas of the disease for experimental studies, which only began with the work of Barnicot[12] and Walker.[150,151] Much of what we have learned in animals has been confirmed and used clinically in children, but there are unique features of some human osteopetroses not found in animals to date. These points warrant a brief review

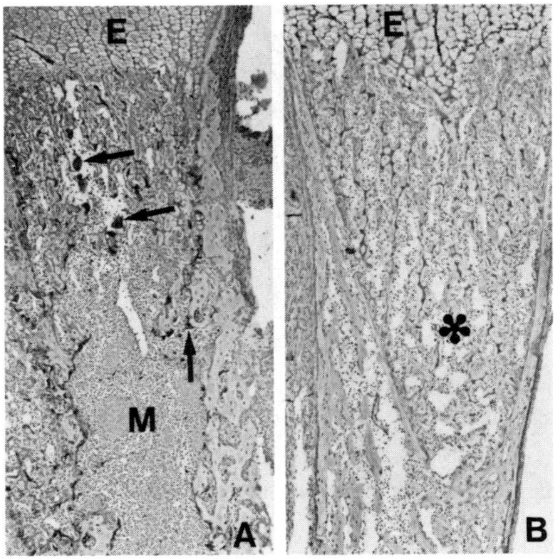

FIGURE 2. Photomicrographs of the proximal tibial metaphysis of a 1-d-old normal rat (A) and its osteopetrotic littermate (B) stained for tartrate-resistant acid phosphatase (TRAP) and counterstained with toluidine blue. A marrow cavity (M), the product of bone resorption, is present in normal rats (A), but not in those with osteopetrosis (B). The persistance of endochondral (*) and periosteal bone produces the characteristic "bone-within-a-bone" appearance in radiographs or histologic sections of long bones of osteopetrotic mutants (B). In this particular mutation (microphthalmia), one of four in rats, osteoclasts (arrows) are more numerous in normal rats (A) and stain more intensely for TRAP than in mutants (B). Marrow spaces eventually develop in this mutation, but metaphyses remain sclerotic throughout life. E, proximal epiphyseal plate. (Original magnification × 50.)

to emphasize the heterogeneity of all the osteopetroses and the unique features of some human mutations.

Human osteopetroses comprise a spectrum of diseases of recessive and dominant inheritance. These exhibit variations in osteoclast parameters[129,130,149] and response to bone marrow transplantation.[17,26,55,135] In children, blood levels of 1,25$(OH)_2$D tend to be elevated,[38,59,95] and some, but not all, show a skeletal response to high-dose 1,25$(OH)_2$D therapy.[57,59] A transient form of neonatal osteopetrosis has recently been described.[95] In adults, osteopetrosis is usually of dominant inheritance, and two distinct forms of autosomal dominant osteopetrosis have been described by radiographic and biochemical criteria.[18,19] It is becoming increasingly appreciated that human osteopetroses exhibit considerable clinical and pathogenetic heterogeneity.[16,18,50,52,53,67] In addition, a recessively inherited form of osteopetrosis occurs in children, the result of a deficiency of carbonic anhydrase II (CA II), the only isoenzyme in osteoclasts.[132,143,146] But even this illustration of the importance of CA II in osteoclast function exhibits considerable variation in clinical presentation.[137] Thus, the hallmark of osteopetrosis in laboratory animals and human beings is heterogeneity.

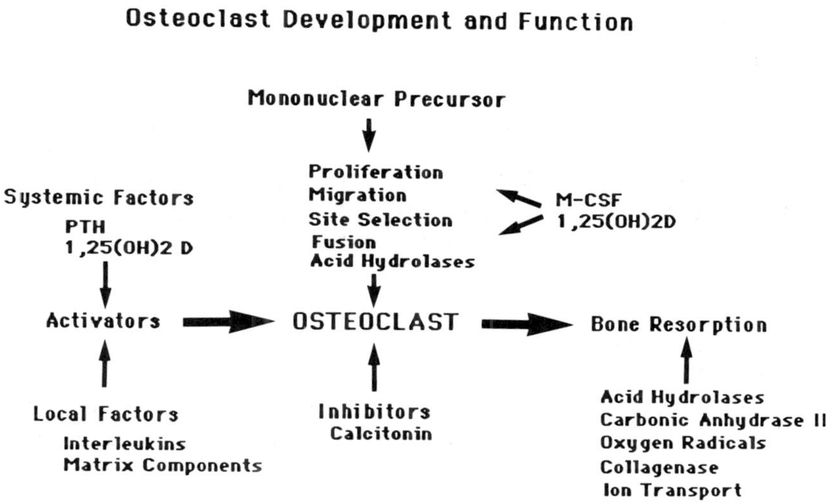

FIGURE 3. Summary of osteoclast development and function. M-CSF, macrophage colony-stimulating factor; $1,25(OH)_2D$, 1,25-dihydroxyvitamin D; PTH, parathyroid hormone.

III. OSTEOCLAST DEVELOPMENT AND FUNCTION

The origin and regulation of osteoclasts are the main considerations in this volume, and have recently been reviewed.[13,76,78,145] Nevertheless, an overview of these processes and some of their regulators will be a useful departure for discussion of the osteopetroses, and is provided in Figure 3. Osteoclasts arise from hemopoietic stem cells whose progeny migrate via vascular pathways to the skeleton, where they fuse with each other or existing osteoclasts to form multinucleated cells on bone surfaces. Certain cytokines and hormones, e.g., macrophage colony-stimulating factor (M-CSF) and $1,25(OH)_2D$, accelerate this developmental sequence *in vitro* by enhancing proliferation and fusion. Mononuclear precursors acquire the ability to synthesize characteristic acid hydrolases (tartrate-resistant acid phosphatases) late in development. Osteoclasts in bone must be activated for resorption to occur. These activators are produced by local cells in response to systemic (PTH and $1,25(OH)_2D$) or local factors (bone matrix components themselves, or cytokines and growth factors released from local cells). These factors cause bone resorption by enhancing formation of ruffled borders, release of acid hydrolases into the ruffled border region, and the production of superoxide (oxygen radicals). CA II, located in the ruffled border region, facilitates delivery of H^+ to this area, and a proton pump in this membrane produces acidification of the bone surface below the ruffled border. This, in turn, aids in demineralization of bone and the release and activation of procollagenase in the bone matrix. Products of bone resorption are taken up into the cell as phagosomes, digested further in secondary lysosomal vacuoles, and released extracellularly on the opposite side of the cell next to a vascular sinus. Bone resorption is inhibited by calcitonin. In each osteopetrotic mutation, this cellular and biochemical cascade leading to bone resorption appears to have been interrupted at a different site.

IV. CONGENITAL OSTEOPETROSIS AS AN ILLUSTRATION OF COMPROMISED OSTEOCLAST FUNCTION

Congenital osteopetrosis occurs in a number of mammalian species. We shall emphasize the four different mutations in the mouse (grey-lethal, *gl*; microphthalmia, *mi*; osteosclerosis, *oc*; and osteopetrosis, *op*), the four in the rat (incisors absent, *ia*; toothless, *tl*; osteopetrosis, *op*; and microphthalmia, *mib*), and the single mutation in the rabbit (osteopetrosis, *os*), because these have been studied most extensively.[126] Each is inherited as an autosomal recessive.[72] Osteopetrosis is lethal within 5 weeks of birth in *gl* and *oc* mice, the *op* rat, and the *os* rabbit. Bone resorption is reduced only slightly in *ia* and *mib* rats, and these mutations develop marrow spaces within 1 to 2 weeks after birth. However, skeletal sclerosis persits in caudal vertebrae and most metaphyses throughout life. The bones of *mi* mice and *tl* rats remain sclerotic throughout life.

A. OSTEOCLAST ONTOGENY

Studies of osteopetrotic mutations have made significant contributions to our understanding of the origin of osteoclasts. The hemopoietic origin of osteoclasts alluded to in earlier work[34,51] was substantiated by the pioneering work of Walker,[153-156] which revealed that precursors arising from stem cell proliferation in extraskeletal hemopoietic tissues migrate to skeletal sites via vascular pathways and fuse with one another or preexisting osteoclasts. These cells then differentiate and become bone resorbers. Earlier studies had overlooked the possibility of separate lineages for osteoblasts and osteoclasts.[106,161] The cure of osteopetrosis in mice undergoing temporary parabiosis with a normal littermate,[153] and later the transplantation of normal spleen cells, bone marrow,[154,156] or stem cells,[114] provided direct evidence for the hemopoietic origin of the osteoclast. This was confirmed by successful cures of osteopetrosis in animals[65,69,89] and children.[9,26,135] In one study,[26] the donor origin of osteoclasts in an osteopetrotic child was proved by fluorescent Y-body analysis. However, not all animal mutations[70,84,123] or children[10,17,135] are cured by this procedure. These observations serve to exemplify the complexity of osteoclast ontogeny and the remarkable heterogeneity of the osteopetroses. The mononuclear precursor to osteoclasts has yet to be identified, although investigations[14,148] have shown that mononuclear cells exist with certain enzymatic features of osteoclasts (tartrate-resistant acid phosphatase) and that the osteoclasts develop from a common stem cell or one related to monocytes and macrophages.[117]

Among the osteopetrotic mutations, osteoclasts are reduced in number in the *gl, op*, and *mi* mutations in the mouse, the *tl, mib*, and *op* mutations in the rat, and in the osteopetrotic *(os)* rabbit.[126] This indicates that in these mutations, osteoclast development is compromised. Osteoclasts are increased in *ia* rats and *oc* mice and in type II adult human osteopetrosis.[19] Reductions in osteoclast numbers are indicative of problems with the sequence of osteoclast development (Figure 3), and increases in osteoclast populations are believed to reflect a compensatory hyperplasia in response to reduced bone resorption. In *mi* mice, osteoclasts are small,[139] and in *op* rats and *os* rabbits,[77,83] they are large, suggesting problems with the regulation of cell fusion. The *mi* and *gl* mutations in the mouse and the *op* mutation in the rat are cured by bone marrow transplantation, which has been interpreted to mean that the defect in these mutations is one of stem cells or their progeny. The induction of osteopetrosis in normal animals of these stocks by transplantation of stem (spleen) cells from osteopetrotic littermates supports this conclusion.[73] On the other hand, the *op* mutation in the mouse, the *tl* mutation in the rat, and the *os* mutation in the rabbit are not cured by stem cell transplantation, which indicates that the problem lies elsewhere, perhaps in the skeletal matrix itself or in other cells in the skeletal or hemopoietic microenvironment.[126]

A number of factors influence the development of osteoclasts and, while all aspects are not known, one can appreciate its complexity. In other developing systems,[45,142] the local environment has been shown to influence the proliferation, migration, differentiation, and activation of cells, and this is likely to be the case for the skeleton.[71] Some or all stages of osteoclast ontogeny may be dependent upon or coordinated by extracellular signals in the local environment. For example, the mutant skeletal environment of the *tl* rat is unable to support the development and differentiation of normal osteoclast precursors,[100] suggesting the absence of a promoter or the presence of an inhibitor in this mutation. The paucity of osteoclasts in *op* mice has recently been shown to result from an absence of M-CSF.[159,160] Injections of M-CSF increase osteoclast numbers, restore resorption, and cure the skeletal sclerosis in *op* mice.[33] Untreated *op* mice have few macrophages and osteoclasts, and injections of M-CSF are much more effective in increasing osteoclasts than macrophages.[33] These data suggest a differential effect of this growth factor on osteoclasts and macrophages, and show that M-CSF is required for the development of osteoclasts.

Osteoclast development from cultured bone marrow cells requires $1,25(OH)_2D$.[36] The coexistence of high blood levels of $1,25(OH)_2D$ and significant variations in osteoclast numbers among osteopetrotic mutations suggest a resistance to the osteoclastogenic effects of $1,25(OH)_2D$ in some mutations. There may be additional factors affecting osteoclast ontogeny within the skeletal microenvironment. Studies of isolated mononuclear cells have shown that various products of bone resorption are chemoattractants for these cells[63,66,94,96] and presumably for osteoclasts and/or their precursors.

B. OSTEOCLAST ACTIVATION

The active metabolite of vitamin D, $1,25(OH)_2D$ and PTH are potent stimulators of osteoclast resorption. The administration of either factor gives a rapid activation of preexisting osteoclasts, as evidenced by increased resorption,[47] increased numbers of osteoclasts,[61] and increased nuclei per osteoclast. The effects of $1,25(OH)_2D$ are also independent of PTH.[49] Osteopetrotic mutations are resistant to the hypercalcemic effects of both PTH and $1,25(OH)_2D$,[74] and blood levels of $1,25(OH)_2D$ are elevated in osteopetrotic children[59] and animals.[162] This suggests a skeletal resistance to both hormones, a hypothesis most thoroughly explored for $1,25(OH)_2D$. Treatment of children with high-dose $1,25(OH)_2D$[59] produced evidence of increased bone resorption, but no significant clinical improvement. Similar studies in the *os* rabbit, a mutation with features found in osteopetrotic children, caused an increase in osteoclast number and staining for characteristic enzymes,[40] but no significant effects on skeletal sclerosis.[104] These data suggest that osteoclasts in these osteopetrotic mutations can respond to $1,25(OH)_2D$ with respect to osteoclastogenesis and resorption, but that each response is quantitatively insufficient to produce a cure. Investigation of one of the autosomal dominant forms of osteopetrosis in humans has shown reduced levels of $1,25(OH)_2D$.[20]

Receptors for PTH and $1,25(OH)_2D$ are found in osteoblasts and not osteoclasts. This led to the theory that osteoclast function is osteoblast-directed,[110] i.e., that neighboring osteoblasts may release paracrine factors which indirectly activate osteoclasts. There is increasing evidence that osteoblasts expose bone surfaces (collagenase) for osteoclast attachment. Indeed, molecular analyses of osteoblast gene expression in osteopetrosis suggest that in some mutations, particularly the *tl* rat, decreased osteoclast activity is related to osteoblast dysfunction.[128] The potential sites for interception of osteoclast function could involve either recruitment or activation.

Similar roles may be played by other local microenvironment factors such as cytokines, growth factors, and prostaglandins. Numerous osteoclast-activating factors (IL-1, IL-2, TNF-α, TNF-β) stimulate bone resorption *in vitro* and may act synergistically with PTH to couple the bone resorption and formation required for normal skeletal development.[39,140,141] A reduced level of IL-2 has been reported in one patient with osteopetrosis.[57] Lymphocytes harvested from this patient revealed a decrease in natural killer (NK) cell activity *in vitro* which was reversed by the addition of IL-2 to the cultures. IL-2 is known to increase the acid production and bone-resorbing activity of osteoclasts *in vitro*, and both effects are inhibited by calcitonin.[108] IL-1 stimulates bone resorption *in vivo*.[22] It is not known if osteoclasts are directly or indirectly stimulated by cytokines. Evidence[140] suggests that osteoblasts are required for osteoclasts to resorb bone in the presence of IL-1.

C. OSTEOCLAST INHIBITION

Calcitonin is a hypocalcemic and hypophosphatemic peptide produced by cells derived from the ultimobranchial body. The osteoclast has receptors for calcitonin, a known inhibitor of bone resorption in both animals[8,24] and humans.[99] Calcitonin will directly decrease the number of osteoclasts, decrease the number of nuclei per osteoclast, flatten the ruffled border, and separate the osteoclast from the underlying bone surface.[54] Osteoclast-like multinucleated cells which are formed from mononuclear precursor cells in response to 1,25$(OH)_2$D possess calcitonin receptors.[138]

Osteopetrosis in *gl* mice has been postulated to result from excess calcitonin[80,97] because of hypocalcemia, hypophosphatemia, and a significant increase in C-cells of the thyroid, the cellular source of calcitonin in mammals. This is supported by the demonstration that intravenous injection of plasma from *gl* mice into normal mice produces a profound but transient hypocalcemia.[98] In *mi* mice, basal and stimulated blood levels of calcitonin do not differ from those in normal mice,[56] but *mi* osteoclasts exhibit a greatly increased binding of ^{125}I-calcitonin,[111] and the ruffled border is poorly developed or absent.[48,82] These features suggest an increased sensitivity to calcitonin in *mi* mice. It is unlikely that calcitonin hypersecretion or hypersensitivity is the method by which bone resorption is reduced in some animal mutations. Calcitonin does not appear to be part of human osteopetroses. Whether it contributes indirectly to the osteopetrotic syndrome remains to be established.

D. BONE RESORPTION

The ruffled border, the ultrastructural hallmark of an active osteoclast, is the site on a bone surface at which the agents responsible for resorption leave the cell and at which the products of resorption reenter the cell for further processing and secretion. Active osteoclasts exhibit prominent cytoplasmic vacuoles next to bone centrally (ruffled border), with peripheral small areas devoid of vacuoles (clear zone) tightly adherent to the bone surface. For resorption to occur, osteoclast products are emptied at the ruffled border into the confined extracellular space bounded by the clear zone. An acid environment is provided through the action of carbonic anhydrase and the Na^+,K^+ pump. The products of bone resorption are then taken up into digestive vacuoles with further degradation to secondary lysosomes, which are released into nearby vascular spaces. Bone resorption is reduced in any circumstances in which the exocytic and endocytic functions of the ruffled border are impaired. These include interruptions in the release of acid hydrolases, reduction in the size of the ruffled border, minimal cytoplasmic vacuolization, failure to acidify the confined extracellular space of the ruffled border, and detachment of an osteoclast from the bone surface. Some mutations in children and the majority of animal mutants have osteoclasts which are unable to elaborate a ruffled border.[21,38,79,129] Osteoclasts in children with osteopetrosis are described as large, numerous, and lacking cytoplasmic vacuolization next to bone surfaces.[21,38,129] The ruffled border is absent in the *os* rabbit, *mi* mouse, *ia* rat, and *oc* mouse, normal in the grey lethan mouse, and exaggerated in the *op* mouse.[79] In addition to the formation of ruffled borders, bone

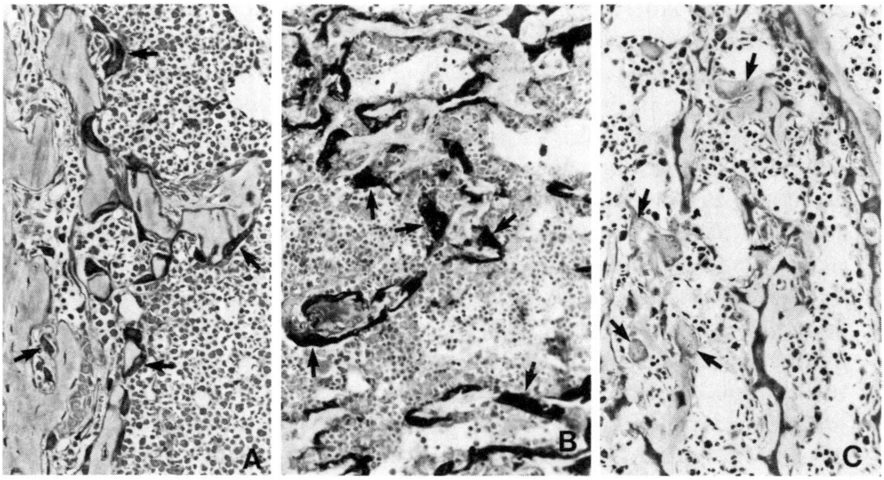

FIGURE 4. Photomicrographs of osteoclasts (arrows) from a normal rat of *mib* stock (A), a mutant of *ia* stock (B), and a mutant of *mib* stock (C), stained histochemically for tartrate-resistant acid phosphate (TRAP) and counterstained lightly with toluidine blue. The ability of osteopetrotic osteoclasts to synthesize TRAP varies considerably among mutations. The extremes shown here, *ia* (B) and *mib* (C) rats, represent two of the milder forms of the disease. (Original magnification × 200.)

resorption requires synthesis and release of acid hydrolases, acidification of the extracellular space below the ruffled border, production of superoxide, and activation of collagenase.

Acid hydrolases are synthesized by osteoclasts, and this ability is one sign of a mature cell. The osteoclastic enzymes are characterized by a resistance to tartrate and inhibition by vanadate.[44] Recent studies have shown that these tartrate-resistant acid phosphatases (TRAPs) are more precisely described as an acid ATPase[5-7] which is acquired late in osteoclast ontogeny.[64] Aberrations in the acid hydrolase content of osteoclasts from osteopetrotic mutations have been known for some time,[81] and have recently been extended to tartrate-resistant ATPase.[29] The spectrum of abnormalities is illustrated in Figure 4. Histochemical staining for TRAP is excessive in *ia* rats (Figure 4B) and deficient in *mi, gl*, and *op* mice, *tl, op*, and *mib* (Figure 4C) rats, and *os* rabbits. The excessive staining seen in *ia* rats is presumed to result from normal synthesis and an inability to secrete (absence of a ruffled border). Reduced staining in other mutations is presumed to reflect synthesis.

Carbonic anhydrase isoenzyme II (CA II) is located in the ruffled border[4] in osteoclasts,[143] and its inhibitors reduce bone resorption.[93] This enzyme catalyzes the conversion of CO_2 and H_2O to bicarbonate and hydrogen ion, which is translocated across the ruffled border to acidify that environment. Both 1,25$(OH)_2$D and PTH have stimulatory effects on this enzyme.[42,43] Treatment with calcitonin has been shown to reduce CA II activity and thus decrease bone resorption.[4] The demonstration that the congenital absence of CA II in children causes osteopetrosis with renal tubular acidosis and cerebral calcification[132] established the central role of CA II in osteoclast function.

Recently, the role of oxygen-derived free radicals, specifically superoxide, in osteoclast function has been investigated.[37,60] Production of oxygen-derived free radicals by osteoclasts is inhibited by calcitonin. It is detectable in bone-resorbing systems and is generated adjacent to bone surfaces *in vivo*.[37] It is not clear how superoxide production affects bone resorption. In some patients with autosomal recessive osteopetrosis, the deficiency in bone resorption has been associated with a defect in superoxide generation by leukocytes.[15,57,107] Thus, studies of osteopetrosis could help clarify the role of superoxide in osteoclast biology.

The osteoclast ruffled border membrane has been shown to contain both Na^+,K^+-[14] and H^+-ATPases.[144] While the precise role of each potential ion transport mechanism in resorption has not been established, it appears that one or both may play an important role in bone resorption. Defects in this pump could contribute to some osteopetrotic mutations.

Collagenase is believed to act in at least two parts of the resorption cycle. This enzyme is synthesized and secreted by osteoblasts as an inactive precursor, procollagenase. The enzyme can be activated extracellularly by osteoblasts and osteoclasts. Activation by osteoblasts occurs during osteoclast recruitment and serves to expose a mineralized surface for osteoclast attachment. This occurs when osteoblasts are stimulated by bone-resorbing agents.[46] Prior to resorption, bone is covered with a layer of osteoid,[105,147] which must be removed before the osteoclast can resorb bone mineral.[25] Osteoclastic activation of the procollagenase in bone matrix occurs during bone resorption by various proteases and the lysosomal cathepsin B released by osteoclasts into the confines of the ruffled border.[27,28] The active enzyme, together with cysteine proteases produced by the osteoclast, degrade the collagen in bone matrix.[145] While not yet pursued, aberrations in the procollagenase content of bone or production of cystein proteases by osteoclasts might be involved in the pathogenesis of certain osteopetroses.

V. UNRESOLVED ISSUES IN OSTEOPETROSIS

It is clear that except for the M-CSF deficiency in *op* mice[32,33,160] and the CA II deficiency in some osteopetrotic children,[133] the precise genetic defect remains to be established for each osteopetrotic mutation (Figure 5). In addition to the challenges of each mutation, the osteopetroses also present opportunities to understand the relationships between osteoclast development and function in at least three broad areas: the hemopoietic and immune systems, tissue-specific resistance to vitamin D, and neuronal development.

While initial reports of direct thymic involvement in osteopetrosis[87-89] have been largely unsubstantiated,[30,101,115] there is abundant evidence that

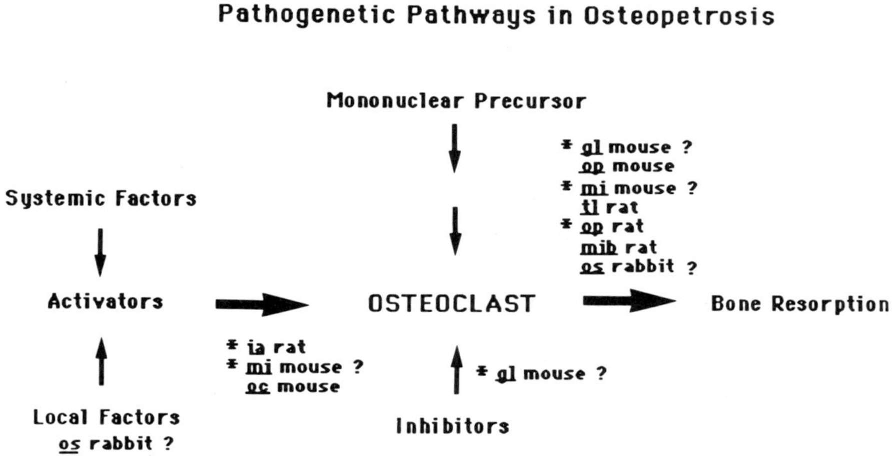

FIGURE 5. Pathogenetic pathways for osteopetrosis in laboratory animals. Each mutation is positioned with respect to osteoclast development and function according to the evidence to date. An asterisk indicates that this mutation is cured by bone marrow (stem cell) transplantation from a normal littermate. A question mark indicates evidence for more than one potential defect.

certain host defense mechanisms are deficient. These include reduced bactericidal capacity in monocytes and neutrophils in osteopetrotic children,[15,107] *ia* rats,[116,121] and *mi* mice,[92] reduced NK cell function in *mi* mice and *ia, tl,* and *op* rats,[102,119,120,136] and a deficiency in IL-2 in children[57] and *ia* rats.[118] The relationship of these immune aberrations to the osteoclast dysfunction in osteopetrosis is unknown. The ability of IL-2 infusions to restore NK cell function[118] and to elevate bone resorption in *ia* rats[116] suggests a direct relationship. However, $1,25(OH)_2D$ is known to reduce IL-2 levels,[3] and the IL-2 and NK cell defects could be due to elevated blood levels of $1,25(OH)_2D$ in osteopetrosis.[162] The anemia seen in many osteopetrotic children is not a feature of osteopetrosis in most animals, presumably because considerable splenic hemopoiesis persists in adult mice and rats.[122] However, a severe monocytopenia and osteoclastopenia coexist in *op* mice.[75,157,158] Treatment with M-CSF restores resorption and elevates monocytes and peritoneal macrophages, but only to one fourth the number in normal littermates.[33] Thus, osteoclasts and macrophages appear to respond differently to injections of M-CSF. Whether there are other abnormal hematological parameters and their relationship to bone resorption are questions worth pursuing.

In osteopetrotic animals and some human mutations, blood levels of $1,25(OH)_2D$ are markedly elevated.[59,162] This has been interpreted as a vitamin D-resistant condition. However, the intestinal response to these high levels of vitamin D is appropriate at least with respect to production of the vitamin D-dependent, calcium-binding protein in *oc* mice and *tl* rats.[124,125] Treatment of children[58,59] and osteopetrotic animals[103,104] increases osteoclast numbers,[40] but does not elevate resorption sufficiently to qualify as a cure. Thus, the osteopetroses are likely to exhibit selective skeletal resistance to $1,25(OH)_2D$[31,85,134] and represent excellent opportunities to explore the relationships of this hormone to bone resorption during skeletal modeling.

Neurological complications of osteopetrosis in children typically include optic atrophy. Since this has been assumed to be due to compression of the optic nerve within the optic canal, early surgical decompression at this site has been recommended.[1,41,109] However, reports of optic atrophy in the absence of nerve compression and the presence of a more generalized neuronal degeneration and cerebral atrophy[2,35,62,112] suggest more generalized, degenerative neuronal changes which appear to be unrelated to the occasional hydrocephalus, at least in some children. These associations of a neuronal storage disease and increased lysosomal storage in osteoclasts (no ruffled border) in these children suggest a common etiology as a lysosomal storage disease. Study of the osteoclastic and neuronal aberrations in these children may provide clues about the potential interdependence of skeletal and neuronal development and function.

Investigations of these areas, together with delineation of the specific defect in osteoclast function in each osteopetrotic mutation, can be expected to provide new, specific insights into osteoclast biology.

ACKNOWLEDGMENTS

The authors' work was supported by NIH Grants DE 07444 and AR 39737, and a research fellowship from the Department of Orthopedics and Physical Rehabilitation, University of Massachusetts Medical Center. We thank C. Dunshee and E. Larson for superb technical assistance.

REFERENCES

1. **Al-Mefty, O., Fox, J. L., Al-Rodhan, N., and Dow, J. H.,** Optic nerve decompression in osteopetrosis, *J. Neurosurg.*, 68, 80, 1988.
2. **Ambler, M. W., Trice, J., Grauerholz, J., and O'Shea, P. A.,** Infantile osteopetrosis and neuronal storage disease, *Neurology,* 33, 437, 1983.
3. **Amento, E. P.,** Vitamin D and the immune system, *Steroids,* 49, 507, 1987.
4. **Anderson, R. E., Schraer, H., and Gay, C. V.,** Ultrastructural immunocytochemical localization of carbonic anhydrase in normal and calcitonin-treated chick osteoclasts, *Anat. Rec.*, 204, 9, 1982.
5. **Andersson, G. N., Ek-Rylander, B., and Hammarstrom, L. E.,** Purification and characterization of a vanadate-sensitive nucleotide tri- and diphosphatase with acid pH optimum from rat bone, *Arch. Biochem. Biophys.*, 228, 431, 1984.
6. **Andersson, G. N., Ek-Rylander, B., Hammarstrom, L. E., Lindskog, S., and Toverud, S. U.,** Immunocytochemical localization of a tartrate-resistant and vanadate-sensitive acid nucleotide tri- and diphosphatase, *J. Histochem. Cytochem.*, 34, 293, 1986.

7. **Andersson, G., Lindunger, A., and Ek-Rylander, B.,** Isolation and characterization of skeletal acid ATPase — a new osteoclast marker?, *Connect. Tissue Res.*, 20, 151, 1989.
8. **Arnett, T. R. and Dempster, D. W.,** Effect of pH on bone resorption by rat osteoclasts *in vitro, Endocrinology,* 119, 119, 1986.
9. **Ballet, J., Griscelli, C., Courtris, C., Milhaud, G., and Maroteaux, P.,** Bone marrow transplantation in osteopetrosis, *Lancet,* 2, 1137, 1977.
10. **Ballet, J. J. and Griscelli, C.,** Lymphoid cell transplantation in human osteopetrosis, in *Mechanisms of Localized Bone Loss,* Horton, J. E., Tarpley, T. M., and Davis, W. F., Eds., Information Retrieval, Washington, D.C., 1978, 399.
11. **Banco, R., Seifert, M. F., Marks, S. C., Jr., and McGuire, J. L.,** Rickets and osteopetrosis: the osteosclerotic (*oc/oc*) mouse, *Clin. Orthop. Relat. Res.*, 201, 238, 1985.
12. **Barnicot, N. A.,** Studies on the factors involved in bone absorption. I. The effect of subcutaneous transplantation of bones of the grey-lethal house mouse into normal hosts and of normal bones into grey-lethal hosts, *Am. J. Anat.*, 68, 497, 1941.
13. **Baron, R.,** Molecular mechanisms of bone resorption by the osteoclast, *Anat. Rec.*, 224, 317, 1989.
14. **Baron, R., Neff, L., Roy, C., Boisvert, A., and Caplan, M.,** Evidence for a high specific concentration of (Na^+,K^+)ATPase in the plasma membrane of the osteoclast, *Cell,* 46, 311, 1986.
15. **Beard, C. M., Key, L., Newburger, P. E., Ezekowitz, A. B., Arcesi, R., Miller, B., Proto, P., Ryan, T., Anast, C., and Simons, E. R.,** Neutrophil defect associated with malignant infantile osteopetrosis, *J. Lab. Clin. Med.*, 108, 498, 1986.
16. **Beighton, P., Hamersma, H., and Cremin, B. J.,** Osteopetrosis in South Africa. The benign, lethal and intermediate forms, *S. Afr. Med. J.*, 55, 659, 1979.
17. **Blazar, B., Fallon, M., Teitelbaum, S., Ramsay, N., and Brown, D.,** Calcitriol for congenital osteopetrosis, *N. Engl. J. Med.*, 311, 55, 1984.
18. **Bollerslev, J.,** Autosomal dominant osteopetrosis: bone metabolism and epidemiological, clinical and hormonal asepcts, *Endocr. Rev.*, 10, 45, 1989.
19. **Bollerslev, J. and Andersen, P. E., Jr.,** Radiological, biochemical and hereditary evidence of two types of autosomal dominant osteopetrosis, *Bone,* 9, 7, 1988.
20. **Bollerslev, J., Neilsen, H. K., Storm, T., and Mosekilde, L.,** Serum levels of vitamin D metabolites and nuclear uptake of 1,25-dihydroxyvitamin D_3 in monocytes from patients with autosomal dominant osteopetrosis, *Bone,* 9, 355, 1988.
21. **Bonucci, E., Sartori, E., and Spina, M.,** Osteopetrosis fetalis, *Virchows Arch. A,* 368, 109, 1975.
22. **Boyce, B. F., Aufdemorte, T. B., Garrett, I. R., Yates, A. J. P., and Mundy, G. R.,** Effects of interleukin-1 on bone turnover in normal mice, *Endocrinology,* 125, 1142, 1989.
23. **Cecchini, M. G. and Fleisch, H.,** Bisphosphonates *in vitro* specifically inhibit, among the hematopoietic series, the development of the mouse mononuclear phagocytic lineage, *J. Bone Miner. Res.*, 5, 1019, 1990.
24. **Chambers, T. J.,** The patholobiology of the osteoclast, *J. Clin. Pathol.*, 38, 241, 1985.
25. **Chambers, T. J., Revell, P. A., Fuller, K., and Athanasou, N. A.,** Resorption of bone by isolated rabbit osteoclasts, *J. Cell Sci.*, 66, 383, 1984.
26. **Coccia, P. F., Krivit, W., Cervenka, J., Clawson, C., Kersey, J. H., Kim, T. H., Nesbit, M. E., Ramsay, N. K. C., Warkentin, P. I., Teitelbaum, S. L., Kahn, A. J., and Brown, D. M.,** Successful bone-marrow transplantation for infantile malignant osteopetrosis, *N. Engl. J. Med.*, 301, 701, 1980.
27. **Delaisse, J.-M., Eeckhout, Y., and Vaes, G.,** Bone-resorbing agents affect the production and distribution of procollagenase as well as the activity of collagenase in bone tissue, *Endocrinology,* 123, 264, 1988.

28. **Eeckhout, Y., Delaisse, J.-M., and Vaes, G.,** Direct extraction and assay of bone tissue collagenase and its relation to parathyroid-hormone-induced bone resorption, *Biochem. J.*, 239, 793, 1986.
29. **Ek-Rylander, B., Marks, S. C., Jr., Hammarstrom, L. E., and Andersson, G. N.,** Osteoclastic acid ATPase: biochemical and histochemical studies of the osteopetrotic mutations in the rat, *Bone Miner.*, 5, 309, 1989.
30. **Evans, S. F., Nisbet, N. W., Marshall, M. J., and Catty, D.,** A secondary immune deficiency in the fatty/Orl-*op* rat, *Br. J. Exp. Pathol.*, 66, 251, 1985.
31. **Feldman, D. and Malloy, P. J.,** Hereditary 1,25-dihydroxyvitamin D resistant rickets: molecular basis and implications for the role of $1,25(OH)_2D_3$ in normal physiology, *Mol. Cell. Endocrinol.*, 72, C57, 1990.
32. **Felix, R., Cecchini, M. G., Hofstetter, W., Elford, P. R., Stutzer, A., and Fleisch, H.,** Impairment of macrophage colony-stimulating factor production and lack of resident bone marrow macrophages in the osteopetrotic *op/op* mouse, *J. Bone Miner. Res.*, 5, 781, 1990.
33. **Felix, R., Cecchini, M. G., and Flesich, H.,** Macrophage colony stimulating factor restores *in vivo* bone resorption in the *op/op* osteopetrotic mouse, *Endocrinology*, 127, 2592, 1990.
34. **Fischman, D. A. and Hay, E. D.,** Origin of osteoclasts from mononuclear leucocytes in regenerating newt limbs, *Anat. Rec.*, 143, 329, 1962.
35. **Fitch, N., Carpenter, S., and Lachance, R. C.,** Prenatal axonal dystrophy and osteopetrosis, *Arch Pathol.*, 95, 298, 1973.
36. **Fuller, K. and Chambers, T. J.,** Generation of osteoclasts in cultures of rabbit bone marrow and spleen cells, *J. Cell. Physiol.*, 132, 441, 1987.
37. **Garrett, I. R., Boyce, B. F., Oreffo, R. O. C., Bonewald, L., Poser, J., and Mundy, G. R.,** Oxygen-derived free radicals stimulate osteoclastic bone resorption in rodent bone *in vitro* and *in vivo, J. Clin. Invest.*, 85, 632, 1990.
38. **Glorieux, F. H., Petifor, J. M., Marie, P. J., Delvin, E. E., Travers, R., and Shepard, N.,** Induction of bone resorption by parathyroid hormone in congenital malignant osteopetrosis, *Metab. Bone Dis. Relat. Res.*, 3, 143, 1981.
39. **Gowen, M., Wood, D. D., Ihrie, E. J., McGuire, M. K. B., and Russell, R. G. G.,** An interleukin 1 like factor stimulates bone resorption *in vitro, Nature (London)*, 306, 378, 1983.
40. **Grise, M. A., Marks, S. C., Jr., MacKay, C. A., and Popoff, S. N.,** The effects of 1,25-dihydroxyvitamin D on osteoclast number and cytochemistry in normal and osteopetrotic *(os)* rabbits, *Am. J. Anat.*, 189, 261, 1990.
41. **Haines, J. S., Erickson, D., and Wirtschafter, J.,** Optic nerve decompression of osteopetrosis in early childhood, *Neurosurgery*, 23, 470, 1988.
42. **Hall, G. E. and Kenny, A. D.,** Role for carbonic anhydrase in bone resorption induced by 1,25-dihydroxyvitamin D_3 *in vitro, Calcif. Tissue Int.*, 37, 134, 1985.
43. **Hall, G. E. and Kenny, A. D.,** Role of carbonic anhydrase in bone resorption: effect of acetazolamide on basal and parathyroid hormone-induced bone metabolism, *Calcif. Tissue Int.*, 40, 212, 1987.
44. **Hammarstrom, L. E., Hanker, J. S., and Toverud, S. U.,** Cellular differences in acid phosphatase isoenzymes in bone and teeth, *Clin. Orthop.*, 78, 151, 1971.
45. **Hay, E.,** *Cell Biology of Extracellular Matrix,* Plenum Press, New York, 1981.
46. **Heath, J. K., Atkinson, S. J., Meikle, M. C., and Reynolds, J. J.,** Mouse osteoblasts synthesize collagenase in response to bone resorbing agents, *Biochim. Biophys. Acta*, 802, 151, 1984.
47. **Holtrop, M. E. and Raisz, L. G.,** Comparison of the effects of 1,25-dihydroxycholecalciferol, prostaglandin E_2, and osteoclast-activating factor with parathyroid hormone on the ultrastructure of osteoclasts in cultured long bones of fetal rats, *Calcif. Tissue Int.*, 29, 201, 1979.

48. **Holtrop, M. E., Cox, K. A., Eilon, G., Simmons, H. A., and Raisz, L. G.,** The ultrastructure of osteoclasts in microphthalmic mice, *Metab. Bone Dis. Relat. Res.*, 3, 123, 1981.
49. **Holtrop, M. E., Cox, K. A., Clark, M. B., Holick, M. F., and Anast, C. S.,** 1,25-dihydroxycholecalciferol stimulates osteoclasts in rat bones in the absence of parathyroid hormone, *Endocrinology,* 108, 2293, 1981.
50. **Horton, W. A., Schimke, R. N., and Iyama, T.,** Osteopetrosis: Further heterogeneity, *J. Pediatr.*, 97, 580, 1980.
51. **Jee, W. S. S. and Nolan, P. D.,** Origin of osteoclasts from the fusion of phagocytes, *Nature (London),* 200, 225, 1963.
52. **Johnston, C. C., Lavy, N., Lord, T., Vellios, F., Merritt, A. D., and Deiss, W. P.,** Osteopetrosis, *Medicine,* 47, 149, 1968.
53. **Kahler, S. G., Burns, J. A., and Aylsworth, A. S.,** A mild autosomal recessive form of osteopetrosis, *Am. J. Med. Genet.*, 17, 451, 1984.
54. **Kallio, D. M., Garant, P. R., and Minkin, C.,** Ultrastructural effects of calcitonin on osteoclasts in tissue cultures, *J. Ultrastruct. Rev.*, 39, 205, 1972.
55. **Kaplan, F. S., August, C. S., Fallon, M. D., Dalinka, M., Axel, L., and Haddad, J. G.,** Successful treatment of infantile malignant osteopetrosis by bone-marrow transplantation, *J. Bone J. Surg. Am. Vol.*, 70, 617, 1988.
56. **Kent, G. N. and Cohn, D. V.,** Blood levels of calcitonin in microphthalmic (*mi*/*mi*) osteopetrotic mouse cannot account for the resistance of bone to this hormone, *Metab. Bone Dis. Relat. Res.*, 5, 151, 1981.
57. **Key, L.,** Osteopetrosis: a genetic window into osteoclast function, *CPC Ser. Cases Metab. Bone Dis.*, 2, 1, 1987.
58. **Key, L. L. and Baron, R.,** High-dose calcitriol reversed malignant osteopetrosis, *J. Bone Miner. Res.*, 1, 111A, 1986.
59. **Key, L., Carnes, D., Cole, S., Holtrop, M., Bar-Shavit, Z., Shapire, F., Arceci, R., Steinberg, J., Gundberg, C., Kahn, A., Teitelbaum, S., and Anast, C.,** Treatment of congenital osteopetrosis with high-dose calcitriol, *N. Engl. J. Med.*, 310, 409, 1984.
60. **Key, L. L., Ries, W. L., Taylor, R. G., Hays, B. D., and Pitzer, B. L.,** Oxygen derived free radicals in osteoclasts: the specificity and location of the nitroblue tetrazolium reaction, *Bone,* 11, 115, 1990.
61. **King, G. J., Holtrop, M. E., and Raisz, L. G.,** The relation of ultrastructural changes in osteoclasts to resorption in bone cultures stimulated with parathyroid hormone, *Metab. Bone Dis. Relat. Res.*, 1, 67, 1978.
62. **Lehman, R. A. W., Reeves, J. D., Wilson, W. B., and Wesenberg, R. L.,** Neurological complications of infantile osteopetrosis, *Ann. Neurol.*, 2, 378, 1977.
63. **Lian, J. B., Tassinari, M., and Glowacki, J. C.,** Resorption of implanted bone prepared from normal and warfarin-treated rats, *J. Clin. Invest.*, 73, 1223, 1984.
64. **Lindunger, A., MacKay, C. A., Ek-Rylander, B., Andersson, G., and Marks, C. S., Jr.,** Histochemistry and biochemistry of tartrate-resistant acid phosphatase (TRAP) and tartrate-resistant acid adenosine triphosphatase (TrATPase) in bone, bone marrow and spleen: implications for osteoclast ontogeny, *Bone Miner.*, 10, 109, 1990.
65. **Loutit, J. F. and Sansom, J. M.,** 1976 Osteopetrosis of microphthalmic mice — a defect of the hematopoietic stem cell?, *Calcif. Tissue Res.*, 20, 251, 1976.
66. **Malone, J. D., Teitelbaum, S. L., Griffin, G. L., Senior, R. M., and Kahn, A. J.,** Recruitment of osteoclast precursors by purified bone matrix constituents, *J. Cell Biol.*, 92, 227, 1982.
67. **Manzke, E., Gruber, H., Baylink, D., and Hiness, R.,** Different radiological, histomorphometrical and biochemical findings in two cases of adult osteopetrosis, *Radiology,* 11, 175, 1982.
68. **Marks, C. R., Seifert, M. F., and Marks, S. C., III,** Osteoclast populations in congenital osteopetrosis: additional evidence of heterogeneity, *Metab. Bone Dis. Relat. Res.*, 5, 259, 1984.

69. **Marks, S. C., Jr.,** Osteopetrosis in the *ia* rat cured by spleen cells from a normal littermate, *Am. J. Anat.,* 146, 331, 1976.
70. **Marks, S. C., Jr.,** Osteopetrosis in the toothless *(tl)* rat: presence of osteoclasts but failure to respond to parathyroid extract or to be cured by infusion of spleen or bone marrow cells from normal littermates, *Am. J. Anat.,* 149, 289, 1977.
71. **Marks, S. C., Jr.,** The origin of osteoclasts: the evidence, clinical implications and investigative challenges of an extraskeletal source, *J. Oral Pathol.,* 12, 226, 1983.
72. **Marks, S. C., Jr.,** Congenital osteopetrotic mutations as probes of the origin, structure and function of osteoclasts, *Clin. Orthop.,* 189, 239, 1984.
73. **Marks, S. C., Jr.,** Osteopetrosis — multiple pathways for the interception of osteoclast function, *Appl. Pathol.,* 5, 172, 1987.
74. **Marks, S. C., Jr.,** Osteoclast biology: lessons from mammalian mutations, *Am. J. Med. Genet.,* 34, 43, 1989.
75. **Marks, S. C., Jr. and Lane, P. W.,** Osteopetrosis, a new recessive skeletal mutation on chromosome 12 of the mouse, *J. Hered.,* 67, 11, 1976.
76. **Marks, S. C., Jr. and Popoff, S. N.,** Bone cell biology: the regulation of development, structure and function in the skeleton, *Am. J. Anat.,* 183, 1, 1988.
77. **Marks, S. C., Jr. and Popoff, S. N.,** Osteoclast biology in the osteopetrotic *(op)* rat, *Am. J. Anat.,* 186, 325, 1989.
78. **Marks, S. C., Jr. and Popoff, S. N.,** Ultrastructural biology and pathology of the osteoclast, in *Ultrastructure of Skeletal Tissues. Bone and Cartilage in Health and Disease,* Bonucci, E. and Motta, P. M., Eds., Kluwer Academic, Boston, 1990, 239.
79. **Marks, S. C., Jr. and Seifert, M. F.,** The development and structure of osteoclasts in osteopetrosis, in *The Chemistry and Biology of Mineralized Tissues,* Butler, W. T., Ed., EBSCO Media, Birmingham, 1985, 408.
80. **Marks, S. C., Jr. and Walker, D. G.,** The role of the parafollicular cell of the thyroid gland in the pathogenesis of congenital osteopetrosis in mice, *Am. J. Anat.,* 126, 299, 1969.
81. **Marks, S. C., Jr. and Walker, D. C.,** Mammalian osteopetrosis: a model for studying cellular and humoral factors in bone resorption, in *The Biochemistry and Physiology of Bone,* Vol. 4, 2nd ed., Bourne, G., Ed., Academic Press, New York, 1976, 227.
82. **Marks, S. C., Jr. and Walker, D. G.,** The hematogenous origin of osteoclasts: experimental evidence from osteopetrotic (microphthalmic) mice treated with spleen cells from beige mouse donors, *Am. J. Anat.,* 161, 1, 1981.
83. **Marks, S. C., Jr., MacKay, C. A., and Seifert, M. F.,** The osteopetrotic rabbit — skeletal cytology and ultrastructure, *Am. J. Anat.,* 178, 300, 1987.
84. **Marks, S. C., Jr., Seifert, M. F., and McGuire, J. L.,** Congenitally osteopetrotic *op/op* mice are not cured by transplants of spleen or bone marrow cells from normal littermates, *Metab. Bone Dis. Relat. Res.,* 5, 183, 1984.
85. **Marx, S. J. and Barsony, J.,** Tissue-selective 1,25-dihydroxyvitamin D_3 resistance: novel applications of calciferols, *J. Bone Miner. Res.,* 3, 481, 1988.
86. **McGuire, J. L. and Marks, S. C., Jr.,** The effects of parathyroid hormone on bone cell structure and function, *Clin. Orthop.,* 100, 392, 1974.
87. **Milhaud, G. and Labat, M.-L.,** Thymus and osteopetrosis, *Clin. Orthop.,* 135, 260, 1978.
88. **Milhaud, G. and Labat, M.-L.,** Osteopetrosis reconsidered as a curable immune disorder, *Biomedicine,* 30, 71, 1979.
89. **Milhaud, G., Labat, M.-L., Graf, B., Juster, M., Balmain, N., Moutier, R., and Toyama, K.,** Demonstration cinetique, radiographique et histologique de la guerison de l'osteopetrose congenitale du Rat, *C. R. Acad. Sci. Paris,* 280, 2485, 1975.
90. **Milhaud, G., Labat, M.-L., Parant, M., Damais, C., and Chedid, L.,** Immunologic defect and its correction in the osteopetrotic mutant rat, *Proc. Natl. Acad. Sci. U.S.A.,* 74, 339, 1977.

91. **Miller, S. C. and Jee, W. S. S.,** The effect of dichloromethylene diphosphonate, a pyrophosphate analog, on bone and bone cell structure in the growing rat, *Anat. Rec.*, 193, 439, 1979.
92. **Minkin, C.,** Defective macrophage chemotaxis in osteopetrotic mice, *Calcif. Tissue Int.*, 33, 677, 1981.
93. **Minkin, C. and Jennings, J. M.,** Carbonic anhydrase and bone remodeling: sulfonamide inhibition of bone resorption in organ culture, *Science,* 176, 1031, 1972.
94. **Minkin, C., Posek, R., and Newbrey, J.,** Mononuclear phagocytes and bone resorption: identification and preliminary characterization of a bone-derived macrophage chemotactic factor, *Metab. Bone Dis. Relat. Res.*, 2, 363, 1981.
95. **Monaghan, B. A., Kaplan, F. S., August, C. S., Fallon, M. D., and Flannery, D. B.,** Transient infantile osteopetrosis, *J. Pediatr.*, 118, 252, 1991.
96. **Mundy, G. R., Altman, A. J., Gondek, M. D., and Bandelin, J. G.,** Direct resorption of bone by human monocytes, *Science,* 196, 1109, 1977.
97. **Murphy, H. M.,** A review of inherited osteopetrosis in the mouse, *Clin. Orthop.*, 65, 97, 1969.
98. **Murphy, H. M.,** Calcitonin-like activity in the circulation of osteopetrotic grey-lethal mice, *J. Endocrinol.*, 53, 139, 1973.
99. **Murrills, R. J., Shane, E., Lindsay, R., and Dempster, D. W.,** Bone resorption of isolated human osteoclasts *in vitro*: effects of calcitonin, *J. Bone Miner. Res.*, 4, 259, 1989.
100. **Osier, L. K., Popoff, S. N., and Marks, S. C., Jr.,** Osteopetrosis in the toothless rat: failure of osteoclast differentiation and function, *Bone Miner.*, 3, 35, 1987.
101. **Popoff, S. N., Jackson, M. E., and Marks, S. C., Jr.,** Thymic development and function in the osteopetrotic *(op)* rat, *J. Bone Miner. Res.*, 5 (Abstr.) (Suppl. 2), 384, 1990.
102. **Popoff, S. N., Jackson, M. E., Koevary, S. B., and Marks, S. C., Jr.,** Coexistence of reduced function of natural killer cells and osteoclasts in two distinct osteopetrotic mutations in the rat, *J. Bone Miner. Res.*, 6, 263, 1991.
103. **Popoff, S. N., Marks, S. C., Jr., and McGuire, J. L.,** Treatment of osteopetrotic mutations with high-dose 1,25 dihydroxyvitamin D, *J. Bone Miner. Res.*, 1 (Abstr.) (Suppl. 1), 191, 1986.
104. **Popoff, S. N., McGuire, J. L., Zerwekh, J. E., and Marks, S. C., Jr.,** Treatment of congenital osteopetrosis in the rabbit with high-dose 1,25-dihydroxyvitamin D, *J. Bone Miner. Res.*, 4, 57, 1989.
105. **Raina, V.,** Normal osteoid tissue, *J. Clin. Pathol.*, 25, 229, 1972.
106. **Rasmussen, H. and Bordier, P.,** *The Physiological and Cellular Basis of Metabolic Bone Disease,* Williams & Wilkins, Baltimore, 1974.
107. **Reeves, J. D., August, C. L., Humbert, J. R., and Weston, W. L.,** Host defense in infantile osteopetrosis, *Pediatrics,* 64, 202, 1979.
108. **Ries, W. L., Seeds, M. C., and Key, L. L.,** Interleukin-2 stimulates osteoclastic activity: increased acid production and radioactive calcium release, *J. Periodont. Res.*, 24, 242, 1989.
109. **Riser, R. O.,** Marble bones and optic atrophy, *Am. J. Ophthalmol.*, 24, 874, 1941.
110. **Rodan, G. A. and Martin, T. J.,** Role of osteoblasts in hormonal control of bone resorption — a hypothesis, *Calcif. Tissue Int.*, 33, 349, 1981.
111. **Rouleau, M. F., Rarshawsky, H., Marks, S. C., Jr., and Goltzman, D.,** Calcitonin receptor binding as a marker of osteoclast heterogeneity in osteopetrotic rodents, *J. Bone Miner. Res.*, 1, 543, 1986.
112. **Ruben, J. B., Morris, R. J., and Judisch, G. F.,** Chorioretinal degeneration in infantile malignant osteopetrosis, *Am. J. Opthalmol.*, 110, 1, 1990.
113. **Schmidt, E. V., Crapo, J. D., Harrelson, J. M., and Smith, R. E.,** A quantitative histologic study of avian osteopetrotic bone demonstrating normal osteoclast numbers and increased osteoblastic activity, *Lab. Invest.*, 44, 164, 1981.

114. **Schneider, G. B.,** Cellular specificity of the cure for osteopetrosis: isolation of and treatment with pluripotent hemopoietic stem cells, *Bone,* 6, 241, 1985.
115. **Schneider, G. B. and Marks, S. C., Jr.,** Immunological competence in osteopetrotic (microphthalmic) mice, *Exp. Cell Biol.,* 51, 327, 1983.
116. **Schneider, G. B., Kelly, J. D., and Relfson, M.,** Effects of interleukin-2 on bone resorption in *(ia)* osteopetrotic rats, *J. Bone Miner. Res.,* 5 (Abstr.) (Suppl. 2), 433, 1990.
117. **Schneider, G. B., Relfson, M., and Nicholas, J.,** Pluripotent hemopoietic stem cells give rise to osteoclasts, *Am. J. Anat.,* 177, 505, 1986.
118. **Schneider, G. B., Relfson, M., and Ellis, T.,** Natural killer cell activity in *(ia)* osteopetrotic rats: effects of rIL-2, *Calcif. Tissue Int.,* 4 (Abstr.) (Suppl.), 185, 1988.
119. **Schneider, G. B., Relfson, M., and Ellis, T. M.,** An evaluation of natural killer cells in osteopetrotic rats by flow cytometry, *J. Bone Miner. Res.,* 4 (Abstr.) (Suppl. 2), 187, 1989.
120. **Seaman, W. E., Gindhart, T. D., Greenspan, J. S., Blackman, M. A., and Talal, N.,** Natural killer cells, bone and the bone marrow: studies in estrogen-treated mice and in congenitally osteopetrotic (*mi/mi*) mice, *J. Immunol.,* 122, 2541, 1979.
121. **Seifert, M. F. and Dixit, P. K.,** Impaired macrophage migration in incisors absent *(ia)* osteopetrotic rats, *Bone Miner.,* 4, 167, 1988.
122. **Seifert, M. F. and Marks, S. C., Jr.,** Morphological evidence of reduced bone resorption in osteosclerotic *(oc)* mice, *Am. J. Anat.,* 172, 141, 1985.
123. **Seifert, M. F. and Marks, S. C., Jr.,** Congenitally osteosclerotic (*oc/oc*) mice are resistant to cure by transplantation of bone marrow or spleen cells from normal littermates, *Tissue Cell,* 19, 29, 1987.
124. **Seifert, M. F., Gray, R. W., and Bruns, M. E.,** Elevated levels of vitamin D-dependent calcium-binding protein (Calbindin-D9k) in the osteosclerotic *(oc)* mouse, *Endocrinology,* 122, 1067, 1988.
125. **Seifert, M. F., Gray, R. W., and Bruns, M. E.,** Elevated 1,25-dihydroxyvitamin D_3 and intestinal calbindin-D9k in the toothless rat, *Am. J. Physiol.,* 258, E377, 1990.
126. **Seifert, M. F., Popoff, S. N., Jackson, M. E., MacKay, C. A., Cielinski, M., and Marks, S. C.,** Experimental studies of osteopetrosis in laboratory animals, *Clin. Orthop. Relat. Res.,* in press.
127. **Selye, H.,** On the stimulation of new bone formation with parathyroid extract and irradiated ergosterol, *Endocrinology,* 16, 547, 1932.
128. **Shalhoub, V., Jackson, M. E., Lian, J. B., Stein, G. S., and Marks, S. C., Jr.,** Gene expression during skeletal development in three osteopetrotic rat mutations: evidence for osteoblast abnormalities, *J. Biol. Chem.,* 266, 9847, 1991.
129. **Shapiro, F., Glimcher, M. J., Holtrop, M. E., Tashjian, A. H., Brickley-Parsons, D., and Kenzora, J. E.,** Human Osteopetrosis. A histological, ultrastructural, and biochemical study, *J. Bone J. Surg. Am. Vol.,* 62, 384, 1980.
130. **Shapiro, F., Key, L. L., and Anast, C.,** Variable osteoclast appearance in human infantile osteopetrosis, *Calcif. Tissue Int.,* 43, 67, 1988.
131. **Simpson, C. F. and Sanger, V. L.,** A review of avian osteopetrosis: comparisons with other bone diseases, *Clin. Orthop.,* 58, 271, 1968.
132. **Sly, W. S., Hewett-Emmett, D., Whyte, M. P., Lu, Y.-S. L., and Tashian, R. E.,** Carbonic anhydrase II deficiency identified as the primary defect in the autosomal recessive syndrome of osteopetrosis with renal tubular acidosis and cerebral calcification, *Proc. Natl. Acad. Sci. U.S.A.,* 80, 2752, 1983.
133. **Sly, W. S., Whyte, M. P., Sundaram, V., Tashian, R. E., Hewett-Emmett, D., Guibaud, P., Vainsel, M., Baluarte, H. J., Gruskin, A., Al-Mosawi, M., Sakati, N., and Ohlsson, A.,** Carbonic anhydrase II deficiency in 12 families with the autosomal recessive syndrome of osteopetrosis with renal tubular acidosis and cerebral calcification, *N. Engl. J. Med.,* 313, 139, 1985.

134. **Sone, T., Marx, S. J., Liberman, U. A., and Pike, J. W.,** A unique point mutation in the human vitamin D receptor chromosomal gene confers hereditary resistance to 1,25-dihydroxyvitamin D_3, *Mol. Endocrinol.*, 4, 623, 1990.
135. **Sorell, M., Kapoor, N., Kirkpatrick, D., Rosen, J., Chagnanti, R., Lopez, C., Dupont, B., Pollack, M., Terrin, B., Harris, M., Vine, D., Rose, J., Goossen, C., Good, R., and O'Reilly, R. J.,** Marrow transplantation for juvenile osteopetrosis, *Am. J. Med.*, 70, 1280, 1981.
136. **Stechschulte, D., Sharma, R., Dileepan, K. N., Simpson, K. M., Aggarwal, N., Clancy, J., and Jilka, R. L.,** Effect of the *mi* allele on mast cells, basophils, natural killer cells and osteoclasts in C57B1/6J mice, *J. Cell. Physiol.*, 132, 565, 1987.
137. **Strisciuglio, P., Sartorio, R., Pecoraro, C., Lotito, F., and Sly, W. S.,** Variable clinical presentation of carbonic anhydrase deficiency: evidence for heterogeneity?, *Eur. J. Pediatr.*, 149, 337, 1990.
138. **Takahashi, N., Akatsu, T., Saski, T., Nicholson, G. C., Moseley, J. M., Martin, T. J., and Suda, T.,** Induction of calcitonin receptors by 1α,25-dihydroxyvitamin D_3 in osteoclast-like multinucleated cells formed from mouse bone marrow cells, *Endocrinology*, 123, 1504, 1988.
139. **Thesingh, C. W. and Scherft, J. P.,** Fusion disability of embryonic osteoclast precursor cells and macrophages in the microphthalmic osteopetrotic mouse, *Bone*, 6, 43, 1985.
140. **Thomson, B. M., Saklatvala, J., and Chambers, T. J.,** Osteoblasts mediate interleukin 1 stimulation of bone resorption by rat osteoclasts, *J. Exp. Med.*, 164, 104, 1986.
141. **Thomson, B. M., Mundy, G. R., and Chambers, T. J.,** Tumor necrosis factors α and β induce osteoblastic cells to stimulate osteoclastic bone resorption, *J. Immunol.*, 138, 775, 1987.
142. **Trelstad, R.,** *The Role of Extracellular Matrix in Development*, Alan R. Liss, New York, 1984.
143. **Väänänen, H. K. and Parvinen, E.-K.,** High active isoenzyme of carbonic anhydrase in rat calvaria osteoclasts, *Histochemistry*, 78, 481, 1983.
144. **Väänänen, H. K., Karhukorpi, E.-K., Sundquist, K., Wallmark, B., Roininen, I., Hentunen, T., Tuukkanen, J., and Lakkakorpi, P.,** Evidence for the presence of a proton pump of the vacuolar H^+-ATPase type in the ruffled borders of osteoclasts, *J. Cell Biol.*, 111, 1305, 1990.
145. **Vaes, G.,** Cellular biology and biochemical mechanism of bone resorption. A review of recent developments on the formation, activation, and mode of action of osteoclasts, *Clin. Orthop. Relat. Res.*, 231, 239, 1988.
146. **Vainsel, M., Fondu, P., Cadranel, S., Rocmans, C. L., and Gepts, W.,** Osteopetrosis associated with proximal and distal tubular acidosis, *Acta Paediat. Scand.*, 61, 429, 1972.
147. **Vanderwiel, C. J.,** An ultrastructural study of the components which make up the resting surface of bone, *Metab. Bone Dis. Relat. Res.*, 2 (Suppl.), 109, 1980.
148. **Van de Wijngaert, F. P. and Bruger, E. H.,** Demonstration of tartrate-resistant acid phosphatase in un-decalcified, glycomethacrylate-embedded mouse bone: a possible marker for (pre)osteoclast identification, *J. Histochem. Cytochem.*, 34, 1317, 1986.
149. **van Tran, P., Dryll, A., Lansaman, J., Naveau, B., Trefe, R., Miravet, L., and Ryckewaert, A.,** Osteoclast abnormalities in idiopathic osteopetrosis, *Virchows Arch. Pathol. Anat.*, 408, 269, 1985.
150. **Walker, D. G.,** Elevated bone collagenolytic activity and hyperplasia of parafollicular light cells of the thyroid gland in parathormone-treated grey-lethal mice, *Z. Zellforsch.*, 72, 100, 1966.
151. **Walker, D. G.,** Counteraction to parathyroid therapy in osteopetrotic mice as revealed in the plasma calcium level and ability to incorporate ^{3}H-proline into bone, *Endocrinology*, 79, 836, 1966.
152. **Walker, D. G.,** Congenital osteopetrosis in mice cured by parabiotic union with normal siblings, *Endocrinology*, 91, 916, 1972.

153. **Walker, D. G.,** Experimental osteoporosis, *Clin. Orthop.,* 97, 158, 1973.
154. **Walker, D. G.,** Bone resorption restored in osteopetrotic mice by transplants of normal bone marrow and spleen cells, *Science,* 190, 784, 1975.
155. **Walker, D. G.,** Spleen cells transmit osteopetrosis in mice, *Science,* 190, 785, 1975.
156. **Walker, D. G.,** Control of bone resorption by hematopoietic tissue, *J. Exp. Med.,* 142, 651, 1975.
157. **Wiktor-Jedrzejczak, A., Skelly, R. R., and Ahmed, A.,** Hematopoietic stem cell differentiation and its role in osteopetrosis. Immunologic implications, in *Immunological Defects in Laboratory Animals,* Vol. 1, Gershwin, M. E. and Merchant, B., Eds., Plenum Press, New York, 1981, 51.
158. **Wiktor-Jedrzejczak, W., Ahmed, A., Szczylik, C., and Skelly, R. R.,** Hematological characterization of congenital osteopetrosis in *op/op* mouse, *J. Exp. Med.,* 156, 1516, 1982.
159. **Wiktor-Jedrzejczak, W., Bartocci, A., Ferrante, A. W., Jr., Ahmed-Ansari, A., Sell, K. W., Pollard, J. W., and Stanley, E. R.,** Total absence of colony-stimulating factor 1 in the macrophage-deficient osteopetrotic (*op/op*) mouse, *Proc. Natl. Acad. Sci. U.S.A.,* 87, 4828, 1990.
160. **Yoshida, H., Hayashi, S.-I., Kinisada, T., Ogawa, M., Nishikawa, S., Okamura, H., Sudo, R., Shultz, L. D., and Nishikawa, S.-I.,** The murine mutation osteopetrosis is in the coding region of the macrophage colony stimulating factor gene, *Nature (London),* 345, 442, 1990.
161. **Young, R. W.,** Cell proliferation and specialization during endochondral osteogenesis in young rats, *J. Cell Biol.,* 14, 357, 1962.
162. **Zerwekh, J. E., Marks, S. C., Jr., and McGuire, J. L.,** Elevated serum 1,25 dihydroxyvitamin D in osteopetrotic mutations in three species, *Bone Miner.,* 2, 193, 1987.

Chapter 21

THE OSTEOCLAST IN PAGET'S DISEASE OF BONE

G. David Roodman

TABLE OF CONTENTS

I. CLINICAL FEATURES

Paget's disease of bone was first described more than 100 years ago by Sir James Paget, a British surgeon. He named this deforming, painful disorder of bone "osteitis deformans",[1] because he believed it was due to an inflammatory condition. The patients had deformity or enlargement of the skull, clavicles, or long bones that was usually focal, with one or more bones involved. The clinical description of Paget's disease has changed little since then. Deformity or enlargement of the bones by Paget's disease can result in significant morbidity to the patients, including bone pain, fracture, neurologic complications due to nerve root compression, dental abnormalities, and deafness. In a very small minority of the patients (less than 1%), osteogenic sarcomas or giant cell tumors of bone can develop in pagetic bone.[2,3]

Paget's disease is the second most common metabolic bone disease in the world, surpassed only by osteoporosis. It affects approximately 3% of the population over age 40 in the Western Hemisphere and Australia.[4-6] with about 2 to 3 million patients in the U.S. The prevalence of the disease rises with age, with about 8% of the population over 75 years of age being affected by Paget's disease. Paget's disease is rarely seen in patients under the age of 40. There is a marked geographical distribution for Paget's disease. In the U.K., Paget's disease is most prevalent in Lancashire.[6] About 7% of those over 55 years of age are affected. In contrast, Paget's disease is extremely rare in the Scandinavian countries, Africa, the Far East, and the Middle East. There is an increased incidence of Paget's disease in some families, and there are several reports of Paget's disease affecting identical twins, although the overall concordance among family members is unknown.[7]

The primary cellular abnormality in Paget's disease resides in the osteoclast (Table 1). Pagetic osteoclasts are increased in number and size in these bone lesions, and this increased activity of osteoclasts produces localized areas of bone destruction. Osteoblast activity is also increased in Paget's disease, but the osteoblasts appear to be normal.[8] The constant destruction and formation of bone results in a loss of normal bone architecture. Bone formed in response to the increased osteoclastic bone resorption seen in Paget's disease is of poor quality, leading to the complications noted above, such as fractures, degenerative arthritis, and neurologic impairment.[9] The resorbed bone is replaced by coarse, dense trabecular bone which is organized in a haphazard fashion. The disorganized and very rapid deposition of new bone in Paget's disease results in irregular cement lines responsible for the characteristic mosaic pattern of pagetic bone seen on X-rays. Many Paget's disease patients are asymptomatic and are often diagnosed on X-rays that are done in the course of a medical or dental evaluation.[2] The most common symptoms of these patients are pain and deformity. The pain is generally mild to moderate, often persistent, and frequently due to osteoarthritis rather than to Paget's disease. Due to the increased vascularity of the pagetic bone, the skin

TABLE 1
Factors Suggesting that the Osteoclast is the Abnormal Cell in Paget's Disease

1. Osteoclast numbers, nuclei per osteoclast, and osteoclast size are markedly decreased in Paget's disease.
2. Viral inclusions are found only in osteoclast in Paget's bone biopsies and not in other bone cells.
3. The initial event in Paget's disease is intense focal bone resorption.
4. Therapies such as calcitonin or bisphosphonates that are directed against osteoclast improve the disease and decrease alkaline phosphatase levels.

over pagetic lesions, especially over the tibia, has an increased temperature. In addition to these local bone problems, the patients can develop hypercalciuria and hypercalcemia, hyperuricemia, and increased cardiac output.[7] On laboratory examination, the patients have increased serum alkaline phosphatase levels, reflecting the high rates of bone turnover.[10] Urinary hydroxyproline levels may also be increased due to increased collagen turnover.[10,11]

Radiologic evaluation of the patients often shows localized osteolytic lesions, especially in the skull and pelvis.[7] Osteoporosis circumscripta is a term used to describe these early lesions, which are usually discreet, round or oval, circumscribed areas of reduced bone density in the frontal or occipital bones of the skull. The patients often show local areas of osteolysis in the long bones, with an advancing v-shaped radiolucent area present at one end of the diaphysis that spreads at a constant rate toward the other end of the long bone.[12] Dormant lesions of Paget's disease are always osteosclerotic. Sequential long-term radiologic studies of these patients show that the osteoblastic or sclerotic phase of Paget's disease is most likely preceded by an osteolytic phase that frequently can go undetected due to the asymptomatic nature of early Paget's disease. Incomplete transverse fissues are often found in the cortex along the converse side of long bones in patients affected with Paget's disease. These occur during the osteoblastic phase and are easily distinguished from osteomalacia. Pathologic fractures may develop through these transverse fissures. Bone scans show diffuse, increased uptake of the radionuclide over the pagetic lesions, and early lesions that are not detected radiographically can be seen on the bone scan.[3]

Because the primary abnormality in Paget's disease resides in the osteoclast, treatment of Paget's disease is directed at the osteoclast. Treatment is indicated when bone pain is unresponsive to analgesics, to prevent medical complications such as hypercalcemia and congestive heart failure, to treat or prevent the neurologic complications of Paget's disease, in preparation for orthopedic surgery, and to prevent fractures or skull deformity in patients with osteolytic disease.[3,7-9] The primary modes of treatment include calcitonin and bisphosphonates. Pharmacologic doses of calcitonin rapidly inhibit osteoclastic bone resorption, and when administered for long periods of time, reduce the number of osteoclasts present in the bone. In contrast to patients with hypercalcemia of malignancy, patients with Paget's disease can respond to calcitonin for long periods of time.[7] In patients who become resistant to calcitonin, bisphosphonates can be used. The clinical effects of bisphosphonates are similar to those of calcitonin. However, unlike calcitonin, bisphosphonate therapy seldom heals osteolytic lesions despite improvement of the biochemical parameters of bone turnover in these patients.[3,9] In patients who are unresponsive to both calcitonin and bisphosphonates, mithramycin, a cytotoxic antibiotic, has been used in a small number of patients with Paget's disease. This agent has potent suppressive effects on osteoclasts, but can be marrow suppressive. In addition, renal dysfunction and liver dysfunction may be serious side effects of mithramycin therapy.

II. PATHOLOGY AND BONE HISTOMORPHOMETRY

Pathologic studies of bones from patients with Paget's disease were the first studies to suggest that Paget's disease is a disease of the osteoclast. Pagetic osteoclasts in bone biopsies contain more nuclei and are increased in size compared to normal osteoclasts.[13] Initially, there is an intense phase of bone resorption, followed by increased bone formation.[14] However, the bone formed in Paget's disease may be woven rather than normal lamellar bone, which most probably reflects the greatly increased rate of bone turnover rather than a primary abnormality of the osteoblast. These repeated cycles of bone resorption followed by new bone formation result in the deposition of cement lines, which produce the mosaic pattern that is a characteristic feature seen in the histologic sections of pagetic bone. There is a loss of definition between the cortical and trabecular bone, with areas of trabecular thickening and other areas in which the trabeculae are completely resorbed.[15] The trabeculae are often numerous and thick, and trabecular bone volume is markedly increased in patients with Paget's disease compared to normals. The increase in trabecular bone volume in Paget's patients is about 40% above that seen in normals.[15] Overall, the trabecular bone volume in patients with Paget's disease is increased at least twofold compared to normal bone.

Histomorphometric studies have shown that total resorption surfaces are markedly increased in pagetic bone. The number of resorption surfaces is about seven times greater than that seen in normal bone and is at least three times greater than that seen in patients with primary hyperparathyroidism.[15] Osteoclast numbers are increased about tenfold in pagetic bone compared to normal bone (3.18 ± 2 in pagetic bone vs. 0.33 ± 0.23 in nonpagetic areas).[15] Osteoid volume is moderately but significantly increased in pagetic bone compared to normal controls, and osteoid surfaces are increased about four times above normal.[15] In contrast, thickness of osteoid borders is significantly lower in pagetic bone than in normal bone ($18.8 + 4\%$ in normal bone vs. $15.2 \pm 3.6\%$ in pagetic bones).[15] Overall pagetic bone has extensive but thin osteoid borders, suggesting that the calcification rate in pagetic bone is slightly faster than the osteoblastic appositional rate. In addition, marrow fibrosis is seen in the pagetic bone, which decreases when patients are treated with bisphosphonates or other therapies to decrease osteoclast numbers.[16,17]

Of interest is that in patients with Paget's disease, bone not clinically involved with Paget's disease also appears to show evidence of increased bone remodeling. Approximately 45% of uninvolved bone shows significantly increased trabecular bone resorption and a modest increase in the number of osteoclasts.[15] Meunier and co-workers[15] and Siris et al.[18] attributed this increased bone remodeling in uninvolved bone from patients with Paget's disease to increased levels of parathyroid hormone rather than to subclinical involvement with Paget's disease. However, immunoreactive parathyroid hormone levels were found to be increased in only 13 of 109 untreated pagetic patients with normal renal function. In Meunier's study,[15] only 7 of 39 patients had increased parathyroid hormone levels, and 18% had increased PTH levels in the study by Siris et al.[18] Thus, at present there is no comprehensive explanation for the increased bone remodeling seen in uninvolved bone from patients with Paget's disease. Factors that increase osteoclast activity could be produced by bone cells in the pagetic lesion, either the osteoblast or osteoclast, and released systemically. This possibility will be discussed further in another section of this chapter.

III. OSTEOCLAST ULTRASTRUCTURE

Scanning electron microscopic studies[19] of bone from patients with Paget's disease have confirmed most of the results shown by histomorphometry. These studies have shown that osteocondensation due to large, fused trabeculae and structural and textural abnormalities such as woven bone are present in pagetic bone. Increased amounts of bone were seen in a random array, explaining the fragility of pagetic bone. The resting surfaces of the bone appeared as an irregular patchwork compared to normal bone.[20] These studies have shown that the resorption surfaces are very abundant and extend irregularly in multiple directions compared to normal bone. These resorption lacunae are often unusually deep. Boyde and Jones[21] showed that the volume of resorption lacunae in Paget's bone was more than ten times greater than resorption lacunae in normal bone. In addition, irregularities of the resorption lacunae were also found.[19]

Transmission electron microscopic studies have revealed several unique features that differentiate pagetic osteoclasts from normal osteoclasts. Rebel and co-workers[22] were the first to report several cytological abnormalities present in osteoclasts from patients with Paget's disease. The osteoclasts were irregular in shape, with multiple extensions and invaginations. The plasma membrane showed an external coat of dense particles consistent with membranes involved in active transport. Occasionally, masses of small, electron-dense, glycogen-like particles were found in the cytoplasm, and the cytoplasm also contained numerous microfilaments of various dimensions. The most striking feature of these cells was their nuclei. Large numbers of nuclei were polymorphic; some were smooth and ovoid, while others were badly deformed, with multiple indentations. The paranuclear space was dilated and contained clear vesicles. The nuclei had a peripheral distribution of dense chromatin and large nucleoli. In each of these osteoclasts, several nuclei contained microcylindrical inclusion bodies that were filamentous structures about 150 Å thick. Transverse sections of these inclusions showed a clear center surrounded by dense structures about 50 Å in diameter. Occasionally, these filaments were closely packed in a paracrystalline array, with the interspace reduced to about 50 Å. The filaments were organized in bundles. These nuclear inclusions in Paget's osteoclasts have never been reported in normal osteoclasts. Nuclear inclusions were present in all cells. The nuclear inclusions present in pagetic osteoclasts were similar to nuclear inclusions seen in glial cells from patients with progressive multifocal leukoencephalopathy, a disease most probably due to a papava virus.[23,24] In addition, patients who had subacute sclerosing panencephalitis virus infections also showed similar nuclear inclusions.[25] These investigators suggested that Paget's disease may be due to an "external agent". Mills and Singer[26] confirmed that nuclear inclusions were present in osteoclasts of patients with Paget's disease. These investigators studied 18 patients with Paget's disease. Nuclear inclusions similar to those reported by Rebel and co-workers[22] were present in 20 to 40% of osteoclasts and in about one fourth of the nuclei. Nuclear inclusions were present in all biopsy specimens from patients with Paget's disease. These nuclear inclusions differed from nuclear bodies, which are nuclear organelles associated with cellular hyperactivity.[27] They described these nuclear inclusions as most closely resembling the viral nucleocapsids of a measles-type virus. In addition to the nuclear inclusions, these pagetic osteoclasts also contained cytoplasmic inclusions which were similar to nuclear inclusions from patients with measles virus. Harvey and co-workers[28] found nuclear inclusions in 56 to 100% of osteoclasts present in Paget's bone biopsy sections viewed by transmission electron microscopy. These nuclear inclusions occupied 15 to 75% of the nuclear cross-sectional area. Intracytoplasmic inclusions also were seen in 30 to 40% of osteoclasts in Paget's bone biopsies. Of interest is that treatment of patients with calcitonin or bisphosphonates, although reducing the number of osteoclasts, did not affect the morphology or prevalence of the nuclear inclusions present in these cells.

Howatson and Fornasier[29] compared the nuclear inclusion microfilaments associated with Paget's disease of bone with the nucleocapsids of measles virus and respiratory syncytial virus. The authors confirmed that only osteoclasts contained these nuclear inclusions, and the inclusions were present in all biopsy specimens from Paget's patients. Cytoplasmic inclusions were also detected in eight of ten patients. The distribution of microfilament inclusions in pagetic osteoclasts closely paralleled that of measles nucleocapsids in monkey cells, but the dimensions of the pagetic microtubules were significantly different from those of measles virus. They also noted that the microtubules in the nuclear inclusions of Paget's osteoclasts were indistinguishable in dimension from nucleocapsides from respiratory syncytial virus, but were dissimilar in conformation and location. These data supported the hypothesis that Paget's disease of bone was a slow viral disease and suggested that the respiratory syncytial virus rather than measles virus may be the cause of Paget's disease. Similar ultrastructural studies by Bonucci and co-workers[30] further suggested a viral etiology for Paget's disease. Typical nuclear inclusions in pagetic osteoclasts are shown in Figure 1.

IV. PATHOGENESIS

Ultrastructural studies, which demonstrated viral-like nuclear inclusions in pagetic osteoclasts, suggests that Paget's disease may be a slow viral disease. For the last 10 years, investigators have pursued the identification of this virus. Basle et al.[31] and Rebel et al.,[32,33] using indirect immunofluorescence techniques with anti-measles polyclonal antibodies, showed reactivity of pagetic osteoclasts with these anti-measles antibodies. These investigators showed that most of the osteoclasts, but none of the other bone cells (e.g., osteoblasts or osteocytes), reacted with the anti-measles antibodies. These positive reactions with the polyclonal anti-measles antibodies could be abolished with specific absorption of the antibodies on measles virus-infected vero cells, but not by absorption on uninfected vero cells. No reactivity of normal osteoclasts with anti-measles antibodies could be detected. Similar results were obtained either by immunofluorescence or immunoperoxidase techniques. Mills and co-workers[34,35] reported evidence for both respiratory syncytial virus and measles virus antigens in osteoclasts from patients with Paget's disease of bone. Using indirect immunofluorescent antibody assays, bone biopsy specimens from 28 of 29 Paget's patients showed reactivity with antibodies against respiratory syncytial virus, and 11 of 22 patients had positive immunofluorescence for measles virus. Using antiviral monoclonal antibodies, Basle et al.[36] showed that antigens of measles virus, simian virus 5 (SV5), and human parainfluenza virus type 3 could be detected in pagetic osteoclasts. Measles and SV5 nucleoprotein and hemagglutinin-neuraminidase antigens were also present in all cases of Paget's bone examined. These studies further supported the hypothesis that a paramyxovirus may be responsible for Paget's disease of bone. Recently, other investigators suggested that canine distemper virus rather than other paramyxoviruses may play a role in the etiology of Paget's disease.[37,38] *In situ* hybridization studies carried out by Basle and co-workers[39] have shown that measles nucleocapsid protein mRNA was expressed in the cytoplasm and nuclei of 80 to 90% of pagetic osteoclasts. In addition, the tritiated, labeled DNA probes also hybridized with 30 to 40% of the mononuclear cells, including osteoblasts, osteocytes, fibroblasts, and

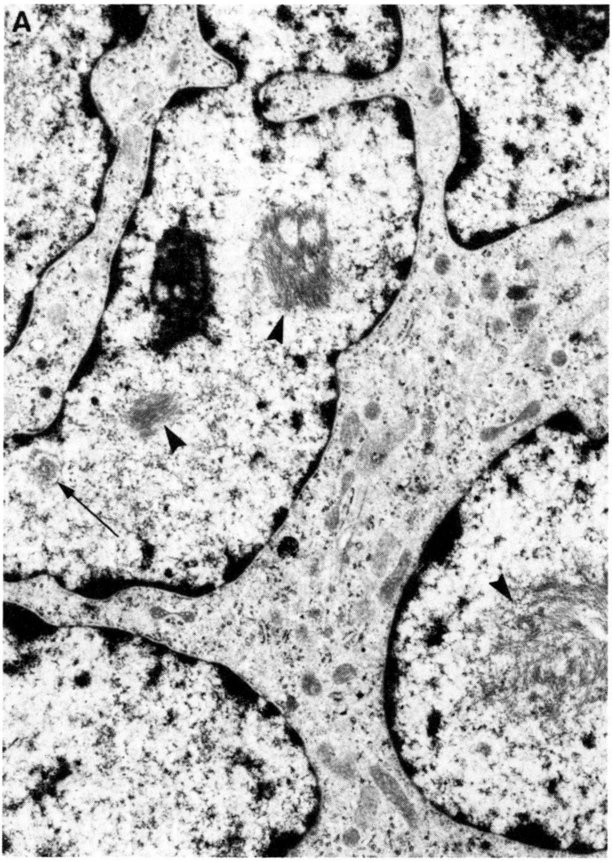

FIGURE 1. Ultrastructure of a pagetic osteoclast. Pagetic osteoclasts were obtained from bone biopsy specimens and processed for electronmicroscopy. (A) The cells contain polymorphic nuclei, and the nuclei contain large nuclear inclusions (arrowhead). The nuclear inclusions are typical of paramyxoviral inclusions. In addition, a nuclear body can be seen indistinctly in one of the nuclei (arrow). (Original magnification × 1600.) (B) Higher magnification of the viral nuclear inclusions present in the pagetic osteoclast shown in A. The large microcylindrical nuclear inclusions present in these cells are characteristic of paramyxoviral inclusions of measles or respiratory syncytial virus. (Original magnification × 105,000.) (These photomicrographs were generously provided by Dr. Barbara Mills.)

lymphocytes. *In situ* hybridization studies with hepatitis B cDNA were negative. These data suggest that a wide range of cells contained measles virus mRNA, but only in the osteoclast was this message being translated into mature virus. Thus, the presence of viral inclusions in all cases of Paget's disease has led to the suggestion that Paget's disease is a slow viral disease analogous to subacute sclerosing panencephalitis and canine distemper. However, investigators have failed to demonstrate complement-fixing antiviral

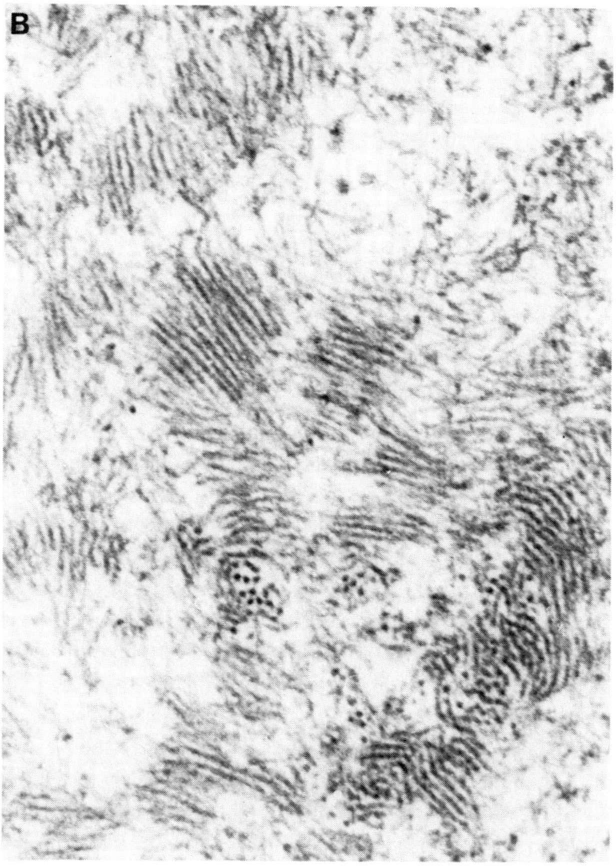

FIGURE 1B.

antibodies for measles in serum from patients with Paget's disease.[40] In contrast, these anti-measles antibodies are present in the serum of patients with subacute sclerosing panencephalitis. In addition, attempts to isolate and transfer the virus have also been unsuccessful.[41,42] Furthermore, the geographic prevalence of acute measles does not match the geographic distribution of Paget's disease. For example, Paget's disease is very rare in Scandinavia but common in Lancashire, although the two regions have an equal incidence of measles. Other etiologic factors could be active in Paget's disease in addition to the virus. Further, viral inclusions have been found in some patients with giant cell tumors of bone[43] as well as in rare patients with osteopetrosis,[44] suggesting that these viral inclusions are not unique to Paget's disease. Until the virus can be isolated and osteoclasts infected with the virus with subsequent development of Paget's disease in the bone, the viral etiology of Paget's disease will remain in question. Table 2 summarizes the evidence for a viral etiology for Paget's disease.

TABLE 2
Evidence for a Viral Etiology in Paget's Disease

1. Nuclear and cytoplasmic viral inclusions are present in osteoclasts.
2. Viral transcripts have been detected in osteoclasts and other bone cells.
3. Immunocytological studies show that paramyxovirus proteins are expressed in pagetic osteoclasts.
4. Clinical features of Paget's disease resemble slow viral infections of the central nervous system.

V. MODEL SYSTEMS OF PAGET'S DISEASE

Presently, there are no *in vivo* model systems for Paget's disease. Attempts have been made to develop animal models for Paget's disease by injecting pagetic osteoclasts into nude mice,[64] but to date none have been successful. Singer and co-workers[44,46] have isolated pagetic cells from bone biopsy specimens and cultured these cells *in vitro*. Although these cells contained viral inclusions, they did not release a virus into the media, nor could the cells survive for long periods of time. Conditioned media from these cultures could not infect normal bone cells. Of interest is that some mononuclear cells in these cultures expressed paramyxovirus antigens. Basle and co-workers[47] have isolated pagetic osteoclasts and other cells from pagetic bone tissue by treating bone samples from eight patients with Paget's disease with ethylene diamine tetraacetic acid. This technique yields variable numbers of osteoclasts, ranging from 10 to 70 osteoclasts per slide. The cells had between 20 to 50 nuclei per cell, and the isolated osteoclasts failed to exclude trypan blue. The cells stained positively for nonspecific esterase that was inhibitable by fluoride, succinic dehydrogenase, and acid phosphatase. Thus, these preparations did not yield viable osteoclasts, but were useful for examining the cytochemistry of isolated osteoclasts from Paget's bone.

Recently, Kukita and co-workers[48] have used long-term human bone marrow cultures as a model system for osteoclast formation in patients with Paget's disease. Long-term marrow culture systems have been used as a model system for normal osteoclast formation.[49-55] Multinucleated cells formed in these human marrow cultures express the osteoclast phenotype (multinucleation, contains tartrate-resistant acid phosphatase, cross-reacts with the 23c6 monoclonal antibody which preferentially binds to osteoclasts, expresses calcitonin receptors, and forms resorption lacunae on calcified matrices.[50,56-58] These workers have cultured bone marrow mononuclear cells from active lesions from patients with Paget's disease. Multinucleated cells which differ from normal marrow-derived multinucleated cells are similar to the pagetic cells isolated from bone formed in these cultures (Table 3). The multinucleated cells formed in long-term marrow cultures of Paget's disease contained large numbers of nuclei (7 to 90 nuclei per cell), were much larger than multinucleated cells formed in normal marrow cultures, and formed between 10 to

TABLE 3
Differences Expressed by Multinucleated Cells (MNC) Formed in Long-Term Cultures of Paget's Marrow Compared to Marrow Cultures from Normals

1. Increased numbers of MNC
2. Increased numbers of nuclei per MNC
3. Increased tartrate-resistant acid phosphatase activity per cell
4. Increased rate of MNC formation
5. Increased sensitivity of MNC to 1,25-dihydroxyvitamin D_3
6. Ultrastructural differences from normal MNC
7. Differences in MNC surface phenotype from normal MNC
8. Only Paget's MNC express paramyxovirus proteins

100 times the number of multinucleated cells formed in long-term human marrow cultures (Figure 2). These multinucleated cells contained tartrate-resistant acid phosphatase, cross-reacted with the 23c6 monoclonal antibody, formed resorption lacunae on calcified matrices, and expressed ultrastructural features which were similar to agetic osteoclasts. They contained pleomorphic nuclei, large amounts of granular material in the cytoplasm, many filamentous structures, and centrioles. In addition, nuclear inclusions were found in these pagetic, marrow-derived, multinucleated cells. However, the nuclear inclusions were not viral inclusions, but represented nuclear bodies,[27] consistent with the hyperactivity of these cells. The pagetic marrow-derived, multinucleated cells had increased levels of tartrate-resistant acid phosphatase per cell and an increased sensitivity to 1,25-dihydroxyvitamin D_3. Concentrations as low as 10^{-11} *M* 1,25$(OH)_2D_3$ stimulated formation of multinucleated cells formed in pagetic marrow cultures. In contrast, concentrations of 10^{-10} *M* 1,25$(OH)_2D_3$ or higher were required for normal marrow multinucleated cell formation (Figure 3). The ultrastructural features of multinucleated cells formed in long-term marrow cultures of marrow derived from Paget's patients are shown in Figures 4 and 5. More recently, Mills and co-workers[59] have tested pagetic marrow-derived multinucleated cells formed in long-term marrow cultures for expression of viral antigens for paramyxoviruses. These investigators found that Paget's marrow-derived multinucleated cells expressed the nucleocapsid antigens for both respiratory syncytial virus and measles virus, while normal marrow-derived multinucleated cells did not express these viral proteins. Thus, the long-term cultures of Paget's marrow may be a model system for examining the factors controlling formation of osteoclasts in Paget's patients. Interestingly, marrow derived from uninvolved bones of Paget's patients also showed similar increases in the number of multinucleated cells formed in long-term marrow cultures.[48] These data suggest that there may be a systemic factor present in Paget's patients that enhances osteoclast formation at sites not clinically involved with Paget's disease. These data are similar to the results of Meunier and co-workers[15] and Siris et al.,[18] who also found increased osteoclast activity in bones not clinically involved with Paget's disease.

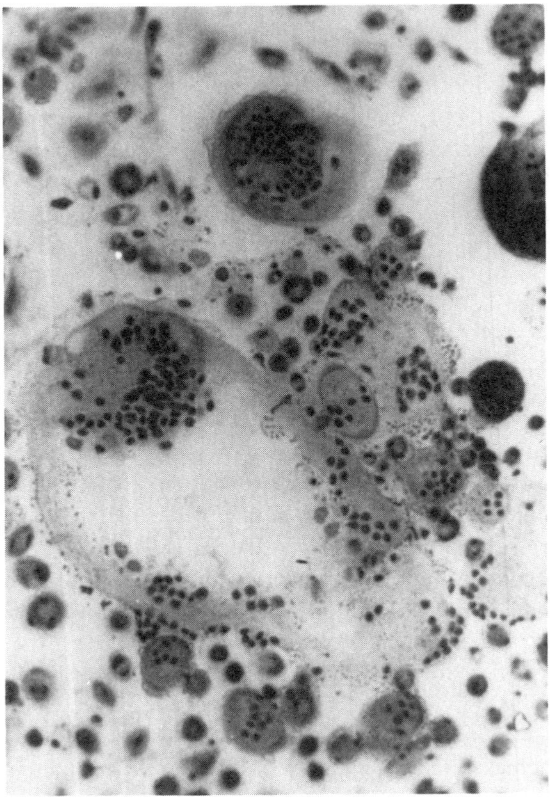

FIGURE 2. Multinucleated cells formed in long-term cultures of Paget's marrow mononuclear cells. Bone marrow mononuclear cells were obtained from patients with active Paget's disease and cultured in the presence of 1,25-dihydroxyvitamin D_3 for 2 weeks. Large numbers of multinucleated cells formed in these cultures. The multinucleated cells were markedly increased in size and contained many nuclei. (Original magnification × 400.)

VI. IL-6 AS A POTENTIAL AUTOCRINE/PARACRINE FACTOR IN PAGET'S DISEASE

At present, it is unclear which factors are responsible for the increased osteoclast formation and bone resorption seen in Paget's disease of bone. Recent studies by our group[60] have suggested that interleukin-6 (IL-6) may play a role in the increased osteoclast formation seen in Paget's patients. Bone marrow samples were obtained from patients with Paget's disease and placed in long-term marrow culture. Conditioned media from these long-term marrow cultures were then added to normal marrow cultures and tested for their capacity to increase normal osteoclast-like multinucleated cell formation in normal marrow cultures. The Paget marrow culture-conditioned media increased normal multinucleated cell formation. One of the factors present in

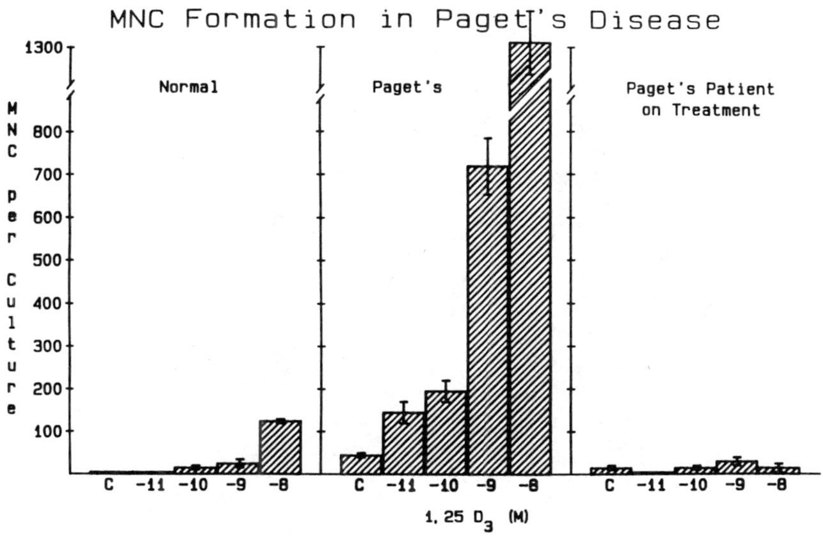

FIGURE 3. Formation of multinucleated cells in normal and Paget's marrow cultures. Bone marrow mononuclear cells from normals or Paget's patients were cultured in the presence of varying concentrations of 1,25-dihydroxyvitamin D_3 for 3 weeks. At the end of the culture period, the cells were stained with Wright's Geimsa stain, and the number of cells with three or more nuclei were counted. Few, if any, multinucleated cells formed in long-term marrow cultures from normals when 10^{-10} *M* 1,25-dihydroxyvitamin D_3 was added to the cultures. In contrast, significant numbers of multinucleated cells formed in the Paget's marrow cultures at concentrations of 1,25-dihydroxyvitamin D_3 as low as 10^{-11} *M*. Bone marrow obtained from Paget's patients undergoing treatment with calcitonin formed few, if any, multinucleated cells, consistent with the clinical response of these patients to calcitonin treatment.

this conditioned media appeared to be IL-6 because antibodies to IL-6, but not IL-1 granulocyte-macrophase colony stimulating factor (GM-CSF), or tumor necrosis factor-α(TNF-α) GM-CSF, or TNF-α, neutralized the stimulating activity present in the Paget-conditioned media. Further, *in situ* hybridization studies showed that the multinucleated cells present in the Paget marrow cultures were actively transcribing IL-6 mRNA. When bone marrow plasma from patients with Paget's disease was assayed for IL-6, increased levels of IL-6 were found in 3 of 3 patients, while elevated levels of IL-6 were not found in 10 of 10 normal marrow plasma samples. Further, in 8 of 14 patients with Paget's disease, increased levels of IL-6 were found in their peripheral blood plasma, while elevated levels of IL-6 were not found in normal marrow plasma from a similar number of controls. These data suggest that IL-6 may be an autocrine/paracrine factor in Paget's disease of bone (Figure 6). Further studies will be required to demonstrate that IL-6 is actively transcribed by osteoclasts isolated from Paget's bone. Suda and co-workers[61] and others[62] have recently shown that IL-6 may play a critical role in osteoclast formation and activity. IL-6 increased the formation of osteoclasts in bone organ culture systems and also increased osteoclastic bone resorption in new-

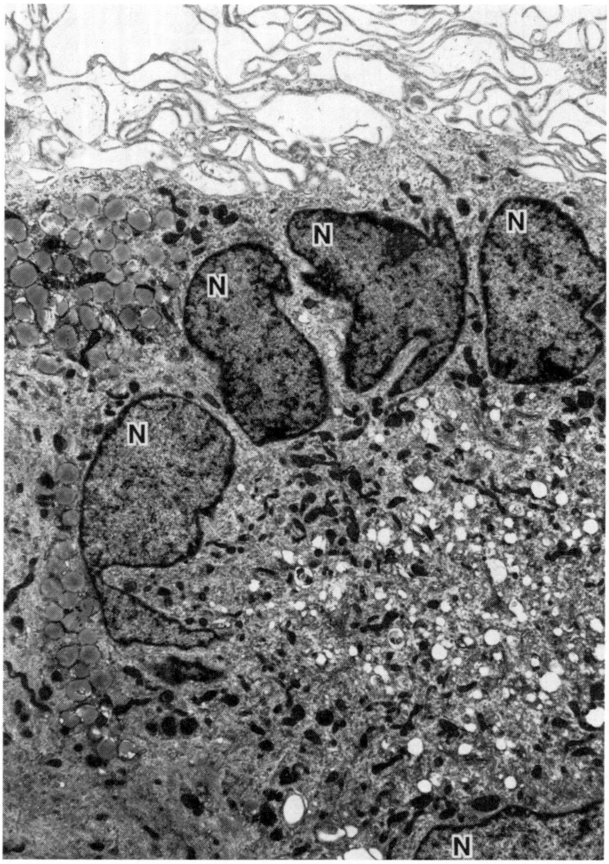

FIGURE 4. Ultrastructure of a multinucleated cell formed in long-term cultures of Paget's bone marrow. Marrow mononuclear cells from active lesions of patients with Paget's disease were cultured with 1,25-dihydroxyvitamin D_3 for 2 weeks. At the end of the culture period, the cells were fixed and processed for electron microscopy. Polymorphic nuclei (N) similar to those in pagetic osteoclast are present in these cells. (Original magnification × 4750.)

born mouse calvarial cultures. In addition, Kurihara et al.[63] have shown that IL-6 stimulates osteoclast-like multinucleated cell formation in normal marrow cultures and increases the percentage of the cells reacting with the 23c6 monoclonal antibody that preferentially reacts with osteoclasts.

VII. SUMMARY

Paget's disease of bone is predominantly a disease of osteoclasts. The osteoclasts are abnormal both in number and size, and show viral nuclear inclusions as well as cytoplasmic inclusions. These inclusions appear to be paramyxoviruses, although the patients do not express neutralizing antibodies

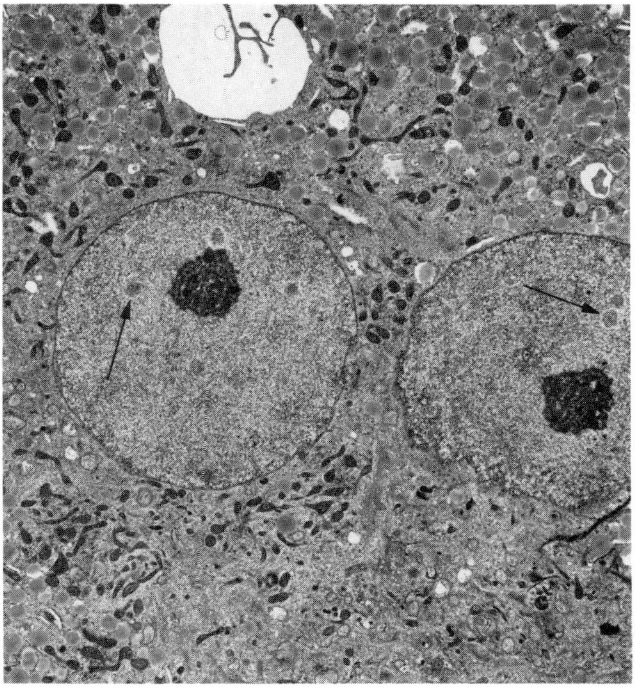

FIGURE 5. Ultrastructure of a multinucleated cell formed in long-term human marrow cultures from patients with Paget's disease. Marrow mononuclear cells from patients with Paget's disease were cultured in the presence of 1,25-dihydroxyvitamin D_3 for 2 weeks. At the end of the culture period, the cells were fixed and processed for electron microscopy. The large nuclei present in this cell contain multiple nuclear inclusions (arrows) which are nuclear bodies. (Original magnification × 4750.)

to the measles or respiratory syncytial virus in their peripheral blood. Ultrastructural and immunocytological studies have suggested that paramyxoviruses may play a role in Paget's disease and that Paget's disease may represent another slow viral disease similar to kuru, systemic sclerosing panencephalitis, and lupus erythematosus seen in NZB mice. However, no studies have demonstrated that this virus can infect other osteoclasts, nor is there an *in vivo* model of Paget's disease that allows studies of the effects of this virus on the expression of Paget's disease in laboratory animals. Long-term marrow cultures are currently being used as model systems to study osteoclast formation in Paget's disease, and may provide further insights into the mechanisms underlying the increased osteoclast formation and bone resorption seen in patients with Paget's disease.

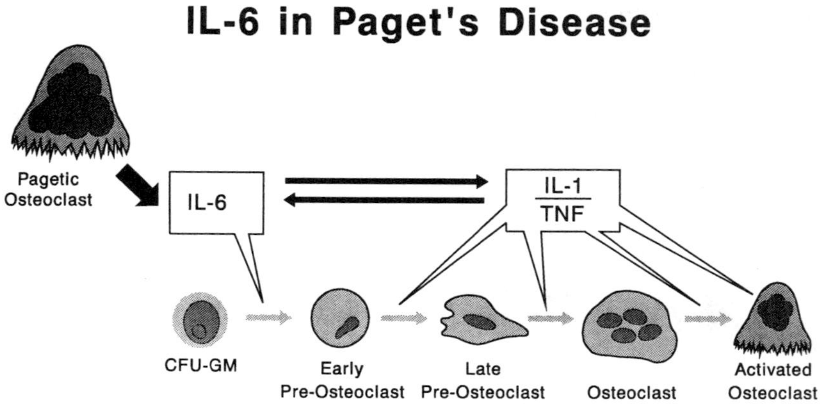

FIGURE 6. A model for the role of IL-6 in Paget's disease. In this model, pagetic osteoclasts actively produce IL-6. IL-6 increases the formation of other osteoclasts from their marrow precursors and induces IL-1 and TNF production by marrow stromal cells present in the bone microenvironment. IL-1 and TNF further induce IL-6 production by the pagetic osteoclasts. These factors (IL-1, TNF, and IL-6) work together to increase osteoclast formation and activity.

REFERENCES

1. **Paget, J.,** On a form of chronic inflammation of bones (osteitis deformans), *Med. Chir. Trans.*, 60, 37, 1877.
2. **Singer, F. R.,** Paget's disease of bone, *Clin. Endocrinol. Calcium Metab.*, 10, 369, 1988.
3. **Hamdy, R. C.,** Paget's disease of bone, *Hosp. Pract.*, (Oct. 30), 33, 1990.
4. **Collins, D. M.,** Paget's disease of bone, incidence and subclinical forms, *Lancet*, 2, 41, 1956.
5. **Schmorl, G.,** Uber osteitis deformans Paget, *Virchows Arch. Pathol. Anat. Physiol.*, 283, 694, 1932.
6. **Barker, D. J. P.,** The epidemiology of Paget's disease, *Metab. Bone Dis. Relat. Res.*, 4&5, 231, 1981.
7. **Yates, A. J. P.,** Paget's disease of bone, *Bailliere's Clin. Endocrinol. Metab.*, 2, 267, 1988.
8. **Hosking, D. J.,** Paget's disease of bone, *Br. Med. J.*, 283, 686, 1981.
9. **Ryan, W. G.,** Pathophysiology and modern management of Paget's disease, *Compr. Ther.*, 9, 64, 1983.
10. **Krane, S. M. and Simon, L. S.,** Metabolic consequences of bone turnover in Paget's disease of bone, *Clin. Orthop.*, 217, 26, 1987.
11. **Dull, T. A. and Henneman, P.H.,** Urinary hydroxyproline as an index of collagen turnover in bone, *N. Engl. J. Med.*, 268, 132, 1963.
12. **Maldague, B. and Malghem, J.,** Dynamic radiologic patterns of Paget's disease of bone, *Clin. Orthop. Relat. Res.*, 217, 126, 1987.
13. **Singer, F. R. and Mills, B. G.,** Primary bone cell dysfunction in Paget's disease of bone, in *Metabolic Bone Disease: Cellular and Tissue Mechanisms*, Tam, C. S., Heersche, J. N. M., and Murray, T. M., Eds., CRC Press, Boca Raton, FL, 1989, 33.

14. **Krane, S.,** Paget's disease of bone, *Calcif. Tissue Int.,* 38, 309, 1986.
15. **Meunier, P. J., Coindre, J. M., Edouard, C. M., and Arlot, M. E.,** Bone histomorphometry in Paget's disease, *Arthritis Rheum.,* 23, 1095, 1980.
16. **Russell, R. G. G., Smith, R., Preston, C., Walton, R. J., and Woods, C. G.,** Diphosphonates in Paget's disease, *Lancet,* 1, 894, 1974.
17. **Khairi, M. R. A., Meunier, P., Edouard, C., Courpron, P., Bernard, J., De Rosa, G. P., and Johnston, C. C.,** Quantitative bone histology in Paget's disease of bone: influence of sodium etidronate therapy, *Calcif. Tissue Res.,* 22, 355, 1977.
18. **Siris, E. S., Clemens, T. P., McMahon, D., Gordon, A., Jacobs, T. P., and Canfield, R. E.,** Parathyroid function in Paget's disease of bone, *J. Bone Miner. Res.,* 4, 75, 1989.
19. **Chappard, D., Alexandre, C., Laborier, J. C., Robert, J. M., and Riffat, G.,** Paget's disease of bone. A scanning electron microscopic study, *J. Submicrosc. Cytol.,* 16, 341, 1984.
20. **Boyde, A.,** Scanning electron microscope studies of bone, in *The Biochemistry and Physiology of Bone,* 2nd ed., Bourne, G. H., Ed., Academic Press, New York, 1972, 1, 259.
21. **Boyde, A. and Jones, J.,** Estimation of the size of resorption lacunae in mammalian calcified tissues using SEM stereophotogrammetry, in *Scanning Electron Microscopy/II,* Johari, O., Ed., AMF O'Hare, 1979, 393.
22. **Rebel, A., Malkani, K., Basle, M., and Bregeon, C. H.,** Osteoclast ultrastructure in Paget's disease, *Calcif. Tissue Res.,* 20, 187, 1976.
23. **Dayan, A. D.,** Progressive multifocal leukoencephalopathy, in *Virus Diseases and the Nervous System,* Whitty, C. W. M., Hughes, J. T. C., and MacCallum, F. O., Eds., Blackwell Scientific, Oxford, 1973, 199.
24. **Zu Rhein, G. M.,** Association of papova-virions with a human demyelinating disease, progressive multifocal leukoencephalopathy, *Progr. Med. Virol.,* 11, 185, 1969.
25. **Raine, C. S., Feldman, L. A., Sheppard, R. D., Barbosa, L. H., and Bornstein, M. B.,** Subacute sclerosing panencephalitis virus, *Lab. Invest.,* 31, 42, 1974.
26. **Mills, B. G. and Singer, F. R.,** Nuclear inclusions in Paget's disease of bone, *Science,* 194, 201, 1976.
27. **Bouteille, M., Kalifat, S. R., and Delarue, J.,** Ultrastructural variations of nuclear bodies in human diseases, *J. Ultrastruct. Res.,* 19, 474, 1967.
28. **Harvey, L., Gray, T., Beneton, M. N. C., Douglas, D. L., Kanis, J. A., and Russell, R. G. G.,** Ultrastructural features of the osteoclasts from Paget's disease of bone in relation to a viral aetiology, *J. Clin. Pathol.,* 35, 771, 1982.
29. **Howatson, A. F. and Fornasier, V. L.,** Microfilaments associated with Paget's disease of bone: comparison with nucleocapsids of measles virus and respiratory syncytial virus, *Intervirology,* 18, 150, 1982.
30. **Gherardi, G., Cascio, V. L. O., and Bonucci, E.,** Fine structure of nuclei and cytoplasm of osteoclasts in Paget's disease of bone, *Histopathology,* 4, 63, 1980.
31. **Basle, M., Rebel, A., Pouplard, A., Kouyoumdjian, S., Filmon, R., and Lepatezour, A.,** Mise en evidence d'antigenes viraux de rougeole dans les osteoclastes de la maladie osseuse de Paget, *C. R. Acad. Sci. Paris,* 289, 225, 1979.
32. **Rebel, A., Basle, M., Pouplard, A., Malkani, K., Filmon, R., and Lepatezour, A.,** Bone tissue in Paget's disease of bone, *Arthritis Rheum.,* 23, 1104, 1980.
33. **Rebel, A., Basle, M., Pouplard, A., Malkani, K., Filmon, R., and Lepatezour, A.,** Towards a viral etiology for Paget's disease of bone, *Metab. Bone Dis. Relat. Res.,* 4, 235, 1981.
34. **Mills, B. G., Singer, F. R., Weiner, L. P., and Holst, P. A.,** Immunohistological demonstration of respiratory syncytial virus antigen in Paget's disease of bone, *Proc. Natl. Acad. Sci. U.S.A.,* 78, 1209, 1981.
35. **Mills, B. G., Singer, F. R., Weiner, L. P., Suffin, S. C., Stabile, E., and Holst, P.,** Evidence for both respiratory syncytial virus and measles virus antigens in the osteoclasts of patients with Paget's disease of bone, *Clin. Orthop. Relat. Res.,* 183, 303, 1984.

36. **Basle, M. F., Russell, W. C., Goswami, K. A., Rebel, A., Giraudon, P., Wild, R., and Filmon, R.,** Paramyxovirus antigens in osteoclasts from Paget's bone tissue detected by monoclonal antibodies, *J. Gen. Virol.,* 66, 2103, 1985.
37. **Barker, D. J. and Detheridge, F. M.,** Dogs and Paget's disease (letter), *Lancet,* 30, 1245, 1985.
38. **Gordon, M., Sharpe, P., and Anderson, D.,** Evidence from *in situ* hybridization that Paget's disease of bone may be due to infection with canine distemper virus, *J. Bone Miner. Res.,* 5, S272, 1990.
39. **Basle, M. F., Rebel, A., Fournier, J. G., Russell, W. C., and Malkani, K.,** On the trail of paramyxoviruses in Paget's disease of bone, *Clin. Orthop. Relat. Res.,* 217, 9, 1987.
40. **Winfield, J. and Sutherland, S.,** Measles antibody in Paget's disease, *Lancet,* 1, 891, 1981.
41. **Singer, F. R. and Mills, B. G.,** Evidence for a viral etiology of Paget's disease of bone, in *Basic Science and Pathology,* Lippincott, New York, 1983, 178, 245.
42. **Mills, B. G. and Singer, F. S.,** Critical evaluation of viral antigen data in Paget's disease of bone, *Clin. Orthop. Relat. Res.,* 217, 16, 1987.
43. **Mirra, J. M., Bauer, F. C. H., and Grant, T. T.,** Giant cell tumor with viral-like intranuclear inclusions associated with Paget's disease, *Clin. Orthop. Relat. Res.,* 158, 243, 1981.
44. **Mills, B. G., Yabe, H., and Singer, F. R.,** Osteoclasts in human osteopetrosis contain viral-nucleocapsid-like nuclear inclusions, *J. Bone Miner. Res.,* 3, 101, 1988.
45. **Mills, B. G., Singer, F. R., Weiner, L. P., and Holst, P. A.,** Cell cultures from bone affected by Paget's disease, *Arthritis Rheum.,* 23, 1115, 1980.
46. **Mills, B. G., Singer, F. R., Weiner, L. P., and Holst, P. A.,** Long-term culture of cells from bone affected by Paget's disease, *Calcif. Tissue Int.,* 29, 79, 1979.
47. **Basle, M. F., Mazaud, P., Malkani, K., Chreitien, M. F., Moreau, M. F., and Rebel, A.,** Isolation of osteoclasts from pagetic bone tissue, morphometry and cytochemistry on isolated cells, *Bone,* 9, 1, 1988.
48. **Kukita, A., Chenu, C., McManus, L. M., Mundy, G. R., and Roodman, G. D.,** Atypical multinucleated cells form in long term marrow cultures from patients with Paget's disease, *J. Clin. Invest.,* 85, 1280, 1990.
49. **Takahashi, N., Yamana, H., Yoshiki, S., Roodman, G. D., Mundy, G. R., Jones, S. J., Boyde, A., and Suda, T.,** Osteoclast-like cell formation and its regulation by osteotropic hormones in mouse bone marrow cultures, *Endocrinology,* 122, 1382, 1988.
50. **Takahashi, N., Kukita, T., MacDonald, B. R., Bird, A., Mundy, G. R., McManus, L. M., Miller, M., Boyde, A., Jones, S. J., and Roodman, G. D.,** Osteoclast-like cells form in long term human bone marrow but not in peripheral blood cultures, *J. Clin. Invest.,* 83, 543, 1989.
51. **MacDonald, B. R., Takahashi, N., McManus, L. M., Holahan, J., Mundy, G. R., and Roodman, G. D.,** Formation of multinucleated cells which respond to osteotropic hormones in human long-term marrow cultures, *Endocrinology,* 120, 2326, 1987.
52. **Roodman, G. D., Ibbotson, K. J., MacDonald, B. R., et al.,** 1,25 $(OH)_2$ Vitamin D_3 causes formation of multinucleated cells with osteoclast characteristics in cultures of primate marrow, *Proc. Natl. Acad. Sci. U.S.A.,* 82, 8213, 1985.
53. **Chenu, C., Pfeilschifter, J., Mundy, G. R., and Roodman, G. D.,** Transforming growth factor beta inhibits formation of osteoclast-like cells in long-term human marrow cultures, *Proc. Natl. Acad. Sci. U.S.A.,* 85, 5683, 1988.
54. **Takahashi, N., MacDonald, B. R., Hon, J., Winkler, M. E., Derynck, R., Mundy, G. R., and Roodman, G. D.,** Recombinant human transforming growth factor alpha stimulates the formation of osteoclast-like cells in long-term human marrow cultures, *J. Clin. Invest.,* 78, 894, 1986.

55. **Takahashi, N., Mundy, G. R., and Roodman, G. D.,** Recombinant human gamma interferon inhibits formation of osteoclast-like cells by inhibiting fusion of their precursors, *J. Immunol.*, 137, 3544, 1986.
56. **Kurihara, N., Gluck, S., and Roodman, G. D.,** Sequential expression of phenotype markers for osteoclasts during differentiation of precursors for multinucleated cells formed in long term human marrow cultures, *Endocrinology,* 127, 3215, 1990.
57. **Kukita, T., McManus, L. M., Miller, M., Civin, C., and Roodman, G. D.,** Osteoclast-like cells formed in long-term human bone marrow cultures express a similar surface phenotype as authentic osteoclasts, *Lab. Invest.*, 60, 532, 1989.
58. **Kurihara, N., Chenu, C., Civin, C. I., and Roodman, G. D.,** Identification of committed mononuclear precursors for osteoclast-like cells formed in long term marrow cultures, *Endocrinology,* 126, 2741, 1990.
59. **Frausto, A., Mills, B., Singer, F., and Roodman, G. D.,** Multinucleated cells (MNC) formed in long term marrow cultures (LTMC) from Paget's patients express viral antigens, *J. Bone Miner. Res.*, 5, 791A, 1990.
60. **Roodman, G. D., Kurihara, N., Ohsaki, Y., Kukita, A., Hosking, D., Demulder, A., and Singer, F. S.,** Interleukin-6: a potential autocrine/paracrine factor in Paget's disease of bone, *J. Clin. Invest.*, submitted.
61. **Ishimi, Y., Miyaura, C., Jin, C. H., Akatsu, T., Abe, E., Nakamura, Y., Yamaguchi, A., Yoshiki, S., Matsuda, T., Hirano, T., Kishimoto, T., and Suda, T.,** IL-6 is produced by osteoblasts and induces bone resorption, *J. Immunol.*, 145, 3297, 1990.
62. **Lowik, C. W. G. M., van der Pluijm, G., Bloys, S., Hoekman, K., Bivoet, O. L. M., Aarden, L. A., and Papapoulos, S. E.,** Parathyroid hormone (PTH) and PTH-like protein (PLP) stimulate interleukin-6 production by osteogenic cells: a possible role of interleukin-6 in osteoclastogenesis, *Biochem. Biophys. Res. Commun.*, 162, 1546, 1989.
63. **Kurihara, N., Bertolini, D., Suda, T., Akiyama, Y., and Roodman, G. D.,** Interleukin-6 stimulates osteoclast-like multinucleated cell formation in long term human marrow cultures by inducing IL-1 release, *J. Immunol.*, 144, 426, 1990.
64. **Singer, S. R.,** personal communication.

INDEX

A

B

C

D

E

F

G

K

L

M

N

O

P

R

S

T

U

V

W–Z